I0478977

Soil Science and Sustainable Agriculture

Soil Science and Sustainable Agriculture

Editor: Henry Wang

www.callistoreference.com

Callisto Reference,
118-35 Queens Blvd., Suite 400,
Forest Hills, NY 11375, USA

Visit us on the World Wide Web at:
www.callistoreference.com

© Callisto Reference, 2018

This book contains information obtained from authentic and highly regarded sources. Copyright for all individual chapters remain with the respective authors as indicated. All chapters are published with permission under the Creative Commons Attribution License or equivalent. A wide variety of references are listed. Permission and sources are indicated; for detailed attributions, please refer to the permissions page and list of contributors. Reasonable efforts have been made to publish reliable data and information, but the authors, editors and publisher cannot assume any responsibility for the validity of all materials or the consequences of their use.

ISBN: 978-1-63239-909-0 (Hardback)

Trademark Notice: Registered trademark of products or corporate names are used only for explanation and identification without intent to infringe.

Cataloging-in-Publication Data

Soil science and sustainable agriculture / edited by Henry Wang.
    p. cm.
Includes bibliographical references and index.
ISBN 978-1-63239-909-0
1. Soil science. 2. Sustainable agriculture. I. Wang, Henry.
S591 .S65 2018
631.4--dc23

Table of Contents

Preface

Soil science studies the texture, classification and formation of soil. The two main branches of this subject are edaphology and pedology. Recent research done in the field of soil science focuses on climate change and resource degradation. Soil quality and fertility play a significant role in agriculture. This book elucidates the role of soil in practicing sustainable agriculture. It aims to serve as a resource guide for students and experts alike and contribute to the growth of the discipline. Coherent flow of topics, student-friendly language and extensive use of examples make it an invaluable source of knowledge.

After months of intensive research and writing, this book is the end result of all who devoted their time and efforts in the initiation and progress of this book. It will surely be a source of reference in enhancing the required knowledge of the new developments in the area. During the course of developing this book, certain measures such as accuracy, authenticity and research focused analytical studies were given preference in order to produce a comprehensive book in the area of study.

This book would not have been possible without the efforts of the authors and the publisher. I extend my sincere thanks to them. Secondly, I express my gratitude to my family and well-wishers. And most importantly, I thank my students for constantly expressing their willingness and curiosity in enhancing their knowledge in the field, which encourages me to take up further research projects for the advancement of the area.

Editor

Transgenic Potatoes for Potato Cyst Nematode Control Can Replace Pesticide Use without Impact on Soil Quality

Jayne Green[⑨¤a], **Dong Wang**[⑨¤b], **Catherine J. Lilley**, **Peter E. Urwin***, **Howard J. Atkinson**

Centre for Plant Sciences, University of Leeds, Leeds, United Kingdom

Abstract

Current and future global crop yields depend upon soil quality to which soil organisms make an important contribution. The European Union seeks to protect European soils and their biodiversity for instance by amending its Directive on pesticide usage. This poses a challenge for control of *Globodera pallida* (a potato cyst nematode) for which both natural resistance and rotational control are inadequate. One approach of high potential is transgenically based resistance. This work demonstrates the potential in the field of a new transgenic trait for control of *G. pallida* that suppresses root invasion. It also investigates its impact and that of a second transgenic trait on the non-target soil nematode community. We establish that a peptide that disrupts chemoreception of nematodes without a lethal effect provides resistance to *G. pallida* in both a containment and a field trial when precisely targeted under control of a root tip-specific promoter. In addition we combine DNA barcoding and quantitative PCR to recognise nematode genera from soil samples without microscope-based observation and use the method for nematode faunal analysis. This approach establishes that the peptide and a cysteine proteinase inhibitor that offer distinct bases for transgenic plant resistance to *G. pallida* do so without impact on the non-target nematode soil community.

Editor: Guy Smagghe, Ghent University, Belgium

Funding: This work was funded by research grant number BB/D001749/1 from the Biotechnology and Biological Sciences Research Council, United Kingdom. (http://www.bbsrc.ac.uk). DW was funded by a Dorothy Hodgkin post-graduate award from the Natural Environment Research Council, United Kingdom. (http://www.nerc.ac.uk) The funders had no role in study design, data collection and analysis, decision to publish, or preparation of the manuscript.

Competing Interests: The authors have declared that no competing interests exist.

* E-mail: p.e.urwin@leeds.ac.uk

⑨ These authors contributed equally to this work.

¤a Current address: Advanced Technologies (Cambridge) Ltd., Cambridge, United Kingdom
¤b Current address: Institute of Genetics and Developmental Biology, Chinese Academy of Sciences, Beijing, China

Introduction

Arable agriculture must remain productive and become more sustainable than in past decades [1]. A key aspect of this is maintaining soil quality for which soil organism abundance, diversity, food web structure or community stability are useful indicators and important contributors to soil function. They are responsive to land management practices [2] and are of value in defining when soil organisms have been exposed to harm. The European Union (EU) seeks to protect soils and their biodiversity [3] for instance by changes to its directive on use of plant protection products [4] to reduce usage of those pesticides that harm soil quality. An example consequence is that the pesticides currently applied to 23% of UK potato fields to control potato cyst nematode (PCN) may be withdrawn either abruptly or progressively thereby doubling the economic cost of this pest, in the UK alone, to £56 m/year [5,6]. This is a serious concern as one of the two species of PCN, *Globodera pallida*, is particularly difficult to control by either crop rotation or the partially resistant potato cultivars available against it [7]. One future economic control option is genetically modified, nematode-resistant (GMNR) potato cultivars but the potential of such crops to provide essential pest management and reduce pesticide usage is inadequately considered in current EU policies. Each transgenic trait/crop developed

must be shown to be both effective and environmentally benign under field conditions before its deployment can be considered. The expression of a plant cysteine proteinase inhibitor (cystatin) in potato roots controls potato cyst nematode (*Globodera* spp) by impairing digestion of its dietary protein [8–10]. Growth of these GMNR plants did not harm above ground organisms [11,12] or soil microarthropods [13] in potato fields. A small effect on the soil microbe community was detected by phospholipid fatty acid analysis in some years [13] but not others [14]. It was insufficient to influence soil function as assessed by litter decomposition and was much less than imposed by seasonal factors like soil moisture content. A second sensitive approach (community level physiological profiles) that measures substrate use by rhizosphere bacteria did not identify changes imposed by the cystatin-expressing transgenic potato plants but readily detected the consequences of growing different conventional crops [15].

Our work to develop GMNR potato plants, test their efficacy and evaluate their possible impact on biotic aspects of soil quality has now advanced in three ways. First, we report the efficacy of a second GMNR trait in providing field control of *Globodera* spp in addition to the cystatin used in previous work. The new potato plants secrete a non-lethal peptide from their roots that disrupts the chemoreception that cyst nematodes require to locate host plants [16]. The peptide was originally obtained by biopanning a phage display library against membrane fractions of *Caenorhabditis elegans* that are rich in nicotinic acetylcholine receptors (nAChR). The peptide was displaced by the anthelmintic levamisole that binds to these receptors [17]. It is a disulphide-constrained 7-mer, termed nAChRbp [18] with the amino acid sequence CTTMHPRLC that inhibits chemoreception of *G. pallida* at 1 µM [19] and reduces parasitism of hairy roots of potato [16]. Fluorescent tagging has shown the peptide is taken up from an aqueous environment by the primary cilia of some nematode chemoreceptive sensilla before undergoing retrograde transport to their neuronal cell bodies. It is then transported to a limited number of connecting neurons and probably exerts its effect at the synapses of cholinergic interneurons [18]. The second new aspect aimed to enhance biosafety of the transgenic approach by restricting expression of the peptide to root tips using a tissue-specific promoter (*MDK4–20*) that we have previously described from *Arabidopsis* [20]. The third development was to improve an approach that detects the direct or indirect impact of the transgenic lines on non-target soil nematodes. Nematodes are worthy of such consideration as the most abundant metazoan taxon with a high abundance in soil where their community participates in many functions at different levels of the soil food web. These nematodes are stable in response to fluctuations in moisture and temperature while responding to land management effects in predictable ways that reflect changes in soil microenvironments [21]. Soil nematode faunal analysis has been based on a weighted abundance for five sub-groups derived from the rapidity of their multiplication in favourable conditions [22,23]. One outcome is an enrichment index (EI) ranging from 0–100 for nematodes that respond rapidly to environmental change and a structural index (SI) with the same score range for those that prefer undisturbed habitats. These indices enable the extent of soil disturbance, enrichment, the decomposition channels, C:N ratio and food web condition to be inferred [23]. One drawback to nematode faunal analysis is the requirement for time-consuming microscopic identification of nematode genera in many samples, a technique that relies on skilled assessment of a range of morphological features. We have overcome this potential obstacle to rapid analysis by deploying a molecular bar-coding approach based on the 18S ribosomal gene [24,25] to recognise nematode genera in mixed samples. Use of that method suggests no impact on the nematode faunal index and hence soil health when GMNR potato plants are deployed to control *Globodera*.

Results

Transgenic potato lines secreting a chemodisruptive peptide

Forty transgenic cv Désirée potato lines were generated that were confirmed by PCR to contain the MDK-peptide construct (results not shown). The peptide has previously been shown to inhibit alkaline phosphatase [16]. The effect is concentration dependent with 1 mM of the synthetic peptide providing almost complete inhibition of the enzyme in the experimental assay conditions (Figure 1A). Root exudates of all transgenic lines were screened for their ability to inhibit alkaline phosphatase in comparison to exudates from wild-type plants. Exudates from several lines caused detectable inhibition and data for a range of these lines are presented in Figure 1B. Comparison with the standards suggests that up to 860 µM of the peptide accumulated in root exudates after 14 days incubation of the plant in 7 ml of water. Three MDK-peptide potato lines (L1–3) were selected for further evaluation on the basis of their level of active peptide secretion and normal growth phenotype.

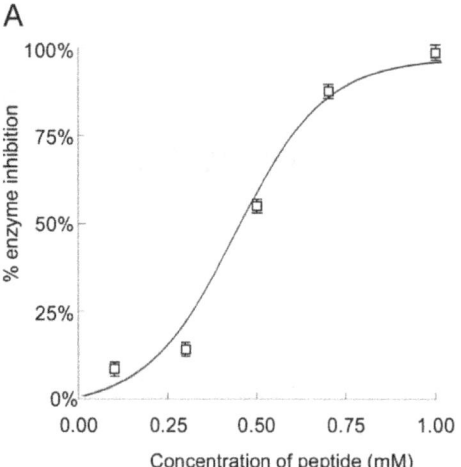

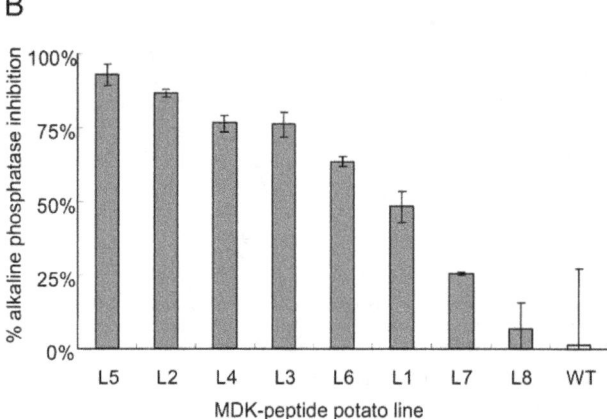

Figure 1. Inhibition of alkaline phosphatase by the behavioural disrupting peptide. (A) The chemically synthesised peptide inhibited alkaline phosphatase in concentration dependent manner. (B) Sufficient peptide could be collected from root exudates of transgenic potatoes to inhibit the enzyme. Values are means ± sem.

Cyst nematode resistance conferred by the secreted peptide

The three selected potato lines were challenged by *G. pallida* in a containment glasshouse trial before promising results led to them being advanced to a replicated field trial. There was concordance between resistance observed in the containment and field trials with transgenic lines supporting significantly reduced nematode multiplication compared to wild type plants in both environments. Up to 77±4% (mean ± sem) resistance was obtained for the most effective line in the field (Figure 2). A general linear model with univariate analysis on transformed arcsin values established significant difference between transgenic and control plants (P = 0.001) but not between the two environments (P = 0.9) and there was no significant interaction between the two factors (P = 0.089). Oneway ANOVA with *a priori* contrasts established each peptide-expressing line differed from the untransformed wildtype plants in containment (P<0.001 in all three cases) and for two lines (L1 and L2) in the field trial (P<0.01 and P<0.05 respectively).

The initial containment trial to assess resistance did not include plants expressing the cystatin OcIΔD86 under control of either the constitutive CaMV35S promoter or the ARSK1 promoter

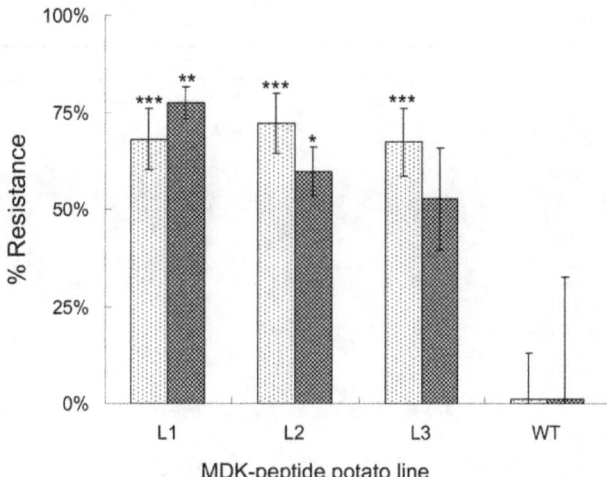

Figure 2. Resistance to *Globodera pallida* provided by transgenic expression of a behavioural disrupting peptide in potato. Resistance is conferred by expression of the peptide under control of a root-tip specific promoter (*MDK4–20*) in trials in both a containment glasshouse (light shading) and the field (darker shading) relative to *G. pallida* eggs in soil after multiplication on untransformed cv Désirée. Values are means ± sem based on 2 replicate samples for each of ten pots per line in containment and pools of nine samples for each of 4 replicate plots per line in the field. The differences between transgenic lines and the corresponding wildtype control are shown (***, $p<0.001$; **, $p<0.01$; *, $p<0.05$; oneway ANOVA with *a priori* contrasts).

expressed preferentially at feeding sites of cyst nematodes. This is because the efficacy of these particular transgenic lines in containment and field trials has been reported previously [9,10]. The level of resistance in the field in the current work was $57\pm13\%$ for ARSK/OcIΔD86 plants and $53\pm7\%$ for CaMV35S/OcIΔD86 plants. Potato cv Santé, which has natural resistance to *G. pallida*, provided a resistance level of $96\pm1\%$ in the field compared to the wild type, fully susceptible cv Désirée. Populations of *G. pallida* vary in their virulence to cv Santé, with lower levels of relative resistance recorded on previous occasions [9].

Development of genus-specific qPCR primers for detection of soil nematodes

The 18S small subunit ribosomal RNA gene (SSU) of nematodes is approximately 1600 bp in length. The 500–600 bp region at the 5′ end contains both conserved stem and more divergent loop structures, accounting for around half the nucleotide variability of the complete gene [25]. Sequencing of this DNA region, amplified using a standard primer pair, enabled individual nematodes to be assigned a likely genus by cross-reference to database sequence entries.

Ninety three nematodes were randomly sampled during an initial survey to identify those soil nematodes present at the field site prior to planting. Each individual was assigned a tentative genus based on morphological identification, although it was not possible to distinguish between the morphologically similar *Pellioditis* and *Pelodera* or between *Acrobeloides* and *Cephalobus*. Sequence of the 18S SSU gene was obtained for 96% of the nematode individuals sampled and this revealed the presence of 21 genera. The preliminary morphological identification of each nematode was confirmed by comparison of the sequence obtained from its PCR product with database entries (Table 1). The sequence analysis allowed discrimination of *Pellioditis* and *Pelodera* so that genus specific primers could be developed for these two nematodes. It was not possible to discriminate *Acrobeloides* and *Cephalobus* individuals in the sample by either observation or sequence comparison but this distinction was unimportant for the current work as they belong to the same functional guild. Each sequence generated was 99–100% identical to database entries for a single genus thus allowing unambiguous genus assignment except for four individuals with sequence that was only 95% identical to database entries for *Aphelenchoides* as the most similar genus. Morphological identification also assigned these nematodes to the *Aphelenchoides* genus and they were considered as such for this work. Eight of the genera identified were plant feeding nematodes that are not included in nematode community index calculations. The sequences of each genus other than those of plant parasites were used to design 11 primer pairs (Table 2) that would each be specific for a particular genus and suitable for use in quantitative polymerase chain reaction (qPCR). The 11 primer pairs allowed detection and discrimination of >90% of the non-plant feeding

Table 1. Free-living soil nematode genera recovered at the field site.

Genus	Functional guild*	Sequence identity (%)	NCBI accession
Acrobeloides/Cephalobus	Ba_2	100	AF430515, AF034390
Anaplectus	Ba_2	99	AY284696
Ceratoplectus	Ba_2	100	AY284706
Eucephalobus	Ba_2	100	AY284667
Aphelenchoides	Fu_2	95	DQ901551
Pellioditis	Ba_1	99	AF430633
Pelodera	Ba_1	99	AF083002
Rhabditella	Ba_1	100	AY284654
Anatonchus	Ca_4	99	AJ966474
Mesodorylaimus	Om_4	100	AJ966488
Aporcelaimellus	Om_5	100	AY284812

*Ba, bacterivore; Fu, fungivore; Ca, carnivore; Om, omnivore.
These genera contributed >90% of individual non-parasitic nematodes at the site. The functional guild and their position on the coloniser-persister scale (1–5, shown as a suffix) are as previously assigned to each genus [22,23]. The sequence identity of the amplified region of 18S small subunit ribosomal RNA gene to that of the most similar genus in the GenBank data set is indicated together with the Accession Number of the most similar database sequence.

Table 2. Primer pairs used in qPCR to identify each nematode genus.

Genus	Primer*	Primer sequence	Product (bp)
Aphelenchoides	AphF	TTGGACTGCCATGGTGTTGA	173
	AphR	ATGTCCGACCTCATAGAGAAC	
Pellioditis	PellF	AAGTTTTCGGCTGCCTTTTAG	161
	PellR	ACAGTTTACGGCCATCGGAA	
Pelodera	PeoF	GAACGCCGTTTCGGTTTTTC	189
	PeoR	AAGCTAACGCTCGGTTTCATA	
Rhabditella	RhaF	TTTTACCTATTCCGAAATCTTATT	198
	RhaR	TGGTTGATAGGGCAGACTCC	
Anatonchus	AnF	TGGTAAGAATTGGTAAACACGA	188
	AnR	CGAGTCCAGTCCGAAGAATT	
Aporcelaimellus	ApoF	GTTACGCCTAGTTCGGAAGA	222
	ApoR	GCACTCATTACAAGCACCTTT	
Ceratoplectus	CplecF	GGTAAACCCCTCAAAATCCTA	190
	CplecR	GAAAACCCCGACAGCAGCA	
Acrobeloides/Cephalobus	GUF	TTACGTCCCTGCCCTTTGTA	121
	AcCR	TCAAATCAGTTTCCAGCGAAC	
Anaplectus	GUF	TTACGTCCCTGCCCTTTGTA	93
	AnaplR	CCCGAAAGCCCCAACAGC	
Mesodorylaimus	GUF	TTACGTCCCTGCCCTTTGTA	95
	MeR	GCACTTTCGTACACCTTAACT	
Eucephalobus	GUF	TTACGTCCCTGCCCTTTGTA	118
	EuR	AGTCAGCTTCCAACGACTCG	

*F, forward primer; R, reverse primer.

nematodes identified at the field site. Sequence divergence between the genera was sufficient to allow two unique primers to be used in seven cases. It was not possible to design unique primer pairs that also satisfied the desired criteria for qPCR for the remaining four genera *Anaplectus, Mesodorylaimus, Eucephalobus* and *Acrobeloides/Cephalobus*. For specific detection of these genera, one common forward primer (GUF2) was used in combination with a unique reverse primer. All primer pairs were tested initially in a standard PCR using template DNA from single nematodes of the corresponding genus and subsequently using DNA prepared from a random pool of 200 nematodes from the field site (Figure 3). A single, unambiguous sequence matching the expected genus was obtained for each product amplified from pooled DNA, confirming the specificity of the primers. In qPCR, dissociation curves exhibited identical single peaks when using template DNA from either single or pooled nematodes. The final verification of the specificity of the primers centred on discrimination achieved by the qPCR primer pairs for *Acrobeloides/Cephalobus* and *Eucephalobus* which have highly similar SSU gene sequences. No cross amplifications were detected when *Eucephalobus* specific primers (GUF2/EuR1) were used for amplification from *Acrobeloides/Cephalobus* genomic DNA, or when *Acrobeloides/Cephalobus* specific primers (GUF2/AcCR2) were used on *Eucephalobus* DNA. In both cases no threshold cycle (Ct) value was recorded using template from non-target species after 40 PCR cycles in contrast to Ct values of <30 cycles for the target sequences (Figure S1).

Once the specificity of the primers for the nematode community under study was determined, it was necessary to

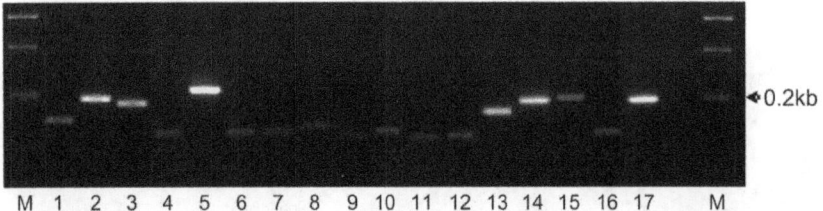

Figure 3. PCR products amplified using genus specific primers. The template for each pair of primers was DNA extracted from two hundred nematodes of mixed genera as recovered from the field site. The primers, detailed in Table 2 for non-parasitic genera, were designed to amplify SSU sequence from (1) *Acrobeloides/Cephalobus* (2) *Anatonchus*, (3) *Aphelenchoides*, (4) *Aporcelaimellus*, (5) *Coslenchus*, (6) *Ditylenchus*, (7) *Eucephalobus*, (8) *Filenchus*, (9) *Globodera*, (10) *Helicotylenchus*, (11) *Mesodorylaimus*, (12) *Pellioditis*, (13) *Pelodera*, (14) *Rhabditella*, (15) *Scutylenchus*, (16) *Ceratoplectus* (17) *Anaplectus*. (M) Hyperladder I marker.

relate the quantity of product from qPCR to that for a single nematode of each genus. This was achieved in two steps by first using the mean DNA content per nematode from qPCR analysis with *Aphelenchoides* as an arbitrary calibrator. A subsequent minor adjustment was then made from relating parallel morphological counts to qPCR-based abundance data of mixed nematodes in replicate soil samples.

Validation of the qPCR measurement of nematode faunal indices

The qPCR-based method of determining nematode faunal indices was validated using soil samples taken from the field trial.site in autumn following a potato crop. Nematodes were extracted from multiple sub-samples from each of three plots for one determination of genera present by microscopic examination of morphology and three to seven replicate analyses by qPCR per plot. The proportion of each genus identified by the two methods is presented in Table 3 and the derived EI and SI values are shown in Table 4. The qPCR approach distinguished between the morphologically very similar *Pellioditis* and *Pelodera* that were grouped together for results of morphological analysis. The two genera belong to the same functional guild so inability to distinguish between them morphologically does not influence the indices derived in this work. The percentage sample compositions recorded by the two approaches were similar for the prevalent genera with the mean from qPCR not differing significantly from the corresponding value from morphology (single sample t tests). *Pellioditis* and *Pelodera* together comprised $74\pm1\%$ of the non-plant parasitic nematodes as determined by morphology compared with a value of $78\pm5\%$ from qPCR analysis. The corresponding values for *Acroboloides/Cephalobus* were $24\pm1\%$ by morphological analysis and $20\pm5\%$ by qPCR. Values for genera present at only a very low abundance are likely to vary between samples by both chance and as a result of a patchy nematode distribution. The full set of 11 primer pairs was used for analysis of these test samples by qPCR, but some genera were not detected either by microscopic examination or by qPCR. *Anaplectus* and *Ceratoplectus* were present in the initial soil sample and so primers had been developed for these genera. They were not recorded in subsequent field samples analysed by either the morphological or the qPCR-based approach. *Rhabditella* was also present in the initial sample and

had a low abundance in soil used for the containment trial but was not detected in subsequent field samples.

Determination of EI and SI values by qPCR during containment and field trials

Soil was collected from the site of the future field trial for use in a containment trial set up to assess effects of the transgenic lines on the soil nematode community. Included in this trial were transgenic potato lines expressing either a cystatin or the chemodisruptive peptide, wild-type cv. Sante and Désirée and oil seed rape (OSR) plants. The nematode faunal analysis for the containment soil prior to planting gave an EI = 48.0 ± 9.8 and SI = 23.1 ± 9.2. Differences (Δ) from these initial values are given for each type of plant, at either flowering time or harvest, in the upper panels of Figure 4. The univariate procedure of the general linear model suggested that ΔEI but not ΔSI varied overall between the flowering and harvest samples (P = 0.016 and P = 0.11) but neither index varied among lines. Further analysis used oneway ANOVA with *a priori contrasts* to compare ΔSI values for untransformed cv. Désirée with those for transgenic lines. This suggested soil that supported growth of OcIΔD86 cystatin and peptide expressing plants had higher SI values at flowering (P = 0.023 and 0.006 respectively) than untransformed cv. Désirée whereas at harvest there were no significant differences between the soil samples (P>0.28 in all cases).

The detection of a difference with sampling time and some differences between lines at one of the two sampling times encouraged us to use this approach in a field trial in the following spring. On this occasion, the values at planting were EI = 80.3 ± 6.1 and SI = 9.3 ± 1.5. Differences (Δ) from these initial values are presented for each type of plant, at either flowering time or harvest, in the lower panels of Figure 4. In this trial, the univariate procedure established a significant difference between flowering and harvest in ΔSI (P = 0.021) but not for ΔEI (P = 0.28). No significant differences were evident by this statistical approach between lines for either index (P = 0.61 and P = 0.22 respectively). Analysis for the field trial was developed further using oneway ANOVA with *a priori contrasts* as in the containment trial. This detected no differences in ΔEI or ΔSI values of transgenic lines relative to untransformed cv. Désirée at either sample time. There were also no significant differences between untransformed

Table 3. Proportions of nematode genera determined by both morphology and qPCR for field soil samples.

Genus	Functional guild	Soil sample 1		Soil sample 2		Soil sample 3	
		morphology	qPCR	morphology	qPCR	morphology	qPCR
Pellioditis	Ba$_1$	73%	$74\pm5\%$	75%	$61\pm15\%$	75%	$47\pm11\%$
Pelodera	Ba$_1$		0%		$8\pm8\%$		$36\pm14\%$
Aphelenchoides	Fu$_2$	1%	0%	1%	0%	1%	0%
Acrobeloides or Cephalobus	Ba$_2$	25%	$25\pm5\%$	22%	$28\pm18\%$	24%	$15\pm5\%$
Eucephalobus	Ba$_2$	0%	$1\pm1\%$	0%	$3\pm1\%$	0%	$1\pm0.4\%$
Anatonchus	Ca$_4$	0%	0%	1%	0%	0%	0%
Mesodorylaimus	Om$_4$	2%	0%	1%	0%	0%	0%
Aporcelaimellus	Om$_5$	0%	$1\pm0.2\%$	0%	0%	0%	0%

The percentage of each nematode genus present was estimated from three field soil samples using in each case one sub-sample for morphological identification and 3–7 sub-samples for qPCR-based analysis. The functional guild and their position on the coloniser-persister scale (1–5, shown as a suffix) are as previously assigned to each genus [22,23]. The percentages are based on the number of nematodes observed (morphological analysis) or on all nematodes (>200) extracted from each 100 g sub-sample used to prepare template for qPCR. *Anaplectus*, *Ceratoplectus* and *Rhabditella* were not detected by either morphology or qPCR in these samples. Values for qPCR represent means ± sem.

Table 4. Values for enrichment (EI) and structural (SI) indices of soil nematode communities determined by both morphology and qPCR.

	Soil sample 1		Soil sample 2		Soil sample 3	
	morphology	qPCR	morphology	qPCR	morphology	qPCR
EI value	92.0	92.1±2.3	92.9	89.8±8.89	92.5	95.2±1.71
SI value	26.7	13.0±2.76	17.4	1.3±1.48	0.0	5.0±1.53
No. of nematodes	175	>200	166	>200	106	>200

The functional guilds of each genus [22] and their standard weightings were used to calculate the components of the food web from which the enrichment (EI) and structural (SI) indices were calculated for each soil sample (see [23] for further details). Values derived from qPCR data represent means ± sem.

Désirée and OSR which was chosen as the non-solanaceous comparison crop because it is grown in rotation with potato at the field site. The lack of any differences between untransformed and transformed cv. Désirée at flowering or harvest establishes little impact of the transgenic plants on the non-target nematode soil community in the field in spite of the level of control of *G. pallida* that was achieved.

Discussion

Potato plants were developed that transgenically expressed a disulphide-constrained peptide (nAChRbp) capable of binding to nematode acetylcholine receptors and inhibiting chemoreception of cyst nematodes. A tissue-specific promoter restricted expression of the peptide to the outer cell layers of the root tip. Exudates from the transgenic potato plants inhibited alkaline phosphatase as

expected if the peptide was successfully expressed and secreted from the roots. However it was uncertain that the quantity of peptide secreted into soil and its stability there would ensure effective resistance when transgenic potato plants were challenged with *G. pallida*. The potato lines trialled did display a level of resistance to *G. pallida* in both a containment glasshouse trial and in the field that establishes the potential of this novel approach.

The results validate the root tip-specific promoter of the *Arabidopsis MDK4–20* gene [20] as a means of delivering effective root protection by the peptide under field conditions. This promoter is active in potato in both the zone of elongation and root border cells even after they detach from the root cap, often decorating the zone of elongation. This may enhance protection of this zone from invading nematodes [20]. PCN normally invades near root tips which slows root extension, particularly by lateral roots. This reduces the volume of soil from which the plant draws

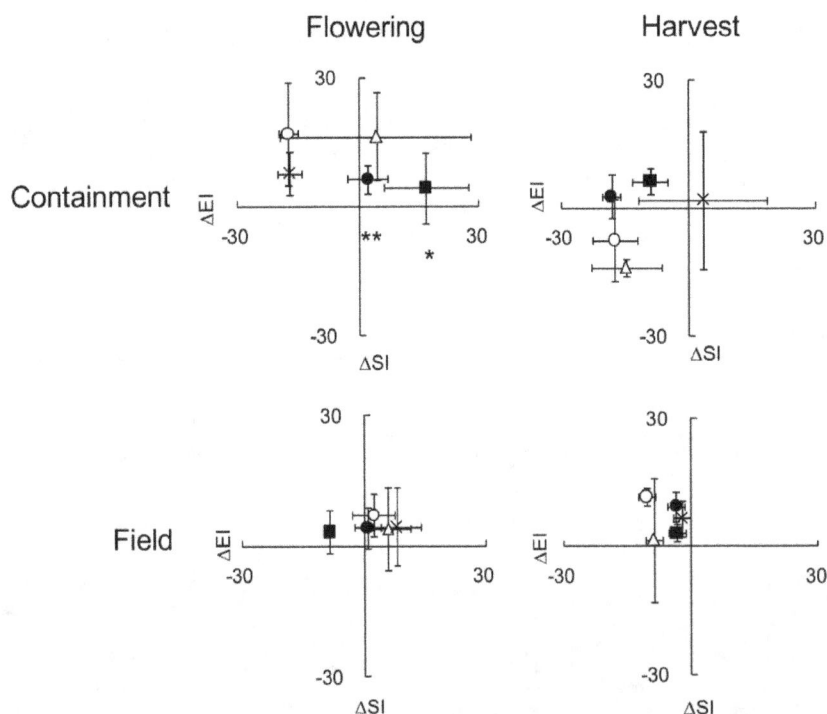

Figure 4. Changes in enrichment and structural indices for soil nematode faunal analysis relative to pre-plant levels. The changes for enrichment (ΔEI) and structural indices (ΔSI) were measured in both containment and the field at potato flowering and immediately before harvest. For each trial, soil was sampled from around the roots of oil seed rape (○), untransformed potato cultivars Santé (△) and Désirée (**X**) plus 2 lines of the latter cultivar expressing a cystatin [9,10] (■) and 3 lines expressing the peptide (●). The containment trial involved three pots for each plant type. For the field trial there were nine pooled samples for each of four replicate plots/line or plant type. Values are means ± sem.

water and nutrients [26]. The peptide's mode of action suppresses this important aspect of the pathology before other defences such as a cystatin could act as an anti-feedant on just those nematodes that establish in roots. This suggests the resistance conferred on potato roots by expressing these different traits should be additive. If so, this is likely to prevent economic damage by *G. pallida*. Both a cystatin [10] and the peptide have provided >75% resistance so if fully additive they should provide circa 95% control. This possibility will be studied in future work.

Plants expressing the peptide-based resistance, or a previously described approach involving transgenic expression of a cystatin in nematode feeding cells, had no impact on standard enrichment or structural indices of the non-target nematode soil community relative to changes caused by non-transgenic potato plants. The relative abundance of nematode genera that contributed to the faunal indices was determined by qPCR analysis of DNA from pooled nematodes extracted from soil samples. This employed genus-specific primers designed from 18S SSU DNA sequence of those nematodes present at the study site. The values obtained by determining EI and SI concurrently for replicate samples by morphology and the qPCR approach established that the molecular technique provided reliable estimates of these indices. This outcome is consistent with nematodes being particularly suitable for a normalised qPCR approach for determining the relative abundance of each genus. Their somatic cells are post-mitotic and growth involves an increase in cell size rather than number [27]. In *C. elegans* the genome copy number rises 2–4 fold as germ line cells increase during adult development [28] before this new level is maintained [29,30] until a post-reproductive variation in DNA content occurs [28]. Errors associated with variation in the relative abundance or reproductive state of adults in soil samples were clearly unimportant in the current work given the good agreement obtained with the estimates based on morphological identification.

Measurements of faunal indices were made at flowering when the root size of potato plants peaks [31] and immediately prior to harvest. The current work emphasised changes (ΔEI and ΔSI) relative to pre-plant values to compare the relative effect that the different plantings imposed rather than absolute values that reflect past agricultural activity. The SI value is primarily determined by omnivorous and predatory nematodes that are sensitive to disturbance. They are often uncommon and variable in frequently tilled arable soils [32] such as those used in this work. The fall in SI value between the two pre-plant samples taken for the containment trial and the later field trial probably reflects the impact of tilling which occurred just before establishing the field trial in spring. It is the more tolerant taxa contributing to SI that are likely to persist in such conditions [33]. In contrast, soil for the containment trial was collected in the summer of the previous year. The ΔSI values associated with the transgenic lines were greater at flowering in the containment trial only and no other differences from soil supporting growth of untransformed Désirée were recorded. The EI value is a good indicator of the amount of N in the soil [34] as it detects changes for those genera that respond to increased bacterial abundance arising from factors like fertiliser application and root turnover. The growth of GMNR lines in soil had no significant adverse effects on the nematodes that contribute to ΔEI values. There is no evidence from this work that the transgenic lines pose an environmental risk to the non-target nematode community. This is in contrast to the impact of the nematicides they could replace [12,15].

The faunal analysis suggests field trials have value for assessing non-target effects on the soil nematode community. Unfortunately EU Directive 2001/18/EC [35] that governs such field trials recognises the need for field releases at the research stage (clause 23) but it does not distinguish between these and large development stage trials in the detailed data and risk assessments it demands. The EU regulations discourage small scale trials and the situation in the UK is a particular concern. The results we report are from a field trial in the UK in 2009 covered by one of only 9 consents issued in the UK from 2002-09. The total for the EU over this period is over 730 [36] in contrast to 16,049 records for USA [37] from the first field test application to the accrued total by February 2010. Both USA and Canadian regulations recognise the need for small scale experimental trials to gain the type of ecological information we have now provided. Canadian regulations identify the need to study the environmental safety and performance of the modified plants in the natural environment rather than a glasshouse [38] which our results support. Without reform, Directive 2001/18/EC will continue to hinder development of transgenic approaches that could help conserve the soil quality of productive arable fields in the EC.

Materials and Methods

Plasmid construction

The MDK-peptide construct comprised the promoter region of the *Arabidopsis MDK4–20* gene [20] (At5g54370) driving expression of the disulphide constrained 7-mer peptide nAChRbp (CTTMHPRLC). The peptide coding sequence was fused to an N-terminal signal sequence from the *Nicotiana plumbaginifolia* calreticulin gene to ensure efficient secretion from the roots cells as in previous work [16]. A DNA fragment encoding the signal sequence and constrained peptide sequence was obtained using a series of splice overlap extension polymerase chain reactions. Restriction enzyme sites were incorporated into the primers to allow cloning as a *Bam* HI-*Sac* I fragment into the binary vector pBI121 (Clontech, California, USA) from which the GUS coding region had been removed. The CaMV35S promoter was then excised as a *Hin*d III-*Bam* HI fragment and replaced with a 924 bp promoter region of the *MDK4–20* gene [20].

Transformation and analysis of transgenic potato lines

The binary construct was introduced into *Agrobacterium tumefaciens* LBA4404 and leaf discs of *Solanum tuberosum* cv Désirée were transformed as described before [39]. Transgenic plantlets were rooted in liquid MS medium supplemented with 0.1 mg l^{-1} NAA before transfer to either soil or a peptide assay system as described below. Untransformed control plants were taken through identical tissue culture procedures with the exception that the *Agrobacterium* did not harbour a construct and kanamycin selection was not included.

Root exudates were collected from plant lines confirmed as transgenic by PCR analysis. Plantlets were rooted in liquid medium then transferred to 15 ml polypropylene tubes containing 7 ml sterile tap water. Plantlets were kept in a growth chamber for 14 days during which time the volume of water in each tube was maintained at 7 ml. The water was removed and used in an assay to measure the inhibition of alkaline phosphatase. For each reaction, 20 µl root exudate and 0.2 units alkaline phosphatase were preincubated for 30 min at room temperature prior to the addition of 100 µl p-nitrophenyl phosphate (1 mg ml^{-1} in 0.2 M Tris buffer). The absorbance at 405 nm was measured immediately and at subsequent 5 min intervals for one hour and the resultant regression lines used to determine enzyme activity. Four replicates were tested for each root exudate sample. Appropriate controls were included that lacked enzyme, substrate or exudate and a standard curve was derived from non-transgenic root exudate spiked with known concentrations of synthetic peptide (GL Biochemistry, Shanghai, China).

Containment trials

The potato plants for resistance trial were grown as before [40] in a sand/loam mix containing cysts of *G. pallida* to provide an initial population of 10 viable eggs/g soil. There were 10 replicate pots for untransformed Désirée and each of three transgenic lines harbouring the MDK-peptide construct. *G. pallida* final population estimates were made for two dry soil samples of 100 g taken post harvest from each pot. Cysts were recovered from each 100 g sample using a Fenwick can, disrupted in 10 ml water and the released eggs in three replicate 1 ml aliquots counted using standard methods [41].

For nematode faunal analysis in containment, all plants were trialled in 25 cm pots containing approximately 9 kg soil taken from the future field trial site. A single potato plant or six oil seed rape (OSR; *Brassica napus*) plants were grown in each pot. Two GMNR lines studied previously that express the engineered rice cystatin OcIΔD86 either constitutively [9] or under control of a root specific promoter [10] were included in this trial. Immediately prior to planting the soil was thoroughly mixed and triplicate 150 g samples removed for nematode faunal analysis. Rooted potato plantlets taken from liquid media were transferred to a compost/Perlite mix in 7 cm pots and established for 42 days prior to transfer into the field soil. OSR was similarly pre-grown for 42 days. There were three replicate pots each for OSR, untransformed Désirée and Santé potato cultivars, two cystatin-expressing transgenic lines and three MDK-peptide transgenic lines. Plants were grown at 20±2°C on 16 h: 8 h light cycle in a containment glasshouse and a 100 g soil sample was withdrawn from the plant rhizosphere of each pot both when the untransformed cv Désirée was in flower and just before harvest. Each sample was collected from 6 positions that were 10 cm deep around a 7 cm radius from the plant stem. Soil nematodes were extracted for analysis using the Seinhorst flask method [41].

Field trial

The field trial was conducted at the University of Leeds Field Station, Headley Hall Farm, Tadcaster, West Yorkshire, UK (DEFRA Consent 07/R31/01). This was a randomised block design with 4 replicate small plots per line using the same range of plants as in containment trial for faunal analysis. Each potato plot contained 9 plants grown from chitted tubers 38 cm apart in the row with rows 77 cm apart. OSR was planted near the site and transplanted at 17 plants/m² into the replicate plots for this crop within the trial, concurrently with potato planting. The trial was surrounded with guard rows of potato cv Santé. The sampling for faunal analysis was before planting, at flowering of cv Désirée and before harvest. On the two occasions after planting, 100 g of soil was recovered from the rhizosphere of each plant and nematodes extracted from pooled samples for the nine plants per replicate plot using the tray method [41] to facilitate handling many samples. Preliminary work established that both this and the Seinhorst flask method of nematode extraction from soil provided similar values for nematode faunal analysis. Following removal of the haulms, soil samples were collected from the base of each plant upon harvest of the tubers. The soil samples from 9 plants in a replicate plot were mixed thoroughly, dried and duplicate 100 g aliquots per plot removed for *Globodera* cyst extraction and egg counts as described above.

Bar coding of nematodes and development of genus specific qPCR primers

Soil nematodes were initially sampled from the future field trial site and extracted using the techniques described above to allow identification of the genera present. Three soil cores (2 cm diameter and 15 cm deep) were removed from each of 12

positions and the samples combined and mixed prior to nematode extraction. Each individual nematode was assigned a tentative identification based upon morphological characteristics observed using a Leica DMRB microscope. It was then digested to release DNA [25] and the lysate was stored at −20°C until required. PCR was carried out on 2 μl lysate using Biotaq Red DNA polymerase (Bioline, London, UK) and primers SSU18A (AAAGATTAAGC-CATGCATG) and SSU26R (CATTCTTGGCAAATGCTT-TCG) designed to conserved sequences of the SSU gene [24]. The PCR reaction conditions were as described previously [25].

PCR products were purified (QIAquick kit; Qiagen, Sussex, UK) then directly sequenced using a Taq Dye Deoxy Terminator Cycle Sequencing system (Applied Biosystems) and an automated sequencer (Applied Biosystems 373A) with primer SSU9R (AGCTGGAATTACCGCGGCTG) [24]. The resulting sequences were used in BLASTN similarity searches of GenBank nucleotide database entries allowing a putative genus to be assigned to each sample (Table 1). Multiple sequence alignments performed in MegAlign within the DNASTAR software suite (Lasergene, Madison, Wisconsin, USA) were used as a basis for designing genus specific qPCR primers (PrimerSelect, DNAS-TAR) to the variable regions within the SSU gene that could discriminate between the different nematode genera present at the trial site (Table 2). BLASTN searches of the GenBank database confirmed the predicted specificity of each primer pair. Genus-specific primer pairs were tested in two SYBR Green qPCR reactions with dissociation curves using DNA template prepared from a) single nematodes and b) a pool of 200 randomly selected worms. In addition, the sequence of products amplified from pools of 200 randomly selected individuals was determined. The primers were validated as specific for the range of nematodes present in the soil samples.

qPCR based method for faunal analysis

The nematodes extracted from 100 g soil samples as described above were incubated in 1 ml of 0.25 M NaOH overnight at 25°C followed by heating at 99°C for 3 minutes. After cooling to room temperature, 200 μl 1 M HCl, 500 μl 0.5 M Tris-HCl (pH 8.0) and 250 μl 2% Triton X-100 were added and the samples heated again at 99°C for 3 min. Lysate was stored at −20°C. Comparative qPCR was performed on a Mx3005P qPCR system (Agilent Technologies, Cheshire, UK) using 5 μl template DNA in a 25 μl reaction volume with 1× iQ SYBR Green Supermix (BioRad, Herts, UK) and 300 nM of the appropriate genus specific primer pair. Reactions, including no template controls, were performed in triplicate. The reaction conditions were 95°C for 3 minutes, then 40 cycles of 95°C for 30 seconds, 60°C for 30 seconds and 72°C for 30 seconds.

Ct values obtained for a series of individuals for each genus were used to normalise mean DNA content per nematode from qPCR results, with *Aphelenchoides* as an arbitrary calibrator. For initial sampling from the field trial site, the relative numbers of nematode genera were validated with tandem measurements using a microscope-based morphological approach of nematode identification. A subsequent minor adjustment to the normalization was made after relating these parallel morphological counts to qPCR measurements of nematodes. Normalised values were used to identify relative number per genus to determine EI and SI values [22,23] of containment and field trial soil samples.

Statistical approaches

All analyses were carried out using SPSS version 16.0 (http://wwwspsscom/?source=homepage&hpzone=nav_bar). For Fig. 2, resistance is expressed as before [9,40]. The number of soil samples in containment trials and the field trial is given above. The arcsin values of the proportions were analysed by the general linear model univariate procedure. This latter procedure and oneway analysis of variance was also used for the data in Fig. 4.

Supporting Information

Figure S1 Specificity of qPCR primer pairs for *Eucephalobus* and *Acrobeloides/Cephalobus*. Fluorescent emission from SYBR green labeled products of individual qPCR reactions using a primer pair designed for either *Eucephalobus* (GUF + EuR) or *Acrobeloides/Cephalobus* (GUF + AcCR) and template DNA from each genus. For each primer pair, amplification occurred only with the correct template DNA and not with DNA

from the other, closely related nematode. Primer sequences are given in Table 2.

Acknowledgments

We thank B. Merry, F. Moulton, M. Fearnehough and J. Hibbard for technical assistance with potato transformation and trials.

Author Contributions

Conceived and designed the experiments: JG DW PEU HJA. Performed the experiments: JG DW. Analyzed the data: CJL PEU HJA. Wrote the paper: CJL HJA. Made and evaluated the transgenic lines: JG. Designed and carried out bar coding and faunal analysis: DW JG CJL PEU.

References

1. The Royal Society (2009) Reaping the benefits: science and the sustainable intensification of global agriculture. RS Policy document 11/09. Available: http://royalsociety.org/uploadedFiles/Royal_Society_Content/policy/publications/2009/4294967719.pdf.
2. Doran JW, Zeiss MR (2000) Soil health and sustainability: managing the biotic component of soil quality. App Soil Ecol 15: 3–11.
3. European Community Commission (2010) Soil biodiversity: functions, threats and tools for policy makers. doi: 10.2779/14571.
4. EC Regulation (2009) No 1107/2009 Concerning the placing of plant protection products on the market and repealing Council Directives 79/117/EEC and 91/414/EEC. Official Journal of the European Union 52: 1–50.
5. Clayton R, Parker B, Ballingall M, Davies K (2008) Impact of reduced pesticide availability on control of potato cyst nematodes and weeds in potato crops. Potato Council Ltd, Oxford, UK. Available: http://www.potato.org.uk/sites/default/files/%5Bcurrent-page%3Aarg%3A%3F%5D/Reduced%20pesticide%20PCN%20case%20report%20Sept%2008_0.pdf.
6. Twining S, Clarke J, Cook S, Ellis S, Gladders P, et al. (2009) Pesticide availability for potatoes following revision of Directive 91/414/EEC: Impact assessments and identification of research priorities. (Report No. 2009/2). Potato Council Ltd, Oxford, UK. Available: http://www.potato.org.uk/sites/default/files/%5Bcurrent-page%3Aarg%3A%3F%5D/20092_Pesticide_availability_Report_R415_0.pdf.
7. Trudgill DL, Elliott MJ, Evans K, Phillips MS (2003) The white potato cyst nematode (*Globodera pallida*) - a critical analysis of the threat in Britain. Annals Appl Biol 143: 73–80.
8. Urwin PE, Atkinson HJ, Waller DA, McPherson MJ (1995) Engineered oryzacystatin-1 expressed in transgenic hairy roots confers resistance to *Globodera pallida*. Plant J 8: 121–131.
9. Urwin PE, Green J, Atkinson HJ (2003) Expression of a plant cystatin confers partial resistance to *Globodera*, full resistance is achieved by pyramiding a cystatin with natural resistance. Mol Breeding 12: 263–269.
10. Lilley CJ, Urwin PE, Johnston KA, Atkinson HJ (2004) Preferential expression of a plant cystatin at nematode feeding sites confers resistance to *Meloidogyne incognita* and *Globodera pallida*. Plant Biotech J 2: 3–12.
11. Cowgill SE, Atkinson HJ (2003) A sequential approach to risk assessment of transgenic plants expressing protease inhibitors: effects on nontarget herbivorous insects. Transgenic Res 12: 439–449.
12. Cowgill SE, Danks C, Atkinson HJ (2004) Multitrophic interactions involving genetically modified potatoes, nontarget aphids, natural enemies and hyperparasitoids. Mol Ecol 13: 639–647.
13. Cowgill SE, Bardgett RD, Kiezebrink DT, Atkinson HJ (2002) The effect of transgenic nematode resistance on non-target organisms in the potato rhizosphere. J Appl Ecol 39: 915–923.
14. Atkinson HJ, Keizenbrink DT (2004) Determining risks to soil organisms associated with a genetically modified (GM) crop expressing a biopesticide in its roots. Available: http://randd.defra.gov.uk/Default.aspx?Menu=Menu&Module=More&Location=None&Completed=0&ProjectID=11999.
15. Celis C, Scurrah M, Cowgill S, Chumbiauca S, Green J, et al. (2004) Environmental biosafety and transgenic potato in a centre of diversity for this crop. Nature 432: 222–225.
16. Liu B, Hibbard JK, Urwin PE, Atkinson HJ (2005) The production of synthetic chemodisruptive peptides *in planta* disrupts the establishment of cyst nematodes. Plant Biotech J 3: 487–496.
17. Qian H, Robertson AP, Powell-Coffman JA, Martin RJ (2008) Levamisole resistance resolved at the single-channel level in *Caenorhabditis elegans*. FASEB J 22: 3247–3254.
18. Wang D, Jones LM, Urwin PE, Atkinson HJ (2011) A synthetic peptide shows retro- and anterograde neuronal transport before disrupting the chemosensation of plant-pathogenic nematodes. PLoS ONE. pp e17475, 1–9.
19. Winter MD, McPherson MJ, Atkinson HJ (2002) Neuronal uptake of pesticides disrupts chemosensory cells of nematodes. Parasitol 125: 561–565.
20. Lilley CJ, Wang D, Atkinson HJ, Urwin PE (2010) Effective delivery of a nematode-repellent peptide using a root cap-specific promoter. Plant Biotech J 9: 151–61.
21. Ingham ER (2000) Nematodes. In: Tugel A, Lewandowski A, Happe-vonArb D, eds. Soil Biology Primer. Iowa: Soil and Water Conservation Society, Available: http://soils.usda.gov/sqi/concepts/soil_biology/nematodes.html.
22. Bongers T (1990) The maturity index: an ecological measure of environmental disturbance based on nematode species composition. Oecologia 83: 14–19.
23. Ferris H, Bongers T, de Goede RGM (2001) A framework for soil food web diagnostics: extension of the nematode faunal analysis concept. Appl Soil Ecol 18: 13–29.
24. Blaxter ML, De Ley P, Garey JR, Liu LX, Scheldeman P, et al. (1998) A molecular evolutionary framework for the phylum Nematoda. Nature 392: 71–75.
25. Floyd R, Abebe E, Papert A, Blaxter M (2002) Molecular barcodes for soil nematode identification. Mol Ecol 11: 839–850.
26. Trudgill DL, Evans K, Phillips MS (1998) Potato cyst nematodes: Damage mechanisms and tolerance in the potato. In: Marks RJ, Brodie BB, editors. Potato cyst nematodes: Biology, distribution and control. Wallingford: CAB International. pp 117–133.
27. Flemming AJ, Shen ZZ, Cunha A, Emmons SW, Leroi AM (2000) Somatic polyploidization and cellular proliferation drive body size evolution in nematodes. Proc Natl Acad Sci U S A 97: 5285–5290.
28. Golden TR, Beckman KB, Lee AHJ, Dudek N, Hubbard A, et al. (2007) Dramatic age-related changes in nuclear and genome copy number in the nematode *Caenorhabditis elegans*. Aging Cell 6: 179–188.
29. Gumienny TL, Lambie E, Hartwieg E, Horvitz HR, Hengartner MO (1999) Genetic control of programmed cell death in the *Caenorhabditis elegans* hermaphrodite germline. Development 126: 1011–1022.
30. Blum ES, Driscoll M, Shaham S (2008) Noncanonical cell death programs in the nematode *Caenorhabditis elegans*. Cell Death Diff 15: 1124–1131.
31. Jefferies RA, Lawson HM (1991) A key for the stages of development of potato (*Solanum tuberosum*). Annals Appl Biol 119: 387–399.
32. Leroy BLM, De Sutter N, Ferris H, Moens M, Reheul D (2009) Short-term nematode population dynamics as influenced by the quality of exogenous organic matter. Nematol 11: 23–38.
33. Sanchez-Moreno S, Minoshima H, Ferris H, Jackson LE (2006) Linking soil properties and nematode community composition: effects of soil management on soil food webs. Nematol 8: 703–715.
34. Sanchez-Moreno S, Smukler S, Ferris H, O'Geen AT, Jackson LE (2008) Nematode diversity, food web condition, and chemical and physical properties in different soil habitats of an organic farm. Biol Fertil Soils 44: 727–744.
35. European Parliament, Council (2001) European Parliament Directive 2001/18/EC on the deliberate release into the environment of genetically modified organisms. Available: http://eur-lex.europa.eu/LexUriServ/LexUriServ.do?uri=CELEX:32001L0018:EN:NOT.
36. European Commission Joint Research Institute of Health and Consumer Protection (2010) Deliberate release into the environment of GMOs for any other purposes than placing on the market (experimental releases). Available: http://gmoinfo.jrc.ec.europa.eu/.
37. ISB (2010) Information Systems for Biotechnology Website. Blacksburg, VA: ISB Project, Virginia Tech. Available: http://www.isb.vt.edu/.
38. Canadian Food Inspection Agency Science Branch (2008) Office of Biotechnology - confined research field trials for plants with novel traits (PNTs). Available: http://www.inspection.gc.ca/english/plaveg/bio/pub/pntvcne.shtml.
39. Dietze J, Blau A, Willmitzer L (1995) *Agrobacterium*-mediated transformation of potato (*Solanum tuberosum*). In: Potrykus I, Spangenberg G, eds. Gene transfer to plants. Heidelberg/Berlin: Springer. pp 24–30.
40. Urwin PE, Troth KM, Zubko EI, Atkinson HJ (2001) Effective transgenic resistance to *Globodera pallida* in potato field trials. Mol Breeding 8: 95–101.
41. Southey JF (1986) Laboratory methods for work with plant and soil nematodes. London: HMSO.

2

Estimation of Wheat Agronomic Parameters using New Spectral Indices

Xiu-liang Jin[1,2], Wan-ying Diao[3], Chun-hua Xiao[3], Fang-yong Wang[4], Bing Chen[4], Ke-ru Wang[1,3]*, Shao-kun Li[1,3]*

1 Institute of Crop Science, Chinese Academy of Agricultural Sciences/Key Laboratory of Crop Physiology and Production Ministry of Agriculture, Beijing, China, 2 Beijing Research Center for Information Technology in Agriculture, Beijing, China, 3 Key Laboratory of Oasis Ecology Agriculture of Xinjiang Construction Crops, Shihezi, China, 4 Institute of Cotton, Xinjiang Academy of Agricultural Reclamation Sciences, Shihezi, China

Abstract

Crop agronomic parameters (leaf area index (LAI), nitrogen (N) uptake, total chlorophyll (Chl) content) are very important for the prediction of crop growth. The objective of this experiment was to investigate whether the wheat LAI, N uptake, and total Chl content could be accurately predicted using spectral indices collected at different stages of wheat growth. Firstly, the product of the optimized soil-adjusted vegetation index and wheat biomass dry weight (OSAVI×BDW) were used to estimate LAI, N uptake, and total Chl content; secondly, BDW was replaced by spectral indices to establish new spectral indices (OSAVI×OSAVI, OSAVI×SIPI, OSAVI×CI$_{red\ edge}$, OSAVI×CI$_{green\ mode}$ and OSAVI×EVI2); finally, we used the new spectral indices for estimating LAI, N uptake, and total Chl content. The results showed that the new spectral indices could be used to accurately estimate LAI, N uptake, and total Chl content. The highest R^2 and the lowest RMSEs were 0.711 and 0.78 (OSAVI×EVI2), 0.785 and 3.98 g/m^2 (OSAVI×CI$_{red\ edge}$) and 0.846 and 0.65 g/m^2 (OSAVI×CI$_{red\ edge}$) for LAI, nitrogen uptake and total Chl content, respectively. The new spectral indices performed better than the OSAVI alone, and the problems of a lack of sensitivity at earlier growth stages and saturation at later growth stages, which are typically associated with the OSAVI, were improved. The overall results indicated that this new spectral indices provided the best approximation for the estimation of agronomic indices for all growth stages of wheat.

Editor: Ive De Smet, University of Nottingham, United Kingdom

Funding: The study is supported by the National Natural Science Foundation of China (grant numbers 31071371, 30760109, 41161068) and the Key Projects in the National Science & Technology Pillar Program during the Twelfth Five-Year Plan Period (grant number 2012BAH27B04). The National Natural Science Foundation of China had role in study design, data collection and analysis, the Key Projects in the National Science & Technology Pillar Program during the Twelfth Five-Year Plan Period had role in decision to publish,or preparation of the manuscript.

Competing Interests: The authors have declared that no competing interests exist.

* E-mail: wkeru01@163.com (KRW); lishk@mail.caas.net.cn (SKL)

Introduction

The development of remote sensing has provided opportunities to quantitatively describe agronomic parameter changes across all growth stages of crops. The application of remote sensing to agronomic problems has created new methods to effectively improve field crop management. Many authors have provided detailed information about the relationships between spectral indices and agronomic parameters, including the leaf area index (LAI), nitrogen (N) uptake, total chlorophyll (Chl) content, and so on.

The LAI is a key variable for the diagnosis and prediction of crop growth and yield. This makes the LAI critical for effective understanding of the biophysical processes of plant canopies and the prediction of plant growth and productivity [1–6]. Rouse et al. suggested that the most well-known and widely used vegetation index was the normalized difference vegetation index (NDVI), and found that NDVI was linearly correlated with the leaf area index (LAI) in crop fields [7]. However, the NDVI does possess certain limitations related to soil background brightness, in that the NDVI tends to be affected by different soil color and moisture conditions [8–10]. To overcome this problem, Rondeaux et al. proposed using an optimized soil-adjustment factor, and obtained an optimized soil-adjusted vegetation index (OSAVI), which mitigated the effects of soil and moisture conditions [11].

Nitrogen is a critically important element that is monitored in an effort to maintain crop health, while there is a good relationship between nitrogen content and chlorophyll content [12–13], therefore chlorophyll content is an important indicator for nitrogen fertilizer applications. Stone et al. detected and predicted N uptake in winter wheat using hand-held sensors [14–15]. Osborne et al. identified the important reflectance wavelengths for the prediction of N concentration changes at different growth stages [16]. Gitelson et al. suggested that the green NDVI (GNDVI) was more sensitive than the NDVI for wheat N uptake over 100 kg/ha [17], but Moges et al. indicated that the NDVI was more robust than the GNDVI for prediction of crop N uptake [18]. Certain researchers have proposed many N indicators for the assessment of crop N change according to the spectral features of chlorophyll in the visible and red-edge bands [19–20]. A good relationship between a combined index (the ratio of modified chlorophyll absorption ratio index and modified triangular vegetation index2, MCARI/MTVI2) and leaf nitrogen concentration was found by Eitel et al. [21]. Fitzgerald et al. reported that a spectral index, the canopy chlorophyll content index (CCCI), could predict canopy N, and specifically found a good relationship between the CCCI and canopy N [22]. Vigneaua et al. used field hyperspectral imaging as a non-destructive method to assess leaf nitrogen content in wheat [23]. And recently, remote sensing methods have been developed and applied for the prediction of Chl content for field crop management [24–30]. Gitelson et al. identified the best vegetation indices for the estimation of Chl content [30]. A good correlation between total Chl content and (R_{NIR} which is in the near infrared band reflectance/$R_{red\ edge}$ which is in the red edge position reflectance)–1, (R_{NIR}/R_{green} which is in the green band reflectance)–1 was found by Gitelson et al. [31], and Gitelson et al. applied the new vegetation index ($R_{NIR}/R_{red\ edge}$–1 and R_{NIR}/R_{green}–1) for the estimation of Chl content and indirectly estimated gross primary production (GPP) [32,33].

However, spectral indices still eixst insensitivity at earlier growth stages, and saturation at later growth stages. Because biomass dry weight (BDW) was accumulated gradually as growth stages progressed, and insensitivity at earlier growth stages and saturation at later growth stages was not observed. Consequently, the objectives of this study were to (1) combine OSAVI and BDW for estimating LAI, N uptake, and total Chl content in wheat; (2) replace BDW with spectral indices; (3) and attempt to propose new spectral indices, in an effort to improve insensitivity at earlier growth stages and saturation at later growth stages.

Materials and Methods

2.1. Design of Experiment

Field experiments were conducted in 2009 and 2010 at the ShiHezi University experiment site (44°20′N, 86°3′E), Xinjiang Province, China. The experiment site had representative soil types and crop management practices for Xinjiang Province, China. The soil was fine-loamy with a total N content of 42.6 mg kg^{-1}, Olsen P of 26.5 mg kg^{-1}, exchangeable K of 139.4 mg kg^{-1}, and organic matter content of 11.6 g kg^{-1} in the 0–30 cm layer. Three local wheat cultivars: Xinchun 6, Xinchun 17, and Xinchun 22, were planted on April 5th 2008 and April 8th 2009. Nitrogen fertilizer as urea was applied at four rates (0, 105, 225, and 345 kg N ha^{-1}) before planting. The N application was distributed at three stages in the growth process in the following percentages: 50% at seeding, 25% at jointing, and 25% at booting. For all treatments, 99 kg ha^{-1} P_2O_5 (as monocalcium phosphate [Ca(H$_2$PO$_4$)$_2$]) and 150 kg ha^{-1} K$_2$O (as KCl) was applied prior to seeding. The experiment was a 2-way factorial arrangement of treatments in a randomized complete block design with three replications for each treatment. Other management practices followed local standard wheat production practices.

2.2. Measurement of Canopy Reectance

Spectral measurements were carried out at the following stages (2009, 2010): tillering (8th May, 6th May), jointing (20th May, 25th May), heading (8th June, 10th June), anthesis (18th June, 20th June), and filling (27th May, 25th May), respectively. All canopy spectral measurements were mounted on the tripod boom and held in a nadir orientation 1.0 m above the canopy. Measurements were taken under clear sky conditions between 10:00 and 14:00 (Beijing local time) using an ASD Field Spec Pro Spectrometer (Analytical Spectral Devices, Boulder, CO, USA). This spectrometer is fitted with a 25° field of view fiber optics, operating in the 350–2500 nm spectral region with a sampling interval of 1.4 nm between 350 and 1050 nm, and 2 nm between 1050 and 2500 nm, and with spectral resolution of 3 nm at 700 nm, and 10 nm at 1400 nm. A 40 cm × 40 cm BaSO4 calibration panel was used for calculating the black and baseline

reectance. To reduce the possible effect of sky and field conditions, spectral measurements were taken at four sites in each plot and were averaged to represent the canopy reflectance of each plot. Vegetation radiance measurement was taken by averaging 16 scans at an optimized integration time, with a dark current correction at every spectral measurement. A panel radiance measurement was taken before and after the vegetation measurement by two scans each time.

2.3. Agronomic Parameters Measurement

Immediately following spectral measurements, the leaf area index (LAI) was measured using the LAI-2000 Plant Canopy Analyzer (LI-COR Inc., Lincoln, NE, USA) with spectrometric measurements at the same position and gained by destructive sampling. The wheat was cut at the ground level and wet weights were recorded in 0.24 m^2. Each plant was then dried at 70°C for 3 d and the dry weight was recorded. Dry plant material was then milled and analyzed in the laboratory. Total N content was estimated using a dry combustion method in a Dumas Elementary Analyser (Macro-N, Foss Heraeus, Hanau, Germany) [34].

The spectral measurements positions of the wheat leaves were collected using a hole puncher (diameter, 0.4 cm). Then, about 0.2 g wheat leaves of each sample was punched off in the laboratory. The selected samples were placed in 95% ethanol or acetone solution and allowed to stand for 24 hr in the dark. Following the 24-hr treatment, the leaves were white-green in color. Finally, leaf pigment density was measured using a colorimetric spectro-photometer. Absorbance of the supernatant was measured at 645, 652 and 663 nm, and chlorophyll a plus chlorophyll b content per unit leaf area was then calculated by the method of McKinney [35].

2.4. Selection of Spectral Indices

This study tested the spectral indices that were considered to be good candidates for estimation of plant chlorophyll content and LAI (Table 1) [11,31,36–37].

2.5. Statistical Analysis

Linear and nonlinear regression analysis was carried out using the biomass dry weight, OSAVI and OSAVI×biomass dry weight (the product of OSAVI and biomass dry weight) as independent variables, and the LAI, total Chl content, and N uptake (leaves yield N concentration of measured leaves) as dependent variables.

Statistically significant (p<0.05 or 0.01) coefficients of determination (R^2) between three crop parameters (LAI, total Chl content, N uptake) and new vegetation indices or biomass dry weight were analyzed using SPSS software (16.0, SPSS, Chicago, IIIinois, USA). The R^2 and root mean square error (RMSE) were used as metrics for quantifying the amount of variation explained by the relationships developed, as well as their accuracy. The performance of the model was evaluated through R^2 and RMSE for the estimation of in-situ measured LAI, total Chl content and N uptake. Generally, the performance of the model was estimated by comparing the differences of the R^2 and RMSE between the measured value and predicted value. The higher the R^2 and the lower the RMSE were, the higher the precision and accuracy of model to predict agronomic parameters was considered to be.

Results

3.1. Leaf Area Index (LAI)

A significant relationship was found between the OSAVI and LAI for the selected data, ranging from all growth stages of wheat across 2009 (Figure 1a). As noted above, the OSAVI was

Table 1. Summary of spectral indices, wavebands and citations for LAI, nitrogen uptake and total chlorophyll content in this paper.

Index	Name	Formula	Developer(s)
OSAVI	Optimized soil-adjusted vegetation index	$1.16 \times (R_{800} - R_{670})/(R_{800} + R_{670} + 0.16)$	Rondeaux et al. (1996) [11]
SIPI	Structure insensitive pigment index	$(R_{800} - R_{445})/(R_{800} - R_{680})$	Penuelas et al. (1995) [36]
$CI_{red\ edge}$	Red edge model	$(R_{800}/R_{700}) - 1$	Gitelson et al. (2005) [31]
$CI_{green\ model}$	Green model	$(R_{800}/R_{550}) - 1$	Gitelson et al. (2005) [31]
EVI2	Enhanced vegetation index 2	$2.5 \times (R_{800} - R_{660})/(1 + R_{800} + 2.4 \times R_{660})$	Jiang et al. (2008) [37]

Note: Ri denotes reectance at band i (nanometer).

calculated and measured for each stage by averaging ASD-2500 sensor readings from taken at the vertical height of the canopy at 1.0 m. Steven showed that the LAI has a strong relationship with the OSAVI [38]. Figure 1a has shown that for the OSAVI index, there was a problem with saturation at later growth stages in wheat, determination coefficient (R^2) value was 0.536. It is mainly for this reason that the OSAVI was not sensitive to the LAI (ranges from 4 to 6) at later growth stages. Possibly, this was because the band combination only considered the visible and near-infrared bands and was thus affected by environmental conditions.

Biomass dry weight and LAI were highly correlated ($R^2 = 0.688$, $P < 0.01$), independent of the area the wheat occupied. At earlier growth stages, the correlation between biomass dry weight and LAI (Figure 1b) was better than that between OSAVI and LAI (Figure 1a), R^2 value was 0.688. The improvement was even greater at later growth stages. This is important because it indicates that biomass dry weight can be used to estimate wheat LAI, and to compensate for the OSAVI's lack of sensitivity to the later growth stages changes in LAI. So, we used the OSAVI× biomass dry weight (BDW) index to estimate LAI changes using the data from all growth stages in wheat. The results showed that the OSAVI×BDW index was a good predictor of LAI, though the biomass dry weight alone was a more accurate predictor (Figure 1c). We performed non-linear regression on the OSA-

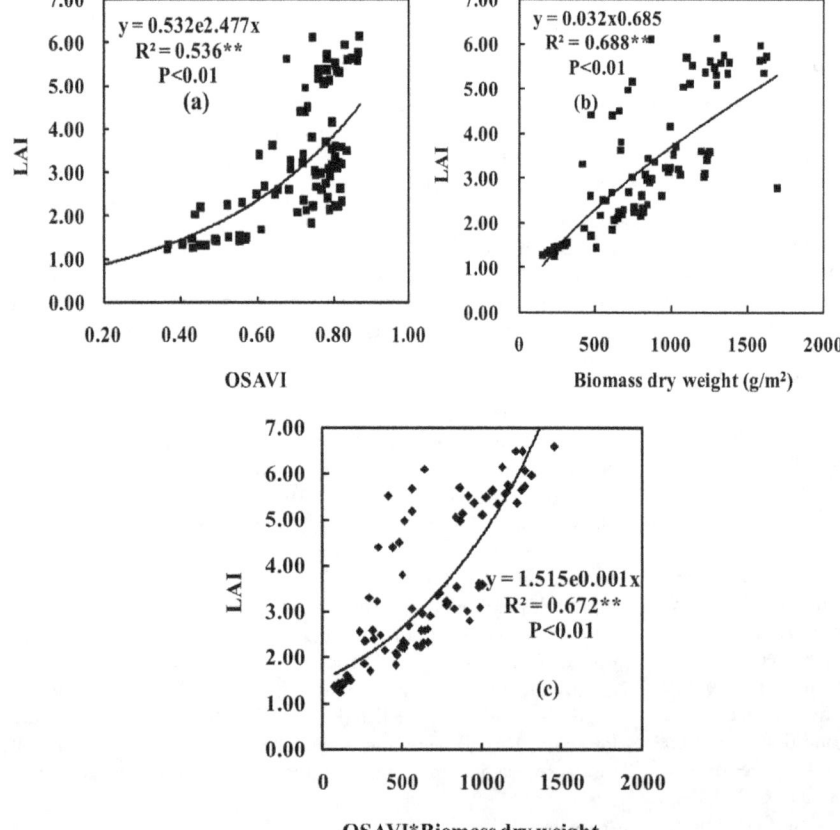

Figure 1. The coefficient of determination (R^2) between OSAVI, biomass dry weight (BDW), BDW×OSAVI and LAI under the different nitrogen treatments for wheat (n = 90). Note: Probability levels are indicated by n.s., * and ** for 'not significant', 0.05, and 0.01, respectively.

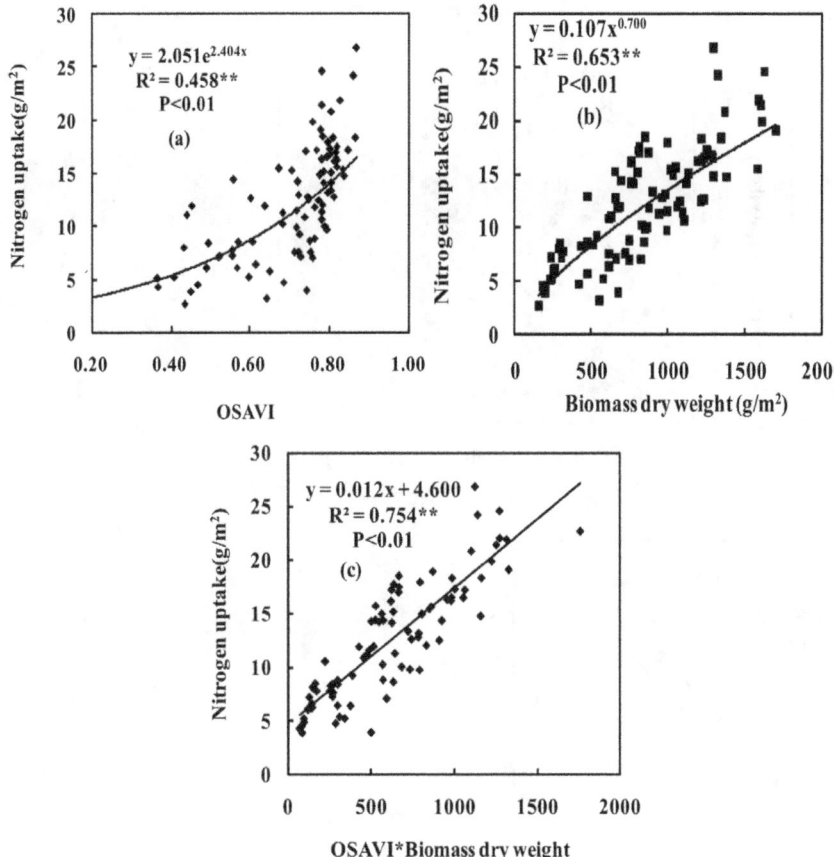

Figure 2. Relationship between OSAVI, biomass dry weight (BDW) and BDW×OSAVI and nitrogen uptake under the different nitrogen treatments for wheat (n = 90). Note: Same as above.

VI×BDW and LAI at all growth stages, and the resulting R^2 was 0.672. These results suggested that the OSAVI×biomass dry weight index was an improvement over the OSAVI index, because the OSAVI alone was not sensitive to LAI changes at later growth stages.

3.2. Nitrogen Uptake

The amount of nitrogen (N) taken up in wheat was correlated with the OSAVI with an R^2 value of 0.458 (Figure 2a). The results showed that the relationship between the OSAVI and N uptake was highly correlated ($P<0.01$) at all growth stages. However, at later growth stages when N uptake ranged from 15 to 27 g/m^2, saturation was observed. A main reason for this problem was that the OSAVI was affected by other environmental factors [11]. Specifically, in this paper, the OSAVI saturation was mainly influenced by the structure of the wheat canopy. For example, the plant height of wheat canopy vary from 80 cm to 120 cm at the later growth stages, the reflectance at the top of wheat canopy (0–40 cm) can be detected by spectrometer (direct light), but the reflectance at the bottom of canopy (40–120 cm) cannot be detected by spectrometer (diffused light). The high-density wheat affected the spectral measurement and plant height. This caused the proportion of the diffused light to increase, thereby increasing the OSAVI saturation. The above problems led to a lack of sensitivity to the changes in N uptake at later growth stages. This suggested that with the increase in wheat biomass, the OSAVI became slightly affected by the surrounding environmental

conditions, and the OSAVI and N uptake were correlated. The correlation between the OSAVI and N uptake could be explained by the ability of OSAVI to detect differences in red absorption. Moges et al. showed a similar finding, their research investigated the relationships of green, red, and near infrared bands to N uptake [16]. The results showed that the correlation between biomass dry weight and N uptake was better than that between OSAVI and N uptake, and the R^2 value was 0.653 (Figure 2b).

The OSAVI×biomass dry weight index (BDW) could be used to estimate N uptake changes, and was a good predictor of N uptake. It performed better than either biomass dry weight or OSAVI alone in the prediction of N uptake for wheat. The results showed that a linear regression performed between the OSAVI×BDW and N uptake at all growth stages resulted in an R^2 value of 0.754 (Figure 2c). This result suggested that the OSAVI×BDW index could be effectively used to improve the OSAVI's prediction accuracy for N uptake.

3.3. Total Chlorophyll Content

The total chlorophyll (Chl) content was correlated with OSAVI, with a coefficient of determination (R^2) of 0.632. The results suggested that the total Chl content in the wheat canopy could be estimated using OSAVI (Figure 3a). Possible explanation could be the red edge absorption waveband (670 nm) is very sensitive to total Chl content [23–24,26,29]. Previous research also showed a good relationship between total Chl content and biomass [39], and biomass changes influenced the spectral reflectance in the near-

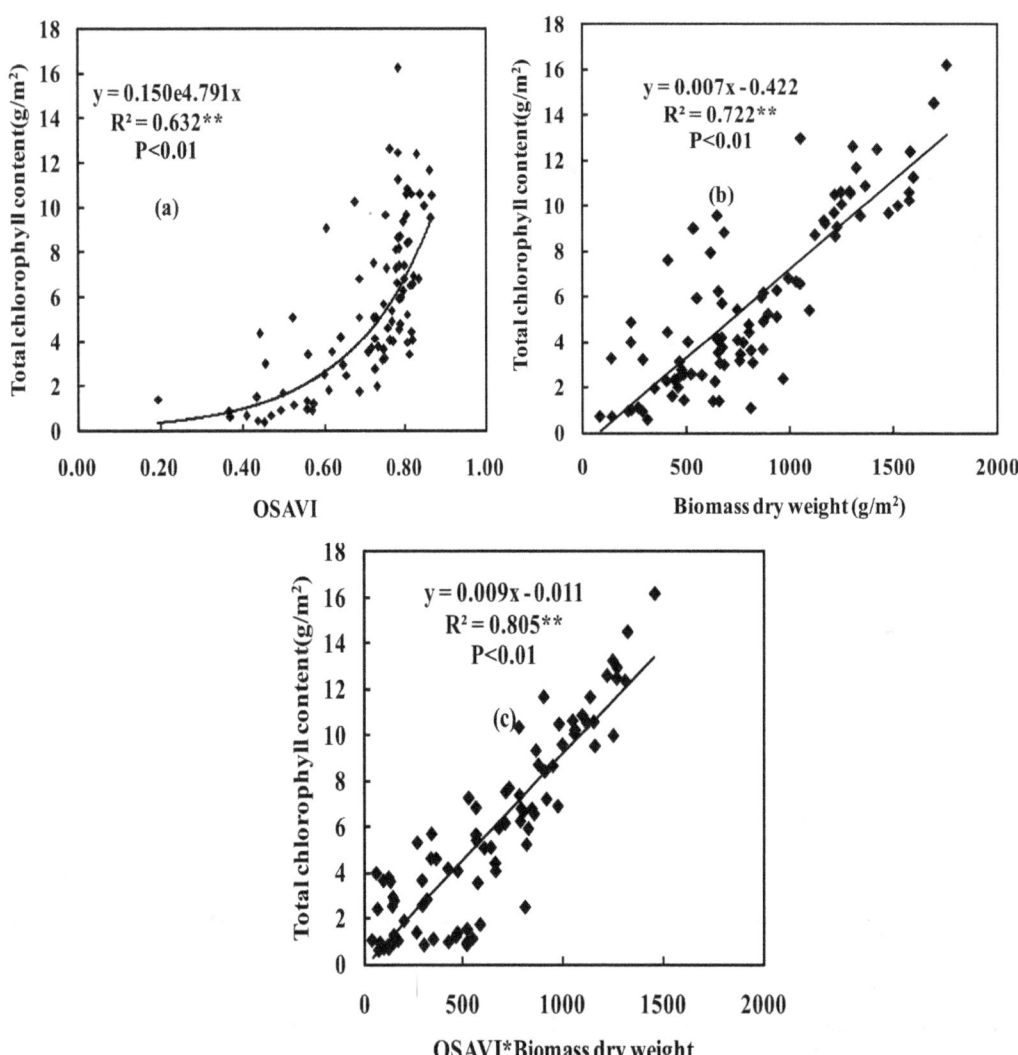

Figure 3. Quantitative relationship between OSAVI, biomass dry weight (BDW) and BDW×OSAVI and total chlorophyll content under the different nitrogen treatments for wheat (n = 90). Note: Same as above.

infrared waveband (800 nm). Both red edge absorption waveband and near-infrared waveband are involved in OSAVI calculation.

Table 2. Relationship between the new spectral indices and LAI under the different nitrogen treatments for wheat (n = 90).

Spectral indices	Simulated equations	Determination coefficient (R^2)
OSAVI×OSAVI	$y = 0.253\ln(x)+0.221$	0.634**
OSAVI×SIPI	$y = 0.174\ln(x)+0.507$	0.563**
OSAVI×Cl$_{red\ edge}$	$y = 3.756\ln(x)+0.337$	0.692**
OSAVI×Cl$_{green\ model}$	$y = 3.534\ln(x)+0.527$	0.706**
OSAVI×EVI2	$y = 0.269\ln(x)+0.151$	0.711**

Note: n = number of pairs of data. x represents spectral indices, y represents LAI. Probability levels are indicated by n.s.,
*and ** for 'not significant', 0.05, and 0.01, respectively.

Therefore, variations in Chl content changes were detected by OSAVI.

Figure 3b shows that biomass dry weight and total chlorophyll content were more highly correlated than were the OSAVI and total Chl content, with an R^2 value of 0.722. Linear regression was performed between biomass dry weight and total Chl content, and between biomass dry weight and total Chl content at the different growth stages. The results showed that the relationship between the OSAVI and total chlorophyll content exhibited similar results to that between OSAVI and N uptake vis-à-vis a lower correlation at earlier growth stages.

The OSAVI×biomass dry weight index could be used to estimate total Chl content changes, and it was a good indicator of total Chl content for wheat. It improved upon the results of the biomass dry weight and OSAVI alone for estimation of the total Chl content for wheat, and the R^2 was 0.805 (Figure 3c). This result indicated that the OSAVI×biomass dry weight index improved upon the prediction accuracy for total Chl content.

Taken together, these results indicated that the OSAVI×biomass dry weight index (BDW) could be used to improve the estimation accuracy of LAI, nitrogen (N) uptake, and total

Table 3. Relationship between the new spectral indices and nitrogen uptake under the different nitrogen treatments for wheat (n = 90).

Spectral indices	Simulated equations	Determination coefficient (R^2)
OSAVI×OSAVI	$y = 0.246\ln(x) - 0.095$	0.542**
OSAVI×SIPI	$y = 0.164\ln(x) + 0.302$	0.641**
OSAVI×CI$_{red\ edge}$	$y = 0.145x^{1.331}$	0.785**
OSAVI×CI$_{green\ model}$	$y = 0.203x^{1.200}$	0.776**
OSAVI×EVI2	$y = 0.274\ln(x) - 0.212$	0.783**

Note: n = number of pairs of data. x represents spectral indices, y represents nitrogen uptake.
Probability levels are indicated by n.s.,
*and ** for 'not significant', 0.05, and 0.01, respectively.

chlorophyll (Chl) content, respectively. The most accurate estimation was gained for total Chl content, the worst for LAI, and the median for N uptake, among three agronomic parameters.

The results showed that biomass dry weight and LAI were highly correlated ($R^2 = 0.688$, $P<0.01$) (Fig. 1b). Most of results indicated that the R^2 and RMSE of the curves fitting was better than the linear fitting [40–43]. But sometimes, the linear fitting was better than the curves fitting, it may be a high related to measurement data and crops physiological mechanism. In this paper, we try to find the best fitting lines by using the curves fitting or the linear fitting, the result indicated that the equations of curves fitting is more than the equations of linear fitting, our results was consistent with previous researches [40–43]. Therefore, we justify the curves fitting was better than the linear fitting according to the R^2 and RMSE of regression model. The results showed that biomass dry weight and LAI were highly correlated (Fig. 1b). It indicated that the size of biomass dry weight represented the size of LAI. It was mainly because LAI growth is synchronized with biomass dry weight according to the certain proportion. The biomass dry weight and nitrogen content were also highly correlated (Fig. 2b). It suggested that the appropriate nitrogen was used to increase the biomass dry weight accumulation. Because the nitrogen was applied for improving crops photosynthesis, and there was a good relationships between photosynthesis and biomass dry weight, thereby biomass dry weight accumulation

Table 4. Relationship between the new spectral indices and total chlorophyll content under the different nitrogen treatments for wheat (n = 90).

Spectral indices	Simulated equations	Determination coefficient (R^2)
OSAVI×OSAVI	$y = 0.264x^{0.400}$	0.652**
OSAVI×SIPI	$y = 0.504x^{0.279}$	0.683**
OSAVI×CI$_{red\ edge}$	$y = 0.137x^{0.769}$	0.846**
OSAVI×CI$_{green\ model}$	$y = 1.307x^{0.742}$	0.824**
OSAVI×EVI2	$y = 0.208x^{0.482}$	0.798**

Note: n = number of pairs of data. x represents spectral indices, y represents total chlorophyll content.
Probability levels are indicated by n.s.,
*and ** for 'not significant', 0.05, and 0.01, respectively.

was increased. Similarly, the appropriate total chlorophyll content could be used to increase crops photosynthesis and biomass dry weight accumulation, thus there was a high correlation between biomass dry weight and total chlorophyll content (Fig. 3b).

A straight linear relationships between biomass and LAI before heading stages, but this relationship between biomass and LAI have changed after heading stages. The rate of increase in the LAI is less than in the biomass (stem and spike weights increase more than leaf weight) before and after flowering. The LAI arrived at the highest values at flag leaf fully expanded, but the biomass dry weight (BDW) is still increased (it is mainly increased from spike weight). Therefore, a curvilinear relationship between wheat BDW and LAI should be better after heading stages. Metabolic physiology exists different before and after flowering. The wheat metabolic physiology was dominated by nitrogen and supplemented by carbon before flowering, canopy leaf nitrogen content change was relatively small, canopy leaf nitrogen content was accurately estimated by more accurately monitoring LAI. However, the wheat metabolic physiology was dominated by carbon and supplemented by nitrogen after flowering, the differences are significant in canopy leaf nitrogen content, nitrogen transfers from lower leaves to the upper leaves and spike in wheat. The nitrogen was not estimated accurately by monitoring LAI because of the lower leaves nitrogen was moved away by nitrogen transfer, resulting in a relative large error will be generated. If biomass factors are taken into account, thereby the transfer of nitrogen will also be included in the monitoring results, therefore it could be used to improve the nitrogen estimation accuracy. These results indicated that the biomass dry weight was highly related to LAI, nitrogen content and total chlorophyll content because biomass dry weight had a close relationship with LAI, nitrogen content and total chlorophyll content in crops physiological mechanism. These results also suggested that the biomass dry weight could be used to estimate the LAI, nitrogen content and total chlorophyll content for wheat. The results showed that a good relationships among the LAI, nitrogen content, total chlorophyll content and OSA-VI×BDW (Figs. 1c, 2c and 3c). It provided a basis for BDW was replaced by spectral indices to establish new spectral indices.

3.4. New Spectral Indices

The new spectral index (OSAVI×BDW) had a good relationship with LAI, N uptake and total Chl content, but BDW was gained by destructive sampling, and required significant time investment for data acquisition. Thus, we attempted to build new spectral indices by replacing the BDW with others spectral indices. Tables 2, 3 and 4 showed that the new spectral indices: OSAVI×OSAVI, OSAVI×SIPI, OSAVI×CI$_{red\ edge}$, OSAVI× CI$_{green\ model}$ and OSAVI×EVI2, could be used to improve LAI, N uptake and total Chl content estimation accuracy. For LAI, the lowest and highest determination coefficient (R^2) observed were OSAVI×SIPI and OSAVI×EVI2 with R^2 values of 0.563 and 0.711, respectively (Table 2); similarly, the lowest and highest R^2 were OSAVI×CI$_{green\ model}$ and OSAVI×OSAVI, for which R^2 were 0.542 and 0.785 for nitrogen uptake, respectively (Table 3); the lowest and highest R^2 were OSAVI×CI$_{red\ edge}$ and OSAVI×OSAVI, with R^2 values of 0.652 and 0.846 for total Chl content, respectively (Table 4).

3.5. Validation Model and Comparison

To validate the model accuracy, we compared the predicted values (the predicted values were gained by the LAI, N uptake and total Chl content of regression equations in 2009) with the actual values (the actual values were gained by the LAI, N uptake and total Chl content field measurement data in 2010). A good

Table 5. The relationships the measured value and predicted value for LAI, nitrogen uptake and total chlorophyll content under the different nitrogen treatments for wheat.

LAI		Nitrogen uptake		Leaf chlorophyll content	
	RMSE (g/m^2)		RMSE (g/m^2)		RMSE (g/m^2)
OSAVI	1.41	OSAVI	7.93	OSAVI	3.42
biomass dry weight	1.02	biomass dry weight	4.45	biomass dry weight	3.12
OSAVI×BDW	1.12	OSAVI×BDW	4.23	OSAVI×BDW	2.11
OSAVI×OSAVI	1.18	OSAVI×OSAVI	6.46	OSAVI×OSAVI	3.36
OSAVI×SIPI	1.36	OSAVI×SIPI	4.85	OSAVI×SIPI	3.02
OSAVI×CI$_{red\ edge}$	0.98	OSAVI×CI$_{red\ edge}$	3.98	OSAVI×CI$_{red\ edge}$	0.65
OSAVI×CI$_{green\ model}$	0.84	OSAVI×CI$_{green\ model}$	4.12	OSAVI×CI$_{green\ model}$	0.79
OSAVI×EVI2	0.78	OSAVI×EVI2	4.03	OSAVI×EVI2	1.02

correlation between the predicted values and the actual values was observed for the following indices: OSAVI, biomass dry weight, and OSAVI×biomass dry weight (BDW). The corresponding root mean square errors (RMSEs) were 1.42, 1.02, and 1.12 for LAI; 7.93 g/m^2, 4.45 g/m^2, and 4.23 g/m^2 for nitrogen uptake; 3.42 g/m^2, 1.85 g/m^2, and 2.23 g/m^2 for leaf chlorophyll content, respectively (Table 5). These data indicated that OSAVI×biomass dry weight could be used to improve the estimation accuracy of LAI, nitrogen uptake, and leaf chlorophyll content. The new spectral indices were proposed by replacing BDW with spectral indices, and then obtained OSAVI×OSAVI, OSAVI×SIPI, OSAVI×CI$_{red\ edge}$, OSAVI×CI$_{green\ model}$ and OSAVI×EVI2. The results showed that the new spectral indices were better than OSAVI alone for estimating LAI, nitrogen uptake and total chlorophyll content (Tables 2, 3, 4 and 5 and Figs. 1a, 2a and 3a). The results indicated that the products of spectral indices and OSAVI could be used to improve the LAI, N uptake and total Chl content estimation accuracy.

Discussion

Crop leaf area index (LAI), total chlorophyll (Chl) content and nitrogen (N) uptake was estimated using new spectral indices. The results showed that the new spectral indices could be used to improve the LAI, total Chl content and N uptake estimation accuracy. We used the OSAVI×biomass dry weight (BDW) index (the product of OSAVI and biomass dry weight) to improve the relationship between the LAI and spectral indices. The results indicated that the OSAVI×BDW index was better than the OSAVI at estimating wheat LAI. The OSAVI×BDW index was more sensitive to LAI due to the further decreased effects of soil at earlier growth stages. For example, the spectrometer was sensitive to the soil color and moisture, thus reducing the wheat canopy spectral information detection at earlier growth stages, and leading decreased sensitive to canopy spectral reflectance changes. The OSAVI×BDW ameliorated the saturation at later growth stages because of the addition of biomass dry weight, which was increased gradually with progression through the wheat growth stages. Overall, the saturation problem observed for the spectral indices was mainly influenced by the structure of the wheat canopy at later growth stages (see section 3.2). The OSAVI×BDW index was used to improve the estimation of nitrogen, and the results demonstrated that it was better than OSAVI alone for nitrogen assessment. The reason for this was similar to that explaining the reasonable estimation of LAI, the results were similar to those of

previous research [14–19]. We proposed that the OSAVI×BDW index could be used to effectively improve the estimation accuracy of total Chl content indirectly, and the results demonstrated that it could accomplish the proposed tasks. Again, the reasons for such are similar to those explaining the reasonable estimation of N uptake.

The new spectral indices in Tables 2, 3 and 4 are derived from BDW×OSAVI. Biomass dry weight (BDW) was gained by destructive sampling, but required a greater time investment for data acquisition. To quickly obtain data and estimate leaf area index (LAI), total chlorophyll (Chl) content and nitrogen (N) uptake, we attempted to build new spectral indices by replacing BDW with other spectral indices. We obtained the OSAVI×O-SAVI, OSAVI×SIPI, OSAVI×CI$_{red\ edge}$, OSAVI×CI$_{green\ model}$ and OSAVI×EVI2. The results suggested the new spectral indices were better than OSAVI alone for estimating LAI, total Chl content and N uptake (Tables 2, 3, 4 and 5 and Figs. 1a, 2a and 3a). The main reason is that the SIPI, CI$_{red\ edge}$, CI$_{green\ model}$ and EVI2 include the 800 nm band is sensitive to LAI changes. The products of OSAVI and SIPI, CI$_{red\ edge}$, CI$_{green\ model}$ and EVI2 will further improve the OSAVI sensitivity in LAI changes. Therefore, these new spectral indices could be used for improving the LAI. For N uptake, the products of OSAVI and SIPI, CI$_{red\ edge}$, CI$_{green\ model}$ and EVI2, making OSAVI increase the sensitive bands of the chlorophyll content changes (445 nm, 550 nm, 660 nm and 680 nm), these new spectral indices are sensitive to detect chlorophyll content changes. Thus, the N uptake estimation was improved by using the new spectral indices. For total Chl content, it was similar to those explained reasonable estimation of N uptake.

Taken together, the results indicated that it was feasible to use new methods to improve agronomic parameters (LAI, N uptake and total Chl content) assessment accuracy. This paper evaluated the estimation accuracy of agronomy parameters by multiplying the spectral indices in OSAVI, it showed the OSAVI is unnecessary and able to be replaced by others spectral indices. For example, if you want to better estimate total Chl content or N uptake, you could be selected to the product of two spectral indices are highly related to chlorophyll. Further, we used these indices to improve the accuracy of these predictions for all crop growth stages. In future studies, we will try to multiply two spectral indices that are highly related to LAI, total Chl content or N uptake to estimate agronomic parameters of different crops.

Conclusions

Building upon previous studies, wheat leaf area index (LAI), nitrogen uptake, and total chlorophyll content were predicted using the products spectral indices and spectral indices methods, and the main results and conclusions follow. The results suggested that the LAI, nitrogen uptake and total Chl content could be accurately predicted using the products of OSAVI and biomass dry weight (OSAVI×biomass dry weight (BDW) index), the corresponding determination coefficient (R^2) and root mean square errors (RMSEs) were 0.672 and 1.12, 0.754 and 4.23 g/m^2, 0.805 and 2.11 g/m^2, respectively. We obtained the new spectral indices by using spectral indices replace BDW. The

relationships between LAI, nitrogen uptake and total Chl content and new spectral indices, OSAVI×EVI2, OSAVI×CI$_{red\ edge}$ and OSAVI×CI$_{red\ edge}$ had the highest R^2 and the lowest RMSEs for LAI, nitrogen uptake and total Chl content, respectively, R^2 and RMSEs were 0.711 and 0.78, 0.785 and 3.98 g/m^2, 0.846 and 0.65 g/m^2, respectively.

Author Contributions

Conceived and designed the experiments: XLJ CHX WYD FYW BC SKL. Performed the experiments: XLJ WYD KRW FYW. Analyzed the data: XLJ. Contributed reagents/materials/analysis tools: CHX KRW SKL. Wrote the paper: XLJ SKL KRW.

References

1. Daughtry CST, Gallo KP, Goward SN, Prince SD, Kustas WD (1992) Spectral estimates of absorbed radiation and phytomass production in corn and soybean canopies. Remote Sensing of Environment 39: 141–152.
2. Goetz SJ, Prince SD (1996) Remote sensing of net primary production in boreal forest stands. Agricultural and Forest Meteorology 78: 149–179.
3. Liu J, Chen J, Cihlar J, Park WM (1997) A process-based boreal ecosystem productivity simulator using remote sensing inputs. Remote Sensing of Environment 62: 158–175.
4. Moran MS, Inoue Y, Barnes EM (1997) Opportunities and limitations for image-based remote sensing in precision crop management. Remote Sensing of Environment 61: 319–346.
5. Moran MS, Maas SJ, Pinter JPJ (1995) Combining remote sensing and modeling for estimating surface evaporation and biomass production. Remote Sensing of Environment 12: 335–353.
6. Tucker CJ, Holben BN, Elgin IJH, McMurtrey JE (1980) Relationship of spectral data to grain yield variations. Photogrammetric Engineering & Remote Sensing 46: 657–666.
7. Rouse JW, Haas RH, Schell JA, Deering DW, Harlan JC (1974) Monitoring the vernal advancements and retrogradation of natural vegetation. In: NASA/GSFC, Final Report, Greenbelt, MD, USA, 1974, pp.1–137.
8. Bausch WC (1993) Soil background effects on reectance-based crop coefficients for corn. Remote Sensing of Environment 46: 213–222.
9. Elvidge CD, Lyon RJP (1985) Inuence of rock-soil spectral variation on the assessment of green biomass. Remote Sensing of Environment 17: 265–279.
10. Huete AR, Jackson RD, Post DF (1985) Spectral response of a plant canopy with different soil backgrounds. Remote Sensing of Environment 17: 37–53.
11. Rondeaux G, Steven M, Baret F (1996) Optimization of soil adjusted vegetation indices, Remote Sensing of Environment 55: 95–107.
12. Yang JC, Wang ZQ, Zhu QS (1996) Effects of nitrogen nutrition on rice yield and its physiology mechanism under the different status of soil moisture, Chinese Agriculture Science 29: 58–66 (in Chinese).
13. Rhykerd CL, Noller CH (1974). The role of nitrogen in forage production, In D.A. Mays (ed.) Forage fertilization, American Society of Agronomy, Madison, WI, 416–424.
14. Stone ML, Solie JB, Raun WR, Whitney RW, Taylor SL, et al. (1996a) Use of spectral radiance for correcting in-season fertilizer nitrogen deficiencies in winter wheat. Transactions of the American Society Agriculture Engineers 39: 1623–1631.
15. Stone ML, Solie JB, Whitney RW, Raun WR, Lees HL (1996b) Sensors for the detection of nitrogen in winter wheat. Tech. Paper Series No. 961757. SAE, Warrendale, PA.
16. Osborne SL, Schepers JS, Francis DD, Schlemmer MR (2002) Detection of phosphorous and nitrogen deficiencies in corn using spectral radiance measurement. Agronomy Journal 94: 1215–1221.
17. Gitelson AA, Merzlyak MN (1996) Signature analysis of leaf reflectance spectra: algorithm development for remote sensing of chlorophyll. Journal of Plant Physiology 148: 494–500.
18. Moges SM, Raun WR, Mullen RW, Freeman KW, Johnson GV, et al. (2004) Estimating of green, red, and near infrared bands for predicting winter wheat biomass, nitrogen uptake, and final grain yield. Journal of Plant Nutrition 27: 1431–1441.
19. Reyniers M, Walvoort DJJ, De Baardemaaker J (2006) A linear model to predict with a multi-spectral radiometer the amount of nitrogen in winter wheat. International Journal of Remote Sensing 27: 4159–4179.
20. Zhu Y, Yao X, Tian YC, Liu XJ, Cao WX (2008) Analysis of common canopy vegetation indices for indicating leaf nitrogen accumulations in wheat and rice. International Journal of Appllied earth Obsevation and Geoinformation 10: 1–10.
21. Eitel JUH, Long DS, Gessler PE, Smith AMS (2007) Using in-situ measurements to evaluate the new RapidEye satellite series for prediction of wheat nitrogen status. International Journal of Remote Sensing 28: 4183–4190.
22. Fitzgerald G, Rodriguez D, O'Leary G (2010) Measuring and predicting canopy nitrogen nutrition in wheat using a spectral index-The canopy chlorophyll content index (CCCI). Field Crops Research 116: 318–324.

23. Vigneaua N, Ecarnotb M, Rabatel G, Roumetb P (2011) Potential of field hyperspectral imaging as a non destructive method to assess leaf nitrogen content in wheat. Field Crops Research 122: 25–31.
24. Buschmann C, Nagel E (1993) In vivo spectroscopy and internal optics of leaves as basis for remote sensing of vegetation. International Journal of Remote Sensing 14: 711–722.
25. Gitelson AA, Merzlyak MN (1994a) Quantitative estimation of chlorophyll a using reflectance spectra: Experiments with autumn chestnut and maple leaves. Journal of Photochemistry and Photobiology Biology 22: 247–252.
26. Gitelson AA, Merzlyak MN (1994b) Spectral reflectance changes associated with autumn senescence of Aesculus hippocastanum andAcer platanoides leaves. Spectral features and relation to chlorophyll estimation. Journal of Plant Physiology 143: 286–292.
27. Markwell J, Osterman JC, Mitchell JL (1995) Calibration of the Minolta SPAD-502 leaf chlorophyll meter. Photosynthesis Research 46: 467–472.
28. Gamon JA, Surfus JS (1999) Assessing leaf pigment content and activity with a reflectometer. New Phytologist 143: 105–117.
29. Gitelson AA, Merzlyak MN, Chivkunova OB (2001) Optical properties and non-destructive estimation of anthocyanin content in plant leaves, Journal of Photochemistry Photobiology 74: 38–45.
30. Gitelson AA, Zur Y, Chivkunova OB, Merzlyak MN (2002) Assessing carotenoid content in plant leaves with reectance spectroscopy. Journal of Photochemistry Photobiology 75: 272–281.
31. Gitelson AA, Vina A, Ciganda V, Rundquist DC, Arkebauer TJ (2005) Remote estimation of canopy chlorophyll content in crops. Geophysical Research Letters 32: L08403, doi:10.1029/2005 GL022688.
32. Gitelson AA, Vina A, Verma SB, Rundquist DC, Arkebauer TJ, et al. (2006) Relationship between gross primary production and chlorophyll content in crops: Implications for the synoptic monitoring of vegetation productivity. Geophysical Research Letters 111: D08S11, doi:10.1029/2006JD006017.
33. Peng Y, Gitelson AA, Keydan G, Rundquist DC, Moses W (2011) Remote estimation of gross primary production in maize and support for a new paradigm based on total crop chlorophyll content. Remote Sensing of Environment 115: 978–989.
34. Schepers JS, Francis DD, Thompson MT (1989) Simultaneous determination of total C, total N and 15N on soil and plant material. Communications in Soil Science and Plant Analysis 20: 949–959.
35. McKinney G (1941) Absorption of light by chlorophyll solutions. Journal of Biology Chemistry 140: 315–322.
36. Penuelas J, Baret F, Filella I (1995) Semiempirical indexes to assess carotenoids chlorophyll-a ratio from leaf spectral reectance. Photosynthetica 31: 221–230.
37. Jiang Z, Huete AR, Didan K, Miura T (2008) Development of a two-band enhanced vegetation index without a blue band. Remote Sensing of Environment 112: 3833–3845.
38. Steven MD (1998) The densitivity of the OSAVI vegetation Index to observational parameters, Remote Sensing of Environment 63: 49–60.
39. Cheng JP, Cao CG, Cai ML, Yuan BZ, Zhai J (2008) Effect of different nitrogen nutrition and soil water potential on the physiology parameters and yield of hybrid rice. Plant Nutrition and Fertilizer Science 14: 199–206 (in Chinese).
40. Gitelson AA, Gurlin D, Moses WJ, Barrow T (2009) A bio-optical algorithm for the remote estimation of the chlorophyll-a concentration in case 2 waters. Environment Research Letters 045003, doi:10.1088/1748-9326/4/4/045003.
41. Houborg R, Anderson MC, Daughtry CST, Kustas WP, Rodell M (2011) Using leaf chlorophyll to parameterize light-use-efficiency within a thermal-based carbon, water and energy exchange model. Remote Sensing of Environment 115: 1694–1705.
42. Sakamoto T, Gitelson AA, Wardlow BD, Arkebauer TJ, Verma SB, et al. (2012) Application of day and night digital photographs for estimating maize biophysical characteristics. Precision Agriculture 13: 285–301.
43. Gurlin D, Gitelson AA, Moses WJ (2011) Remote estimation of chl-a concentration in turbid productive waters-Return to a simple two-band NIR-red model? Remote Sensing of Environment 115: 3479–3490.

3

Tracking Fungal Community Responses to Maize Plants by DNA-and RNA-Based Pyrosequencing

Eiko E. Kuramae[1,2]*, **Erik Verbruggen[2]**, **Remy Hillekens[1]**, **Mattias de Hollander[1]**, **Wilfred F. M. Röling[3]**, **Marcel G. A. van der Heijden[4,5]**, **George A. Kowalchuk[1,5]**

1 Department of Microbial Ecology, Netherlands Institute of Ecology (NIOO-KNAW), Wageningen, The Netherlands, 2 Department of Ecological Science, VU University Amsterdam, Amsterdam, The Netherlands, 3 Department of Molecular Cell Physiology, VU University Amsterdam, Amsterdam, The Netherlands, 4 Research Station ART, Agroscope Reckenholz Tänikon, Zurich, Switzerland, 5 Department of Biology, Utrecht University, Utrecht, The Netherlands

Abstract

We assessed soil fungal diversity and community structure at two sampling times (t1 = 47 days and t2 = 104 days of plant age) in pots associated with four maize cultivars, including two genetically modified (GM) cultivars by high-throughput pyrosequencing of the 18S rRNA gene using DNA and RNA templates. We detected no significant differences in soil fungal diversity and community structure associated with different plant cultivars. However, DNA-based analyses yielded lower fungal OTU richness as compared to RNA-based analyses. Clear differences in fungal community structure were also observed in relation to sampling time and the nucleic acid pool targeted (DNA versus RNA). The most abundant soil fungi, as recovered by DNA-based methods, did not necessary represent the most "active" fungi (as recovered via RNA). Interestingly, RNA-derived community compositions at t1 were highly similar to DNA-derived communities at t2, based on presence/absence measures of OTUs. We recovered large proportions of fungal sequences belonging to arbuscular mycorrhizal fungi and *Basidiomycota*, especially at the RNA level, suggesting that these important and potentially beneficial fungi are not affected by the plant cultivars nor by GM traits (Bt toxin production). Our results suggest that even though DNA- and RNA-derived soil fungal communities can be very different at a given time, RNA composition may have a predictive power of fungal community development through time.

Editor: Mari Moora, University of Tartu, Estonia

Funding: This work was supported by the the Ecology Regarding Genetically Modified Organisms - Netherlands Organisation for Scientific Research (ERGO-NWO) program. The funders had no role in study design, data collection and analysis, decision to publish, or preparation of the manuscript.

Competing Interests: The authors have declared that no competing interests exist.

* E-mail: e.kuramae@nioo.knaw.nl

Introduction

Numerous soil processes are primarily, and some even exclusively, carried out by soil fungi. Two of the most important processes are the facilitation of nutrient uptake and degradation of crop remains [1,2]. The fungi responsible for these key soil functions are in direct contact with plants and plant materials. Therefore, they may be especially vulnerable to alterations in e.g. plant defenses and carbohydrate composition and availability [3,4], as generated by Genetic Modification (GM) or other plant variety specific differences.

Recently, methodological advances have allowed enhanced screening of soil microbial communities, revealing high diversity and temporal turnover based on sequencing of 18S rRNA gene variants [5]. In addition, high-throughput sequencing approaches, such as 454 pyrosequencing, have greatly expanded the power of such nucleic acid-based assessments of complex fungal communities [6,7]. Nuclear SSU rRNA genes have proven to be useful markers for assessing fungal community structure across a range of habitats, including soils [7,8]. However, it has been demonstrated that non-active or dormant fungi may persist in sampled DNA pools, potentially masking the dynamics of the more active constituents of fungal communities [9–11]. As an alternative approach, environmental 18S rRNA can be targeted by RT-PCR amplification from soil RNA extracts [12–14] in order to focus on the active components of fungal communities [12,15,16]. The combination of DNA- and RNA- based assessments of fungal communities is therefore thought to provide a more complete picture of fungal community dynamics [16]. However, there is currently no consensus on whether DNA, RNA or both should be targeted to obtain the most meaningful assessment, and how communities obtained by analysing either nucleic acid type relate to each other.

In this study, we examined the potential impact of four maize plant cultivars, consisting of two genetically modified cultivars (GM) and two near-isogenic non-GM cultivars, on soil-borne fungal communities in a pot-based experiment. GM plants were incorporated into the study to address concerns that some transgenic crops may adversely affect soil processes, including important groups of beneficial soil fungi. To date, most studies on this topic have been focused on the effects of Bt (*Bacillus thuringiensis* toxin coding gene introduced in maize) crops on arbuscular mycorrhizal fungi (AMF). Some of these studies have reported no consistently significant impacts on AMF [17–21], although others have detected significant negative effects of Bt plants on AMF [22,23].

To provide an in-depth and comprehensive account of fungal communities, we targeted 18S rRNA genes and reverse-transcribed 18S rRNA at two sampling times (t1 = 47 and t2 = 104

Table 1. Estimators of sequences diversity, evenness, richness and coverage given by environmental DNA and RNA in soils with two maize cultivars (M = Monumental; D = DKC3420) and their respective genetically modified lines (M-GM = event MON810; D-GM = DKC3421YG) at different sampling times (t1 = 47 days; t2 = 104 days) after sowing.

	Number of OTU[1][2]	Singletons[2]	Shannon (diversity)[2]	Evenness[2]	Chao-1 (richness)[2]	Good's coverage estimator[2]
Cultivar						
M (DNA-t1)	80[3]±11a	36±28	3.30±0.04a	0.35±0.06a	110±44a	94±5
M (DNA-t2)	85±2a	24±6	3.43±0.37a	0.38±0.13a	98±2a	96±1
M-GM (DNA-t1)	94±9a	40±1	3.21±0.32a	0.27±0.06a	126±2a	93±0
M-GM (DNA-t2)	76±10a	17±19	3.48±0.27a	0.44±0.13a	89±28a	97±3
M (RNA-t1)	95±19a	37±14	3.32±0.08a	0.30±0.05a	126±35a	94±2
M (RNA-t2)	86±10a	38±15	3.24±0a	0.30±0.03a	133±43a	94±2
M-GM (RNA-t1)	78±5a	36±6	2.79±0.18a	0.21±0.05a	127±20a	94±1
M-GM (RNA-t2)	94±15a	48±13	2.85±0.36a	0.19±0.03a	155±48a	92±2
D (DNA-t1)	87±8a	27±2	3.45±0.21a	0.37±0.04a	103±8a	96±0
D (DNA-t2)	80±12a	21±6	3.22±0.22a	0.32±0.02a	90±3a	97±1
D (RNA-t1)	78±13a	41±9	2.68±0.44a	0.19±0.05a	137±24a	93±2
D (RNA-t2)	89±6a	38±7	3.20±0.26a	0.28±0.05a	131±24a	94±1
D-GM (DNA-t1)	80±15a	29±5	3.13±0.45a	0.30±0.08a	111±13a	95±1
D-GM (DNA-t2)	85±13a	35±15	3.07±0.11a	0.26±0.03a	120±41a	94±2
D-GM (RNA-t1)	93±12a	50±7	2.75±0.21a	0.17±0.02a	181±3a	92±1
D-GM (RNA-t2)	97±14a	49±18	2.83±0.37a	0.19±0.08a	175±61a	92±3
P value	ns	ns	ns	ns	ns	
Non-GM/GM						
Non-GM	85±6a	33±8	3.23±0.24a	0.31±0.06a	116±18a	95±1
GM	87±8a	38±11	3.01±0.25a	0.25±0.09a	135±32a	94±2
P value	ns	ns	ns	ns	ns	
Sampling time						
t1	85±7a	37±7	3.08±0.29a	0.27±0.07a	128±24	94±1
t2	86±7a	34±12	3.16±0.23a	0.29±0.09a	124±31	94±2
P value	Ns	ns	ns	ns	ns	
Nucleic acid type						
DNA	82±6a	28a ±7	3.28±0.16a	0.33±0.06	105±12	95a ±1
RNA	88±7a	42b ±6	2.96±0.25a	0.23±0.06	146±22	93b ±1
P value	Ns	*	ns	ns	*	

[1]Operational Taxonomic Unit.
[2]The values are mean of replicates (n = 2–3).
[3]Values with the same letters were not significantly (ns) different (*P*<0.05); *P<0.05 Significant comparisons at p<0.05.

days of plant age) by 454 pyrosequencing, and the relative impacts of plant cultivar, plant age and nucleic acid pool were assessed. A previous study using the same system [21] focused on solely AMF (average of 12% total sequence reads) and related their diversity in pot experiments to natural AMF community variation in the field. In the current study, we assess the full fungal community to examine the potential impact of GM-cultivars on soil fungal diversity and community structure, and specifically determine whether DNA and RNA generally reflect similar community dynamics and how these two different nucleic acid pools relate to each other.

Materials and Methods

Plant Cultivars and Experimental Setup

The four different maize (*Zea mays* L.) cultivars used in this study consisted of "Monumental MON810" (GM event MON810) and

"DK 3421YG" (GM event MON810) and two near isogenic non-GM cultivars of these lines ("Monumental" and "DKC 3420", respectively). For brevity, different cultivars are abbreviated by letters, Monumental = M, DKC = D, and GM cultivars are indicated with "GM" (i.e. M - GM, and D – GM). Both GM cultivars had been transformed to express the *CryIAb* gene (an insecticidal endotoxin produced by *Bacillus thuringiensis* that is active against, among others, the European corn borer *Ostrinia nubilalis*).

Intact pot-size soil cores from an organically managed agricultural field were transferred to pots in order to maintain natural stratification and integrity of fungi inhabiting the soil. No specific permits were required for this field sampling. This field is not a protected area and does not involve endangered or protected species. Soil cores were collected randomly from within a homogeneous 10×10 m plot in September 2009. The standing crop was a grass-clover mixture (*Trifolium pratense* L. and *Lolium*

Table 2. Results of two-way PERMANOVA testing the effect of "GM" (Bt *vs.* non-Bt maize) and "cultivar" (each of the two types of parental cultivars).

Nucleic acid	Time		Df	F	R^2	P
DNA	t1	GM	1	0.77	0.09	0.72
		Cultivar	1	0.89	0.10	0.55
		GM*Cultivar	1	1.01	0.12	0.41
		Residuals	6	0.69		
	t2	GM	1	1.55	0.14	0.08
		Cultivar	1	**1.87**	**0.17**	**0.02**
		GM* Cultivar	1	1.41	0.13	0.13
		Residuals	6	0.55		
RNA	t1	GM	1	1.53	0.16	0.12
		Cultivar	1	1.25	0.13	0.24
		GM* Cultivar	1	0.84	0.09	0.58
		Residuals	6	0.62		
	t2	GM	1	2.09	0.20	0.05
		Cultivar	1	1.65	0.16	0.11
		GM* Cultivar	1	0.59	0.06	0.83
		Residuals	6	0.58		
			Df	F	R^2	P
DNA		Time	1	**6.32**	**0.26**	**<0.001**
		Residuals	18	0.74		
RNA		Time	1	**3.30**	**0.16**	**0.002**
		Residuals	18	0.84		

Tests were performed separately for each nucleic acid type at each time point (t1 = 47 days and t2 = 105 days of plant age). Significant values are indicated in bold. The bottom part of the table represents separate PERMANOVA's for DNA and RNA in response to the factor "sampling time".

perenne L.), which had been sown after maize in the autumn of 2007 and mown twice a year. Soil chemical properties were pH (CaCl$_2$-extractable) = 5.8, P (CaCl$_2$-extractable) = 5.1 mg kg^{-1}, N = 1.36 g kg^{-1}, OM = 1%. In each pot (containing approximately 6 kg of soil, diameter = 20 cm, height = 18 cm), one of four different maize (*Zea mays* L.) cultivars was grown.

The experimental design was 2 GMs×2 non-GMs×3 replicate pots×2 sampling times×2 types of nucleic acids. Two seeds were sown into each pot on October 1st 2009, and pots kept in a greenhouse (16/8 hours light/dark). One seedling was kept per pot after two weeks. Hoagland solution (½ strength P; 250 ml per pot) was applied twice during the first month of plant growth. On November 24th, after 47 days (t1) of plant growth, soil samples were taken in the following way: one core (diameter 1 cm) per pot was taken and the part originating from 5–11 cm depth was immediately transferred to dry ice and subsequently stored at −80°C. Cores were taken 5–6 cm from the edge, which was approximately halfway between the edge of the pot and the stem of the plant. On January 20th 2010, after 104 days (t2) of growth, samples were again taken as above, but the position of cores was shifted 45° in relationship to the first core to minimize potential disturbances from the first sampling event. At the end of the experiment (plant age 130 days; at full maturity of the ears), total above- and belowground plant biomass was harvested.

Nucleic Acid Extraction and cDNA Preparation

RNA and DNA were simultaneous extracted from two grams of fresh bulk soil (soil without plant roots) per sample using the RNA PowerSoil kit and the DNA Elution accessory kit (MO BIO Laboratories inc., Carlsbad, CA, USA). The DNA from total RNA samples was digested by DNase I (RNase-free DNase set Qiagen 79254) according to manufacturer recommendations. The total RNA was measured with a ND-1000 spectrophotometer (Nanodrop Technology, Wilmington, DE, USA), and the quality of the total RNA was checked with Experion (BioRad). The cDNA was synthesised from the total RNA using random hexamer primers and the superscript double-stranded cDNA synthesis kit (Invitrogen, Life Technologies) exactly as described in Verbruggen et al. 2012.

Nucleic Acid Amplification and Sequencing of 18S rRNA Gene Fragments

DNA and cDNA templates derived from each of the soil pots at t1 = 47 days and t2 = 104 days of plant age were utilized for nucleic acid amplification and subsequent 454 pyrosequencing. In these procedures, a fungal-specific primer set was used, consisting of the FR1 primer [24] and a modified version of the FF390 primer ([24]; 5′-CGWTAACGAACGAGACCT-3′), to allow for inclusion of the *Glomeromycota* (Arbuscular Mycorrhizal Fungi). Thermocycling conditions were: denaturing at 94°C for 30 s (after initial denaturation of 4 min. initial annealing temperature was 55°C (1 min.), and every two cycles the annealing temperature was lowered by 2°C until 47°C was reached, which was the annealing temperature used for the final twenty cycles (thus, 29 PCR cycles in total). Extension conditions were 68°C for 2 min. for all cycles. Reactions contained about 25 ng of DNA or cDNA template added to a standard PCR mix. Four PCR reactions of 25ul each per biological replicate of a sample were carried out. Primers contained 12 different multiplex tags for sample identification were sequenced in 1/8 lane plate of Roche 454 automated sequencer and GS FLX system using titanium chemistry (454 Life Sciences, Branford, CT, USA). Sequence analysis was performed using QIIME 1.2.1 scripts [25] incorporated into the Galaxy interface [26]. We generally used DNA and cDNA (RNA) of three biological replicates per sample; however, eight samples [M (DNA-t1), M (DNA-t2), M (RNA-t1, M (RNA-t2), D (DNA-t1), D (DNA-t2), D (RNA-t1), D (RNA-t2)] had two replicates due to technical issues. All sequencing reads were checked for containing the right forward and reverse MID-tags and assigned to samples accordingly. Then, barcodes and tags were removed, and sequences were denoised using Denoiser 0.91 [27] and clustered at 97% similarity using the UCLUST 1.2.21 algorithm [28]. The resulting "operational taxonomic units" (OTUs) were assigned to eukaryote families through BLAST searches against the QIIME-compatible version of the Silva 104 release [29]. Singleton OTUs were removed for statistical analyses. The OTU table was rarefied to 906 reads using "single rarefaction" QIIME script since this number was the lowest number of reads in any single sample after singleton removal. This rarefied OTU table was used for all subsequent statistical analyses. Shannon (diversity), evenness and richness (Chao1) were calculated in the PAST program [30], and the percentage of coverage was calculated by Good's method [31]. Good's coverage estimator is a method of estimating what percent of the total species, in this case OTUs, is represented in a sample, which estimator equation gives a good idea of how their limited sampling relates to the entire sampled community. Two-way PERMANOVAs were separately performed for testing GM (Bt versus non-Bt maize) and cultivar (each of the two parental cultivars) effects and for DNA and RNA in response to sampling

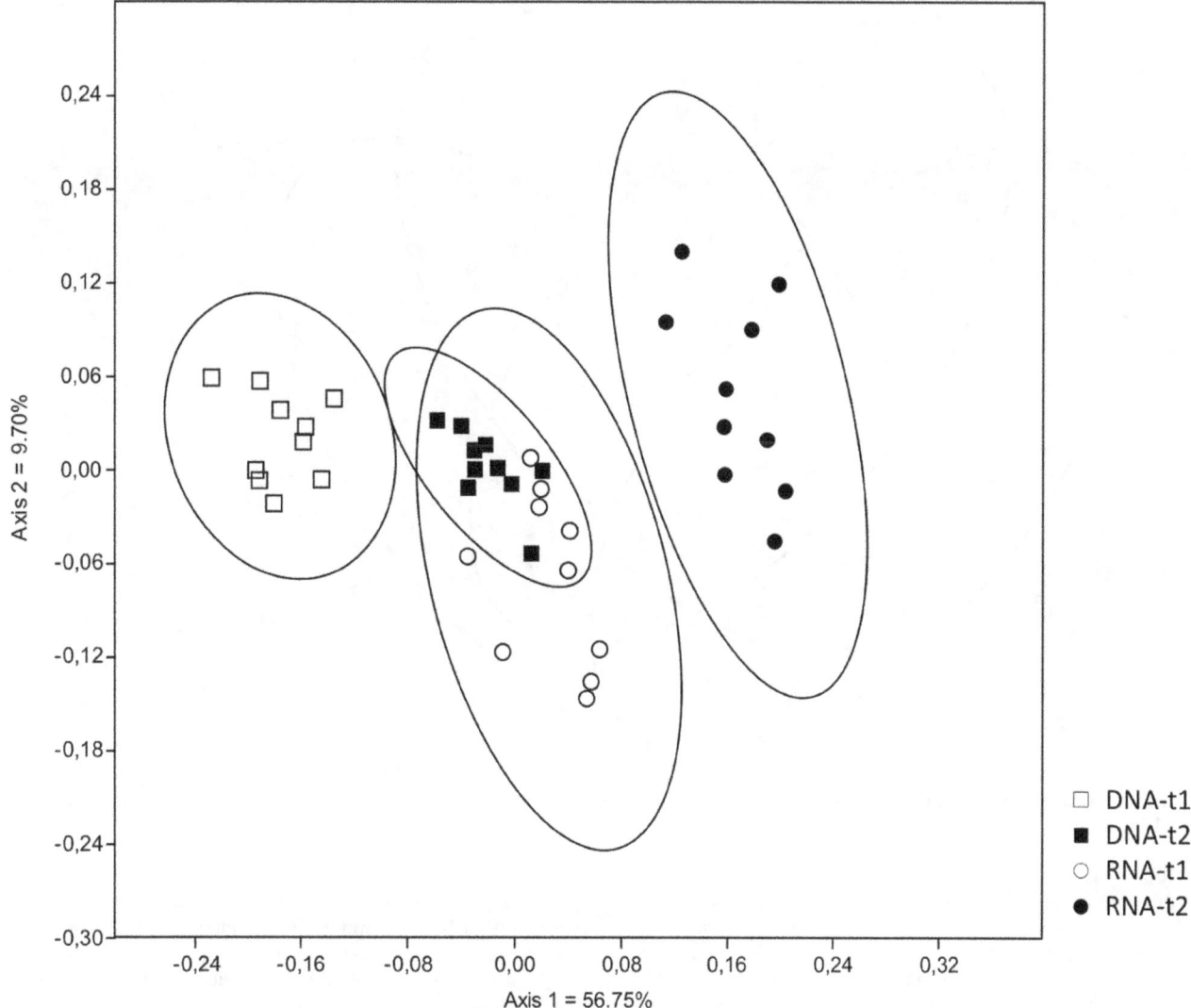

Figure 1. Non-metric multidimensional scaling (NMDS) analysis of presence and absence of fungal OTUs based on Jaccard-index of similarity with 95% confidence intervals shared between sampling times (t1 = 47 days; t2 = 104 days) and nucleic acid type (DNA, RNA).

time using the Vegan package [32] in R (The R Foundation for Statistical Computing). Jaccard's similarity was calculated based on OTU presence and absence in PAST program [30].

Results

Diversity of Fungal Communities

After denoising, chimera detection and removal of non-fungal reads (34% DNA; 24% RNA), a total of 60,845 reads were used for analyses. The average length of reads of fungal 18S rRNA was 375 bp. The numbers of reads of fungal DNA-based ranged from 1,025–6,970 while from RNA-based ranged from 2,448–9,000. Although we sought to mix PCR products in equal amounts prior to sequencing, product amounts were quantified using a spectrophotometer. Such measures may not be entirely accurate, and may have caused undesired variations in template amounts and subsequently sequence numbers.

In order to compare samples, we discarded singletons and rarefied the OTU table to the lowest numbers of reads obtained for any single sample. Doing this, we ended up with 906 reads per sample, yielding between 76 and 97 OTUs per sample, and OTU coverage ranged from 92 to 97% as determined by Good's coverage estimator (Table 1). OTU number comparisons between non-GM versus GM plants, different sampling times (t1 = 47 days; t2 = 104 days) or nucleic acid type were not significantly different (Table 1).

Fungal diversity (Shannon), evenness and richness (Chao-1) indices per plant cultivar were not different (Table 1). Similar results were also observed in soils with different non-GM and GM plants and across the sampling times, t1 and t2. However, the Chao-1 richness index between nucleic acid types, DNA versus RNA, was significantly different. The DNA-based (t1, t2) richness (Chao-1) was lower than that derived from RNA templates (t1, t2) index (Table 1).

Comparison of Fungal Communities

Differences were detected in soil fungal community composition associated with the different plant genetic backgrounds examined

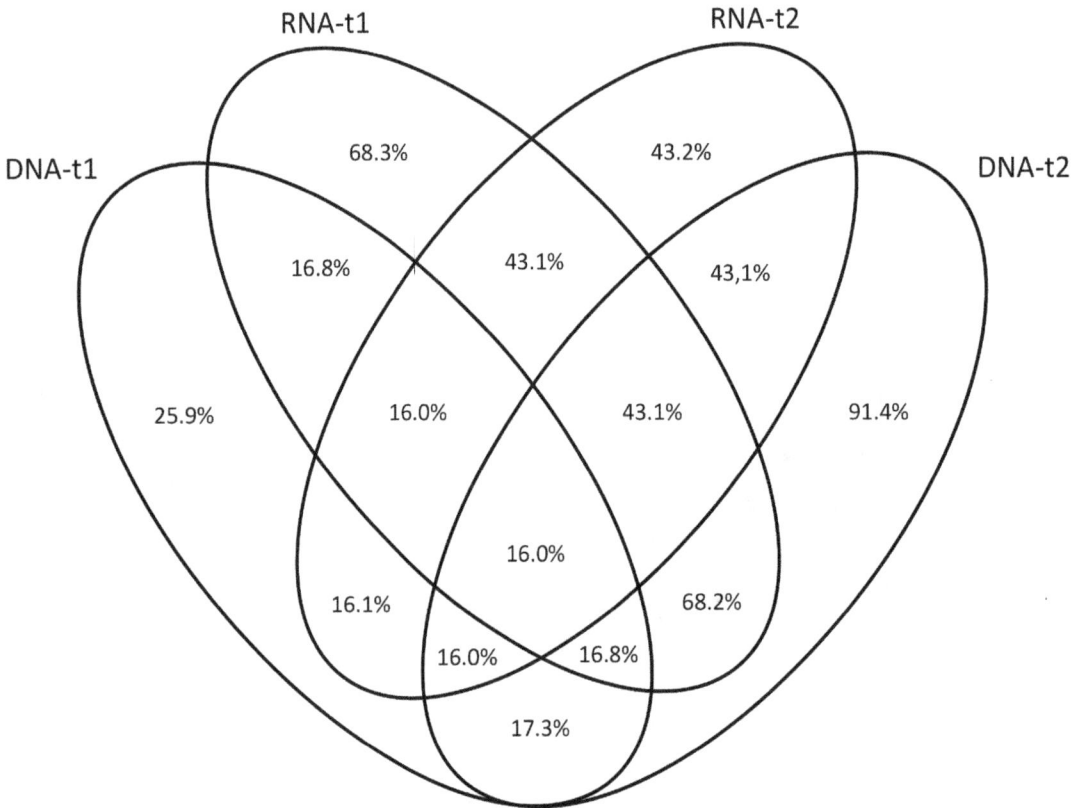

Figure 2. Percentage of fungal OTUs shared between nucleic acid type (DNA; RNA) in different sampling times (t1 = 47 days; t2 = 104 days).

in the experiment. There is a significant effect of plant cultivar in DNA-based analysis for sampling time t2 and no significant effect of GM for both DNA and RNA-based analysis at the same sampling time t2 (Table 2). There is a significant difference between DNA and RNA-microbial community based analysis (Table 2).

When OTU occurrence between sampling times was compared using a presence/absence measure of similarity (Jaccard), a very clear separation of treatments was observed. All nucleic acid type/time combinations formed distinct clusters, with the exception of the RNA-based OTUs at sampling time t1 and the DNA-derived OTUs at sampling time t2, which grouped very close together and were overlapping (Figure 1). Moreover, these two groups shared a large number of OTUs (Figure 2), despite the separation by nucleic acid type and time observed based on relative abundance measures (Figure 3). This indicates that taxa relative read number within treatment is rather dynamic, but that taxa occurrence at a given sampling time represented by either DNA or RNA is relatively more predictable.

There were different patterns in fungal phyla obtained by DNA-t1 and DNA-t2-based analysis, and by RNA-t1 and RNA-t2-based analysis. In the DNA-based fungal community, *Ascomycota* was the most abundant phylum independent of the sampling time, while in the RNA-based fungal community, *Basidiomycota* was the most abundant phylum for both time points. The *Glomeromycota* were more abundant in the RNA-derived datasets, and increased in relative abundance with plant age (t2) (Figure 4).

The relative abundance of members within different fungal phyla from the DNA- and RNA-based analyses were also significant different. The relative abundances of "environmental

fungi" (sequences only known from direct sequencing on nucleic acids isolated from environmental samples), environmental *Chytridiomycota*, *Ascomycota* (environmental, *Sordariomycetes*, *Saccharomycetales*) and *Glomeromycota* (*Glomerales*) were significantly higher in the DNA-derived dataset, while the relative abundance of *Basidiomycota* (*Agaricales*, environmental) was higher in the RNA dataset (Table 3).

In the DNA-based analysis, the relative abundances of members of "environmental fungi", and *Glomeromycota* (*Glomerales*, *Paraglomerales*) significantly increased with plant age, while members of the *Ascomycota* (environmental, mitosporic, *Dothideomycetes*) decreased (Table 3). For RNA-derived sequences, the relative abundance of *Ascomycota* (*Dothideomycetes*) and *Glomeromycota* (*Glomerales*, *Paraglomerales*) significantly increased with plant age, while *Ascomycota* (*Sordariomycetes*) and *Basidiomycota* (*Agaricales*) significantly decreased (Table 3).

At the same time point of sampling (t1 or t2), the relative abundances of fungal groups based on DNA and RNA analyses were not the same. At sampling time t1, members of the *Ascomycota* (environmental *Ascomycota*, *Dothideomycetes*), *Basidiomycota* (*Agaricales*) and *Glomeromycota* (*Glomerales*) were significantly different between DNA and RNA-based analysis (Table 3). Also, at time point t2, members of environmental fungi, *Ascomycota* (environmental, *Dothideomycetes*, *Sordariomycetes*) and *Glomeromycota* (*Diversisporales*, *Glomerales*) were significantly different between DNA and RNA-fungal community based analysis (Table 3).

Discussion

Despite having to restrict our analysis to relatively small numbers of sequence reads per sample, we still recovered 92 to

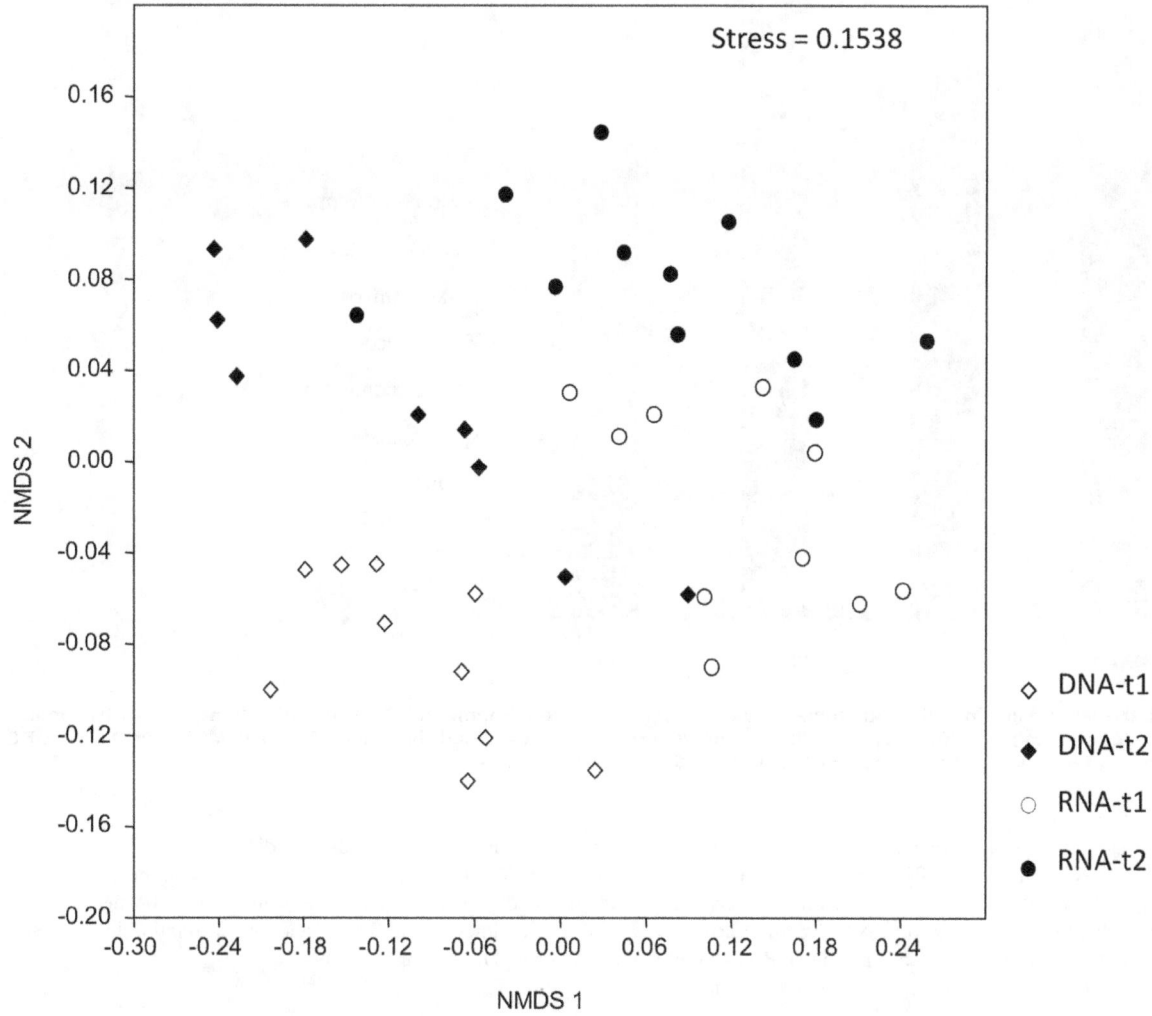

Figure 3. NMDS analysis based on relative abundance of fungal OTUs based on environmental DNA and RNA both at sampling times t1-47 days and t2-104 days of plant age.

97% of the total OTU-level fungal diversity in the soil samples studies as determined by Good's coverage estimator. The numbers of OTUs obtained in this study were rather low compared to other soils i.e. forest soils [6,7]. However, caution must be used in comparing diversity estimates using different markers. In the current study, we targeted a rather conserved region of the 18S rRNA gene, which only provides a taxonomic discrimination to approximately the order level. The use of a more variable marker, such as the Internal Transcribed Spacer (ITS) regions [33], which can discriminate to the subspecies level, would no doubt yield larger OTU numbers than we recovered.

Under the greenhouse conditions used in this study, no differences in soil fungal diversity (Shannon), evenness and richness (Chao-1) indices could be detected with respect to the GM nature of the plant. However, there was a difference in fungal richness depending on the type of nucleic acid (DNA or RNA) and the time point of soil sampling. The fungal community assessed by DNA-based analysis was richer than fungi community assessed by RNA-based analysis.

The fungal community structure was also not affected by non-GM and GM plants. This result is in agreement with previous studies that have examined the colonization and community structure of mycorrhizal fungi in response to GM lines of maize [17,21], soybean [34], cotton [18] and tobacco [35]. Our inability to detect shifts in total fungal community structure is in contrast with the observations of Tan et al. [34], who observed distinct cultivar effects on *Glomus*-like fungi, as determined by PCR-DGGE community profiling. Although these authors did observe differences between GM plants and their nearly isogenic parental lines, these differences were not greater that those observed between different non-GM cultivars. These previous studies have zoomed in specifically on AMF, yet relatively few studies have examined complete fungal community responses to GM plants [36–39], however none of these studies applied next generation sequencing approach as addressed in the current study. Our results indicate that the observed lack of a GM-related effects in our experiment was not caused by a low resolution of our assessment, but rather by weak effects of plant cultivar compared to other community-structuring processes because i) fungal communities of 47 and 104 days of plant growth were clearly distinct, and plant age was the most important explanatory factor of fungal community composition, and ii) we observed clear differences in the fungal community structures as recovered from DNA versus RNA templates. Regarding this last observation, these differences do

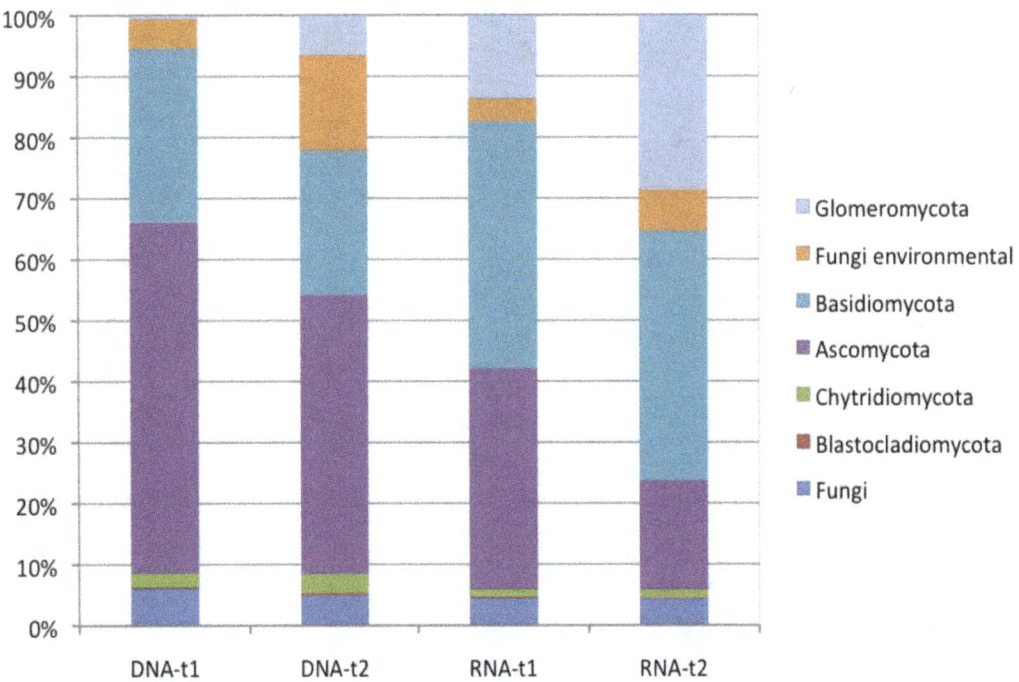

Figure 4. Relative abundance of different fungal phyla recovered from environmental DNA and RNA in soils with two maize cultivars (M = Monumental; D = DKC 3420) and their respective genetically modified lines (M-GM = event MON810; D-GM = DKC 3421YG) at two different sampling times (t1 = 47 days; t2 = 104 days) of plant age.

not seem to be caused by a fundamental difference between the taxa recovered as DNA compared to RNA (in that some are very active but low in abundance or vice-versa), but rather indicate a dynamic system where taxa measured by RNA give an indication of taxa found as DNA later in the season (*i.e.* the t1 RNA-based community appears to foreshadow the t2 DNA-based community).

It should be noted, however, that we did not examine any samples in the intervening time between the two sampling points, and additional observations and time points would be necessary to confirm the time-scale and generality of this relationship between RNA versus DNA template pools.

Table 3. Relative abundance (%) of soil fungal groups based on pyrosequencing analysis of environmental DNA and RNA in soils with two maize cultivars and their respective genetically modified lines at different sampling times (t1 = 47 days; t2 = 104 days) of plant age.

	DNA-t1	DNA-t2	RNA-t1	RNA-t2	DNA	RNA	DNA *vs.* RNA	DNA-t1 *vs.* DNA-t2	RNA-t1 *vs.* RNA-t2
Fungi environmental	4.8±0.04	15.6±0.59	3.7±0.08	7.1±0.60	10.2±0.58	5.40±0.45	**	***	ns
Chytridiomycota (environmental)	1.2±0.06	1.7±0.05	0.8±0.03	0.8±0.03	1.5±0.05	0.80±0.03	**	ns	ns
Ascomycota (environmental)	3.5±0.01	1.8±0.09	1.1±0.02	0.7±0.03	2.65±0.11	0.90±0.03	***	**	ns
Ascomycota (mitosporic)	1.8±0.02	1.1±0.04	1.4±0.05	1.3±0.05	1.45±0.03	1.35±0.05	ns	**	ns
Ascomycota (Dothideomycetes)	2.0±0.03	1.2±0.06	1.2±0.06	2.1±0.06	1.60±0.05	1.65±0.07	ns	**	**
Ascomycota (Sordariomycetes)	37.7±0.89	30.2±0.48	27.1±0.79	13.2±0.79	33.95±0.72	20.15±0.71	***	ns	**
Ascomycota (Saccharomycetales)	9.8±0.35	8.3±0.32	1.0±0.04	0.5±0.04	9.05±0.32	0.75±0.05	***	ns	ns
Basidiomycota (Agaricales)	1.3±0.07	1.4±0.05	2.2±0.04	1.7±0.04	1.35±0.06	1.95±0.04	**	ns	**
Basidiomycota (environmental)	14.5±0.44	16.4±0.26	31.4±0.86	29.6±0.86	15.45±0.33	30.50±1.05	***	ns	ns
Glomeromycota (Glomerales)	0.5±0.02	4.9±0.13	13.1±0.32	25.0±0.32	2.70±0.15	19.05±0.73	***	***	ns
Glomeromycota (Paraglomerales)	0.5±0.01	4.9±0.08	13.1±0.07	25.0±0.07	0.85±0.07	2.50±0.13	ns	***	ns

Numbers are mean of replicates.
Non-Parametric MANOVA (NPMANOVA) test between samples based on Bray-Curtis distance measure. Significance levels: ns: $P>0.05$;
**$P<0.005$;
***$P<0.0005$.
Relative abundances lower than 0.1% is not shown.

Although DNA and RNA clearly give rise to potentially complementary windows of observation, it should be kept in mind that ribosome number (*i.e.* 18S rRNA content) may be differentially correlated with activity for different species [40], and both DNA and RNA measurements have specific biases (DNA; e.g. copy number differences between species, RNA; different ribosome numbers per cell). Thus, while RNA-based analyses may provide additional insight into the generally active populations in environmental samples, one cannot extract precise activity estimates based upon such approaches.

In conclusion, soil fungal community proved to be dynamic with changes over time across nucleic acid pools, but not revealing any detectable impact of maize cultivar or the GM nature of the plant.

Accession Numbers

Nucleotide sequences were deposited in GenBank-SRA under the Study Accession No. ERP002065 for agriculture soil samples.

Acknowledgments

We want to thank Monsanto for providing maize seeds used in the experiment. Publication number 5435 of the NIOO-KNAW, Netherlands Institute of Ecology.

Author Contributions

Conceived and designed the experiments: EEK EV RH WFMR. Performed the experiments: EEK EV RH. Analyzed the data: EEK MdH. Contributed reagents/materials/analysis tools: EEK MGAvdH GAK. Wrote the paper: EEK EV MdH WFMR MGAvdH GAK.

References

1. Martinez AT, Speranza M, Ruiz-Duenas FJ, Ferreira P, Camarero S, et al. (2005) Biodegradation of lignocellulosics: microbial chemical, and enzymatic aspects of the fungal attack of lignin. International Microbiology 8: 195–204.

2. de Boer W, Folman LB, Summerbell RC, Boddy L (2005) Living in a fungal world: impact of fungi on soil bacterial niche development. Fems Microbiology Reviews 29: 795–811.

3. Graham JH, Duncan LW, Eissenstat DM (1997) Carbohydrate allocation patterns in citrus genotypes as affected by phosphorus nutrition, mycorrhizal colonization and mycorrhizal dependency. New Phytologist 135: 335–343.

4. Kiers ET, van der Heijden MGA (2006) Mutualistic stability in the arbuscular mycorrhizal symbiosis: Exploring hypotheses of evolutionary cooperation. Ecology 87: 1627–1636.

5. Dumbrell AJ, Ashton PD, Aziz N, Feng G, Nelson M, et al. (2011) Distinct seasonal assemblages of arbuscular mycorrhizal fungi revealed by massively parallel pyrosequencing. New Phytologist 190: 794–804.

6. Buee M, Reich M, Murat C, Morin E, Nilsson RH, et al. (2009) 454 Pyrosequencing analyses of forest soils reveal an unexpectedly high fungal diversity. New Phytologist 184: 449–456.

7. Lim YW, Kim BK, Kim C, Jung HS, Kim BS, et al. (2010) Assessment of soil fungal communities using pyrosequencing. Journal of Microbiology 48: 284–289.

8. Kowalchuk GA (1999) New perspectives towards analysing fungal communities in terrestrial environments. Current Opinion in Biotechnology 10: 247–251.

9. Agnelli A, Ascher J, Corti G, Ceccherini MT, Pietramellara G, et al. (2007) Purification and isotopic signatures (delta C-13, delta N-15, Delta C-14) of soil extracellular DNA. Biology and Fertility of Soils 44: 353–361.

10. Anderson IC, Parkin PI, Campbell CD (2008) DNA- and RNA-derived assessments of fungal community composition in soil amended with sewage sludge rich in cadmium, copper and zinc. Soil Biology & Biochemistry 40: 2358–2365.

11. Hirsch PR, Mauchline TH, Clark IM (2010) Culture-independent molecular techniques for soil microbial ecology. Soil Biology & Biochemistry 42: 878–887.

12. Girvan MS, Bullimore J, Ball AS, Pretty JN, Osborn AM (2004) Responses of active bacterial and fungal communities in soils under winter wheat to different fertilizer and pesticide regimens. Applied and Environmental Microbiology 70: 2692–2701.

13. Lueders T, Wagner B, Claus P, Friedrich MW (2004) Stable isotope probing of rRNA and DNA reveals a dynamic methylotroph community and trophic interactions with fungi and protozoa in oxic rice field soil. Environmental Microbiology 6: 60–72.

14. Rangel-Castro JI, Killham K, Ostle N, Nicol GW, Anderson IC, et al. (2005) Stable isotope probing analysis of the influence of liming on root exudate utilization by soil microorganisms. Environmental Microbiology 7: 828–838.

15. Mahmood S, Paton GI, Prosser JI (2005) Cultivation-independent in situ molecular analysis of bacteria involved in degradation of pentachlorophenol in soil. Environmental Microbiology 7: 1349–1360.

16. Jumpponen A (2011) Analysis of ribosomal RNA indicates seasonal fungal community dynamics in Andropogon gerardii roots. Mycorrhiza 21: 453–464.

17. de Vaufleury A, Kramarz PE, Binet P, Cortet J, Caul S, et al. (2007) Exposure and effects assessments of Bt-maize on non-target organisms (gastropods, microarthropods, mycorrhizal fungi) in microcosms. Pedobiologia 51: 185–194.

18. Knox OGG, Nehl DB, Mor T, Roberts GN, Gupta V (2008) Genetically modified cotton has no effect on arbuscular mycorrhizal colonisation of roots. Field Crops Research 109: 57–60.

19. Liu WK (2010) Do genetically modified plants impact arbuscular mycorrhizal fungi? Ecotoxicology 19: 229–238.

20. Tan FX, Wang JW, Chen ZN, Feng YJ, Chi GL, et al. (2011) Assessment of the arbuscular mycorrhizal fungal community in roots and rhizosphere soils of Bt corn and their non-Bt isolines. Soil Biology & Biochemistry 43: 2473–2479.

21. Verbruggen E, Kuramae EE, Hillekens R, de Hollander M, Kiers ET, et al. (2012) Testing Potential Effects of Maize Expressing the Bacillus thuringiensis Cry1Ab Endotoxin (Bt Maize) on Mycorrhizal Fungal Communities via DNA-and RNA-Based Pyrosequencing and Molecular Fingerprinting. Applied and Environmental Microbiology 78: 7384–7392.

22. Turrini A, Sbrana C, Nuti MP, Pietrangeli BM, Giovannetti M (2004) Development of a model system to assess the impact of genetically modified corn and aubergine plants on arbuscular mycorrhizal fungi. Plant and Soil 266: 69–75.

23. Castaldini M, Turrini A, Sbrana C, Benedetti A, Marchionni M, et al. (2005) Impact of Bt corn on rhizospheric and on beneficial mycorrhizal symbiosis and soil eubacterial communities iosis in experimental microcosms. Applied and Environmental Microbiology 71: 6719–6729.

24. Vainio EJ, Hantula J (2000) Direct analysis of wood-inhabiting fungi using denaturing gradient gel electrophoresis of amplified ribosomal DNA. Mycological Research 104: 927–936.

25. Caporaso JG, Kuczynski J, Stombaugh J, Bittinger K, Bushman FD, et al. (2010) QIIME allows analysis of high-throughput community sequencing data. Nature Methods 7: 335–336.

26. Goecks J, Nekrutenko A, Taylor J (2010) Galaxy: a comprehensive approach for supporting accessible, reproducible, and transparent computational research in the life sciences. Genome Biology 11: 13.

27. Reeder J, Knight R (2010) Rapidly denoising pyrosequencing amplicon reads by exploiting rank-abundance distributions. Nature Methods 7: 668–669.

28. Edgar RC (2010) Search and clustering orders of magnitude faster than BLAST. Bioinformatics 26: 2460–2461.

29. Pruesse E, Quast C, Knittel K, Fuchs BM, Ludwig WG, et al. (2007) SILVA: a comprehensive online resource for quality checked and aligned ribosomal RNA sequence data compatible with ARB. Nucleic Acids Research 35: 7188–7196.

30. Hammer Ø, Harper DAT, Ryan PD (2001) PAST: Paleontological Statistics Software Package for Education and Data Analysis. Palaeontologia Electronica 4.

31. Good IJ (1953) The population frequencies of species and the estimation of population parameters. Biometrika 40: 237–264.

32. Dixon P (2003) VEGAN, a package of R functions for community ecology. Journal of Vegetation Science 14: 927–930.

33. Mello A, Napoli C, Murat C, Morin E, Marceddu G, et al. (2011) ITS-1 versus ITS-2 pyrosequencing: a comparison of fungal populations in truffle grounds. Mycologia 103: 1184–1193.

34. Powell JR, Gulden RH, Hart MM, Campbell RG, Levy-Booth DJ, et al. (2007) Mycorrhizal and rhizobial colonization of genetically modified and conventional soybeans. Applied and Environmental Microbiology 73: 4365–4367.

35. Vierheilig H, Alt M, Lange J, Gutrella M, Wiemken A, et al. (1995) Colonization of transgenic tabacco constitutively expressing pathogenesis-related proteins by the vesicular-arbuscular mycorrhizal fungus *Glomus mosseae*. Applied and Environmental Microbiology 61: 3031–3034.

36. Hur M, Lim YW, Yu JJ, Cheon SU, Choi YI, et al. (2012) Fungal community associated with genetically modified poplar during metal phytoremediation. Journal of Microbiology 50: 910–915.

37. Hannula SE, Boschker HTS, de Boer W, van Veen JA (2012) 3C pulse-labeling assessment of the community structure of active fungi in the rhizosphere of a genetically starch-modified potato (Solanum tuberosum) cultivar and its parental isoline. New Phytologist 194: 784–799.

38. Hannula SE, de Boer W, van Veen J (2012) A 3-year study reveals that plant growth stage, season and field site affect soil fungal communities while cultivar and GM-trait have minor effects. PLoS ONE 7: e33819.

39. Hannula SE, de Boer W, van Veen JA (2010) In situ dynamics of soil fungal communities under different genotypes of potato, including a genetically modified cultivar. Soil Biology & Biochemistry 42: 2211–2223.

40. Janssen PH (2006) Identifying the dominant soil bacterial taxa in libraries of 16S rRNA and 16S rRNA genes. Applied and Environmental Microbiology 72: 1719–1728.

4

Vertical Variation of Nonpoint Source Pollutants in the Three Gorges Reservoir Region

Zhenyao Shen*, Lei Chen, Qian Hong, Hui Xie, Jiali Qiu, Ruimin Liu

State Key Laboratory of Water Environment Simulation, School of Environment, Beijing Normal University, Beijing, P.R. China

Abstract

Nonpoint source (NPS) pollution is considered the main reason for water deterioration, but there has been no attempt to incorporate vertical variations of NPS pollution into watershed management, especially in mountainous areas. In this study, the vertical variations of pollutant yields were explored in the Three Gorges Reservoir Region (TGRR) and the relationships between topographic attributes and pollutant yields were established. Based on our results, the pollutant yields decreased significantly from low altitude to median altitude and leveled off rapidly from median altitude to high altitude, indicating logarithmic relationships between pollutant yields and altitudes. The pollutant yields peaked at an altitude of 200–500 m, where agricultural land and gentle slopes (0–8°) are concentrated. Unlike the horizontal distributions, these vertical variations were not always related to precipitation patterns but did vary obviously with land uses and slopes. This paper also indicates that altitude data and proportions of land use could be a reliable estimate of NPS yields at different altitudes, with significant implications for land use planning and watershed management.

Editor: Vishal Shah, Dowling College, United States of America

Funding: The study was supported by the National Science Foundation for Distinguished Young Scholars (No. 51025933), the National Science Foundation for Innovative Research Group (No. 51121003) and the National Basic Research Program of China (973 Project, 2010CB429003).The funders had no role in study design, data collection and analysis, decision to publish, or preparation of the manuscript.

Competing Interests: The authors have declared that no competing interests exist.

* E-mail: zyshen@bnu.edu.cn

Introduction

After decades of working to reduce emissions from point sources, problems regarding nonpoint source (NPS) pollution have been highlighted, with agriculture being the largest contributor [1]. The three main forms of NPS pollutants are sediments, nutrients and pesticides [2], the effects of which are well documented [3–5]. Researchers have revealed that NPS pollution may come from a wide range of dispersed sources though a complex combination of physical, chemical and biological processes [6]. From an environmental point of view, there is a dire need to gain insights into the spatial variations of NPS pollution, which are essential for analyzing these complex problems in drainage basins.

In general, the spatial distributions of NPS pollution can be quantified by monitoring or modeling methods. In respect to monitoring strategy, detailed measured data are collected and the spatial variations of water quality can be analyzed by comparing those measured data. However, water quality degradation often results from multiple sources and separating the impacts by monitoring methods is very difficult and costly, especially for a large basin [7]. Watershed models can facilitate in identifying individual sources of NPS pollution and evaluating the decision schemes for watershed management. Up to now, many models have been developed for identifying the spatial distributions of NPS pollution [8]. Some of these models, such as Export Coefficient Model [9], are based on empirical equations and cannot always provide sufficient explanations for those complex watershed processes [10]. By contrast, those physically-based models can simulate the hydrologic and water quality responses at varying scopes and locations [11]. In addition, those physically-based models are usually coupled with the geographic information system (GIS) which can compile extensive input database and visualize the model results. Information of watershed characteristics can be extracted and analyzed with convenience of GIS techniques which are usually integrated into these watershed models. The most commonly-used watershed models are the Soil and Water Assessment Tool (SWAT) model [12], Agricultural Nonpoint Source pollution (AGNPS) model [13], Annualized Agricultural Nonpoint Source pollution (AnnAGNPS) model [14], and Hydrological Simulation Program - Fortran (HSPF) [15].

Currently, the GIS techniques have provided a reliable platform for integrating vertical and horizontal information within a basin. Within such a framework, altitude data are generally extracted from a Digital Elevation Model (DEM) [16], and this 3-dimensional information has been widely applied in studies on atmospheric pollution [17,18]. However, there is current interest in integrating the GIS platform to project large volumes of meteorological and geophysical data into horizontal information to study NPS pollution [19], regardless of whether vertical variations occur. Indeed, altitude is the key attribute of topography and has a direct impact on physical parameters such as precipitation, solar radiation, temperature and soil chemistry [20,21]. Altitude is also essential in other environmental factors, including slope length, slope degree and other properties [22,23]. Researchers have reported that altitude has an impact on geomorphologic processes such as surface runoff, soil erosion and landslides in hilly regions [24,25]. Land use changes, landscape dynamics and other human activities are therefore related to the terrain and altitude [26], especially in the mountainous areas. Therefore, those projected horizontal distributions of NPS pollution is a consequence but may not be the cause of NPS pollution because this information should make reference to specific vertical processes. As far as we know, there has been no attempt to incorporate vertical variations for the analysis of NPS pollution, which should draw increasing attention due to continued hilly urbanization, increased deforestation, and changed precipitation with global warming [27].

The objective of this paper is to contribute new insights into vertical variations to capture the complex features of NPS pollution. The study was performed in the Three Gorges Reservoir Region (TGRR) by: 1) exploring the spatial distributions of sediment, nitrogen (N), and phosphorus (P) using the Soil and Water Assessment Tool (SWAT); 2) establishing the relationships between land use, slope and altitude; and 3) characterizing the vertical variations of sediment, N and P yields in the TGRR.

Materials and Methods

Watershed Description

The Three Gorges Reservoir, which is by far the world's largest hydropower project, completed its first filling stage in 2003 and reached its maximum designed water level in 2008. Geographically, the TGRR, with a total area of approximately 58,000 km^2, is located in the transitional zone from the Tibetan Plateau in the west to the east rolling hills and plains of China between latitudes $28°10'$ and $32°13'$N and longitudes $105°17'$ and $110°11'$E (Fig. 1). The topography is complex, with over 74% of the landscape being mountainous and 21.7% being low hills. The land uses include cropland (39%), grassland (13%) and forest (46%), while the main soils are purplish soils (48%), limestone earths (34%) and yellow (16%) earths. The average precipitation is approximately 1400 mm, 80% of which occurs from April to October. The highest and lowest annual temperature ranges from approximately 27°C to 29°C and 6°C to 8°C, respectively.

When water levels were driven up by the Three Gorges Reservoir, hilly reclamation and deforestation continued to increase above the 175-m inundation line [28,29]. Due to the special geography and structure of the agriculture in the TGRR, the soil loss is serious and the eco-environment is vulnerable. Additionally, after the water was cut off, the water velocity was reduced and the retention time of pollutants prolonged. The water quality challenge has never been greater than now, as indicated by the soil erosion in the uplands and algal blooms in the aquatic environment [29].

Model Description and Preparation

Model description. The ArcSWAT model, developed by Arnold et al. [12], was used to develop the necessary input files. The SWAT components include weather generation, hydrology, soil erosion, crop growth, nutrient leaching and agricultural management [30]. The hydrology calculation was based on the curve number method and the Green-Ampt infiltration method [31]. The sediment yield was estimated by the modified soil loss equation [32]. Runoff, sediments and nutrients were calculated for each Hydrologic Response Unit (HRU) and then routed in stream using the QUAL2E model [33]. More information about the SWAT model can be found in Douglas et al. [30] (Methods S1).

Data description. Taking into account the study needs and data availability, the digital layers for altitude, land use and soil were constructed. DEM data at a scale of 1:25,000 published by the Institute of Geographical and Natural Resources Research, China, were used. Land-use data were interpreted from a 1:100,000 Thematic Mapper image and the proportions of land uses were treated as constants during the simulation period. A soil map at a scale of 1:1,000,000 and the related physical data were obtained from the Institute of Soil Science, Chinese Academy of Sciences. The daily precipitation, relative humidity, solar radiation, wind speed and air temperature data, measured by 49 weather stations from 1980 to 2010, were obtained from the State Meteorological Data Sharing Service System (http://cdc.cma.gov.cn). Crop information, including tillage, irrigation and the amount

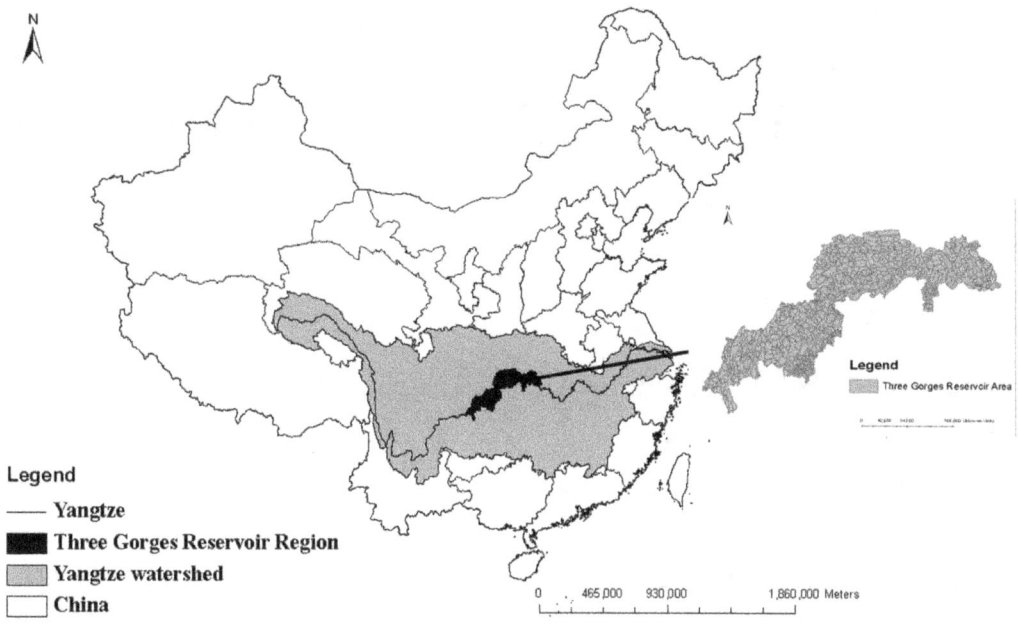

Figure 1. The location of the Three Gorges Reservoir Region.

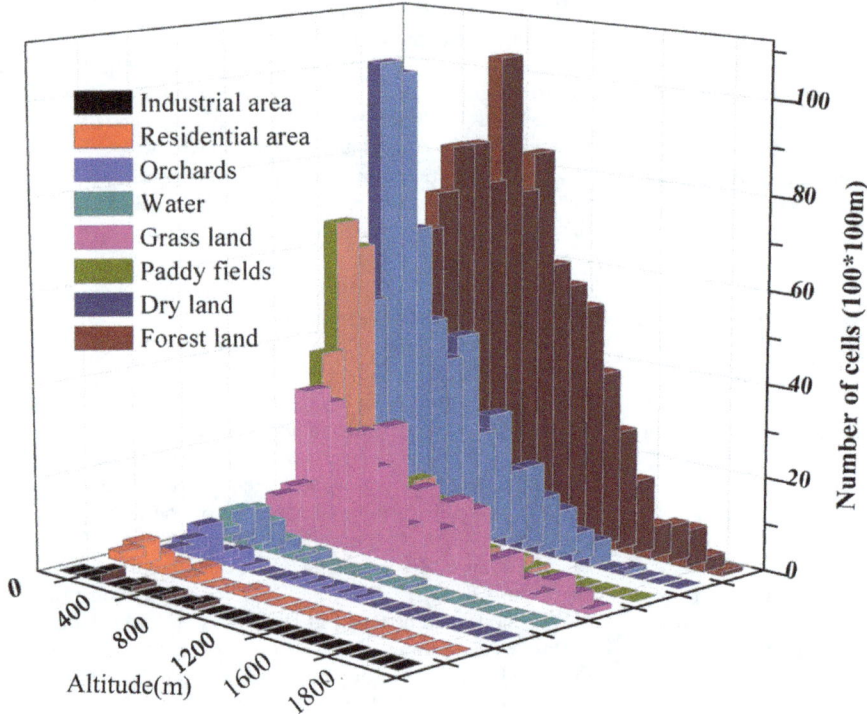

Figure 2. The vertical variations of land use.

of fertilizer used, was based on statistical data from local bureaus as well as field investigations in several local watersheds. Rice, potato, sweet potato and corn were selected as the main crops due to the cultivated areas and the amount of fertilizer. However, no clear records of management practices were available. To compensate for this lack, the average fertilizer rates in each village were calculated for cultivated crops and all agricultural areas were assumed to be tile drained.

Model preparation. In this study, the TGRR was delineated into 613 sub-watersheds interconnected by a stream network, and each sub-watershed was divided further into HRUs by setting 0% thresholds of land use, soil type and slope to accurately capture even small areas. In our previous study [34], we introduced a small-scale watershed extended method (SWEM) for parameter calibration in the TGRR (Methods S2). The detailed processes involve: 1) model calibration- a process of generating model

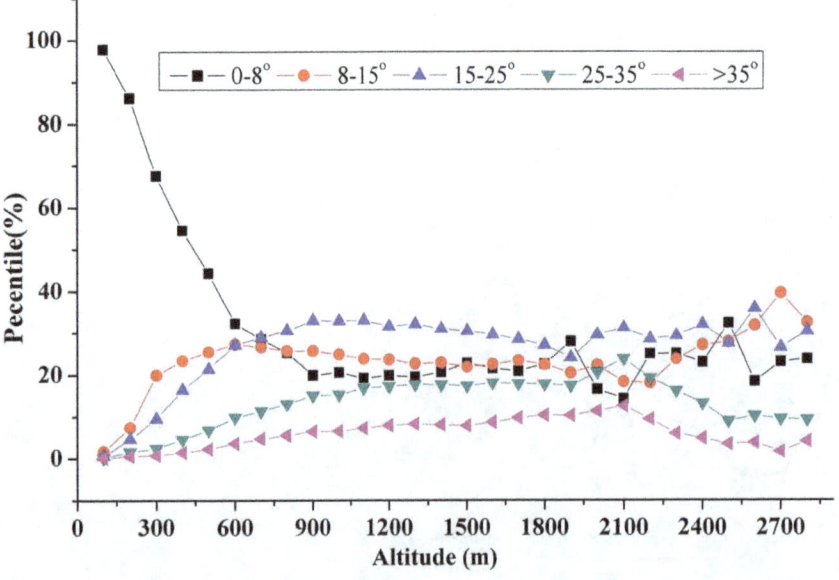

Figure 3. The vertical variations of slope.

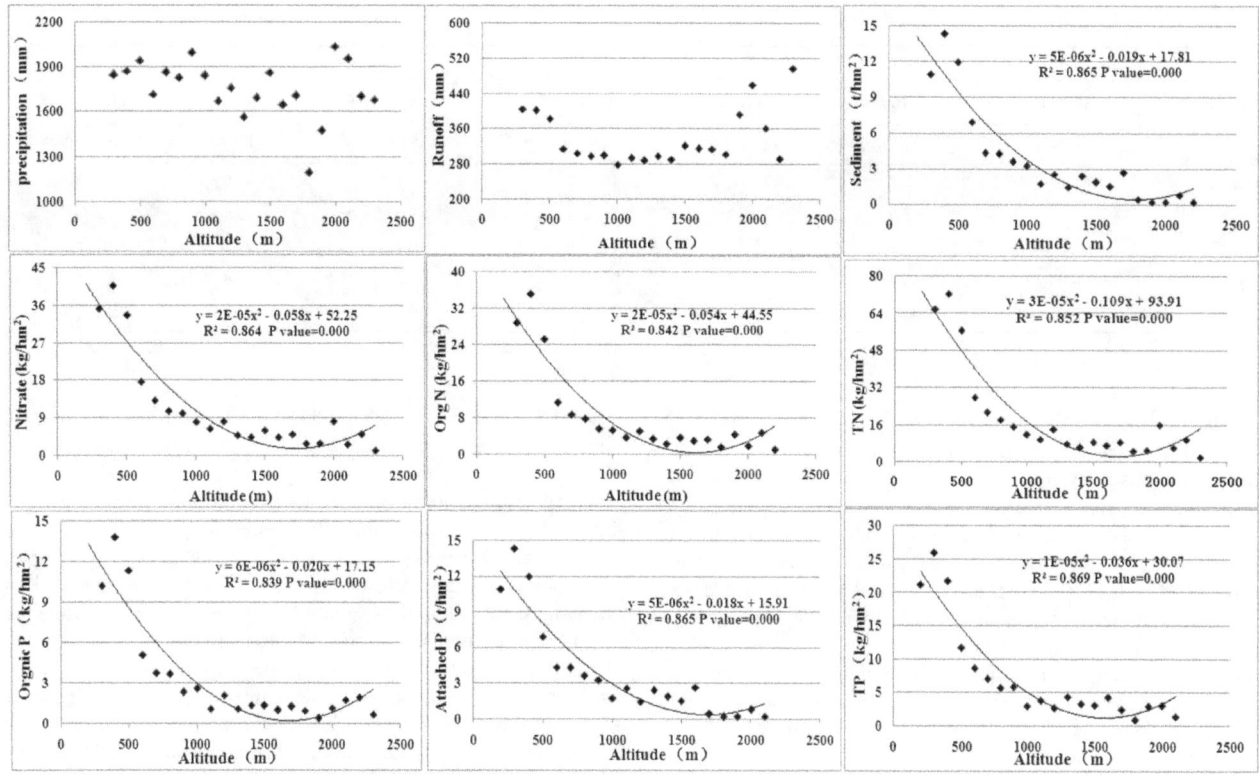

Figure 4. The vertical variations of precipitation and pollution yield.

parameter groups for representing different parts of the TGRR (in terms of the watersheds of the Yulin, Xiaojiang, Daning and Xiangxi); 2) extended modeling- running the well-calibrated models in the corresponding parts of the TGRR (Figure S1). The measured flow and water quality data were obtained from the Changjiang Water Resources Commission and the parameter groups were generated by the SWAT-CUP [35]. The Nash-Sutcliffe efficiency coefficient (E_{NS}) [36] was used to quantify the

degree of fit between the simulated data and the measured data.

$$E_{NS} = 1 - \frac{\sum_{i=1}^{n}(Q_{sim,i} - Q_{mea,i})^2}{\sum_{i=1}^{n}(Q_{mea,i} - \overline{Q}_{mea})^2} \qquad (1)$$

Where, $Q_{mea,i}$ is the ith observation for the constituent being

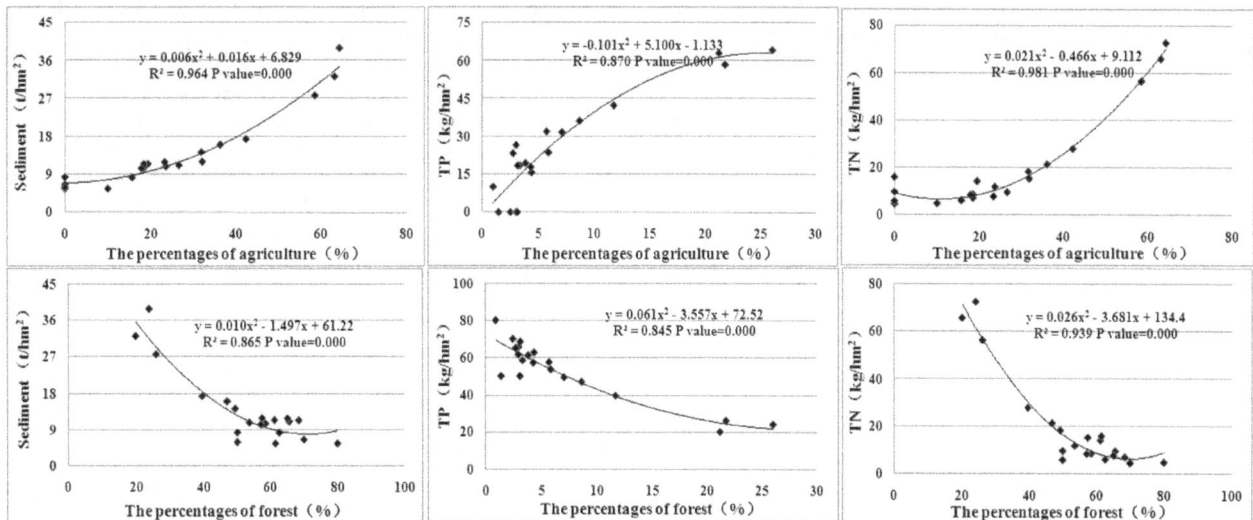

Figure 5. The relationship between pollutant yield and land uses in different altitudes.

Table 1. The load intensities of pollutants at different altitudes.

Slope	Sediment	Nitrate	Orgnic N	Attached P	Solute P	TN	TP
	t/hm^2	kg/hm^2	kg/hm^2	kg/hm^2	kg/hm^2	kg/hm^2	kg/hm^2
0–8°	11.9	28.7	21.0	1.5	1.1	49.8	4.1
8–15°	23.0	17.0	23.4	2.0	0.6	40.5	4.1
15–25°	18.3	10.7	12.4	1.0	0.4	23.1	2.1
25–35°	15.8	5.7	6.1	0.7	0.2	11.8	1.3
>35°	16.2	3.8	3.7	0.5	0.2	7.5	0.9

evaluated, $Q_{sim,i}$ is the predicted value for the constituent being evaluated, $\overline{Q}_{mea}$ is the mean value of observed data for the constituent being evaluated, and n is the total number of observations.

The values of E_{ns} in the respective sub-watersheds ranged from 0.53 to 0.94 for the stream flow, 0.53–0.94 for sediment, 0.60–0.84 for total P (TP), 0.47–0.80 for nitrate-N and 0.41–0.81 for NH$_4$-N. The detailed processes of the model calibration and validation can be found in our previous study [34,37]. With the groups of calibrated parameters, the extended simulation was conducted by running the well-calibrated SWAT models in the entire TGRR.

Data analysis. Following calibration, a 10-year (2000–2009) simulation was performed to isolate variability of climate, land use, crop rotations and runoff regime which may mask the effect of vertical variation [38]. The DEM and land use maps were divided into two matrixes of 4500 rows by 5500 columns, with a cell size of 100*100 m. A mask layer was applied to all cells to avoid noise in the statistical processing of the data, and 6,200,000 cells remained after this selection. Topographic data such as altitude, slope and land use were generated for each cell, and the flow, sediment and nutrient yield were calculated from SWAT outputs. The total yields were defined as the summary of corresponding cells.

Results and Discussion

The Vertical Variations of Land Use and Slope

Fig. 2 illustrates the vertical variations of the land uses at different altitudes. According to Fig. 2, the landscape area increases significantly from 0 m to 400 m and levels off rapidly from 400 m to 1600 m, while only slight declines can be observed when the altitude varies from 1600 m to 2100 m. Specifically, the land from 200 m to 1000 m was dominant in the TGRR, covering more than 71% of the entire area. The landscape area accounted for only 2% from 0 m to 200 m, 13% from 1000 m to 1500 m and 2% from 1500 m to 2100 m of the total area. Among the different land uses, agriculture (paddy field and dry land), forest and grassland were dominant in all altitudes but the vertical variations of these land uses were different. As illustrated in Fig. 2, the proportions of agriculture show obvious declines as the altitude increases, while those of forest and grassland show increasing trends. In particular, agricultural areas were concentrated among the altitudes between 200 m and 800 m. This vertical variation could be explained by most low-altitude areas below the 175-m inundation line having been submerged when the water levels were driven up by the Three Gorges Dam [29,38]. These vertical variations of land uses were also indicated by other studies showing that farmers resettled in the low hilly areas and 80% of the arable farmland is distributed in the low hilly areas or valley terraces [39,40]. In this study, the slope degrees were categorized into 0–

8°, 8–15°, 15–25°, 25–35° and 35~90°. The vertical variations of these slope degrees are analyzed in Fig. 3. In the TGRR, gentle slope (0–8°) made up the largest proportion (31%) of the entire area, while the land on median slope (8–25°) and steep slope (25–90°) accounted for 50% and 19%, respectively.

Fig. 3 also illustrates that the vertical variations of slope degree show obvious trends. Below the altitude of 800 m, the proportions of gentle slope gradually decreased as altitude increased, while the proportions of median and steep slopes increased slightly. Between the altitudes of 800 m and 2000 m, the proportions of gentle slope and median slope remained stable, while the proportions of steep slope continued to increase. Above the altitude of 2000 m, the landscape area was again dominated by gentle and median slopes. This result can be explained by the widely held view that the rock strength at high altitudes is normally high and that such altitudes usually have weathered rocks or rocks whose shear strength is much higher [25], indicating the high altitude areas in the TGRR are generally flat and have a convergent terrain.

The Vertical Variations of NPS Pollution

Fig. 4 further characterizes the vertical variation of precipitation, sediment, N and P yields. Fig. 4 illustrates that precipitation did not vary significantly with altitude. This result is inconsistent with previous studies [16,41] that demonstrated that precipitation increased with altitude due to the orographic effect, which lifted the air vertically and the condensation occurred due to adiabatic cooling. There are two probable reasons for this inconsistent variation. First, the climate in the TGRR is subtropical, with the annual mean temperature being 17°C, so adequate illumination may compensate for the orographic effect in the mountainous terrain. Second, a 10-year period was considered in the current study to represent the climatic variations. As rainfall is irregular in occurrence, duration and magnitude, this long period is equally true for a flattening effect of precipitation [41]. This paper indicates that NPS pollution did vary with altitude, even in the absence of different precipitation patterns related to altitude.

As illustrated in Fig. 4, the load intensities of all pollutants showed obvious declines from low altitude to high altitude. All variables peaked at the low altitude (200–500 m), where frequency of human actives is the highest. Specifically, soil erosion (above 500 t/(km^2 a)) occurred in over 90% of the altitudes of the TGRR, while 33% of the areas were heavily eroded below the altitude of 500 m, with an erosion coefficient greater than 4,000 t/(km^2 a). This result can be explained by the rock strength and vegetation cohesion making the high altitudes pollutant sinks, while the low altitude areas were normally prone to environmental vulnerability due to human disturbances [24,25]. The logarithmic lines were generated to demonstrate the correlation between the pollutant yields and altitude (Fig. 4). As shown in Fig. 4, the regression results are significant, with the regression correlations being larger than 0.74.

The relationships among land use, slope and NPS yields were also explored. As illustrated in Fig. 5, the proportion of agricultural area was positively correlated with pollution yields, while that of forest was negatively correlated. For every 1% reduction in forest area, the load intensity increased by $0.01\sim11.34$ t/km^2 for sediment, $0.15\sim2.83$ kg/km^2 for TP and $0.40\sim14.00$ kg/km^2 for total nitrogen (TN). The main reason for this result is that forest plants generally have a higher capability to hold and fix soil, while agricultural soil is either regularly over-fertilized or highly vulnerable to erosion [42]. In the TGRR, the agricultural area shrank at a high rate due to the Three Georges Reservoir, and there was no alternative but to rely on greater applications of fertilizer to ensure high productivity for the huge and growing population [29,38,43]. Specifically, the sediment yield increased slightly when the proportions of agriculture changed from 0% to 40%, and it showed a jump when the agriculture varied from 40% to 60%. This phenomenon could also be observed in P and N yields, from which the jumping points were obtained at proportions of 10% and 40%, respectively. For forest, the load intensity of sediment, TN and TP remained stable outside a relevant domain of 40%, 10% and 40%, respectively, and any change inside this proportion domain would have a greater impact on NPS yields. This phenomenon may be explained by the spatial distributions of the converted landscape pattern, which may mitigate certain discharges and may not always intensify the NPS pollution [44].

As shown in Table 1, the load intensity of sediment, organic N and attached P progressively rises with relief up to the median slope. The pollutant yield then declines toward steep slope. The most severe pollution risks typically occurred on the slopes between 8° and 15°, which is inconsistent with the widely held view that the downhill force is highest on steeper slopes [25]. However, the TGRR is major region for national environmental protection, and many projects, such as 'Grain for Green Project', have been implemented in steep slope areas to strengthen soil and reduce NPS yields. Additionally, human practices, including rotation, irrigation and tillage, have been conducted at the gentle slope and low altitude areas; therefore, gentler slopes, particularly in deeply human-impacted slopes, may increase hydrological connectivity and nutrient leaking, resulting in greater efficiency of delivery of P and N to surface waters [42].

Conclusions

In this paper, the vertical variations of land use, slope and NPS yields were estimated and used for studying the behavior of pollutants in the TGRR. Based on our results, the NPS pollution showed an obvious decline from low to high altitude, with all variables peaking at the low altitude (200–500 m), where the frequency of human actives was the highest. The watershed manager can gain insight into vertical dynamics to develop site-specific policies using this spatial information. This paper indicates that the vertical variations of NPS pollution were not related to precipitation patterns but did vary with vertical variations of land uses and slopes. Therefore, altitude data and proportions of land uses can be regarded as a reliable estimate of NPS load intensity, especially in the mountainous areas. However, uncertainty of modeling outcomes must be estimated to establish the reliability of the simulated outputs. In the future, more detailed data should be used and more pollutants, such as pesticide, heavy metal and Polychlorinated Biphenyl, should be incorporated into the list of analysts.

Acknowledgements

The authors wish to express their gratitude to *Plos One*, as well as to the anonymous reviewers who helped to improve this paper though their thorough review.

Author Contributions

Conceived and designed the experiments: ZYS. Performed the experiments: LC QH HX JLQ RML. Analyzed the data: ZYS LC QH HX JLQ. Contributed reagents/materials/analysis tools: ZYS LC QH HX. Wrote the paper: LC ZYS QH.

References

1. Xu Z, Wan S, Ren H, Han X, Li M-H, et al. (2012) Effects of water and nitrogen addition on species turnover in temperate grasslands in northern china. Plos One 7: e39762.

2. Dowd BM, Press D, Los Huertos M (2008) Agricultural nonpoint source water pollution policy: The case of California's Central Coast. Agr Ecosyst Environ 128: 151–161.

3. Nelson JL, Zavaleta ES (2012) Salt marsh as a coastal filter for the oceans: changes in function with experimental increases in nitrogen loading and sea-level rise. Plos One 7: e38558.

4. Keatley BE, Bennett EM, MacDonald GK, Taranu ZE, Gregory-Eaves I (2011) Land-Use Legacies Are Important Determinants of Lake Eutrophication in the Anthropocene. Plos One 6: e15913.

5. Somura H, Takeda I, Arnold JG, Mori Y, Jeong J, et al. (2012) Impact of suspended sediment and nutrient loading from land uses against water quality in the Hii River basin, Japan. J Hydrol 450: 25–35.

6. Short JS, Ribaudo M, Horan RD, Blandford D (2012) Reforming Agricultural Nonpoint Pollution Policy in an Increasingly Budget-Constrained Environment. Environ Sci Tech 46: 1316–1325.

7. Shen ZY, Chen L, Chen T (2012) The influence of parameter distribution uncertainty on hydrological and sediment modeling: a case study of SWAT model applied to the Daning watershed of the Three Gorges Reservoir Region, China. Stochc Env Res Risk A. 27: 235–251.

8. Shen ZY, Liao Q, Hong Q, Gong YW (2011) An overview of research on agricultural non-point sources pollution modelling in China. Sep Purif Technol. 9: 595–604.

9. Ding X, Shen ZY, Hong Q, Yang ZF, Wu X, et al. (2010) Development and test of the Export Coefficient Model in the Upper Reach of the Yangtze River. J Hydrol. 383: 233–244.

10. Shen ZY, Chen L, Hong Q, Ding XW, Liu RM, et al. (2013) Long-term variation (1960–2003) and causal factors of non-point source nitrogen and phosphorus loads in the Upper Reach of Yangtze River. J Hazard Mater. 252–253: 45–56.

11. Woznicki S, Nejadhashemi A, Smith C (2011) Assessing best management practice implementation strategies under climate change scenarios. T ASABE. 54: 171–190.

12. Arnold JG, Srinivasan R, Muttiah RS, Williams JR (1998) Large area hydrologic modeling and assessment - Part 1: Model development. J Am Water Resour As 34: 73–89.

13. Young RA, Onstad C, Bosch D, Anderson W (1989) AGNPS: A nonpoint-source pollution model for evaluating agricultural watersheds. J Soil Water Conserv. 44: 168–173.

14. Bingner R, Theurer F, Yuan Y (2001) AnnAGNPS Technical Processes: Documentation Version 2. Unpublished Report, USDA-ARS National Sedimentation Laboratory, Oxford, Miss.

15. Bicknell B, Imhoff J, Kittle Jr J, Donigian Jr A, Johanson R (1993) Hydrologic Simulation Program-FORTRAN (HSPF): User's Manual for Release 10. Rep. No. EPA/600/R-93/174. US EPA Environmental Research Lab, Athens, Ga.

16. Lin K, Zhang Q, Chen X (2010) An evaluation of impacts of DEM resolution and parameter correlation on TOPMODEL modeling uncertainty. J Hydrol 394: 370–383.

17. Moreau-Guigon E, Motelay-Massei A, Harner T, Pozo K, Diamond M, et al. (2007) Vertical and temporal distribution of persistent organic pollutants in Toronto. 1. Organochlorine pesticides. Environ Sci Tech 41: 2172–2177.

18. Hauck M, Zimmermann J, Jacob M, Dulamsuren C, Bade C, et al. (2012) Rapid recovery of stem increment in Norway spruce at reduced SO_2 levels in the Harz Mountains, Germany. Environ Pollut 164: 132–141.

19. Wong MS, Nichol JE, Lee KH (2009) Modeling of Aerosol Vertical Profiles Using GIS and Remote Sensing. Sensors 9: 4380–4389.

20. Bryan BA (2003) Physical environmental modeling, visualization and query for supporting landscape planning decisions. Landscape and Urban Plan. 65: 237–259.

21. Wu S, Li J, Huang GH (2007) Characterization and Evaluation of Elevation Data Uncertainty in Water Resources Modeling with GIS. Water Resour Manage 22: 959–972.

22. Livne E, Svoray T (2011) Components of uncertainty in primary production model: the study of DEM, classification and location error. Int J Geogr Inf Sci 25: 473–488.

23. Lin K, Zhang Q, Chen X (2010) An evaluation of impacts of DEM resolution and parameter correlation on TOPMODEL modeling uncertainty. J Hydrol 394: 370–383.

24. Pourghasemi HR, Mohammady M, Pradhan B (2012) Landslide susceptibility mapping using index of entropy and conditional probability models in GIS: Safarood Basin, Iran. Catena 97: 71–84.

25. Ghimire M (2011) Landslide occurrence and its relation with terrain factors in the Siwalik Hills, Nepal: case study of susceptibility assessment in three basins. Nat Hazards56: 299–320.

26. Orgiazzi A, Lumini E, Nilsson RH, Girlanda M, Vizzini A, et al. (2012) Unravelling Soil Fungal Communities from Different Mediterranean Land-Use Backgrounds. Plos One 7: e34847.

27. Guo SL, Wang JX, Xiong LH, Ying AW, Li DF (2002) A macro-scale and semi-distributed monthly water balance model to predict climate change impacts in China. J Hydrol 268: 1–15.

28. Shi ZH, Ai L, Fang NF, Zhu HD (2012) Modeling the impacts of integrated small watershed management on soil erosion and sediment delivery: A case study in the Three Gorges Area, China. J Hydrol 438–439: 156–167.

29. Zhang Q, Lou Z (2011) The environmental changes and mitigation actions in the Three Gorges Reservoir region, China. Environ Sci Policy 14: 1132–1138.

30. Douglas-Mankin KR, Srinivasan R, Arnold JG (2010) Soil and Water Assessment Tool (SWAT) model: Current developments and applications. T ASABE 53: 1423–1431.

31. USDA-SCS (1972) Hydrology Sect. 4, Soil Conservation Service National Engineering Handbook; Washington, DC.

32. Williams JR (1976) Flood routing with variable travel time or variable storage coefficients. T ASABE 12: 100–103.

33. Brown LC, Barnwell TO (1987) The Enhanced Stream Water Quality Models QUAL2E and QUAL2E-UNCAS: Documentation and User Manual; Athens.

34. Hong Q, Sun Z, Chen L, Liu R, Shen Z (2012) Small-scale watershed extended method for non-point source pollution estimation in part of the Three Gorges Reservoir Region. Int J Environ Sci Tech 9: 595–604.

35. Abbaspour KC (2008) SWAT-CUP2: SWAT calibration and uncertainty programs - a user manual; Department of Systems Analysis, Integrated Assessment and Modelling (SIAM), Eawag, Swiss Federal Institute of Aquatic Science and Technology: Duebendorf.

36. Nash J, Sutcliffe J (1970) River forecasting using conceptual models, 1. A discussion of principles. J Hydrol 10: 282–290.

37. Shen ZY, Chen L, Chen T (2012) Analysis of parameter uncertainty in hydrological and sediment modeling using GLUE method: a case study of SWAT model applied to Three Gorges Reservoir Region, China. Hydrol Earth Syst Sci 16: 121–132.

38. Ma X, Li Y, Zhang M, Zheng F, Du S (2011) Assessment and analysis of non-point source nitrogen and phosphorus loads in the Three Gorges Reservoir Area of Hubei Province, China. Sci Total Environ 412–413: 154–161.

39. Li Q, Yu M, Lu G, Cai T, Bai X, et al. (2011) Impacts of the Gezhouba and Three Gorges reservoirs on the sediment regime in the Yangtze River, China. J Hydrol 403: 224–233.

40. Wu J, Cheng X, Xiao H, Wanga H, Yang LZ, et al. (2009) Agricultural landscape change in China's Yangtze Delta, 1942–2002: A case study. Agr Ecosyst Environ 129: 523–533.

41. Shen Z, Chen L, Liao Q, Liu R, Hong Q (2012) Impact of spatial rainfall variability on hydrology and nonpoint source pollution modeling. J Hydrol 472–473: 205–215.

42. Shen Z, Hong Q, Yu H, Niu JF (2010) Parameter uncertainty analysis of non-point source pollution from different land use types. Sci Total Environ 408: 1971–1978.

43. Chen Q, Hooper DU, Lin S (2011) Shifts in Species Composition Constrain Restoration of Overgrazed Grassland Using Nitrogen Fertilization in Inner Mongolian Steppe, China. Plos One 6: e16909.

44. Ouyang W, Skidmore AK, Hao F, Wang T (2010) Soil erosion dynamics response to landscape pattern. Sci Total Environ 408: 1358–1366.

5

Predicting Greenhouse Gas Emissions and Soil Carbon from Changing Pasture to an Energy Crop

Benjamin D. Duval[1,2¤a], Kristina J. Anderson-Teixeira[1¤b], Sarah C. Davis[1¤c], Cindy Keogh[3], Stephen P. Long[1,2,4], William J. Parton[3], Evan H. DeLucia[1,2,4]*

1 Energy Biosciences Institute, University of Illinois at Urbana-Champaign, Urbana, Illinois, United States of America, **2** Global Change Solutions, Urbana, Illinois, United States of America, **3** Natural Resource Ecology Laboratory, Fort Collins, Colorado, United States of America, **4** Department of Plant Biology, University of Illinois at Urbana-Champaign, Urbana, Illinois, United States of America

Abstract

Bioenergy related land use change would likely alter biogeochemical cycles and global greenhouse gas budgets. Energy cane (*Saccharum officinarum* L.) is a sugarcane variety and an emerging biofuel feedstock for cellulosic bio-ethanol production. It has potential for high yields and can be grown on marginal land, which minimizes competition with grain and vegetable production. The DayCent biogeochemical model was parameterized to infer potential yields of energy cane and how changing land from grazed pasture to energy cane would affect greenhouse gas (CO_2, CH_4 and N_2O) fluxes and soil C pools. The model was used to simulate energy cane production on two soil types in central Florida, nutrient poor Spodosols and organic Histosols. Energy cane was productive on both soil types (yielding 46–76 Mg dry mass·ha^{-1}). Yields were maintained through three annual cropping cycles on Histosols but declined with each harvest on Spodosols. Overall, converting pasture to energy cane created a sink for GHGs on Spodosols and reduced the size of the GHG source on Histosols. This change was driven on both soil types by eliminating CH_4 emissions from cattle and by the large increase in C uptake by greater biomass production in energy cane relative to pasture. However, the change from pasture to energy cane caused Histosols to lose 4493 g CO_2 eq·m^{-2} over 15 years of energy cane production. Cultivation of energy cane on former pasture on Spodosol soils in the southeast US has the potential for high biomass yield and the mitigation of GHG emissions.

Editor: Chenyu Du, University of Nottingham, United Kingdom

Funding: This research was funded by the Energy Bioscience Institute. The funders had no role in study design, data collection and analysis, decision to publish, or preparation of the manuscript.

Competing Interests: Please note that Duval, Long and DeLucia are affiliated with a company, Global Change Solutions LLC. This affiliation does not represent a "competing interest".

* E-mail: delucia@illinois.edu

¤a Current address: Dairy Forage Research Center, United States Department of Agriculture, Agricultural Research Service, Madison, Wisconsin, United States of America
¤b Current address: Conservation Ecology Center, Smithsonian Conservation Biology Institute, National Zoological Park, Front Royal, Virginia, United States of America
¤c Current address: Voinovich School for Leadership and Public Affairs, Ohio University, Athens, Ohio, United States of America

Introduction

Land use has a pervasive influence on atmospheric greenhouse gas (GHG) concentrations and thereby on climate [1,2,3]. Carbon emissions from land use change, often to make way for agriculture, have contributed substantially to anthropogenic increases in the atmospheric CO_2 concentration [2]. For example, C emissions from tropical deforestation have been estimated at 10.6 ± 1.8 Pg CO_2 per year between 1990 and 2007, equal to $\sim$40% of global fossil fuel emissions [3]. Likewise, it is estimated that 40–52 Pg CO_2 have been released by plowing high-C native prairie soils [4]. Agricultural practices are important to global GHG budgets, with agroecosystems contributing $\sim$14% of global anthropogenic GHG emissions [1]. Agricultural practices can also reduce GHG emissions and enhance soil carbon, and have the potential to mitigate climate change [4,5].

Land use and land management changes associated with the emerging bioenergy industry are likely to have substantial impacts on global GHG budgets [6,7,8]. A change from fossil fuels to an energy economy more reliant on plant-derived biofuels has the potential to reduce GHG emissions [9]. The prospect of lowering emissions is one factor leading to the United States' mandate to produce 136 billion liters of renewable fuel by 2022 [10]. However, meeting this mandate will require substantial land area [11,12], which implies potentially major changes to regional biogeochemical cycling [12,13].

Corn grain (*Zea mays*) is the dominant crop used for ethanol production in the US [14]. However, the ability of corn ethanol to reduce GHG emissions is questionable [15,16], and corn production exacerbates nitrogen pollution and other environmental problems [17–19]. Of particular concern is the possibility that diversion of corn for ethanol production will increase global grain prices and trigger agricultural expansion and deforestation elsewhere in the world [7]. The emerging commercial technology to convert ligno-cellulose to ethanol could redress the reliance on corn grain as an ethanol feedstock [20]. This could be particularly beneficial if cellulosic biofuel crops are grown on land that is not important for food production, while having lower GHG emissions than traditional row-crop agriculture [21,22]. Therefore, considerable research has focused on understanding the soil C and greenhouse gas consequences of replacing traditional agriculture used for bioenergy with perennial grasses like switchgrass (*Panicum virgatum* L.), Miscanthus (*Miscanthus x giganteus* J. M. Greef & Deuter ex Hodk. & Renvoize), or restored prairie cropping systems in the Midwestern United States [17,23–27].

The Southeastern United States holds particular potential for cultivation of second-generation biofuel crops [28,29]. In comparison with the corn-soy and wheat belts of the Midwestern US, this region's longer growing season, high precipitation and relatively lower land costs make it attractive for biofuel crop production. However, far less is known about the biogeochemical consequences of land-use change to biofuel crop production in this region.

Energy cane, a promising crop for ligno-cellulosic fuel production, is a variety of sugarcane (*Saccharum officinarum*) that is higher yielding, more cold tolerant and has lower sucrose content than commercially produced sugarcane [28]. Because of its lower sugar concentration, it has not been widely cultivated, but has been of interest commercially as a genetic stock for improving cold tolerance in higher sucrose sugarcane strains [28]. With the development of ligno-cellulosic ethanol conversion technologies, sucrose concentration is less important for ethanol production, and energy cane could become an important biofuel feedstock as yields are high, ranging from 25–74 $Mg \cdot ha^{-1} \cdot yr^{-1}$ dry mass (Table 1).

Florida is the largest sugarcane producing state in the US and is therefore a likely location for large-scale energy cane production [30]. Currently, 466,000 hectares of land in Florida are used for low-intensity grazing, and converting some portion of this land could provide an option for growing energy cane [31,32]. However, it is unknown if converting pasture to cultivated land will affect GHG exchange with the atmosphere and soil carbon storage. More frequent soil disturbance and the presence of larger quantities of litter from growing energy cane could increase CO_2 efflux to the atmosphere [33,34], while removing cattle from the landscape will displace methane (CH_4) efflux [35]. If fields are fertilized, nitrous oxide (N_2O) emissions may increase because of greater substrate availability for denitrifying microbes [36], and

indeed, high rates of N_2O efflux have been measured from sugarcane grown on highly fertilized soils in Australia [37]. However, considering the entire suite of greenhouse gasses, there may be an overall reduction in GHG flux due to the offset provided by greater atmospheric carbon uptake into the crop.

The region of Florida where energy cane is likely to be grown has two distinct soil types. The most common soils are Spodosols, which are low nutrient and low organic matter sands requiring significant fertilizer to maintain agricultural productivity [38]. Substantial sugarcane production in Florida also occurs on Histosols, which are high organic matter "mucks" that are not typically fertilized, as production on these soils can be maintained by N mineralization from organic matter [39]. The cultivation of Histosols began by draining swamplands, where organic matter had accumulated under anaerobic conditions. Drainage accelerates decomposition and further cultivation of these organic soils is associated with rapid oxidation of organic matter, resulting in significant soil C loss and emissions of CO_2 and N_2O to the atmosphere [9,40,41].

Theoretical [13,25] and empirical research [42,43] indicate that the conversion of land in the rain-fed Midwest currently used to produce corn for ethanol to perennial biofuel feedstocks such as switchgrass or Miscanthus (a close relative of sugarcane) would greatly reduce or reverse the emission of GHG to the atmosphere and rebuild depleted carbon stocks in the soil. Prior studies with Miscanthus in Europe have measured substantial decreases in nitrogen use, and large increases in soil biomass and organic matter relative to other agricultural land uses [44,45]. There have been no experimental studies that address how changing a landscape to cultivate energy cane will impact GHG emissions and soil C stocks. This is addressed here by using the process-based biogeochemical model DayCent to run *in silico* experiments to ask how land use change from pasture to energy cane production changes ecosystem GHG flux and soil C storage. We test the hypotheses that converting pastures to energy cane will lead to reductions in GHG flux to the atmosphere and increase soil C stocks, and that soil type is an important modulator of that change.

Methods

Plant and Soil Analyses

To parameterize the DayCent model, plants and soils were collected on private land in Highlands County, Florida (27° 21' 49" N, 81° 14' 56" W) in May 2011. Paired 4-m^2 plots (n = 3) were randomly located in energy cane fields that had been recently (<2 months) converted from pasture and in adjacent non-cultivated pasture on both Spodosols (hyperthermic Arenic Alaquods) and Histosols (hyperthermic Histic Glossaqualfs). We harvested all aboveground biomass from each plot. Soil samples were taken from the pastures in areas not yet under energy cane cultivation. Three soil cores to a depth of 1 m were extracted from each plot with a 1.75-cm diameter wet sampling tube (JMC product # PN010, Newton, IA). Soil cores were separated by depth (0–30 cm, >30 cm). Plant material and soils were oven dried at 65°C (plant material) and 105°C (soils) until they reached constant mass. Dried soils were coarse ground with a mill (model F-4, Quaker City, Phoenixville, PA), and then fine ground with a coffee grinder (Sunbeam Products Inc., Boca Raton, FL). Total C and N content may have been slightly underestimated from the dried Histosols due to volatilization, but the values we measured (Table 1) fall well within the range reported by NRCS Web Soil Survey [46]. Plant material was ground to pass a 425-μm mesh (Wiley mill, Thomas Scientific, Swedesboro, NJ, USA). Plant and soil subsamples

Table 1. Input parameters (mean and one standard error of the mean; SEM) for carbon and nitrogen concentration of energy cane and soils collected from the Highlands Ethanol farm, Highlands County, Florida.

		%C		%N	
		Mean	SEM	Mean	SEM
Energy cane	Live leaves	43.68	0.22	1.80	0.18
	Dead leaves	39.77	0.22	0.52	0.03
	Stalks	41.18	0.33	0.87	0.10
Soils	Soil Depth				
Histosols	0–60 cm	7.77	2.48	0.50	0.20
	60–100 cm	7.77	2.48	0.50	0.20
Spodosols	0–30 cm	0.77	0.17	0.04	<0.01
	30–60 cm	0.36	0.03	0.02	0.01
	60–100 cm	0.36	0.03	0.02	0.01

When site-specific data were not available, plant information was used from reference [65], and soil data were collected from the NRCS Web Soil Survey (http://websoilsurvey.nrcs.usda.gov/).

Table 2. Site information for studies used in DAYCENT model validation.

Site	Lit. Yield	Model Yield	Max. Temp.	Min. Temp.	Precipitation	Latitude	Longitude	Reference
Auburn, AL	26.1	25.4	24.2	9.8	1160	32.67	−85.44	Woodard and Prine, 1993
Belle Glade, FL	25.0	28.3	27.8	16.4	1378	26.68	−80.67	Korndorfer, 2009
EREC, FL	51.3	43.5	29.1	17.7	1181	26.65	−80.63	Gilbert et al., 2006
Gainesville, FL	35.6	27.3	27.0	13.7	1123	29.68	−82.27	Woodard and Prine, 1993
Hendry, FL	39.2	55.9	28.4	18.3	1362	27.78	−82.15	USDA, 2011
Hillsboro, FL	60.7	62.5	28.5	18.3	1547	27.90	−82.49	Gilbert et al., 2006
Houma, LA (1st ratoon)	36.6	38.2	25.2	14.8	500	29.57	−90.65	Legendre and Burner, 1995
Houma, LA (2nd ratoon)	34.9	37.6	25.2	14.8	500	29.57	−90.65	Legendre and Burner, 1995
Hundley, FL	73.5	62.0	28.5	18.1	1457	26.30	−80.16	Gilbert et al., 2006
Jay, FL (plant cane)	35.8	33.6	26.9	16.4	1321	28.65	−80.82	Woodard and Prine, 1993
Jay, FL (1st ratoon)	27.8	32.8	26.9	16.4	1321	28.65	−80.82	Woodard and Prine, 1993
Lakeview, FL	71.3	62.5	28.5	17.4	1275	26.30	−80.15	Gilbert et al., 2006
Hidalgo, TX	34.6	42.5	28.7	18.1	576	26.17	−97.93	Weidenfeld, 1995
Ona, FL (1st ratoon)	40.5	38.1	28.6	16.0	1160	27.48	−81.92	Woodard and Prine, 1993
Ona, FL (2nd ratoon)	30.2	31.6	28.6	16.0	1160	27.48	−81.92	Woodard and Prine, 1993
Pahokee, FL	60.5	65.5	28.4	17.5	1269	26.82	−80.66	Glaz and Ulloa, 1993
Palm Beach, FL	32.3	35.4	29.0	16.6	851	26.67	−80.15	USDA, 2011
Quincy, FL (1st ratoon)	26.3	25.1	25.8	12.9	1445	30.59	−84.58	Woodard and Prine, 1993
Quincy, FL (2nd ratoon)	27.8	21.7	25.8	12.9	1445	30.59	−84.58	Woodard and Prine, 1993
Shorter, AL	26.4	25.4	24.9	10.9	1119	32.40	−85.94	Sladden et al., 1991
Sundance, FL	42.1	43.5	28.6	17.5	1303	26.60	−80.87	Gilbert et al., 2006

Yield values from the literature and modeled yields for energy cane and sugarcane represent total aboveground biomass expressed as Mg ha^{-1} on a dry mass basis. Climate variables include mean annual maximum and minimum temperature (°C) and mean annual precipitation (mm).

within each plot were combined, and C and N concentrations were measured for depth-stratified soil samples (Table 1) and total above ground biomass with a flash combustion chromatographic separation elemental analyzer (Costech 4010 CHNSO Analyzer, Costech Analytical Technologies Inc. Valencia, CA). The instrument was calibrated with acetanilide obtained from Costech Analytical Technologies, Inc. Other physical soil attributes, including texture, bulk density and water holding capacity were obtained from the NRCS Web Soil Survey [46] for Highlands County, Florida.

The DayCent Model

The DayCent model [47,48] was developed to simulate ecosystem dynamics for agricultural, forest, grassland and savanna ecosystems [49–51]. The model is a daily time step version of the Century model [52,53], using the same soil carbon and nutrient cycling submodels to simulate soil organic matter dynamics (C and N) and nitrogen mineralization. DayCent uses more mechanistic submodels than Century to simulate daily plant production, plant nutrient uptake, trace gas fluxes (N_2O, CH_4), NO_3 leaching, and soil water and temperature [48,54–58].

The DayCent soil organic matter model is widely used to simulate the impacts of management practices on soil carbon dynamics and nutrient cycling. Specifically, the soil organic matter submodel has been used to simulate the impacts of soil tillage practices; no-tillage, minimum tillage and conventional tillage [59,60], crop rotations [59], and biofuel crops; woody biomass, switchgrass (*Panicum virgatum*), Miscanthus (*Miscanthus* X *giganteus*), and sugarcane [13,61] on soil carbon dynamics for agricultural systems. These studies test model performance against observed

data and demonstrate general success in simulating changes in soil carbon levels associated with management practices.

The soil trace gas submodel has been extensively tested using observed soil CH_4 and N_2O data sets from agricultural and natural ecosystems, and once parameterized with plant production data, provides accurate predictions of trace gas fluxes. Specifically, DayCent has successfully simulated the observed impacts of N fertilizer additions and cropping systems [50,58,59] on soil N_2O and CH_4 fluxes. The model results and observed data sets demonstrate that increasing N fertilizer levels increases soil N_2O fluxes and that soil N_2O fluxes are much lower for perennial crops as compared to annual crops.

The DayCent model has been used extensively to simulate grassland and crop yields [50,58,59,62], and to evaluate the environmental impacts of growing crops. Adler et al. [63] used the DayCent model to simulate net greenhouse gas fluxes (soil C status and soil CH_4 and N_2O fluxes) associated with the use of corn, soybeans, alfalfa, hybrid popular, reed canary grass and switchgrass for biofuel energy production in Pennsylvania. Davis et al. [25] used the DayCent model to simulate the environmental impacts of growing switchgrass and Miscanthus in Illinois and compared simulated plant production for switchgrass and Miscanthus with observed yield data. The authors also compared the net soil greenhouse gas fluxes (soil C changes and soil CH_4 and N_2O fluxes) associated with growing switchgrass and Miscanthus and growing corn and soybeans. Davis et al. [13] recently used the DayCent model to simulate the environmental impact of replacing the corn currently grown for ethanol production in the Corn Belt with perennial grasses (Miscanthus and switchgrass) for second-generation biofuel production. The authors found that the

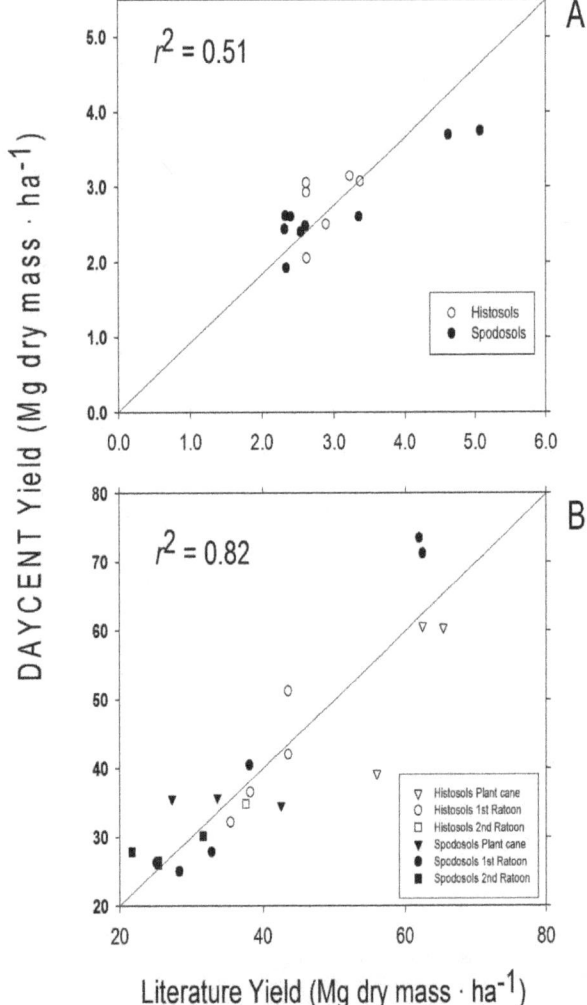

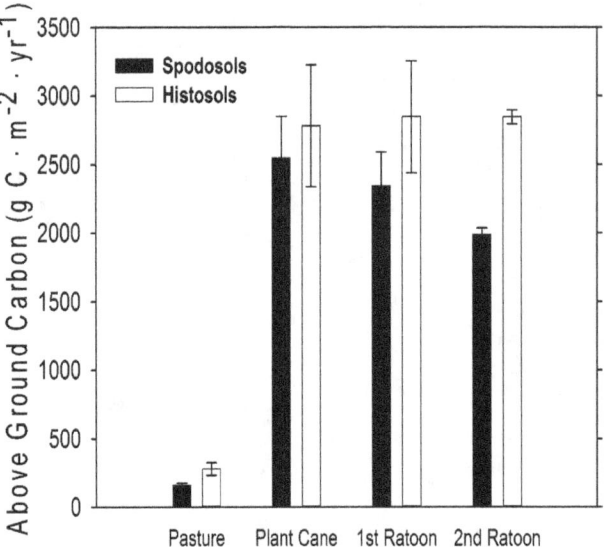

Figure 2. Modeled above ground production of grazed pasture and energy cane in Highlands County, Florida. Values are mean above ground carbon (g C·m^{-2}·yr^{-1}, ± SD) for 15 years in pasture, and for 5 × 3-year ratoon cycles in energy cane (each bar represents the average of 5 values, one for each year for each stage in the planting cycle).

Figure 1. Regression analysis used in DayCent model validation. Model output for dry mass yield was compared to literature values for A) pasture yield values from the USDA-NASS database, B) sugarcane and energy cane dry mass yield. Data points are compared to a 1:1 line.

DayCent model successfully predicted corn, Miscanthus, and switchgrass biomass production for U.S. sites with multiple N fertilizer levels. They also showed that the DayCent model successfully simulated observed annual soil N_2O fluxes from corn and switchgrass grown with multiple N fertilizer levels and showed that soil N_2O fluxes are much lower for fertilized switchgrass than for corn.

Furthermore, the basis for the DayCent model, Century, has been used to simulate sugarcane production in Brazil [61,64] and Australia [65]. These authors show that the Century/DayCent soil organic matter sub-model can correctly simulate the impacts of fertilizer, and organic matter additions on soil carbon levels and surface litter decay.

Model Parameterization

Energy cane is a variety of sugarcane, and thus parameterizing DayCent for this crop required only minor changes to the previously published input data used for sugarcane [61,65]. Energy cane differs from sugarcane in that it has increased cold

tolerance, decreased sucrose content, and higher cellulose content. We adjusted parameters based on direct measurement of energy cane tissue traits described above. The principal changes from sugarcane to energy cane were reducing the minimum C:N ratio of leaves from 28.6 to 22.1 and changing the C:N of stems from 160 to 30.5. Because of this change in C:N, the parameter for C allocation to stems in DayCent also was modified (from 60% to 40%), to reflect the lower C content of stems relative to N for energy cane versus the previously modeled sugarcane parameters [65]. Bahiagrass (*Paspalum notatum* Flueggé) pasture was simulated using the existing DayCent model parameters for warm season grasses [66,67].

Histosols are challenging to model as organic matter and C content typically are uniform throughout the soil profile [68], but they also are known to subside because of oxidation of the highly labile organic C pools characteristic of these soils [69]. This subsidence was calculated from the modeled rate of organic matter loss and bulk density. DayCent simulates soil C flux to a depth of 30 cm [47], so as soil was lost with subsidence new soil and organic matter became part of this upper 30 cm column from below. This assumes that loss only occurred in the upper 30 cm, which is reasonable since this is the disturbed and aerated part of the soil. The C and N added from low in the soil profile was calculated from the rate of subsidence and the measured elemental contents and bulk density of the soil that was below 30 cm, when sampled, which is at time zero in our model. However, model output calculates GHG and soil C to soil depths to 30 cm.

Model Validation

Literature values of aboveground production (dry mass) for grazed pasture, sugarcane and energy cane (Table 2) were used to validate DayCent. Validation focused on aboveground biomass production because this variable has been measured widely across a range of sites. While there were insufficient data on trace gas flux or changes in soil C in sugarcane or energy cane for validation of

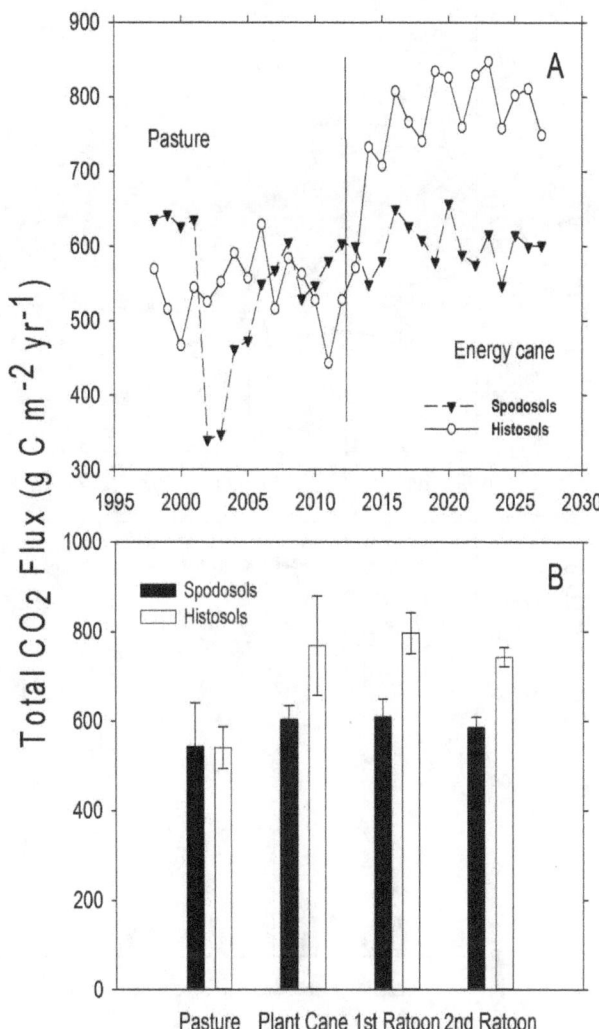

Figure 3. Modeled total soil CO₂ flux from pasture and land converted to energy cane in Highlands County, Florida. A) Total annual soil CO_2 flux (expressed as g C·m^{-2}). Dashed line represents year of land use conversion from pasture to energy cane. B) Mean total soil CO_2 flux (g C·m^{-2}·yr^{-1}, ± SD) for 15 years in pasture, and for 5, 3-year ratoon cycles in energy cane (each bar represents the average of 5 values, one for each year for each stage in the planting cycle).

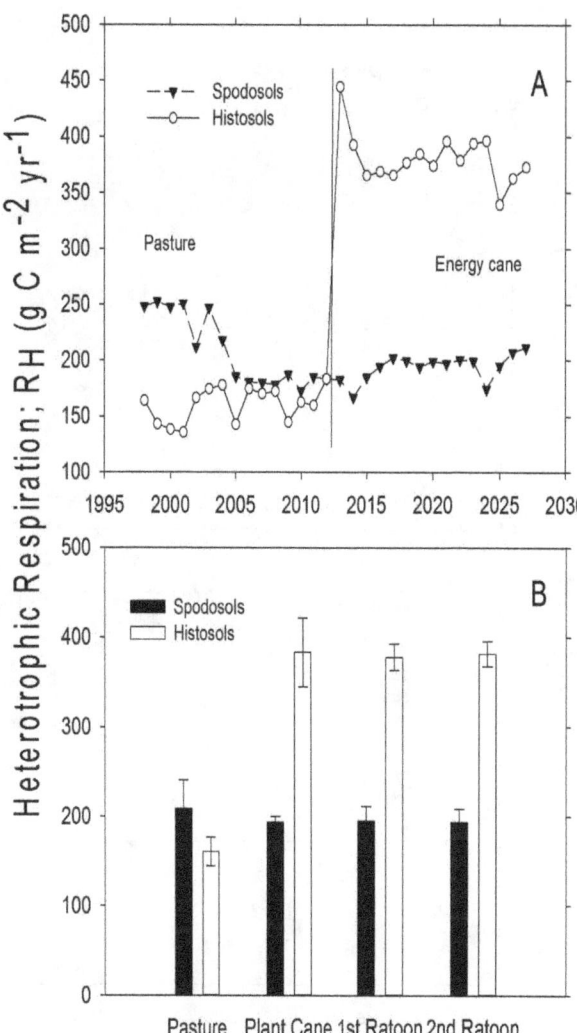

Figure 4. Modeled heterotrophic respiration (R_H) from pasture and land converted to energy cane in Highlands County, Florida. A) Total annual heterotrophic respiration (g C·m^{-2}). Dashed line represents year of land use conversion from pasture to energy cane. B) Mean heterotrophic respiration (g C·m^{-2}·yr^{-1}, ± SD) for 15 years in pasture, and for 5, 3-year ratoon cycles in energy cane (each bar represents the average of 5 values, one for each year for each stage in the planting cycle).

these variables, validation based on productivity for other crops reliably predicts trace gas flux [59,60,63,70–72].

We compiled a literature database of 17 sites that had reliable data on sugarcane and energy cane yield. There were also pasture productivity data for 15 of those sites [73]. In some instances we were able to contact researchers directly to access unpublished data (Table 2). The geographic range of sites represents the breadth of sugarcane production in the continental United States, and the potential range of energy cane production on currently grazed pastures. For all sites, daily weather data inputs (minimum and maximum temperature, daily precipitation) from 1980 to 2002 were obtained from the DayMet database [74]. The model was run using the DayCent growing degree-day subroutine to determine plant emergence, senescence and death, based on plant phenological characters and daily weather data. Soil data for the validation sites were obtained from the NRCS Web Soil Survey [75]. Using the same schedule of management events used for the

in silico experiments (described below), DayCent was run with site-specific soil and weather data for each sites. The fit of modeled to measured above ground dry mass production (Mg dry matter ha^{-1}) of our simulations of grazed pasture and energy cane were separately tested via linear regression, using the linear model function in R [76].

Initial Simulation Conditions

A "spin-up" period in DayCent based on historical land use and vegetation type was used to set initial soil conditions. The dominant, historic vegetation type for this area of south-central Florida was savanna, with a mixture of grasses and several species of scrub-oak, or sawgrass for the swamp areas [77]. A mix of perennial C_3 grasses species and symbiotic N_2 fixing plants, were used as initial conditions for the savanna simulation (initial vegetation type "savanna" in DayCent). A period of 2000 years was simulated to obtain an initial soil C and N conditions prior to

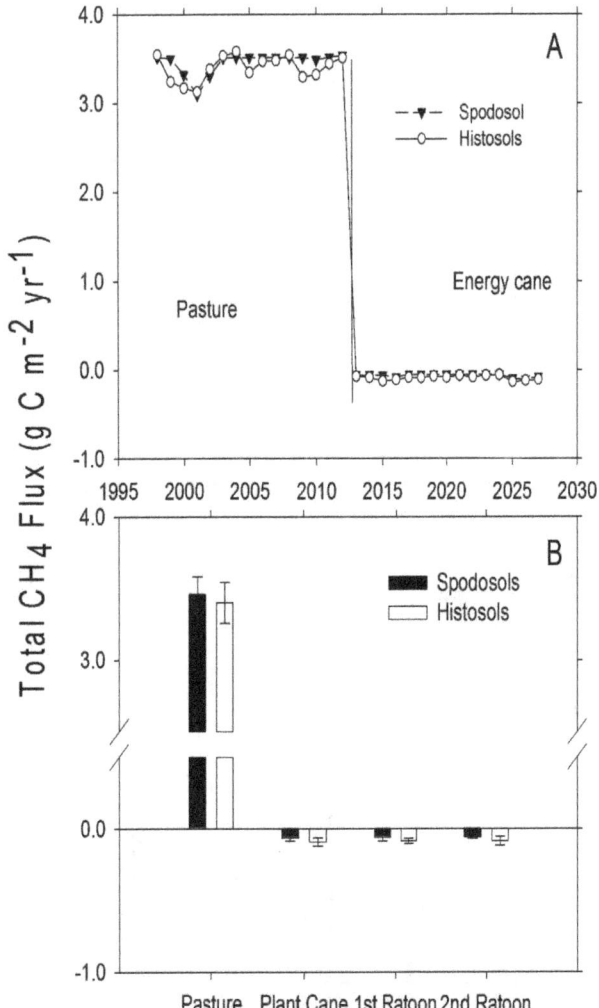

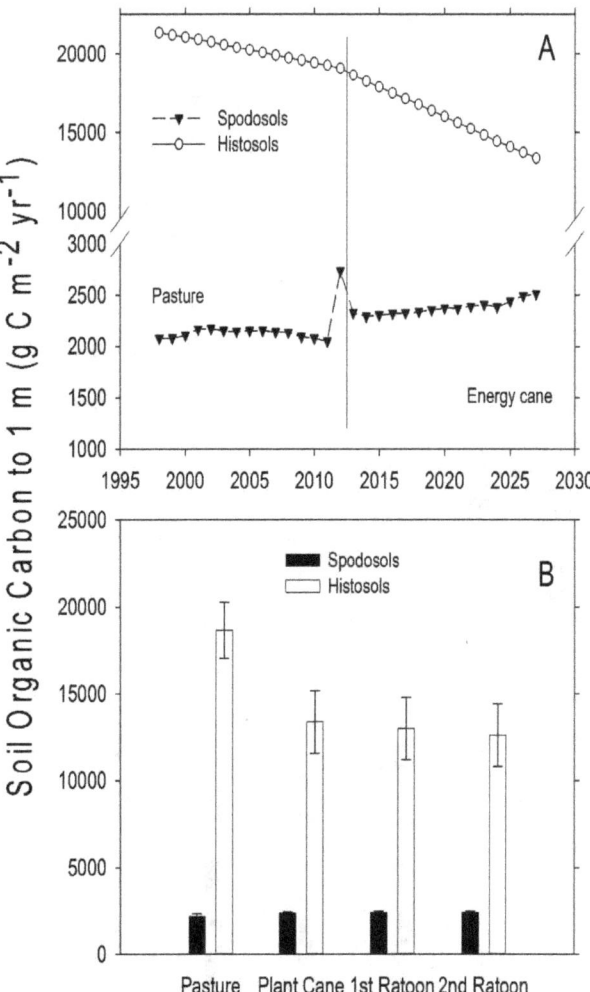

Figure 5. Modeled CH₄ flux from pasture and land converted to energy cane in Highlands County, Florida. A) Total annual CH_4 flux (g $C \cdot m^{-2}$). The solid vertical line represents year of land use conversion from pasture to energy cane, positive values indicate CH_4 efflux and negative values indicate CH_4 uptake. B) Mean CH_4 flux (g $C \cdot m^{-2} \cdot yr^{-1}$, ± SD) for 15 years in pasture, and for 5, 3-year ratoon cycles in energy cane (each bar represents the average of 5 values, one for each year for each stage in the planting cycle).

Figure 6. Changes in total soil organic C from pasture and land converted to energy cane in Highlands County, Florida. A) Total annual SOC flux (g $C \cdot m^{-2}$). The solid vertical line represents year of land use conversion from pasture to energy cane. B) Mean SOC flux (g $C \cdot m^{-2} \cdot yr^{-1}$, ± SEM) for 15 years in pasture, and for 5, 3-year ratoon cycles in energy cane (each bar represents the average of 5 values, one for each year for each stage in the planting cycle).

our *in silico* experiments. The model was run for spin ups and all subsequent experiments using the growing degree-day sub-routine.

In silico Experiments

Model simulations were then run to determine the GHG soil-atmosphere exchange and change in soil C predicted for conversion of pasture to energy cane on the two dominant soil types, Spodosols and Histosols. We used daily weather data inputs (minimum and maximum temperature, daily precipitation) from 1951 to 2002, which was the longest time period available for Highlands County, Florida obtained from the DayMet database [74]. This weather file is used by DayCent to create a mean and standard deviation of weather parameters, thus the more weather data available for a given site, the more accurately the variability of a site will be captured by the model.

To initiate the experimental simulations, in 1998 we converted the savanna by removing all above ground biomass and plowing to a depth of 30 cm. A landscape conversion to a grazed Bahiagrass (*Paspalum notatum* Flueggé) ecosystem was then simulated. Bahia-grass is a common forage grass for this part of Florida that would be considered "improved pasture", although usually not fertilized or irrigated [78]. We simulated grazing in our modeling experiment by annually removing 10% of live shoot and 1.0% of standing dead shoots. Prior to planting energy cane, another plow event to 30 cm was initiated to remove the pasture vegetation and simulate the physical land use change.

The simulated cycle of energy cane planting and harvest was based on the sugarcane literature [65,79,80] and discussions with University of Florida and USDA sugarcane agronomists [81,82]. In the simulations, energy cane was planted in January of the first year (2013), followed by a two-year ratoon (crop regenerated from remaining biomass) from which 80% of the above ground biomass was harvested in December. At the end of the second ratoon, the crop was removed and the land plowed before planting a new

Table 3. Modeled ecosystem carbon, nitrogen and greenhouse gas fluxes after converting pasture to energy cane on nutrient poor Spodosols and organic matter rich Histosols.

	Spodosols			Histosols		
	Pasture	Energy cane	Δ	Pasture	Energy cane	Δ
SOC (g C·m^{-2})	2736	2513	−224	16087	10373	−5715
Nitrogen Mineralization (g N·m^{-2})	134	203	69	216	293	77
Heterotrophic Respiration (g C·m^{-2})	3130	2913	−218	2413	5715	3302
Total Soil CO$_2$ Efflux (g C·m^{-2})	8148	8993	845	8111	11540	3429
CH$_4$ (g CO$_2$eq·m^{-2})	2980	−33	−3013	2958	−46	−3004
N$_2$O (g CO$_2$eq·m^{-2})	214	649	435	6713	1742	−4970
Total System C Flux (g CO$_2$eq·m^{-2})	−1159	−2812	−1653	−1367	924	2291
Total Greenhouse Gas Flux (g CO$_2$eq·m^{-2})	2035	−2196	−4231	8304	2620	−5684

Greenhouse gas and N mineralization values are the sum of values from pasture 15 years prior to conversion to energy cane and the sum values for 15 years following the conversion to energy cane. Positive values indicate a flux to the atmosphere and negative values indicate uptake from the atmosphere by the ecosystem. Soil organic matter values are the differences between the last year of energy cane production and the last year of pasture. Total GHG values are the sums of CH$_4$, N$_2$O and total system C flux (calculated in DayCent as the difference between all C uptake and storage versus efflux from respiration) expressed as CO$_2$e. Differences (Δ) represent the values for energy cane minus pasture.

plant crop. This cycle of ratooning and planting was repeated in the simulation for fifteen years following conversion from pasture; i.e. five cycles of three years each. This three-year planting cycle is typical for sugarcane production in Florida [83,84].

Irrigation events were scheduled every month throughout the dry season, and every two months during the rainy season to maintain soil water at field capacity. Fertilizer (NH$_4^+$ – NO$_3^-$) was applied in mid February and mid June of each year of the simulation, at a rate of 102 kg N· ha^{-1} per fertilization event for Spodosols. No fertilizer was added to the organic rich Histosols. This fertilization schedule was based on studies that suggest that a split fertilization regime at this rate maximizes sugarcane yield, and that fertilizing above this level does not increase yield but increases N$_2$O efflux [38,65,85]. The input files used to drive DayCent (e.g. schedule files, plant input parameters, and soil input files) are available online [86].

Calculations and Statistical Analyses

We summed daily GHG and soil C fluxes from DayCent to calculate yearly fluxes and report those in g C or N·m^{-2} yr^{-1}, with the exception of total GHG values which are reported as CO$_2$ equivalents [87] and factored by warming potential (CO$_2$ = 1, CH$_4$ = 23, N$_2$O = 296; ref. 85). Total ecosystem C flux was calculated as the annual change in total ecosystem C storage between the beginning and end of a year and represents the net ecosystem carbon balance expressed in CO$_2$eq [88,89].

Because the model experiments were performed using the same site with the same weather data, but controlled for soil type, the simulations had the structure of a paired design where each year was a replicate [90]. We therefore used paired t-tests to determine differences between soil types within a plant type (n = 15) and between plant types within a soil type (n = 15). The variation reported with mean annual values represents inter-annual variation in the predicted variables. Heteroscedasticity was examined with the Fligner-Killeen test, and output data distributions, which did not meet variance assumptions, were compared with the Wilcoxon rank-sum test. The routines t.test (paired = TRUE) and wilcox.test were performed using R [76,90].

Because of the large number of pair-wise comparisons of our model results, the False Discovery Rate (FDR) test was used to account for multiple comparisons. The FDR test is less conser-

vative than a *P*-value adjustment such as the Bonferroni correction, and determines the probability of a Type I error. We calculated a FDR of 0.024 for our matrix of tests, and therefore justified the use of multiple paired t-tests without *P*-value adjustment [91].

Results

Predicted harvested yields for both pasture and energy cane in our validation sites agreed well with measured values from the literature (Pasture: r^2 = 0.52, Energy cane: r^2 = 0.82, Figure 1A & 1B), indicating that our modeled predictions provided a good representation of the productivity that drives the biogeochemical dynamics of DayCent.

For our modeled site, DayCent estimated a large increase in aboveground plant biomass production after conversion of pasture to energy cane (Figure 2); annual aboveground biomass production increased by a factor of 14 on Spodosols and by a factor of 10 on Histosols, relative to pasture. Energy cane production ranged from 1911–3153 g C m^{-2} yr^{-1} (46–76 Mg dry biomass·ha^{-1}). Predicted energy cane production remained high through the three harvests on Histosols, but declined through the modeled ratoon cycle on Spodosols (Figure 2).

There was considerable temporal variation in predicted soil CO$_2$ efflux from pasture in the 15 years simulated prior to the conversion to energy cane (Figure 3a). This variation was particularly evident for pasture on Spodosols and was driven primarily by variation in precipitation. Total soil CO$_2$ efflux was similar for pasture on both soil types, but significantly increased when averaged over 15 years after conversion to energy cane on the Histosols (Figure 3a; t = 10.65, d.f. = 14, P<0.001). Land use conversion did not increase CO$_2$ efflux on Spodosols (t = 0.58, d.f. = 14, P = 0.57). Following conversion to energy cane CO$_2$ efflux from Histosols was significantly higher than energy cane on Spodosols (Figure 3b; t = 9.56, d.f. = 14, P<0.001).

The conversion of land from pasture to energy cane had no significant effect on the predicted heterotrophic component of soil respiration (R$_H$) on Spodosols (Figure 4a), but caused a large increase in R$_H$ from the Histosols (Figure 4a; t = 31.86, d.f. = 14, P<0.001) and resulted in higher R$_H$ on Histosols than Spodosols following the conversion to energy cane (Figure 4b; t = 23.68, d.f.

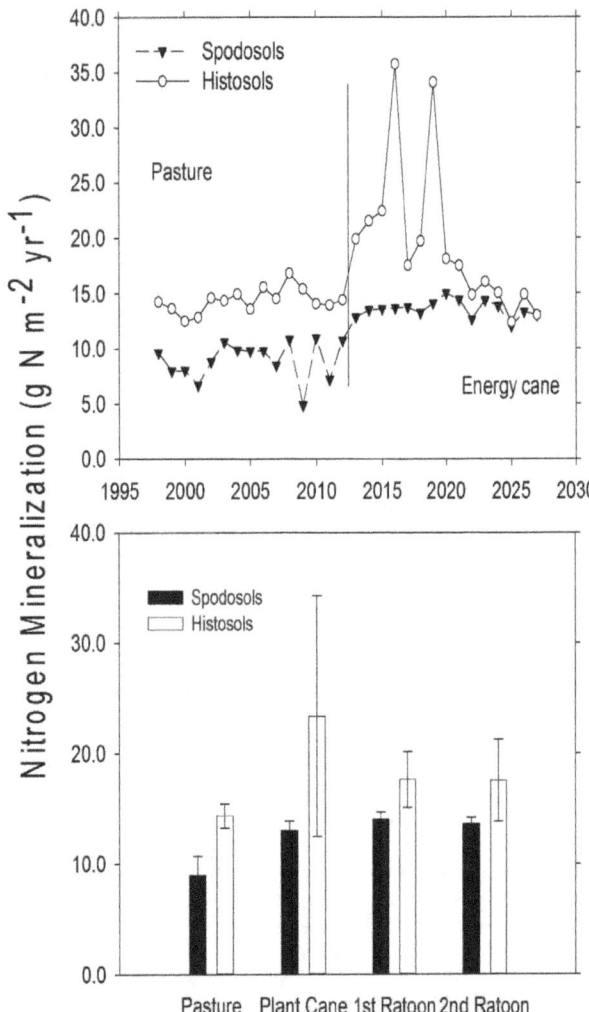

Figure 7. Modeled nitrogen mineralization rates from pasture and land converted to energy cane in Highlands County, Florida. A) Total annual N mineralization rate (g $N \cdot m^{-2}$). The solid vertical line represents year of land use conversion from pasture to energy cane. B) Mean N mineralization rate (g $N \cdot m^{-2} \cdot yr^{-1}$, ± SD) for 15 years in pasture, and for 5, 3-year ratoon cycles in energy cane (each bar represents the average of 5 values, one for each year for each stage in the planting cycle).

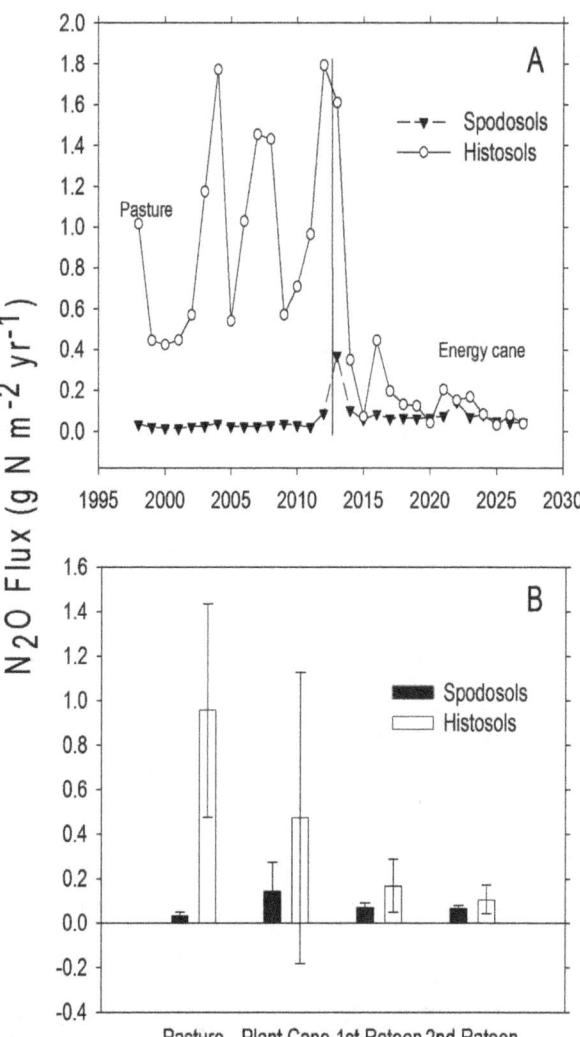

Figure 8. Modeled N_2O flux from pasture and land converted to energy cane in Highlands County, Florida. A) Total annual N_2O flux (g $N \cdot m^{-2}$). The solid vertical line represents year of land use conversion from pasture to energy cane, positive values indicate N_2O efflux and negative values indicate N_2O uptake. B) Mean N_2O flux (g $N \cdot m^{-2} \cdot yr^{-1}$, ± SD) for 15 years in pasture, and for 5, 3-year ratoon cycles in energy cane (each bar represents the average of 5 values, one for each year for each stage in the planting cycle).

= 14, P<0.001). Prior to the conversion to energy cane, modeled (R_H) was slightly higher in pasture on Spododols than on Histosols (Figure 4; t = 31.86, d.f. = 14, P<0.001).

On both soil types, the removal of cattle associated with the conversion of pasture to energy cane caused a substantial change in predicted CH_4 flux (t = 185, d.f. = 14, P<0.001 on Spodosols; t = 167, d.f. = 14, P<0.001 on Histosols; Figure 5a). Without cattle, pastures were a small CH_4 sink (0.16–0.60 g $C \cdot m^{-2} \cdot yr^{-1}$ uptake in Spodosols, 15 year sum = 112 g $CO_2eq \cdot m^{-2}$, 0.12–0.57 g$C \cdot m^{-2} \cdot yr^{-1}$ uptake for Histosols, 15 year sum = 135 g $CO_2eq \cdot m^{-2}$). Introducing cattle at stocking rates and grazing intensity typical for this region (1 head cattle$\cdot ha^{-1}$: ref. 31), caused pasture on both soil types to be a substantial source of CH_4 to the atmosphere (Figure 5).

Changes in vegetation and management practices altered soil organic carbon (SOC), and these changes were particularly evident on the Histosols (Table 2; Figure 6). Histosols had a larger pool of active C (weekly to monthly turnover) than Spodosols under both pasture and energy cane (pasture, t = 19.25, d.f. = 14, P<0.001; energy cane, t = 14.21, d.f. = 14, P<0.001). Comparing the remaining total SOC pools between the end of pasture and the last year of the energy cane simulation, Histosols lost a large amount of soil organic C; 5714 g $C \cdot m^{-2}$ to 1 m depth (Figure 6; t = 296, d.f. = 14, P<0.001), compared to the SOC loss from Spodosols of 224 g $C \cdot m^{-2}$ to 1 m (Table 3).

Nitrogen mineralization increased after pasture was converted to energy cane on both the fertilized Spodosols (t = 9.02, d.f. = 14, P<0.001) and the non-fertilized Histosols (t = 2.72, d.f. = 14, P = 0.02). After conversion to energy cane, Histosols had higher rates of N mineralization than Spodosols (Figure 7; t = 3.43, d.f. = 14, P = 0.004), and this increase in available N likely accounted for the continued high yields on Histosols.

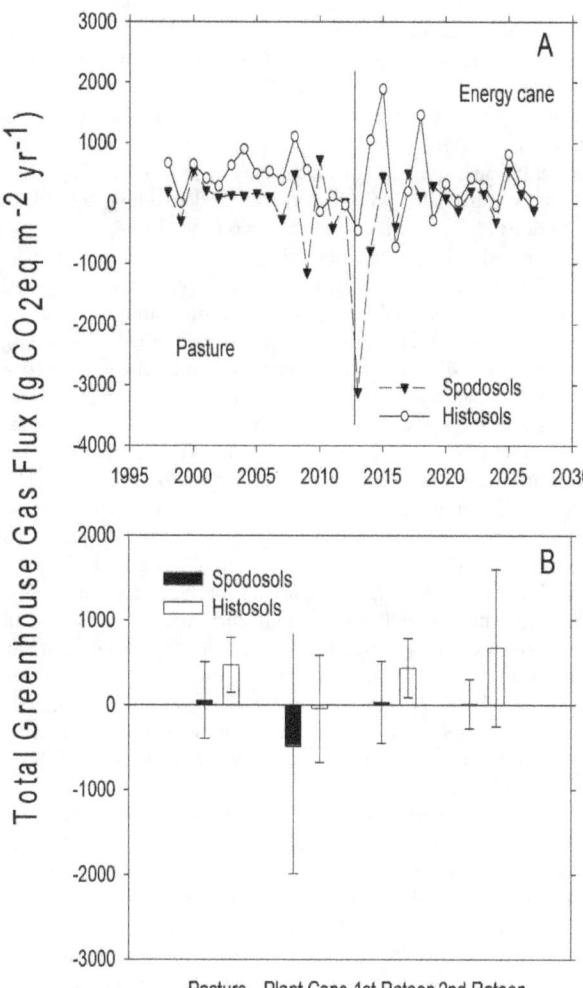

Figure 9. Changes in total greenhouse gas (GHG) from pasture and land converted to energy cane in Highlands County, Florida. Positive values indicate GHG efflux and negative values indicate GHG uptake. A) Total annual GHG flux, reported as CO_2 equivalents converted to account for differences in warming potential (g $CO_2 \cdot m^{-2}$). The solid vertical line represents year of land use conversion from pasture to energy cane. B) Mean greenhouse gas flux in CO_2 equivalents converted to account for differences in warming potential (g $CO_2 e \cdot m^{-2} \cdot yr^{-1}$, ± SD) for 15 years in pasture, and for 5, 3-year ratoon cycles in energy cane (each bar represents the average of 5 values, one for each year for each stage in the planting cycle).

Prior to conversion to energy cane, N_2O efflux was higher in pastures on Histosols compared to Spodosols (Figure 8; Wilcoxon rank sum, W = 225, P<0.001). After conversion to energy cane, Histosols remained greater sources of N_2O than Spodosols (t = 12.15, d.f. = 14, P<0.001). Conversion of pasture to energy cane decreased N_2O efflux on Histosols (Figure 8a; t = 4.30, d.f. = 14, P<0.001), but increased N_2O efflux on Spodosols (Figure 8b; t = 2.87, d.f. = 14, P = 0.01). It is likely that N_2O emission from Histosols decreased following conversion because the increase in productivity resulted in a higher uptake of nitrate that would otherwise be available for denitrification.

Total GHG exchange (global warming potential) was calculated by converting the fluxes of CH_4 and N_2O to CO_2 equivalents based on their warming potential relative to CO_2 [87] and

summing these with total system C flux (Table 2). Variation in weather caused substantial inter-annual variation in total GHG flux, with both pasture and energy cane varying between net GHG sinks and sources (Figure 9); no significant differences in annual GHG flux were resolved on either soil type (Spodosols: t = 1.15, d.f. = 14, P = 0.27; Histosols: t = 0.13, d.f. = 14, P = 0.90). When the cumulative GHG emission were calculated for the fifteen years prior to conversion, pasture was a net source to GHGs to the atmosphere on both soil types, and pasture was a stronger source on Histosols (8304 $gCO_2 eq \cdot m^{-2}$) than on Spodosols (2035 $gCO_2 eq \cdot m^2$; Table 3). Conversion of pasture to energy cane caused the Spodosols to transition from a source to a sink for GHGs and reduced the flux of GHGs to the atmosphere on Histosols. On both soil types, the reduction GHG emission to the atmosphere was associated with a large decrease in CH_4 emissions caused by the elimination of cattle grazing. On the Histosols, the reduction in N_2O emissions to the atmosphere also contributed to reduced emission of GHGs. This analysis of GHG emissions and their corresponding global warming potentials did not account for the displacement of fossil fuel emissions by the biofuel product.

Discussion

Parameterization of the DayCent model for energy cane, an emerging bioenergy crop, successfully simulated biomass production across the southeast United States (Figure 1). Our simulations suggested high yields for energy cane on former pastureland in a subtropical climate when Spodosols are highly fertilized (200 kg $N \cdot ha^{-1} \cdot yr^{-1}$), and when microbial activity in Histosols leads to high rates of N mineralization (rates were 44% higher on Histosols). When integrated over 15 years (Table 3), conversion of pasture to energy cane on Spodosols converted a net source of GHG (due to cattle CH_4 emissions) to a sink driven by the removal of cattle and the increase in C uptake by energy cane. While Histosols were a net GHG source under both pasture and energy cane, the source was reduced by the land use conversion (Table 3). The GHG improvement resulting from this conversion from pasture to energy cane would be even greater if fossil fuel displacement by cellulosic ethanol had been included.

The range of our simulated energy cane yields was 46–76 Mg ha^{-1} dry mass per year on fertilized Spodosols and unfertilized Histosols. Using published values for the conversion efficiency for the production of cellulusoic ethanol [22], a hypothetical energy cane farm of 10,000 ha could therefore produce between 142–236 million liters of ethanol [92]. In comparison, equal areas of land devoted to corn grain and Miscanthus in the Midwest would yield between 25 and 73% this amount of ethanol, respectively, assuming the maximum yields reported by other authors [11,22].

Typically, sugarcane yield declines with ratooning, the repeated harvests of aboveground material generated by vegetative growth [93]. The model reproduced the yield decline for energy cane on Spodosols but not on Histosols, but the model in its current configuration probably failed to capture the mechanisms that would normally cause a decline in yield. Various factors ranging from increases in nematode populations and ratoon stunting disease, to mechanical compaction of the soil have been implicated in ratoon decline [94,95], and these were not accounted for in the model. Although sugarcane in Florida typically is grown for three years and three annual harvests before re-planting, if it were grown for more years between re-planting, we would expect a continuing yield decline on the Spodosols. In contrast, continued mineralization of organic matter on Histosols may sustain high yields beyond the 3-year period simulated in the model. On both soils the GHG benefits would be improved with longer ratoon cycles because of less soil disturbance due to decreased frequency of soil disturbance for replanting.

Organic matter (OM) content of soils is important for sustaining high yields of sugarcane, in part because OM mineralization provides the labile N necessary to sustain plant growth. Spodosols had much less OM than Histosols (Appendix I; Figure 6; [65,96]). The high CO_2 efflux rates from Histosols (Figures 3–4) and the patterns of SOC loss following land use change (Figure 6) correspond to higher rates of OM mineralization. The associated higher rates of N mineralization (Table 3; Figure 7) on Histosols provided additional N to energy cane and improved crop yield (Figure 2). Although energy cane on Spodosols was fertilized to offset the low N content of these soils, rates of nitrification (the process by which NH_4^+ is converted into the highly mobile NO_3^- anion) were higher on these soils. The fertilizer applied to energy cane crops on Spodosols in the simulations was NH_4^+ - NO_3^-, a labile substrate for nitrification [66]. Spodosols had consistently higher nitrification rates than Histosols, and therefore higher NO_3^- content because of fertilization, and it is possible that some fraction of fertilizer was lost before plant uptake [97]. We hypothesize that a combination of NO_3^- leaching from fertilizer before plant uptake, lower initial N content, and lower mineralization rates may have created a stronger N limitation to yield on Spodosols but not on Histosols.

Before land use change, pasture on both soil types was a net source of GHGs to the atmosphere (Table 3). This is consistent with both direct measurements [98] and modeling efforts [99] that have found grazed pastures to be net sources of GHGs, but this is also a function of grass species present and animal stocking density [100]. The model estimated that pastures were sinks for CO_2, with total C uptake of 1159 g CO_2 m^{-2} and 1367 g CO_2 m^{-2} over 15 years on Spodosols and Histosols, respectively (Table 3). In the absence of cattle, both soil types were CH_4 sinks (112 and 135 g CO_2eq, respectively), but including reasonable estimates of CH_4 efflux from cattle (Figure 5) and N_2O efflux from soils (Figure 8) resulted in net GHG emission to the atmosphere on both pasture soils (Table 3). Following conversion to energy cane, the production of N_2O on Spodosols increased (Figure 8) within the range of N_2O flux rates previously reported for Australian sugarcane fertilized at similar rates to this study [37]. The increase in N_2O was offset by uptake of CO_2 and the change from a source to a sink for CH_4 (Figure 5), with the net effect that Spodosols became a net GHG sink (Table 3). Indeed, over 15 years energy cane on Spodosol was a GHG sink of >40 Mg CO_2eq per hectare (Table 3). On Histosols, eliminating grazing following the conversion of pasture to energy cane caused a similar decrease in CH_4 efflux to the atmosphere (Figure 5) and this land use change also reduced N_2O emissions (Figure 8; Table 3). However, following land conversion this system switched from a net CO_2 sink to a source, and this change in total system C prevented energy cane on Histosols from becoming a net sink for GHGs. The driver for GHG production on Histosols was higher R_H, and significant losses of soil organic matter [69] that resulted in total C efflux from these soils (Table 3).

The model successfully simulated energy cane biomass production across a range of sites across the southern United States (Figure 1). Previous studies have shown that DayCent reliably predicts soil biogeochemistry and GHG exchange when parameterized for net primary production [51], suggesting that the estimates of GHG flux and soil C dynamics were reasonable. Eddy-flux measurements of GHG exchange that are now being initiated at this site will provide an independent test of the predictions of GHG effects of conversion made here.

Indirect land use change (ILUC) – the stimulation of deforestation or increased agriculture in other parts of the world driven by diversion of current agricultural land to bioenergy production – potentially poses an environmental risk of bioenergy production [7,101]. Growing energy cane on land converted from low stocking density pasture would be unlikely to trigger significant increases in food price or ILUC in the way that large-scale shifts from corn or soy production in the Midwestern United States would motivate greater production of those crops elsewhere [13]. Indeed, the recommended stocking density for Bahiagrass pasture in this region is ~1 animal·ha^{-1} [31], and cattle and calf operations in Florida account for less than 6% of the state's annual agricultural revenue [102]. The loss in meat production could be redressed with minimal increases in current stocking rates, and would be unlikely to trigger the type of large-scale landscape changes that may occur through the diversion of midwestern agricultural land [7]. However, displacing cattle for energy cane production may potentially increase methane emissions elsewhere, which would negate the local benefit of reduced methane flux to the atmosphere.

The environmental impacts of changing land use from pasture to energy cane were highly dependent on the soil type. Whereas the cultivation of Histosols results in high CO_2 efflux and the reduction of soil carbon (Figures 3, 4, and 6), the model predicted that energy cane crops on Spodosols would act as a net C and GHG sink (Figure 6, Table 3). From both a biofuel and biogeochemical perspective, these results suggest that energy cane grown on nutrient poor soils, as opposed to organic soils, has the potential to be a high-yielding bio-ethanol feedstock that creates a GHG sink in the Southeastern United States.

Acknowledgments

We thank Michael Masters for lab analysis of plant and soil C and N. Lykes Brothers Inc. graciously provided access to energy cane plantations and pastures, which greatly helped in parameterizing our model. We also thank Dr. Robert Gilbert, Dr. Barry Glaz, and Mr. Pedro Korndorfer for sharing their knowledge on sugar and energy cane agronomy and data that aided our modeling effort.

Author Contributions

Conceived and designed the experiments: BDD KJAT WJP EHD. Performed the experiments: BDD SCD CK WJP. Analyzed the data: BDD SCD KJAT WJP SPL EHD. Contributed reagents/materials/analysis tools: CK WJP. Wrote the paper: BDD KJAT WJP SPL EHD.

References

1. IPCC (2007) Climate Change 2007: The Physical Science Basis. Contribution of Working Group I to the Fourth Assessment Report of the Intergovernmental Panel on Climate Change. Solomon S, Qin D, Manning M, Chen Z, Marquis M et al. Eds. Cambridge, UK, Cambridge, UK.
2. Le Quéré C, Raupach MR, Canadell JG, Marland G, Bopp L, et al. (2009) Trends in the sources and sinks of carbon dioxide. Nat Geosci 2: 831–836.
3. Pan Y, Birdsey RA, Fang J, Houghton R, Kauppi PE, et al. (2011) A large and persistent carbon sink in the world's forests. Science 333: 988–993.
4. Lal R (2004) Soil carbon sequestration impacts on global climate change and food security. Science 304: 1623–1627. doi: 10.1126/science.1097396.

5. Tilman D (1998) The greening of the green revolution. Nature 396: 211–212.
6. Fargione JE, Hill JD, Tilman D, Polasky S, Hawthorne P (2008) Land clearing and the biofuel carbon debt. Science 319: 1235–1238.
7. Searchinger T, Heimlich R, Houghton RA, Dong F, Elobeid A, et al. (2008) Use of U.S. croplands for biofuels increases greenhouse gases through emissions from land-use change. Science 319: 1238–1240.
8. Melillo JM, Reilly JM, Kicklighter DW, Gurgel AC, Cronin TW (2009) Indirect emissions from biofuels: How important? Science 326: 1397–1399.
9. Börjesson P (2009) Good or bad bioethanol from a greenhouse gas perspective – What determines this? Applied Energy 86: 589–594.

10. United States Congress (2007) The Energy Independence and Security Act of 2007 (H.R. 6). Available: http://frwebgate.access.gpo.gov/cgibin/getdoc. cgi?dbname = 110_cong_bills&docid = f: h6enr.txt.pdf. Accessed 2010 Dec 19.

11. Heaton EA, Dohleman FG, Long SP (2008) Meeting US biofuel goals with less land: the potential of Miscanthus. Glob Change Biol 14: 2000–2014.

12. Fargione JE, Plevin RJ, Hill JD (2010) The ecological impact of biofuels. Ann Rev Ecol Evol Syst 41: 351–377.

13. Davis SC, Parton WJ, Del Grosso SJ, Keough C, Marx E, et al. (2012) Impacts of second-generation biofuel agriculture on greenhouse gas emissions in the corn-growing regions of the US. Front Ecol Environ 10: 69–74. doi:10.1890/110003.

14. Dien BS, Bothast RJ, Nichols NN, Cotta MA (2002) The U.S. corn ethanol industry: an overview of current technology and future prospects. Int Sugar J 103: 204–211.

15. Davis SC, Anderson-Teixeira KJ, DeLucia EH (2009) Life-cycle analysis and the ecology of biofuels. Trends Plant Sci 14: 140–146.

16. O'Hare M, Plevin RJ, Martin JI, Jones AD, Kendall A, et al. (2009) Proper accounting for time increases crop-based biofuels' greenhouse gas deficit versus petroleum. Environ Res Lett 4: doi:10.1088/1748–9326/4/2/024001.

17. Donner SD, Kucharik CJ (2008) Corn-based ethanol production compromises goal of reducing nitrogen export by the Mississippi River. Proc Natl Acad Sci USA 105: 4513–4518.

18. Hill J, Polasky S, Nelson E, Tilman D, Huo H (2009) Climate change and health costs of air emissions from biofuels and gasoline. Proc Nat Acad Sci USA 106: 2077–2082.

19. Smeets EMW, Bouwman LF, Stehfest E, van Vuuren DP, Posthuma A (2009) Contribution of N_2O to the greenhouse gas balance of first-generation biofuels. Glob Change Biol 15: 1–23.

20. Solomon BD, Barnes JR, Halvorsen KE (2007) Grain and cellulosic ethanol: history, economics, and energy policy. Biomass Bioenergy 31: 416–425.

21. Tilman D, Socolow R, Foley JA, Hill J, Larson E (2009) Beneficial biofuels-the food, energy and environment trilemma. Science 325: 270–271.

22. Somerville C, Youngs H, Taylor C, Davis SC, Long SP (2010) Feedstocks for lignocellulosic biofuels. Science 329: 790–792.

23. Tilman D, Hill J, Lehman C (2006) Carbon-negative biofuels from low-input high-diversity grassland biomass. Science 314: 1598–1600.

24. Anderson-Teixeira KJ, Davis SC, Masters MD, DeLucia EH (2009) Changes in soil organic carbon under biofuel crops. Glob Change Biol Bioenergy 1: 75–96.

25. Davis SC, Parton WJ, Dohleman FG, Smith CM, Del Grosso S, et al. (2010) Comparative biogeochemical cycles of bioenergy crops reveal nitrogen fixation and low greenhouse gas emissions in a Miscanthus x giganteus agro-ecosystem. Ecosystems 13: 144–156.

26. Robertson GP, Hamilton SK, Del Grosso SJ, Parton WP (2011) The biogeochemistry of bioenergy landscapes: carbon, nitrogen, and water considerations. Ecol App 21: 1055–1067. doi: 10.1890/09–0456.1.

27. Zeri M, Anderson-Teixeira KJ, Masters MD, Hickman G, DeLucia EH, et al. (2011) Carbon exchange by establishing biofuel crops in central Illinois. Agric Ecosyst Environ 144: 319–329.

28. Sladden SE, Bransby DI, Aiken GE, Prine GM (1991) Biomass yield and composition, and winter survival of tall grasses in Alabama. Biomass Bioenergy 1: 123–127.

29. Mark T, Darby P, Salassi M (2009) Energy cane usage for cellulosic ethanol: estimation of feedstock costs. Southern Agricultural Economics Association Annual Meeting, Atlanta, Georgia, January 31-February 3, 2009.

30. Baucum LE, Rice RW (2009) An overview of Florida sugarcane. University of Florida IFAS Extension document SS-AGR-232.

31. Hersom M (2005) Pasture stocking density and the relationship to animal performance. Animal Science Department, Florida Cooperative Extension Service, Institute of Food and Agricultural Sciences, University of Florida, Document number AN155.

32. Steiner J (2012) personal communication.

33. Bowden RD, Nadelhoffer KJ, Boone RD, Melillo JM, Garrison JB (1993) Contributions of aboveground litter, belowground litter, and root respiration to soil respiration in a temperate mixed hardwood forest. Can J For Res 23: 1402–1407.

34. Paustian K, Six J, Elliott ET, Hunt HW (2000) Management options for reducing CO_2 emissions from agricultural soils. Biogeochemistry 48: 147–163.

35. DeRamus HA, Clement TC, Giampola DD, Dickison PC (2003) Methane emissions of beef cattle on forages: efficiency of grazing management systems. J Environ Qual 32: 269–277.

36. Mosier A, Kroeze C, Nevison C, Oenema O, Seitzinger S, et al. (1998) Closing the global N_2O budget: nitrous oxide emissions through the agricultural nitrogen cycle. Nutr Cycl Agroecosys 52: 225–248.

37. Thorburn PJ, Biggs JS, Collins K, Probert ME (2010) Nitrous oxide emissions from Australian sugarcane production systems – are they greater than from other cropping systems? Agric Ecosyst Environ 136: 343–350.

38. Rice RW, Gilbert RA, Lentini RS (2002) Nutritional requirements for Florida sugarcane. Florida Cooperative Extension Service. UF/IFAS, Document SS-ARG-228. University of Florida Institute of Food Agricultural Science.

39. Morgan KT, McCray JM, Rice RW, Gilbert RA, Baucum LE (2009) Review of current sugarcane fertilizer recommendations: a report from the UF/IFAS sugarcane fertilizer standards task force. Document SL 295, Soil and Water

Science Department, Florida Cooperative Extension Service, Institute of Food and Agricultural Sciences, University of Florida.

40. Morris DR, Gilbert RA, Reicosky DC, Gesch RW (2004) Oxidation potentials of soil organic matter in Histosols under different tillage methods. Soil Sci Soc Am J 68: 817–826.

41. Stehfest E, Bouwman LF (2006) N_2O and NO emission from agricultural fields and soils under natural vegetation: summarizing available measurement data and modeling of global annual emissions. Nutr Cycl Agroecosys 74: 207–228.

42. Anderson-Teixeira KJ, Masters MD, Black CK, Zeri M, Hussain MZ, et al. (2012) Altered belowground carbon cycling following land use change to perennial bioenergy crops. Ecosystems, in press.

43. Smith CM, David MB, Mitchell CA, Masters MD, Anderson-Teixeira KJ, et al. (2013) Reduced nitrogen losses following conversion of row crop agriculture to perennial biofuel crops. J Env Qual 42: 219–228, doi: 10.2134/jeq2012.0210.

44. Beale CV, Long SP (1997) Seasonal dynamics of nutrient accumulation and partitioning in the perennial C-4-grasses Miscanthus x giganteus and Spartina cynosuroides. Biomass Bioenergy 12: 419–428.

45. Hansen EM, Christensen BT, Jensen LS, Kristensen K (2004) Carbon sequestration in soil beneath long-term Miscanthus plantations as determined by ^{13}C abundance. Biomass Bioenergy 26: 97–105.

46. NRCS Web Soil Survey website. Available: http://websoilsurvey.nrcs.usda. gov/app/HomePage.htm. Accessed 2010 Oct 7.

47. Parton WJ. Hartman MD, Ojima DS, Schimel DS (1998) DAYCENT and its land surface submodel: description and testing. Glob Planet Change 19: 35–48.

48. Parton WJ, Holland EA, Del Grosso SJ, Hartmann MD, Martin RE, et al. (2001) Generalized model for NO_x and N_2O emissions from soils. J Geophys Res-Atmos 106: 17403–17420.

49. DayCent: Daily Century Model website. Available: http://www.nrel.colostate. edu/projects/daycent/. Accessed 2010 Jun 4.

50. Del Grosso SJ, Halvorson AD, Parton WJ (2008a) Testing DayCent model simulations of corn yields and nitrous oxide emissions in irrigated tillage systems in Colorado. J Environ Qual 37: 1383–1389, doi:10.2134/jeq2007.0292.

51. Parton WJ, Hanson PJ, Swanston C, Torn M, Trumbore SE, et al. (2010) ForCent model development and testing using the Enriched Background Isotope Study experiment. J Geophys Res 115: G04001.

52. Parton WJ, Schimel DS, Cole CV, Ojima DS (1987) Analysis of factors controlling soil organic levels of grasslands in the Great Plains. Soil Sci Soc Am J 51: 1173–1179.

53. Parton WJ, Ojima DS, Cole CV, Schimel DS (1994) A general model for soil organic matter dynamics: sensitivity to litter chemistry, texture and management, R.B. Bryant,R.W. Arnoldm, Editors, Quantitative Modeling of Soil Forming Processes, Soil Science Society of America, Madison, WI. Pp. 147–167.

54. Eitzinger J, Parton WJ, Hartman M (2000) Improvement and validation of a daily soil temperature submodel for freezing/thawing periods. Soil Sci 165: 525–534.

55. David MB, Del Grosso SJ, Hu X, McIsaac GF, Parton WJ, et al. (2009) Modeling denitrification in a tile-drained, corn and soybean agroecosystem of Illinois, USA. Biogeochemistry 93: 7–30.

56. Del Grosso SJ, Parton WJ, Mosier AR, Ojima DS, Hartmann MD (2000a) Interaction of soil carbon sequestration and N2O flux with different land use practices. In: van Ham J, Baede APM, Meyer LA, Ybema R (eds.), Non-CO_2 Greenhouse Gases: Scientific Understanding, Control and Implementation. Kluwer Academic Publishers, The Netherlands. 303–311.

57. Del Grosso SJ, Parton WJ, Mosier AR, Ojima DS, Kulmala AE, et al. (2000b) General model for N_2O and N_2 gas emissions from soils due to denitrification. Global Biogeochem Cycles 14: 1045–1060.

58. Del Grosso SJ, Mosier AR, Parton WJ, Ojima DS (2005) DayCent model analysis of past and contemporary soil N_2O and net greenhouse gas flux for major crops in the USA. Soil Tillage Res 83: 9–24.

59. Del Grosso SJ, Ojima DS, Parton WJ, Mosier AR, Peterson GA, et al. (2002) Simulated effects of dryland cropping intensification on soil organic matter and GHG exchanges using the DAYCENT ecosystem model. Environ Pollution 116, S75–S83.

60. Del Grosso SJ, Parton WJ, Ojima DS, Keough CA, Riley TH, et al. (2008b) DAYCENT simulated effects of land use and climate on county level N loss vectors in the USA. Pages 571–595 in: R.F. Follett, and J.L. Hatfield (eds.) Nitrogen in the Environment: Sources, Problems, and Management, 2nd ed. Elsevier Science Publishers, The Netherlands.

61. Galdos MV, Cerri CC, Cerri CEP, Paustian K, Van Antwerpen R (2010) Simulation of sugarcane residue decomposition and aboveground growth, Plant Soil 326: 243–259. DOI 10.1007/s11104–009–004–3.

62. Hartmann MD, Merchant EK, Parton WJ, Gutmann MP, Lutz SM, et al. (2011) Impact of historical land use changes in the U.S. Great Plains, 1883 to 2003. Ecol App 21: 1105–1119.

63. Adler PR, Del Grosso SJ, Parton WJ (2007) Life-Cycle assessment of net greenhouse-gas flux for bioenergy cropping systems. Ecol Appl 17: 675–691.

64. Galdos MV, Cerri CC, Cerri CEP, Paustian K, Van Antwerpen R (2009) Simulation of soil carbon dynamics under sugarcane with the CENTURY Model, Soil Sci Soc Am J 73: 802–811.

65. Vallis I, Parton WJ, Keating BA, Wood AW (1996) Simulation of the effects of trash and N fertilizer management on soil organic matter levels and yields of sugarcane. Soil Tillage Res 38: 115–132.

66. Pepper DA, Del Grosso SJ, McMurtrie RE, Parton WJ (2005) Simulated carbon sink response of shortgrass steppe, tallgrass prairie and forest ecosystems to rising [CO_2], temperature and nitrogen input, Global Biogeochem Cycles 19: GB1004 doi:10.1029/2004GB002226.

67. Kelly RH, Parton WJ, Hartman MD, Stretch LK, Ojima DS, et al. (2000), Intra-annual and interannual variability of ecosystem processes in shortgrass steppe, J Geophys Res 105(D15) 20093–20100 doi:10.1029/2000JD900259.

68. Brady NC, Weil RR (2002) The nature and properties of soils. Prentice Hall, Upper Saddle River, New Jersey. 960 p.

69. Morris DR, Gilbert RA (2005) Inventory, crop use, and soil subsidence of Histosols in Florida. J Food Agr Environ 3: 190–193.

70. Newman Y, Vendramini J, Blount A (2010) Bahiagrass (*Paspalum notatum*): overview and management. University of Florida IFAS Extension. Publication #SS-AGR-332. http://edis.ifas.ufl.edu/ag342. Accessed 2011 Aug 7.

71. Valentine DW, Holland EA, Schimel DS (1994) Ecosystem and physiological controls over methane production in northern wetlands. J Geophys Res 99: 1563–1571.

72. Del Grosso SJ, Parton WJ, Mosier AR, Walsh MK, Ojima DS, et al. (2006) DayCent national scale simulations of N_2O emissions from cropped soils in the USA. J Environ Qual 35: 1451–1460.

73. US Department of Agriculture, National Agricultural Statistics Service website. Available: http://www.nass.usda.gov/Quick_Stats/. Accessed 2010 Feb 25.

74. DAYMET United States Data Center-A source for daily surface weather data and climatological summaries website. Available: www.daymet.org. Accessed 2010 Jun 1.

75. US Department of Agriculture Natural Resources Conservation Service, Web Soil Survey website. Available: http://websoilsurvey.sc.egov.usda.gov/app/HomePage.htm. Accessed 2010 Dec 12.

76. R Development Core Team (2007) R: A language and environment for statistical computing. R Foundation for Statistical Computing, Vienna, Austria.

77. Barbour MG, Billings WD (1988) North American terrestrial vegetation. Press Syndicate of the University of Cambridge. Melbourne, Australia.

78. Pitman WD, Portier KM, Chambliss CG, Kretschmer AE (1992) Performance of yearling steers grazing bahia grass pastures with summer annual legumes or nitrogen fertilizer in subtropical Florida. Trop Grasslands 26: 206–211.

79. Glaz B, Ulloa MF (1993) Sugarcane yields from plant and ratoon sources of seed cane. J Am Soc Sugar Cane Tech 13: 7–13.

80. Wiedenfeld RP, Enciso J (2008) Sugarcane responses to irrigation and nitrogen in semiarid south Texas. Agron J 100: 665–671.

81. Gilbert RA, Shine JM, Miller JD, Rice RW, Rainbolt CR (2006) The effect of genotype, environment and time of harvest on sugarcane yields in Florida, USA. Field Crop Res 95: 156–170.

82. Glaz B (2012) Personal communication.

83. Glaz B, Morris DR (2010) Sugarcane Responses to water-table depth and periodic flood. Agron J 102: 372–380.

84. US Environmental Protection Agency (2011) Florida Sugarcane Metadata. Environmental Protection Agency, Washington DC. Available: http://www.epa.gov/oppefed1/models/water/met_fl_sugarcane.htm. Accessed 2011 August 1.

85. Muchovej RM, Newman PR (2004) Nitrogen fertilization of sugarcane on a sandy soil: II soil and groundwater analysis. J Am Soc Sugar Cane Tech 24: 225–240.

86. University of Illinois, DeLucia Laboratory Public Data Archive website. Available: http://www.life.illinois.edu/delucia/Public%20Data%20Archive/. Accessed 2013 Jun 4.

87. Department of Energy and Climate Change (DECC) and the Department for Environment, Food and Rural Affairs website. Available: https://www.gov.uk/government/publications/2012-guidelines-to-defra-decc-s-ghg-conversion-factors-for-company-reporting-methodology-paper-for-emission-factors. Accessed: 2013 Jul 23.

88. Forster P, Ramaswamy V, Artaxo P, Berntsen T, Betts R, et al. (2007) Climate Change 2007: The Physical Science Basis. Contribution of Working Group I to the Fourth Assessment Report of the Intergovernmental Panel on Climate Change. Cambridge University Press, Cambridge, United Kingdom and New York, NY, USA.

89. Chapin FS, Woodwell G, Randerson J, Rastetter E, Lovett G, et al. (2006) Reconciling carbon-cycle concepts, terminology, and methods. Ecosystems 9: 1041–1050.

90. Crawley MJ (2007) The R Book. John Wiley and Sons, West Sussex, England. 942 p.

91. Storey JD (2003) The positive false discovery rate: a Bayesian interpretation and the q-value. Ann Stat 31: 2013–2035.

92. Graham-Rowe D (2011) Agriculture: beyond food versus fuel. Nature 474: S6–S8. doi: 10.1038/474S06a.

93. Ball-Coelho B, Sampaio EVSB, Tiessen H, Stewart JWB (1992) Root dynamics in plant and ratoon crops of sugar cane. Plant Soil 142: 297–305.

94. Hoy JW, Grisham MP, Damann KE (1999) Spread and increase of ratoon stunting disease of sugarcane and comparison of disease detection methods. Plant Disease 83: 1170–1175.

95. Stirling GR, Blair BL, Pattemore JA, Garside AL, Bell MJ (2001) Changes in nematode populations on sugarcane following fallow, fumigation and crop rotation, and implications for the role of nematodes in yield decline. Australas Plant Pathol 30: 323–335.

96. Yadav RL, Prasad SR (1992) Conserving the organic matter content of the soil to sustain sugarcane yield. Exp Ag 28: 57–62.

97. Chapin FS, Matson PA, Mooney HA (2002) Principals of Terrestrial Ecosystem Ecology. Springer Science, New York, New York, USA.

98. Rowlings D, Grace P, Kiese R, Scheer C (2010) Quantifying N_2O and CO_2 emissions from a subtropical pasture. 19th World Congress of Soil Science, Soil Solutions for a changing World. 1–6 August 2010, Brisbane, Australia.

99. Howden SM, White DH, Mckeon GM, Scanlan JC, Carter JO (1994) Methods for exploring management options to reduce greenhouse gas emissions from tropical grazing systems. Clim Change 27: 49–70.

100. Liebig MA, Gross JR, Kronberg SL, Phillips RL (2010) Grazing management contributions to net global warming potential: A long-term evaluation in the Northern Great Plains. J Environ Qual 39: 799–809.

101. Plevin RJ, O'Hare M, Jones AD, Torn MS, Gibbs HK (2010) Greenhouse gas emissions from biofuels: Indirect land use change are uncertain but may be much greater than previously estimated. Env Sci Tech 44: 8015–8021.

102. Florida Department of Agriculture and Consumer Services website. Available: http://www.fl-ag.com/agfacts.htm. Accessed 2011 Oct 1.

6

Bacterial Indicator of Agricultural Management for Soil under No-Till Crop Production

Eva L. M. Figuerola[1][9], Leandro D. Guerrero[1][9], Silvina M. Rosa[1], Leandro Simonetti[1], Matías E. Duval[2], Juan A. Galantini[2], José C. Bedano[3], Luis G. Wall[4], Leonardo Erijman[1,5]*

1 Instituto de Investigaciones en Ingeniería Genética y Biología Molecular (INGEBI-CONICET) Vuelta de Obligado 2490, Buenos Aires, Argentina, **2** CERZOS-CONICET Departamento de Agronomía, Universidad Nacional del Sur, Bahía Blanca, Argentina, **3** Departamento de Geología, Universidad Nacional de Río Cuarto, Río Cuarto, Córdoba, Argentina, **4** Departamento de Ciencia y Tecnología, Universidad Nacional de Quilmes, Roque Sáenz Peña 352, Bernal, Argentina, **5** Facultad de Ciencias Exactas y Naturales, Universidad de Buenos Aires, Ciudad Universitaria, Pabellón 2, Buenos Aires, Argentina

Abstract

The rise in the world demand for food poses a challenge to our ability to sustain soil fertility and sustainability. The increasing use of no-till agriculture, adopted in many areas of the world as an alternative to conventional farming, may contribute to reduce the erosion of soils and the increase in the soil carbon pool. However, the advantages of no-till agriculture are jeopardized when its use is linked to the expansion of crop monoculture. The aim of this study was to survey bacterial communities to find indicators of soil quality related to contrasting agriculture management in soils under no-till farming. Four sites in production agriculture, with different soil properties, situated across a west-east transect in the most productive region in the Argentinean pampas, were taken as the basis for replication. Working definitions of Good no-till Agricultural Practices (GAP) and Poor no-till Agricultural Practices (PAP) were adopted for two distinct scenarios in terms of crop rotation, fertilization, agrochemicals use and pest control. Non-cultivated soils nearby the agricultural sites were taken as additional control treatments. Tag-encoded pyrosequencing was used to deeply sample the 16S rRNA gene from bacteria residing in soils corresponding to the three treatments at the four locations. Although bacterial communities as a whole appeared to be structured chiefly by a marked biogeographic provincialism, the distribution of a few taxa was shaped as well by environmental conditions related to agricultural management practices. A statistically supported approach was used to define candidates for management-indicator organisms, subsequently validated using quantitative PCR. We suggest that the ratio between the normalized abundance of a selected group of bacteria within the GP1 group of the phylum Acidobacteria and the genus *Rubellimicrobium* of the Alphaproteobacteria may serve as a potential management-indicator to discriminate between sustainable *vs.* non-sustainable agricultural practices in the Pampa region.

Editor: Hauke Smidt, Wageningen University, The Netherlands

Funding: The authors are members of the BIOSPAS consortium (http://www.biospas.org/en). ELMF, JCB, LGW and LE are members of Consejo Nacional de Investigaciones Científicas y Técnicas (CONICET), Argentina. JAG is member of Comisión de Investigaciones Científicas of Buenos Aires Province, Argentina (CIC). MD and SR received fellowships from ANPCyT, Argentina. LDG and LS were fellows of CONICET. The funders had no role in study design, data collection and analysis, decision to publish, or preparation of the manuscript.

Competing Interests: The authors have declared that no competing interests exist.

* E-mail: erijman@dna.uba.ar

[9] These authors contributed equally to this work.

Introduction

Sowing crop into no-till soil is a farming method that has initially been developed as an alternative to conventional tillage practices, with the aims of using less fossil fuels, reducing the erosion of soils, and increasing the soil carbon pool [1]. Soil structure can be significantly modified through reduced-till management practices [2]. Soil aggregates are less subjected to dry and wet cycles in no-tilled soil, compared to conventional-tilled soil, due to the protection exerted by surface residues. Therefore, it appears that reduced-till management reduces the risk of surface runoff, increase soil aggregation, and improve soil hydrological properties [3]. This is particularly true if no-till management is combined with diverse crop rotation [4].

Additional driving forces for no-till agriculture are the lower production costs, the higher yields and the incorporation of less fertile areas into crop production [4]. During the past several decades, no-till agriculture has been increasingly adopted in many areas of the world [5]. In Argentina, this practice has spread steadily in the last 30 years [6], covering presently almost 20 million hectare, which represents 70% of the total cultivated area [4]. Through the adoption of this novel agriculture management, farmers have been gradually incorporated novel technologies for weed, disease and fertilizer management through trial-and-error learning. The combination of these technologies with no-till management led to a farmers' definition of good agricultural practices on the basis of economic yield alongside soil conservation and gain in productivity. Yet this situation rapidly highlighted the need for new working hypotheses to aid in soil quality monitoring.

Driven by the influence of favorable market conditions, a substantial portion of that area is presently dedicated to soybean monoculture, often combined with minimal nutrient restoration. From the noticeable increase in soil born diseases caused by residue- and soil-inhabiting pathogens selected by the previous crop, questions arise about the ability to maintain soil fertility and sustainability if monoculture prevails over the crop rotation [7].

The use of soil quality indicators is important in order to guide land and resource management decisions. Traditionally, soil quality research has focused primarily on soil chemical and physical properties [8]. In general, assessment of soil quality will be influenced by management factors, and by climate and soil type as well. In view of that, different data sets of soil quality indicators have been proposed to discriminate between soil textural classes for different agricultural management systems and a variety of crops [9,10,11,12,13]. Besides the well known chemical and physical parameters used as soil quality indicators, such as soil organic matter and soil structure, there is still no consensus about biological soil indicators of sustainable agricultural systems. The massive adoption of no-till practices in extensive agriculture in Argentina gave rise to many situations, in which improvement of crop yield could not be associated to established quality indicators, but to the history of the soil management, suggesting that additional biological parameters might be necessary to describe changes in soil quality.

By driving crucial soil processes, such as decomposition of organic materials and nutrient cycling, soil bacteria are key players in ecosystem functioning. The structure of the microbial community in soil, the distribution of microbial biomass and enzyme activity may be affected by several factors, such as farming systems [14], plant species [15,16,17], tree species, soil pH [18], soil type [19], tillage and crop rotation [20,21,22,23,24]. This is why it is also important to take into consideration microbiological indicators when evaluating soil quality [25]. Yet, understanding about the influence of bacterial community structure on soil quality, and inversely, revealing the effect of soil characteristics on the structuring of bacterial communities is still scarce. In particular, to our knowledge, no previous study has addressed these issues in the framework of crop productivity, as assessed by farmers' records.

This work is a part of a larger effort to find microbiological indicators of sustainable agriculture in the framework of no-till farming. The project BIOSPAS (http://www.biospas.org/en) is a multidisciplinary research project, in which agricultural soil biology is approached by means of a polyphasic description [26]. We have considered three treatments, which were replicated as blocks in four agricultural sites located across a west-east transect in Argentine Central Pampas, having documented history of no-till management. Two treatments were related to contrasting agricultural management practices under no-till in terms of crop rotation, fertilization, pest management and agrochemical use, which in coincidence to farmers' records of crop yield can be regarded as "Good no-till Agricultural Practices (GAP)" and "Poor no-till Agricultural Practices (PAP)". The third treatment corresponded to non-cultivated soils nearby the agricultural sites, which were used as references for natural environments (NE).

Pyrosequencing of 16S RNA gene using barcoded sequence tags is a high-throughput technique that has the capability to provide sufficient coverage and sequence length to afford an extensive taxonomic description of soil biota, comparing multiple samples in a single run [27]. A highly variable region of the 16S rRNA gene is individually PCR-amplified using primers containing a barcoded sequence (pyrotag) that allows distinction between samples. Tagged amplicons are pooled at equimolar concentration and sequenced in a single reaction. Reads were later assigned to individual samples based on the barcode sequence. Subsequent comparison to databases allows the identification of bacterial taxa and their relative abundance within the community. Here, we used tag-encoded pyrosequencing to deeply sample the 16S rRNA

gene from bacteria residing in soils corresponding to the three treatments at the four sites with the objective of finding out potential candidate bacterial species as indicators of agricultural management. As we have considered soils with varied characteristics, in terms of texture and organic matter, the identification of statistically based soil management-associated taxa can provide useful diagnostic tools for agricultural soil quality across the surveyed region.

Materials and Methods

Study Sites

The management and sites for this study were selected after a thoughtfully discussion between scientists and farmers participants of the BIOSPAS Project (www.biospas.org/en).

Whereas the sites selected may not fulfilled a rigorous definition of replicates, due to slight differences in management (historical crop sequence, years on no-till agriculture, were not the same), the experimental design privileged the perspective of farmers in terms of the relation between soil management and crop productivity. We have therefore followed a working definition of soil management, according to a set of definitions of Certified Agriculture by the Argentine No Till Farmers Association (AAPRESID, www.ac.org.ar/descargas/PyC_eng.pdf) and the Food and Agricultural Organization of the United Nations (FAO, www.fao.org/prods/GAP/index_en.htm).

Three treatments were defined (Table 1): 1) "Good no-till Agricultural Practices" (GAP): Sustainable agricultural management under no-till, subjected to intensive crop rotation (basically wheat/other winter crop/soybean/maize and sometimes including the use of cover crops, such as vicia/triticale), nutrient replacement, minimized agrochemical use (herbicides, insecticides and fungicides) and showing higher yield compared to PAP (Table 1); 2) "Poor no-till Agricultural Practices" (PAP): Nonsustainable agricultural management under no-till with high crop monoculture (soybean), low nutrient replacement, high agrochemical use (herbicides, insecticides and fungicides) and showing lower yields compared to GAP (Table 1); 3) "Natural Environment" (NE): As reference, natural grassland was selected in an area of approximately 1 hectare, close to the cultivated plots (less than 5 km), where no cultivation was practiced for (at least) the last 30 years.

Treatments were replicated 4 times (blocks) in agricultural fields located across a west-east transect in the most productive region in the Argentinean Pampas. Sites of soil sampling were near the following locations in Argentina: Bengolea at Córdoba Province (33° 01′ 31″ S; 63° 37′ 53″ W); Monte Buey at Córdoba Province (32° 58′ 14″ S; 62° 27′ 06″ W); Pergamino at Buenos Aires Province (33° 56′ 36″ S; 60° 33′ 57″ W); Viale at Entre Ríos Province (31° 52′ 59,6″ S; 59° 40′ 07″ W). See Table 2 for a description of soil characteristics.

Sampling

Samples were taken in June 2009 (winter) as triplicate for each treatment-site in three 5 m^2 sampling points separated at least 50 m from each other, taking care not to follow the sowing line in the field. Three additional samplings in the exact same locations were performed in February 2010, September 2010 and February 2011. Samplings were performed at private productive fields, which belong to any of the funders of this work. None of the sampling areas belong to a protected area or land. Permissions were obtained directly from the owners or responsible persons. At Bengolea and Monte Buey locations sampling was allowed by Jorge Romagnoli, from La Lucía SA, at Pergamino sampling was

Table 1. Description of the agricultural management and crop yield, averaged over the five years before the first sampling date, in June 2009 (2005–2009).

| | Bengolea | | Monte Buey | | Pergamino | | Viale | |
	GAP	PAP	GAP	PAP	GAP	PAP	GAP	PAP
% no-tillage	100	80	100	100	100	100	100	100
Soybean/maize ratio[a]	1.5	4	0.67	4	1.5	5	1,5	4
% Winter with wheat[b]	60	40	60	20	40	0	40	20
% Winter cover crops[c]	20	0	40	0	0	0	20	0
Herbicide (L) used[d]	27.7	43.8	25.2	38.9	29.3	46.5	34.5	43.1
Soybean yield (kg.ha^{-1})	3067	2775	3167	2675	2933	2825	3000	1805
Maize yield (kg.ha^{-1})	10500	2700	12550	8000	9500	–[e]	7030	3450

[a]Number of soybean cycles to number of maize cycles over the last 5years.
[b]Percentage of winters that wheat was planted as a winter crop.
[c]Percentage of winters that a cover crop (*Vicia* sp., *Melilotus alba* or *Lolium perenne*) was planted. Cover crops were chemically burned before summer crops are planted.
[d]Calculated as liters of low-toxicity herbicides plus liters of moderate-toxicity herbicides weighted by two. Toxicity was defined according to EPA Toxicity Categories. Unit: total liters over 5 years.
[e]No maize was planted in the last 5 years.

allowed by Gustavo Gonzalez Anta, from Rizobacter Argentina SA, sampling at Viale was allowed by Pedro Barbagelatta, member of Aapresid.

Each sample of the top 10 cm of mineral soil was collected as a composite of 16–20 randomly selected subsamples. Composite soil samples were homogenized in the field and transported to the laboratory at 4°C. Within 3 days after collection, samples were sieved through 2-mm mesh to remove roots and plant detritus. Soils were stored at −20°C until DNA extraction.

Chemical and Physical Soil Properties

Soils were classified according to Soil Taxonomy and INTA (Instituto Nacional de Tecnología Agropecuaria, Argentina) soil map. The main chemical properties of soils were determined by standard methods on samples that were air-dried, crushed and

passed through a 2-mm sieve after removal of plant residues. The pH was determined on mixtures of 1:2.5 sample:water. The contents of organic matter as total organic carbon were measured by dry combustion using a LECO CR12 Carbon analyzer (LECO, St. Joseph, MI, USA). The total nitrogen contents in whole soils were obtained by the Kjeldahl method. Extractable phosphorus was determined by the method of Bray and Kurtz. Data is summarized in Table 2.

DNA Extraction

Further homogenization was performed by careful grounding 10–15 g of each soil sample in a mortar before DNA extraction. DNA was extracted from 0.5 g of soil using FastDNA spin kit for soil extraction kit (Mpbio Inc), following the manufacturer's instructions. In order to reduce the presence of humic substances

Table 2. Soil characteristics according to site and agricultural management at the first sampling date, in June 2009.

| | Bengolea | | | Monte Buey | | | Pergamino | | | Viale | | |
	NE	GAP	PAP	NE	GAP	PAP	NE	GAP	PAP	NE	GAP	PAP
Climate	Temperate Subhumid			Temperate Subhumid			Temperate Humid			Temperate Humid		
MAT[1] (°C)	17			17			16			18		
MAP[2] (mm yr^{-1})	870			910			1000			1160		
Altitude (m)	224	222	223	112	111	108	64	68	65	73	80	81
Slope (%)	0.5	0.75	0.5	0.01	0.5	0.2	0.25	0.5	0.5	0.75	0.75	0.2
Years of no-till		13	5		28	10		6	5		13	9
Soil classification	Entic Haplustoll			Typic Argiudoll			Typic Argiudoll			Argic Pelludert		
Texture	Sandy loam			Silt loam			Silt loam			Silty clay/Silty clay loam		
Carbon %	1.7	1.5	1.1	3.5	2.1	1.7	2.7	1.7	1.8	5	3.5	2.5
Nitrogen %	0.146	0.156	0.125	0.328	0.181	0.132	0.233	0.153	0.136	0.369	0.283	0.179
Extractable P (ppm)	44.3	53.1	17.8	296.5	126.5	20.6	10.5	18	11.9	20.2	40.4	41.8
pH	6.3	6.2	6.2	5.6	5.5	6.2	6.2	6	5.7	6.4	6.7	6.3
Moisture	10.58	7.96	6.32	25.47	21.87	18.03	22.83	22.03	12.73	17.8	25.3	18.2

[1]Mat: Mean annual temperature.
[2]MAP: mean annual precipitation.

that inhibited the subsequent PCR reaction, an additional purification step was performed on the DNA sample using polyvinyl polypyrrolidone (PVPP). Eluted DNA was stored at $-20°C$.

Pyrosequencing

Barcoded pyrosequencing analysis was run on samples from the first sampling date of the BIOSPAS project, in June 2009. DNA samples were diluted to 10 ng/µL and 1.5 µL DNA aliquots of each sample were used for 50 µL PCR reactions. A fragment of the 16S rRNA gene of approximately 525 bp in length was amplified using bacterial primer 27F and universal primer 518R, both containing a unique 10-bp barcode sequence per sample to facilitate sorting of sequences from a single pyrosequencing run. PCR was conducted with 0.3 mM of each forward and reverse barcoded primer, 1.5 µl template DNA, 2X buffer reaction, 0.2 mM of dNTPs, 1 mM $MgSO_4$ and 1U of Platinum *Pfx* DNA polymerase (Invitrogen). Samples were initially denatured at 94°C for 5 min, then amplified using 35 cycles of 94°C for 30s, 50°C for 30 s and 68°C for 30s. A final extension of 10 min at 68°C was added at the end of the program to ensure complete amplification of the target region. Each of the triplicate subsamples was amplified separately and later combined and used as a representative composite of each sample. Amplicons were gel purified using GFX PCR DNA and Gel Band Purification Kit (GE Healthcare), and sent to the Genome Project Division Macrogen Inc. Seoul, Republic of Korea to be run on a Roche Diagnostics (454 Life Science) GS-FLX instrument with Titanium chemistry.

Sequence Data Analysis

Data were processed using MOTHUR v.1.22.2 following the Schloss SOP [28]. Briefly, the 10-bp barcode was examined in order to assign sequences to samples. Sequencing errors were reduced by implementation of the AmpliconNoise algorithm and low-quality sequences were removed (minimum length 200 bp, allowing 1 mismatch to the barcode, 2 mismatches to the primer, and homopolymers no longer than 8 bp). Sequences with ambiguous bases were eliminated as well, as their presence appear to be a strong indication of defective sequences [29]. The choice of these parameters for filtering follows the recommendation of Schloss et al. to reduce the error rate [30].

Chimera were removed with 'chimera.uchime' Mothur command. Sequences were aligned and classified against the SILVA bacterial SSU reference database v 102 [31]. Following the OTU approach [28], sequences from forward primers were clustered according the furthest neighbor-clustering algorithm.

Pyrosequencing raw reads were deposited in the NCBI Short-Read Archive under accession SRA057382. Sequence profile of processed sequences is shown in table S1. Raw and filtered reads per sample are shown in table S2.

Shared tables were created indicating the number of times an OTU appears in each sample. Venn diagrams were constructed with package gplots in R 2.10.1 (http://www.R-project.org/) from shared tables at the 0.05 distance level. Names and sequences from shared phylotypes were retrieved with scripts written in Python.

An indicator value analysis was performed for detecting statistically significant associations between taxa and soil management [32]. The indicator value combines the abundance of the OTU in the target group compared to other groups (specificity), with its relative frequency of occurrence in that particular group (fidelity). The value of the IndVal index was calculated using function IndVal in [R] package 'labdsv' (http://ecology.msu. montana.edu/labdsv/R/labdsv). Data were previously ANOVA filtered to reduce the number of tests and therefore increase the power to detect true differences. In order to perform multiple testing corrections, analysis of false-discovery rate (FDR) of 0.05 of significance were calculated for the complete set of p-values with qvalue.gui() in [R] package 'qvalue' (http://genomics.princeton. edu/storeylab/qvalue/). The FDR estimates the chance of reporting a false-positive result in all the significant results [33].

Real-time PCR Quantification

Bacterial taxa were quantified using the taxon-specific 16S rRNA primers designed in this work. All primers were designed using the PRIMROSE software [34].

For bacteria within the GP1 group of the phylum Acidobacteria, two specific primers were designed using the sequences available at the Ribosomal Database Project (RDP) database (http://rdp.cme.msu.edu/probematch/search.jsp) and the 100 sequences of the selected OTU.

Primers for the genus *Rubellimicrobium* Rub290F (GAGAGGAT-GATCAGCAAC) and Rub547R (CGCGCTTTACGCC-CAGTC) were designed using all the *Rubellimicrobium* sequences available in the RDP database. The specificity of the primers Gp1Ac650R (TTTCGCCACAGGTGTTCC) and SubGp1-143F (CGCATAACATCGCGAGGG) were initially checked by *in silico* analysis against RDP probe match (data set options: good and >1200 bp). The Gp1Ac650R primer matched with 92.4% of the sequences within the class Acidobacteria Gp1 (2268/2454), and with other 394 non-target bacteria. The SubGp1-143F primer matched with 99/100 sequences of the indicator group and only 3 sequences of non-target bacteria in the RDP database. More than 98% of the sequences in the RDP database (105/110) matched the primers combination Rub290F (GAGAGGATGATCAGCAAC) and Rub547R (CGCGCTTTACGCCCAGTC). Only 2 sequences of non-target bacteria in the RDP database matched with this primers combination.

To further test the specificity of the qPCR assays, clone libraries were constructed for each primer set using DNA extracted from GAP soil samples of Pergamino and Monte Buey for Acidobacteria GP1, and PAP soil samples of Bengolea and Monte Buey for *Rubellimicrobium*. Twenty four positive clones of each library (n = 48 for each primer set) were sent to Macrogen Inc. for complete sequencing. Sequences were assigned to taxonomic groups using the RDP classifier program. 100% of the cloned amplicons that could be identified belonged to the correct target groups.

Quantification was based on the increasing fluorescence intensity of the SYBR Green dye during amplification. The qPCR assay was carried out in a 20µL reaction volume containing the SYBR green PCR Master Mix (Applied Biosystems, UK), 0.5 µM of each primer, 0.25 µg/µL of BSA and 10 ng of soil DNA. Primer annealing temperature was optimized for PCR specificity in temperature-gradient PCR assays, utilizing the DNA Engine Opticon 2 System (MJ Research, USA). Optimal conditions for PCR were defined as 10 minutes at 94°C, 35 cycles of 94°C for 30 seconds, 59°C for 20 seconds and 72°C for 30 seconds, for both sets of primers. Standard curves were obtained using at least five ten-fold serial dilutions of a known amount of PCR amplicon mixtures as templates, purified through QIAquick PCR purification columns. Controls with no DNA templates gave null or negligible values.

Statistical Analysis

Patterns of similarity between samples were investigated using Correspondence Analysis (CA) on the relative abundances of $OTUs_{0.05}$. Due to the fact that the number of sequences obtained for NE of Monte Buey was markedly lower than those obtained for all other samples, this sample was excluded from this type of analysis.

Correlation between relative abundances of all significant indicator $OTUs_{0.05}$ and soil environmental gradients was assessed using Canonical Correspondence Analysis (CCA). The model used to explain variability included moisture content, total nitrogen content, total carbon content, ratio of total carbon to total nitrogen content and pH. Abiotic variables were standardized by subtracting the mean and dividing by the standard deviation (z-score standardization), making quantitative variables dimensionless. Significance was assessed using permutation tests. Multivariate analyses were performed in [R] package 'vegan'.

The effect of site and management on the number of copies of 16S rRNA genes for GP1A (the indicator group within the GP1 of the Acidobacteria) and *Rubellimicrobium*, was determined for each sampling date by mixed models. Treatment and seasons (summer, winter) were considered as fixed factors, whereas year, site, and subsample were included in the random structure. For this analysis, data were log transformed to achieve normal distribution of residuals. The mean number of 16S rRNA gene copies of each taxon was compared for the effect of treatment by orthogonal contrasts.

Management indicator value was defined as the logarithm of the ratio of the normalized abundance (i.e. fold-change relative to a non-cultivated soil) of the GP1A and the normalized abundance of *Rubellimicrobium* template. Two-tailed one-sample *t*-tests were performed on mean management indicator values (n = 48 for each type of management) to test the null hypothesis that the mean was equal to 0, and 95% confidence intervals were calculated. Statistical analysis was carried out with InfoStat Plus version 2011 (http://www.infostat.com.ar).

Results

Soil Quality and Productivity According to Sites and Agricultural Practices

The information on the agricultural management and crop yields of the different sites under study are summarized in Table 1. Before the first sampling, all sites had been managed under no-till for at least the preceding five years, with the exception of a single year (2004/2005) in Bengolea, where the PAP site was chisel-plowed. In the four localities, GAP had in average a 62% higher proportion of maize in the crop rotation than the PAP. GAP had in the last five years 50% of the winters with crop, whereas PAP sites had only 20%. In addition, cover crops had been implanted in winter in three of the four GAP localities. Management also differed in terms of the amount of herbicides used, as soils under PAP had used 36% more herbicides than GAP during the previous five years. Soybean yield had been in average 24.7% higher in GAP than in PAP, whereas maize yield had been 149.9% higher in GAP.

Soil chemical and physical properties of the studied sites are presented in Table 2. There is a difference in soil texture among localities, with increasing clay and decreasing sand content from West (Bengolea) to East (Viale). Values of soil organic matter follow the relation NE>GAP>PAP at the different localities, except in Pergamino where the Good no-till Agricultural Practices (GAP) and the Poor no-till Agricultural Practices (PAP) showed similar values. Soil N content also followed the same pattern, with the exception of Bengolea, where GAP had higher values than NE. No clear association was observed between values of extractable P and soil type or management. The pH, which ranged from 5.5 to 6.7, did not appear to correlate with soil type or soil management. A more detailed analysis, comparing physical and chemical soil

properties of the different agricultural management under study, exceeds the purpose of this paper and will be presented elsewhere (Duval et al, unpublished).

Bacterial Community Structure

The structure of bacterial communities related to the agricultural management practices was obtained from the massive parallel sequencing data of the 12 samples, i.e. three management scenarios over the four locations. A total of 210579 sequences with an average read length of 284 bp, were obtained after trimming, sorting, and quality control of the pyrosequencing data (table S2). 80% of these sequences were classified to a known phylum in the domain Bacteria (Fig. S1).

We examined OTU distribution across the pyrosequencing-based data sets. Using 3% sequence variation criterion, the patterns of the rarefaction curves were roughly comparable in all samples and none of the curves reached a plateau (Fig. S2). Therefore it was not possible to establish a trend in the differences of richness as a function of either geographical location or soil management. Since, despite quality filtering, pyrosequencing has a large intrinsic error that could lead to overestimation of rare phylotypes, further estimation of bacterial richness was not attempted.

Taxa Overlap between Soil Samples

Correspondence Analysis (CA) was applied to the data set of relative abundances for taxa defined at 0.05 distance (Fig. 1). Axes 1 and 2 account for 28.8% of the total inertia (16.4% and 12.4%, for axes I and II, respectively). Fitting of environmental factors to ordination indicate that samples were distributed according geographical location (site) with p = 0.001.

The majority of OTUs were unique to the samples in which they were found. The Venn diagrams in Fig. 2 show the number of shared OTUs among the different sample types. When analyzed by soil management, 254 OTUs were common to GAP and PAP samples across all sites (Table S3). Considering only sequences that were found in one type of management, but absent in the other, GAP and PAP samples had respectively 142 and 200 OTUs in common among the four sites, which corresponded to around 1.0% and 1.4% of the total number of sequences (Tables S4 and S5).

Considered by geographical location, the number of common OTUs was around 11% of the total OTUs identified in each location. In these cases, the overlap for the three samples in each sampling location was similar to the numbers of OTUs shared by the pair GAP and NE and the pair GAP and PAP, which in turn were consistently higher than the overlap of OTUs shared by NE and PAP (Fig. S3). We did not detect any OTU that was common to NE and PAP, but absent in GAP, even when the sample of NE from Monte Buey, which had less sequences, was excluded from the analysis. This finding disputes the possibility that the overlaps between groups of samples were due to chance.

Indicator Taxa of Agricultural Management

Table 3 shows the result of the IndVal analysis for indicators containing more than 20 sequences, which identified four significant indicators of GAP and five significant indicators of PAP.

Canonical Correspondence analysis (CCA) was applied on the bacterial taxa identified as indicators using IndVal to study the association of physico-chemical soil properties to sites and taxa (Fig. 3). Bacterial taxa clustered into three well-separated groups associated with the different soil management practices despite the different geographic origin of the data. The first ordination axis was strongly correlated to total nitrogen content (0.96, p<0.05),

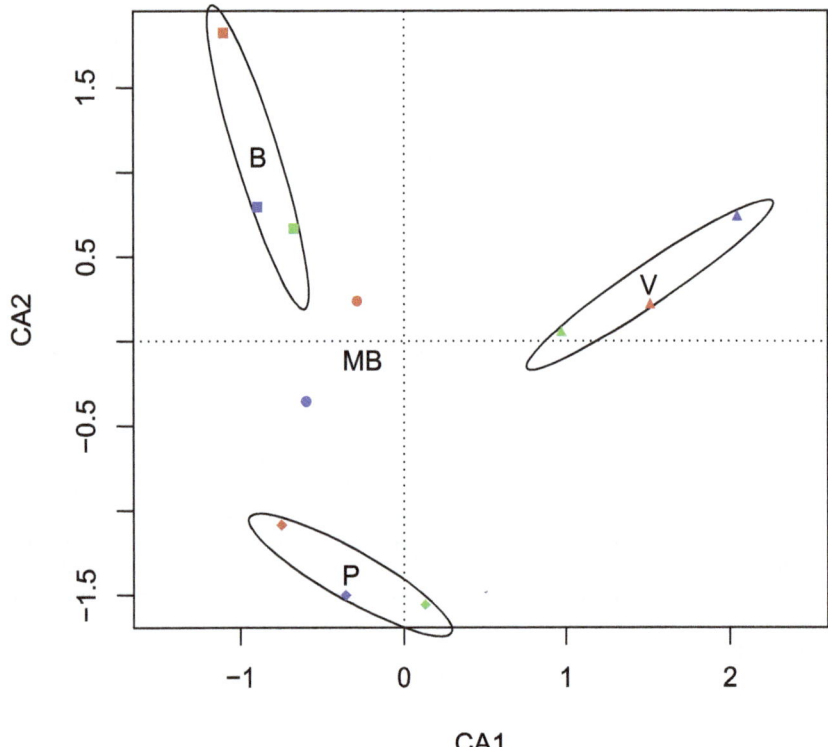

Figure 1. Ordination diagram from the Correspondence Analysis of the relative abundances for taxa defined at 0.05 distance. The 2-D CA diagram account for 29% of inertia. Locations of soils are indicated by squares (Bengolea, B), circles (Monte Buey, MB), diamond (Pergamino, P) and triangles (Viale, V). Colors indicate soil management type: Poor no-till Agricultural Practices in red, Good no-till Agricultural Practices in blue and Natural Environment in green. Standard error ellipses show 95% confidence areas.

meaning the bacterial indicators of natural environments were associated with higher than average nitrogen content. The second canonical axis correlated in descending order to the pH (0.49), moisture (−0.47), carbon to nitrogen ratio (0.45) and total carbon (0.22). Separation between management indicators was influenced by the second canonical axis. Indicators of PAP are located in the positive quadrant. i.e. they occur at sites with higher than average pH, and carbon to nitrogen ratio, and lower than average values of moisture. Inversely, GAP indicators were associated with higher moisture content, lower pH and lower carbon to nitrogen ratio.

Although significant after the application of the false discovery rate, most of the indicators had low abundance across the samples. To obtain meaningful quantitative results, we analyzed only significant indicator containing at least 75 sequences. As a result, we selected single the list's top indicators for GAP and PAP samples respectively, belonging to a taxa within the Acidobacteria Group 1 (GP1A), and to *Rubellimicrobium*, a genus of the order *Rhodobacterales* of the class Alphaproteobacteria (Table 3). Although this threshold can be considered somewhat arbitrary, it was selected on the basis of the fact that those taxa were represented more evenly across all studied locations.

PCR primers were designed to target the sequences detected in the pyrosequence data set of these selected groups. Cloning and sequencing of the PCR products derived from the primer specificity tests confirmed the specificity of these primers (see M&M). Using these newly designed primers, quantitative PCR was conducted to validate the results of the sequence analysis. Primer sets of both Acidobacteria GP1A and the genus *Rubellimicrobium* were calibrated using known concentrations of clones of PCR amplified 16S rRNA genes from the respective

controls. The quantification of a set of Acidobacteria GP1A and of the genus *Rubellimicrobium*, performed over samples from two successive winter-summer seasons (June 2009 to February 2011) are shown respectively in Fig. 4 A and B. Quantitative PCR data revealed that the abundance of both taxa were significantly different among managements (mixed models, p<0.0098 for GP1A and p<0.0001 for *Rubellimicrobium*). Post-hoc contrasts indicated that the number of copies of 16S rRNA genes targeted with GP1A-specific primers were statistically higher in samples of soils defined as GAP (p = 0.005), compared to poorly managed soils. In opposition, the number of copies of 16S rRNA genes targeted with *Rubellimicrobium*-specific set of primers were statistically higher in samples of soils defined as PAP (p = 0.004).

The observation that the GP1A-specific primers target sequences that increase in GAP samples and that the *Rubellimicrobium*-specific primers target sequences that increase in PAP samples, prompted us to evaluate their combined use as potential indicator of agricultural management under no-till regime in the Pampa Region. For that purpose, we calculated the ratio of the normalized abundance (i.e. fold-change relative to a non-cultivated soil in the same geographical location) of the GP1A and the normalized abundance of *Rubellimicrobium* template. By definition, this value is equal to one for NE samples. Log transformation was applied to achieve normal distribution. The resulting indicator value will take therefore a value of zero for NE samples. For all other samples, the sign of the indicator will depend on whether the GP1A and *Rubellimicrobium* abundances increase or decrease relative to NE. The results are indicated in Fig. 5. Despite the high variability of the data, likely due in part to the heterogeneous distribution of bacteria in subsamples within

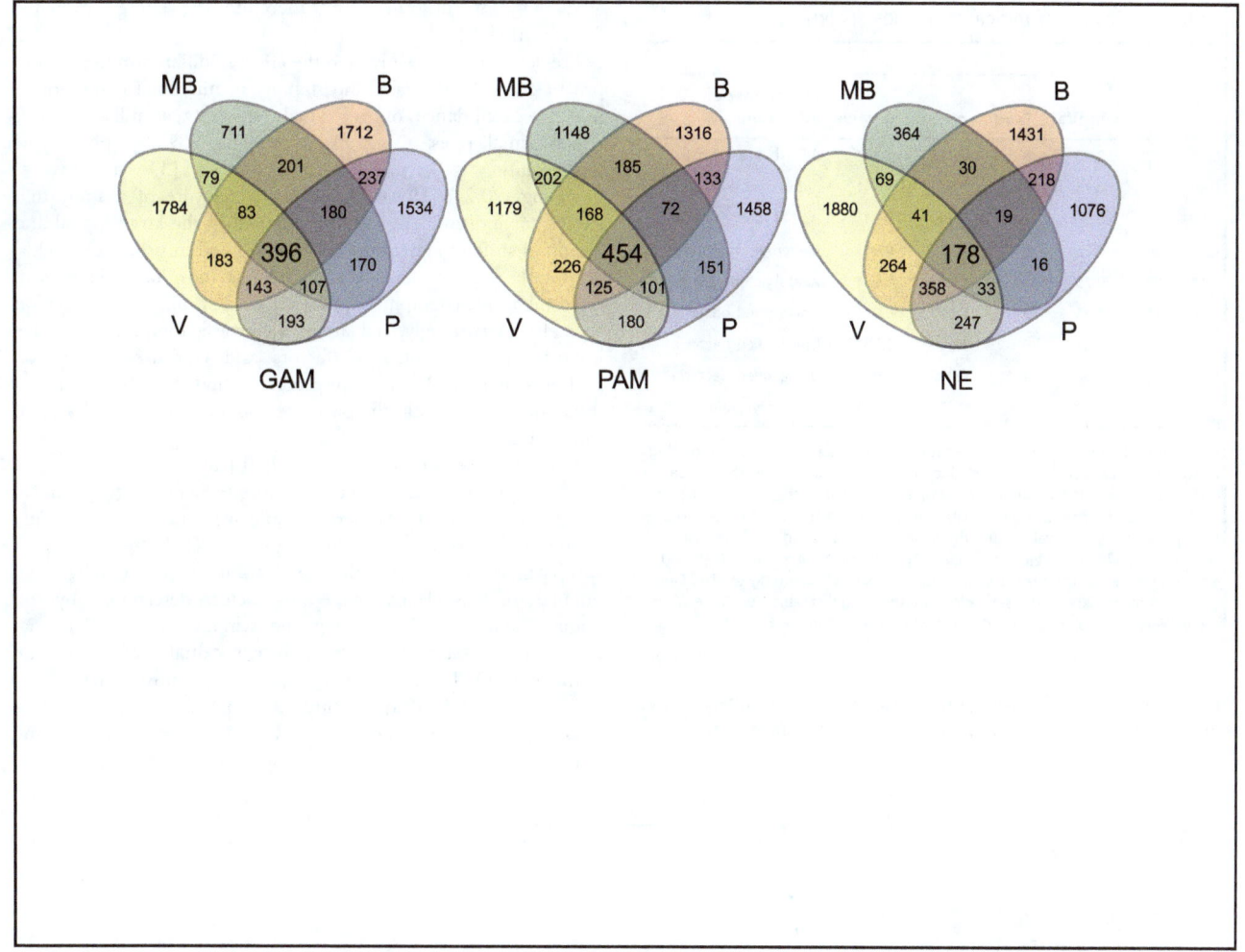

Figure 2. Venn diagram of the overlap of OTUs from the different soil management practices at four geographical locations. The numbers of overlapping tag sequences are indicated in the graph. Management practices are indicated at the bottom of each diagram: GAP: Good no-till Agricultural Practices, PAP: Poor no-till Agricultural Practices; NE: Natural Environment. Location labels are indicated with B (Bengolea), MB (Monte Buey), P (Pergamino) and V (Viale).

each given sample, the mean of the value calculated for all GAP samples across the four sites and four sampling dates (n = 48) was significantly higher than 0 (p = 0.0018), whereas the mean of the value in all PAP samples was significantly lower than 0 (p<0.001).

Discussion

The main hypothesis tested in this study was that the relative abundance of selected soil bacterial taxa could be used as indicator of the impact of agronomic management at a regional scale. Based on massive parallel sequencing and quantitative PCR, we have found that the combined use of the abundance of two bacterial taxa could potentially fulfill this task. The bacteria, belonging to Acidobacteria Group 1, and to the genus *Rubellimicrobium* of the Alphaproteobacteria, were augmented in soils under no-till crop production, managed with sustainable and non-sustainable practices, respectively. What makes our finding more compelling is that the taxa that appeared to be specific of soil management, were present in soils with different physical properties (Table 1) with various crop sequences (Table 2), suggesting that the physiology of these group of bacteria might be affected by nutrient and carbon shifts, and probably different soil microstructure, produced by the

different crop rotation practices: intense crop rotation vs. monoculture practice, the most important characteristic differentiating soil managements, with consistent different yields.

Agricultural soil activities should sustain crop productivity while preserving soil environmental quality. After several years of no-till agriculture and the widespread practice of monoculture, farmers have realized the impact of management on soil quality, which ultimately impacted on crop productivity. This has led to a working definition of "Good no-till Agricultural Practices" (GAP) and "Poor no-till Agricultural Practices" (PAP), according to criteria based on yield, crop rotation, fertilization, pest management and agrochemical use (http://www.ac.org.ar/index_e.asp). In this context, indicators of soil quality are essential tools to evaluate the impact of management on the soil ecosystem. Physical properties, such as soil structure, water storage capacity and soil aeration, as well as soil chemical characteristics are currently used as indicators of soil health. In addition, microbial properties are increasingly regarded as more sensitive and consistent indicators than biochemical parameters for monitoring the effect of management on soil quality [35]. This is because bacteria are in intimate contact with the soil

Table 3. Results of indicator species analysis.

	Size	IndVal	Freq	p value	q value	Phylogenetic affiliation
GAP	100	0.86	8	0.028	0.041	Acidobacteria_Gp1
PAP	76	0.78	9	0.032	0.041	*Rubellimicrobium*
GAP	55	0.91	7	0.050	0.041	Alphaproteobacteria
PAP	34	0.85	8	0.037	0.041	*Micromonosporaceae*
PAP	28	0.75	8	0.043	0.041	Acidobacteria_Gp16
GAP	26	0.85	7	0.014	0.041	Unclassified bacteria
GAP	23	0.83	6	0.044	0.041	Unclassified bacteria
PAP	20	1.00	4	0.009	0.041	Unclassified bacteria
PAP	20	0.80	7	0.038	0.041	Actinomycetales

For each of the taxa, we indicate the total number of sequences corresponding to the OTU that represents the specific groups of samples (size), the Indicator Value index (IndVal), the number of samples that contain the taxon (Freq), the statistical significance of the association (p-value), the chance of reporting a false-positive result (q-value), and the lowest taxonomic rank assigned with a bootstrap confidence greater than 80%. Agricultural managements GAP and PAP are defined in the main text. Results were sorted according to Size. Only OTUs containing 20 or more sequences are shown in this table. See Table S6 for a complete list of significant indicators with IndVal values ≥0.75.

microenvironment. Microorganisms can be directly affected by some toxic effect, or indirectly, e.g. by changes in the

availability of substrates, and therefore the energy available for growth [36].

Previous studies have related the effect of different management practices on the diversity and stability of microbial communities and the abundance of individual taxa, in carefully designed experimental plots, using PLFA profiling [37,38], phenotypic fingerprinting [23], ribosomal fingerprinting [39], and pyrosequencing [20,24]. However, we are not aware of a study that looked for indicators of soil management in the large spatial and temporal scale of agricultural practices tested in productive fields. This task is particularly challenging, as most bacteria present discernible biogeographical patterns, even within a given habitat type [40]. Accordingly, in a large-scale investigation on the relative importance of various soil factors and land-use regimes on soilborne microbial community composition, it was found that the main differences in the bacterial communities were related to soil factors [41].

In our samples we have observed that bacterial communities as a whole appeared indeed to be structured chiefly by geographical proximity, meaning that differences in composition are due mainly to soil characteristics at the landscape scale [42]. Nevertheless, it was particularly interesting that the distribution of certain bacterial populations was clearly shaped by factors determined by soil management as well, opening the window to find bacterial indicators of soil status across a broad spatial scale [24]. The numbers of OTUs, which were found to be common to the four soil locations subjected to similar management practice, was relatively large. We deem unlikely that this overlap was the outcome of chance alone, as for each type of management the

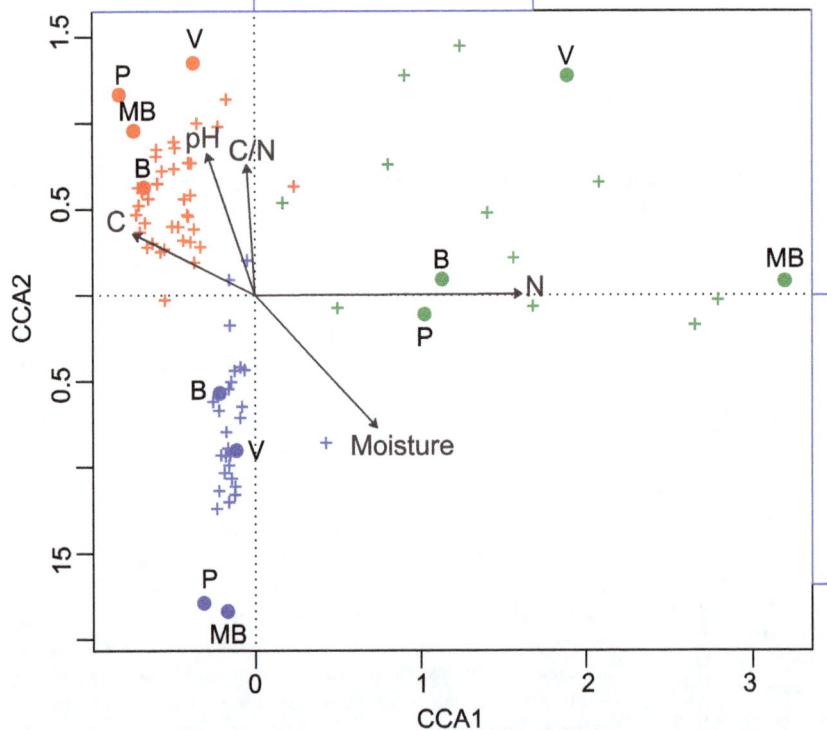

Figure 3. Ordination diagram from Canonical Correspondence Analysis of bacterial taxa identified as indicators using IndVal. Only OTUs identified with IndVal values higher than 0.75 were used in this analysis (Table S6). The 2-D ordination diagram CCA accounts for 66% of inertia. Samples are indicated by circles and site labels. OTUs are indicated by crosses, names are omitted. Arrows for quantitative variables show the direction of increase of each variable, and the length of the arrow indicates the degree of correlation with the ordination axes. Colors indicate soil management type: Poor no-till Agricultural Practices in red, Good no-till Agricultural Practices in blue; Natural Environment in green. Location labels are indicated with: B (Bengolea), MB (Monte Buey), P (Pergamino) and V (Viale).

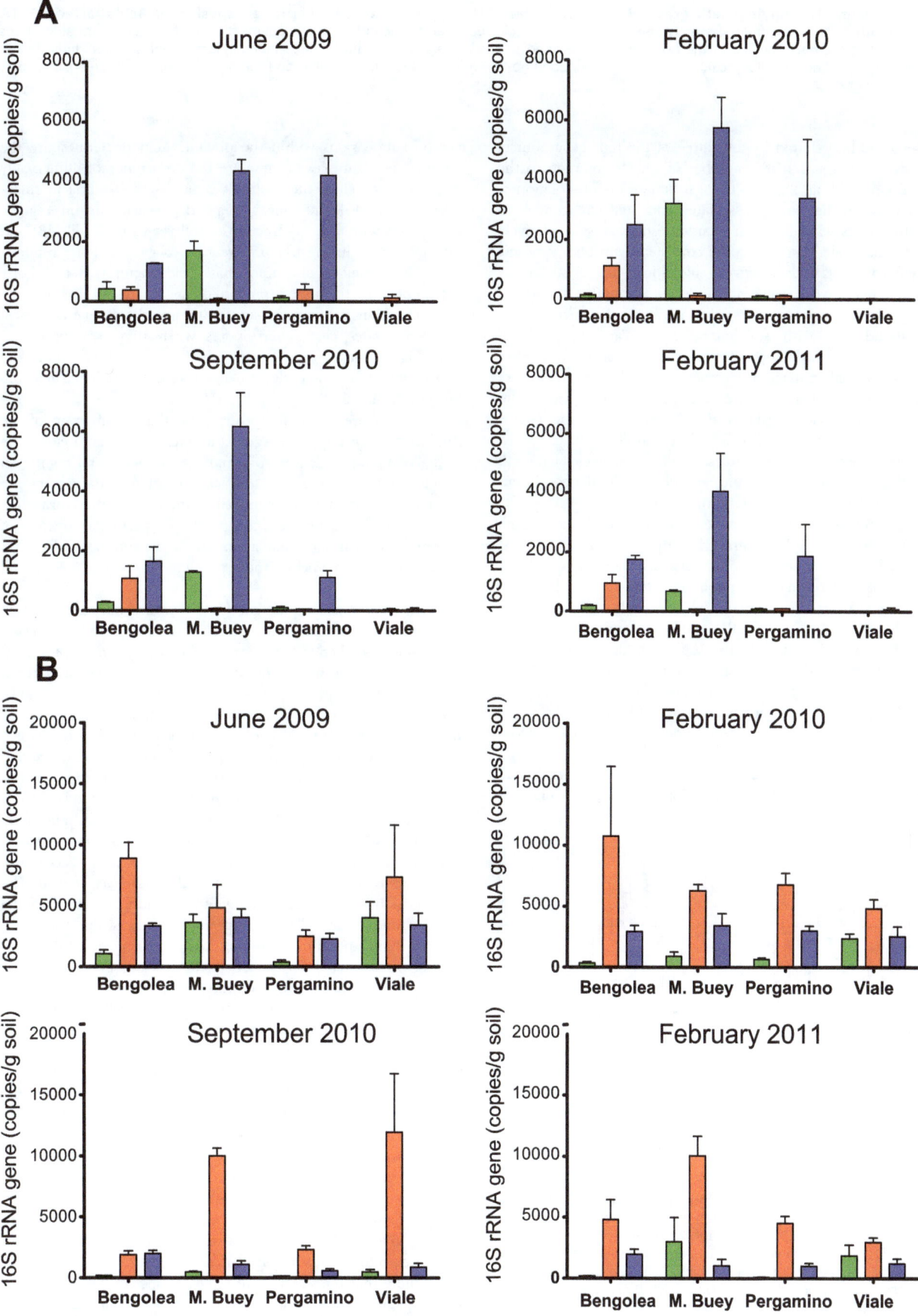

Figure 4. Quantitative phylogenetic group abundance of the OTUs targeted with a set of primers specific for Acidobacteria GP1A (panel A) and *Rubellimicrobium* genus (panel B). Each soil sample subjected to the indicated management in the four geographical locations was sampled at the date showed in the boxes. Bars correspond to the average qPCR data of three independent samples. Colors indicate soil management type: Poor no-till Agricultural Practices in red, Good no-till Agricultural Practices in blue, and Natural Environment in green. Error bars are standard error.

number of OTUs shared by any three of the four geographical locations was lower than the number of OTUs common to the four soil samples (Fig. 2). Even more striking is the observation of bacterial groups that can be associated with soil management in agricultural soils with dissimilar characteristics across a relatively wide regional scale. These data are consistent with both genomic and environmental perspectives suggesting the existence of ecological coherence of bacterial at different taxonomic ranks [43].

The set of indicator taxa were used to evaluate the correlation of their abundances with soil characteristics across sites and management. The ordination illustrated how indicator taxa were responsive to soil management practices. GAP indicators were associated with higher moisture content, and lower carbon to nitrogen ratio and lower pH. Slight changes in pH might have been caused by acidifying reactions (e.g. nitrification). Inversely, the occurrence of PAP indicators at sites was associated with higher than average carbon to nitrogen ratio, i.e. under conditions in which nitrogen becomes a limiting factor.

The phylum Acidobacteria ranked third in abundance in each of the twelve soil samples examined in this study. Acidobacteria constituted an average of 20% of soil bacterial taxa in 16S rRNA gene libraries, according to a published meta-analysis [44] and more recent analysis of agricultural soils indicated that three subgroups (GP4, GP6 and GP1) situate among the five most abundant genera in soils [24,45]. Although the phylum Acidobacteria it has been frequently associated with low nutrient availability [46], its wide global distribution and high diversity led to the proposition that its members are involved in a broad range of metabolic pathways [47]. Several findings point to the fact that

not all subdivisions within the phylum Acidobacteria share the same traits. Examples of these are the occurrence of numerically dominant as well as metabolically active Acidobacteria in rhizospheric soil [45], the lineage-dependent variations in relative abundance within a clay fraction of soil versus bulk soil [48] and the differences in the pH preferences for growth [44]. Interestingly, Mummey et al found that Acidobacteria were poorly represented in the inner fraction of aggregates [49], but were more abundant in soil macroaggregates and the outer fractions of microaggregates, i.e. in coarse pores, where they are supposed to have high turnover rates because of the effect of predation and desiccation events, and due to the transiently high oxygen and nutrient availability [50].

It is therefore not entirely surprising that a subgroup of the Acidobacteria group 1 emerges as a potential bacterial candidate for agronomic practices in soils managed under no till regime, in which carbon conservation and stability of macroagregates are enhanced (Morras et al, personal communication). Unraveling specific details about the ecology of this particular lineage of Acidobacteria through cultivation [51] and genomic studies [52] are needed to gain a better understanding of its involvement in soil processes.

Neither is the natural habitat of Rubellimicrobia currently well characterized. To date, four species of the genus *Rubellimicrobium* had been described. One thermophilic species, *R. thermophilum*, which was isolated from slime deposits on paper machines and a pulp dryer [53] and three mesophilic species, two of which have been isolated from soils: *R. mesophilum* [54] and *R. roseum* [55], and *R. aerolatum*, which was isolated from air samples [56]. It is worth noting that fatty acids profiles of the same soil samples analyzed in this work, show in all PAP treatments that fatty acid $C_{18:1}\omega7c$ is significantly augmented (Ferrari and Wall, unpublished). This is relevant because $C_{18:1}\omega7c$ is one of the major membrane fatty acids in most of the isolates belonging to the genus *Rubellimicrobium* [54,55,56]. Given the limited physiological and ecological information available on the genus *Rubellimicrobium*, it would be too speculative to suggest for it any indicator function in soils at the present time. Nevertheless, it is interesting to note that this genus appeared to respond to the use of the soil in a recent study of the impact of long-term agriculture on desert soil, in which it was shown that *Rubellimicrobium* was among the extremophilic bacterial groups that disappeared from soil after agricultural use [57]. Efforts to isolate *Rubellimicrobium* strains from the soils surveyed in the present study are currently under way in our laboratory, in order to perform a thorough physiological characterization.

The results of this study provide relevant information about the distribution of several groups of numerically abundant taxa in agricultural soils. It was also demonstrated that different taxa of bacteria respond differentially to geographical constraints and contemporary disturbances in no-till agriculture systems, highlighting the potential of high-resolution molecular tools to identify bacterial groups that may serve as potential indicator that might be used to assess the sustainability of agricultural soil management and to monitor trends in soil condition over time.

We note that the selection of indicator species based solely on the frequency of occurrence does not permit conclusions about the processes in which they are involved. In this regard, knowledge on

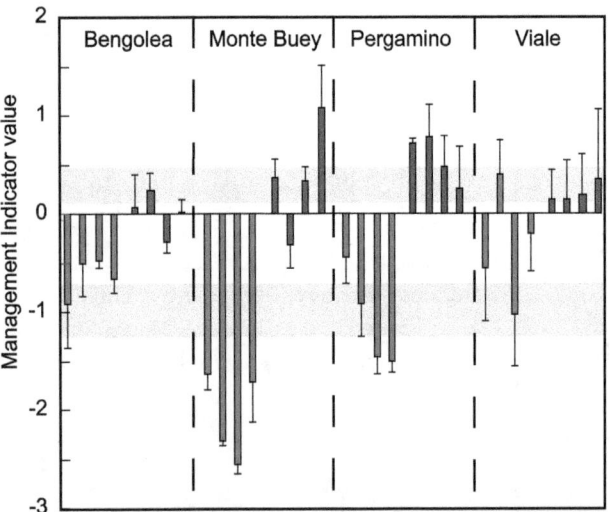

Figure 5. Indicator values for samples of soils under different agricultural management. The geographic sites are indicated in the box. In each site bars are ordered from left to right to the successive sampling dates: June 2009, February 2010, September 2010 and February 2011. PAP, Poor no-till Agricultural Practices (red) and GAP, Good no-till Agricultural Practices (blue). Shadow areas are 95% confidence intervals of indicator of GAP (0.24, 0.71) and PAP (−1.31, −0.41). Error bars are standard error.

their habitat specialization would be important, as this factor is not likely to be influenced by natural variations in environmental conditions [58]. However, considering the scarcity of data regarding the habitat preferences, physiology and in situ activity of Acidobacteria GP1 and *Rubellimicrobium*, a mechanistic link between the factors driving the relative distribution of these taxa and the different soil management is currently not feasible. Thus, although we have initially developed the indicator on a purely phenomenological basis, the understanding of the underlying ecological selection for both groups of taxa depending on the soil management remains a crucial goal for future studies.

Meanwhile, the proposed marker appears to fulfill several of the criteria required for appropriate ecological indicators. It is easily measured, it is sensitive to soil management actions and is integrative, i.e. it provides adequate coverage across a relatively wide range of ecological variables, e.g. soils types, climate, crop sequence, etc. [59]. Based on the data presented here, appropriate tests for simple monitoring can be elaborated to further validate if the proposed candidate biological indicator can be integrated into a minimum dataset, to allow measuring the impact of management practices under no-till at the regional scale.

Supporting Information

Figure S1 Complete set of sequences classified at phylum level against SILVA bacterial SSU refer- ence database v.102 by Bayesian method, with a confidence cutoff of 80% using classify.seqs command in Mothur.

Figure S2 Rarefaction analysis of pyrosequencing tags of the 16S rRNA gene in soils subjected at different agricultural practoces in the four geographic locations. Blue: Good no-till agricultural practices. Red: Poor no-till agricultural practices, Green: Natural environments.

Figure S3 Venn diagram of the overlap of OTUs from the different geographical locations subjected to different soil management practices. The numbers of overlapping tag sequences are indicated in the graph. Locations are indicated at the bottom of each diagram. Management practices are indicated with GAP: Good no-till Agricultural Practices, PAP: Poor no-till Agricultural Practices; NE: Natural Environment.

Table S1 Summary of processed 454-sequencing reads.

Table S2 Filtered and raw (in parenthesis) reads of 454 Pyrosequencing per sample.

Table S3 List of OTUs common to good no-till agricultural practices (GAP) and poor no-till agricultural practices (PAP) in the four locations. Sequences were assigned to taxonomic groups using the RDP classifier (http://rdp.cme.msu.edu/classifier/classifier.jsp). OTUs were sorted by the total number of sequences in the complete data set.

Table S4 List of OTUs only common to good no-till agricultural practices (GAP) in the four locations. Sequences were assigned to taxonomic groups using the RDP classifier (http://rdp.cme.msu.edu/classifier/classifier.jsp). OTUs were sorted by the total number of sequences in the complete data set.

Table S5 List of OTUs common only to poor no-till agricultural practices (PAP) in the four locations. Sequences were assigned to taxonomic groups using the RDP classifier (http://rdp.cme.msu.edu/classifier/classifier.jsp). OTUs were sorted by the total number of sequences in the complete data set.

Table S6 Results of indicator species analysis. For each of the taxa, we indicate the Indicator Value index (IndVal), the number of samples that contain the taxon (Freq), the statistical significance of the association (p-value), the total number of sequences corresponding to the OTU (size), and the lowest taxonomic rank assigned with a bootstrap confidence indicated in parenthesis. Agricultural managements GAP and PAP an NE are defined in the main text. Only OTUs with IndVal higher than 0.75 are shown.

Author Contributions

Conceived and designed the experiments: ELMF LDG LGW LE. Performed the experiments: ELMF LDG. Analyzed the data: ELMF LDG JCB SR LE. Contributed reagents/materials/analysis tools: LS MD JAG. Wrote the paper: ELMF LGW LE. Commented on manuscript: LDG JCB.

References

1. Hobbs PR, Sayre K, Gupta R (2008) The role of conservation agriculture in sustainable agriculture. Phil Trans Royal Soc B: Biol Sci 363: 543–555.
2. Bronick CJ, Lal R (2005) Soil structure and management: a review. Geoderma 124: 3–22.
3. Abid M, Lal R (2009) Tillage and drainage impact on soil quality: II. Tensile strength of aggregates, moisture retention and water infiltration. Soil Till Res 103: 364–372.
4. Derpsch R, Friedrich T, Kassam A, Li H (2010) Current status of adoption of no-till farming in the world and some of its main benefits. Intl J Agric Biol Eng 3: 1–25.
5. Montgomery DR (2007) Soil erosion and agricultural sustainability. Proc Natl Acad Sci USA 104: 13268–13272.
6. Viglizzo EF, Frank FC, Carreño LV, Jobbágy EG, Pereyra H, et al. (2011) Ecological and environmental footprint of 50 years of agricultural expansion in Argentina Global Change Biol 17: 959–973.
7. Cook RJ (2006) Toward cropping systems that enhance productivity and sustainability. Proc Natl Acad Sci USA 103: 18389–18394.
8. Bockstaller C, Guichard L, Makowski D, Aveline A, Girardin P, et al. (2008) Agri-environmental indicators to assess cropping and farming systems. A review. Agron Sustain Dev 28: 139–149.
9. Govaerts B, Sayre KD, Deckers J (2006) A minimum data set for soil quality assessment of wheat and maize cropping in the highlands of Mexico. Soil Tillage Res 87: 163–174.
10. Imaz M, Virto IP, Bescansa P, Enrique A, Fernandez-Ugalde O, et al. (2010) Soil quality indicator response to tillage and residue management on semi-arid Mediterranean cropland. Soil Tillage Res 107: 17–25.
11. Rodrigues de Lima AC, Hoogmoed W, Brussaard L (2008) Soil quality assessment in rice production systems: establishing a minimum data set. J Environ Qual 37: 623–630.
12. Shukla MK, Lal R, Ebinger M (2006) Determining soil quality indicators by factor analysis. Soil Tillage Res 87: 194–204.
13. Yemefack M, Jetten VG, Rossiter DG (2006) Developing a minimum data ser for characterizing soil dynamics in shifting cultivation Systems. Soil Tillage Res 86: 84–98.
14. Hartmann M, Fliessbach A, Oberholzer HR, Widmer F (2006) Ranking the magnitude of crop and farming system effects on soil microbial biomass and genetic structure of bacterial communities. FEMS Microbiol Ecol 57: 378–388.
15. Grayston SJ, Wang S, Campbell CD, Edwards AC (1998) Selective influence of plant species on microbial diversity in the rhizosphere. Soil Biol Biochem 30: 369–378.
16. Marschner P, Crowley D, Yang CH (2004) Development of specific rhizosphere bacterial communities in relation to plant species, nutrition and soil type. Plant and Soil 261: 199–208.
17. Smalla K, Wieland G, Buchner A, Zock A, Parzy J, et al. (2001) Bulk and rhizosphere soil bacterial communities studied by denaturing gradient gel

electrophoresis: plant-dependent enrichment and seasonal shifts revealed. Appl Environ Microbiol 67: 4742–4751.

18. Nacke H, Thürmer A, Wollherr A, Will C, Hodac L, et al. (2011) Pyrosequencing-based assessment of bacterial community structure along different management types in german forest and grassland soils. PloS One 6: 1387–1390.

19. Girvan MS, Bullimore J, Pretty JN, Osborn AM, Ball AS (2003) Soil type is the primary determinant of the composition of the total and active bacterial communities in arable soils. Appl Environ Microbiol 69: 1800–1809.

20. Acosta-Martinez V, Dowd S, Sun Y, Allen V (2008) Tag-encoded pyrosequencing analysis of bacterial diversity in a single soil type as affected by management and land use. Soil Biol Biochem 40: 2762–2770.

21. Lagomarsino A, Moscatelli MC, Di Tizio A, Mancinelli R, Grego S, et al. (2009) Soil biochemical indicators as a tool to assess the short-term impact of agricultural management on changes in organic C in a Mediterranean environment. Ecol Indic 9: 518–527.

22. Laudicina VA, Badalucco L, Palazzolo L (2011) Effects of compost input and tillage intensity on soil microbial biomass and activity under Mediterranean conditions. Biol Fertil Soils 47: 63–70.

23. Lupwayi NZ, Rice WA, Clayton GW (1998) Soil microbial diversity and community structure under wheat as influenced by tillage and crop rotation. Soil Biol Biochem 30: 1733–1741.

24. Yin C, Jones KL, Peterson DE, Garrett KA, Hulbert SH, et al. (2010) Members of soil bacterial communities sensitive to tillage and crop rotation. Soil Biol Biochem 42: 2111–2118.

25. Schloter M, Dilly O, Munch JC (2003) Indicators for evaluating soil quality. Agric Ecosyst Environ 98: 255–262.

26. Wall LG (2011) The BIOSPAS Consortium: Soil Biology and Agricultural Production. In: Bruijn FJd, editor. Handbook of Molecular Microbial Ecology I: Metagenomics and Complementary Approaches. Hoboken, NJ, USA: John Wiley & Sons, Inc. pp. ch. 34.

27. Parameswaran P, Jalili R, Tao L, Shokralla S, Gharizadeh B, et al. (2007) A pyrosequencing-tailored nucleotide barcode design unveils opportunities for large-scale sample multiplexing. Nucl Acids Res 35: e130.

28. Schloss PD, Westcott SL, Ryabin T, Hall JR, Hartmann M, et al. (2009) Introducing mothur: open-source, platform-independent, community-supported software for describing and comparing microbial communities. Appl Environ Microbiol 75: 7537–7741.

29. Huse SM, Huber JA, Morrison HG, Sogin ML, Welch DM (2007) Accuracy and quality of massively parallel DNA pyrosequencing. Genome Biol 8: R143.

30. Schloss PD, Gevers D, Westcott SL (2011) Reducing the effects of PCR amplification and sequencing artifacts on 16S rRNA-based studies. PLoS One 6: e27310.

31. Pruesse E, Quast C, Knittel K, Fuchs BM, Ludwig W, et al. (2007) SILVA: a comprehensive online resource for quality checked and aligned ribosomal RNA sequence data compatible with ARB. Nucl Acids Res 35: 7188–7196.

32. Dufrene M, Legendre P (1997) Species assemblages and indicator species: the need for a flexible asymmetrical approach. Ecol Monogr 67: 345–366.

33. Storey J, Tibshirani R (2003) Statistical significance for genomewide studies. Proc Natl Acad Sci USA 100: 9440–9445.

34. Ashelford KE, Weightman AJ, Fry JC (2002) PRIMROSE: a computer program for generating and estimating the phylogenetic range of 16S rRNA oligonucleotide probes and primers in conjunction with the RDP-II database. Nucleic Acids Res 30: 3481–3489.

35. Garbisu C, Alkorta I, Epelde L (2011) Assessment of soil quality using microbial properties and attributes of ecological relevance. Appl Soil Ecol 49: 1–4.

36. Bending GD, Turner MK, Rayns F, Marx MC, Wood M (2004) Microbial and biochemical soil quality indicators and their potential for differentiating areas under contrasting agricultural management regimes. Soil Biol Biochem 36: 1785–1792.

37. Romaniuk R, Giuffré L, Costantini A, Nannipieri P (2011) Assessment of soil microbial diversity measurements as indicators of soil functioning in organic and conventional horticulture systems. Ecol Indic 11: 1345–1353.

38. Zelles L (1999) Fatty acid patterns of phospholipids and lipopolysaccharides in the characterisation of microbial communities in soil: a review. Biol Fertil Soils 29: 111–129.

39. Wu T, Chellemi DO, Graham JH, Martin KJ, Rosskopf EN (2008) Comparison of soil bacterial communities under diverse agricultural land management and crop production practices. Microb Ecol 55: 293–310.

40. Martiny JB, Bohannan BJ, Brown JH, Colwell RK, Fuhrman JA, et al. (2006) Microbial biogeography: putting microorganisms on the map. Nat Rev Microbiol 4: 102–112.

41. Kuramae E, Gamper H, van Veen J, Kowalchuk G (2011) Soil and plant factors driving the community of soil-borne microorganisms across chronosequences of secondary succession of chalk grasslands with a neutral pH. FEMS Microbiol Ecol 77: 285–294.

42. Ge Y, He J, Zhu Y, Zhang J, Xu Z, et al. (2008) Differences in soil bacterial diversity: driven by contemporary disturbances or historical contingencies? ISME J 2: 254–264.

43. Philippot L, Andersson SG, Battin TJ, Prosser JI, Schimel JP, et al. (2010) The ecological coherence of high bacterial taxonomic ranks. Nat Rev Microbiol 8: 523–529.

44. Sait M, Davis KER, Janssen PH (2006) Effect of pH on isolation and distribution of members of subdivision 1 of the phylum Acidobacteria occurring in soil. Appl Environ Microbiol 72: 1852–1857.

45. Lee SH, Ka JO, Cho JC (2008) Members of the phylum Acidobacteria are dominant and metabolically active in rhizosphere soil. FEMS Microbiol Lett 285: 263–269.

46. Fierer N, Bradford MA, Jackson RB (2007) Toward an ecological classification of soil bacteria. Ecology 88: 1354–1364.

47. Ganzert L, Lipski A, Hubberten HW, Wagner D (2011) The impact of different soil parameters on the community structure of dominant bacteria from nine different soils located on Livingston Island, South Shetland Archipelago, Antarctica. FEMS Microbiol Ecol 76: 476–491.

48. Liles MR, Turkmen O, Manske BF, Zhang M, Rouillard JM, et al. (2010) A phylogenetic microarray targeting 16S rRNA genes from the bacterial division Acidobacteria reveals a lineage-specific distribution in a soil clay fraction. Soil Biol Biochem 42: 739–747.

49. Mummey D, Holben W, Six J, Stahl P (2006) Spatial stratification of soil bacterial populations in aggregates of diverse soils. Microb Ecol 51: 404–411.

50. Chenu C, Hassink J, Bloem J (2001) Short-term changes in the spatial distribution of microorganisms in soil aggregates as affected by glucose addition. Biol Fertil Soils 34: 349–356.

51. Davis KE, Sangwan P, Janssen PH (2011) Acidobacteria, Rubrobacteridae and Chloroflexi are abundant among very slow-growing and mini-colony-forming soil bacteria. Environ Microbiol 13: 798–805.

52. Ward NL, Challacombe JF, Janssen PH, Henrissat B, Coutinho PM, et al. (2009) Three genomes from the phylum Acidobacteria provide insight into the lifestyles of these microorganisms in soils. Appl Environ Microbiol 75: 2046–2056.

53. Denner EB, Kolari M, Hoornstra D, Tsitko I, Kampfer P, et al. (2006) Rubellimicrobium thermophilum gen. nov., sp. nov., a red-pigmented, moderately thermophilic bacterium isolated from coloured slime deposits in paper machines. Int J Syst Evol Microbiol 56: 1355–1362.

54. Dastager SG, Lee JC, Ju YJ, Park DJ, Kim CJ (2008) Rubellimicrobium mesophilum sp. nov., a mesophilic, pigmented bacterium isolated from soil. Int J Syst Evol Microbiol 58: 1797–1800.

55. Cao YR, Jiang Y, Wang Q, Tang SK, He WX, et al. (2010) Rubellimicrobium roseum sp. nov., a Gram-negative bacterium isolated from the forest soil sample. Antonie Van Leeuwenhoek 98: 389–394.

56. Weon HY, Son JA, Yoo SH, Hong SB, Jeon YA, et al. (2009) Rubellimicrobium aerolatum sp. nov., isolated from an air sample in Korea. Int J Syst Evol Microbiol 59: 406–410.

57. Koberl M, Muller H, Ramadan EM, Berg G (2011) Desert farming benefits from microbial potential in arid soils and promotes diversity and plant health. PLoS One 6: e24452.

58. Carignan V, Villard MA (2002) Selecting indicator species to monitor ecological integrity: A review. Environ Monit Assess 78: 45–61.

59. Dale VH, Beyeler SC (2001) Challenges in the development and use of ecological indicators. Ecol indic 1: 3–10.

7

Wheat Seedling Emergence from Deep Planting Depths and Its Relationship with Coleoptile Length

Amita Mohan, William F. Schillinger, Kulvinder S. Gill*

Department of Crop and Soil Sciences, Washington State University, Pullman, Washington, United States of America

Abstract

Successful stand establishment is prerequisite for optimum crop yields. In some low-precipitation zones, wheat (*Triticum aestivum* L.) is planted as deep as 200 mm below the soil surface to reach adequate soil moisture for germination. To better understand the relationship of coleoptile length and other seed characteristics with emergence from deep planting (EDP), we evaluated 662 wheat cultivars grown around the world since the beginning of the 20th century. Coleoptile length of collection entries ranged from 34 to 114 mm. A specialized field EDP test showed dramatic emergence differences among cultivars ranging from 0–66% by 21 days after planting (DAP). Less than 1% of entries had any seedlings emerged by 7 DAP and 43% on day 8. A wide range of EDP within each coleoptile length class suggests the involvement of genes other than those controlling coleoptile length. Emergence was correlated with coleoptile length, but some lines with short coleoptiles ranked among the top emergers. Coleoptiles longer than 90 mm showed no advantage for EDP and may even have a negative effect. Overall, coleoptile length accounted for only 28% of the variability in emergence among entries; much lower than the 60% or greater reported in previous studies. Seed weight had little correlation with EDP. Results show that EDP is largely controlled by yet poorly understood mechanisms other than coleoptile length.

Editor: Randall P. Niedz, United States Department of Agriculture, United States of America

Funding: This research was funded by Washington State University, the Washington Grain Commission, and the National Science Foundation (NSF-BREAD) grant # 0965533 (http://www.nsf.gov/funding/pgm_summ.jsp?pims_id = 503285). The funders had no role in study design, data collection and analysis, decision to publish, or preparation of the manuscript.

Competing Interests: The authors have declared that no competing interests exist.

* E-mail: ksgill@wsu.edu

Introduction

Deep planting of wheat into stored soil moisture is practiced in Mediterranean-like climate regions, including the Pacific Northwest (PNW) of the United States, parts of western and southwestern Australia, central Chile, and several countries surrounding the Mediterranean Sea. Winter wheat in the low-precipitation (<300 mm annual) region of the PNW is planted 100 to 200 mm below the surface of summer-fallowed soils with deep-furrow drills to reach adequate moisture for seedling emergence [1]. Introduction of high-yielding semi-dwarf wheat cultivars in the early 1960s spawned the "green revolution". Height reduction in semi-dwarf cultivars was due to mutations in *Rht-B1* (*Rht1*) or *Rht-D1* (*Rht2*) genes that reduce either the production or the perception of gibberellin (GA), an important plant growth hormone [2]. Semi-dwarf wheat cultivars are lodging resistant, thus allowing application of higher fertilizer inputs, and have improved harvest index compared to standard-height cultivars due to increased partitioning of assimilates to reproductive organs resulting in more fertile florets per spikelet [3–5]. However, semi-dwarf cultivars are not popular among farmers in some low-precipitation Mediterranean regions because of their relatively poor and/or slow emergence from deep planting depths [6–8]. Decreased response to endogenous GA in *Rht-B1b* and/or *Rht-D1b* gene (GA-insensitive) mutants results in reduced cell size and elongation with corresponding reduction in coleoptile length, plant height, and leaf area [9–13]. Compared to GA-insensitive mutants, GA-responsive mutants (*Rht8*) are reported to reduce

plant height without reducing coleoptile length [9,14] and thus have less negative effect on emergence.

Wheat seedling emergence does not have linear relationship with coleoptile length as many other factors have been implicated in the process. Emergence and coleoptile length is reported to be influenced both by genetic background and environmental factors including soil texture, seed-zone water content, temperature, light penetration, and crop residue [15–20]. Increased seed size has been reported to positively affect early seedling vigor in many crop species [21–23], although about an equal number of reports argue against this correlation [18,24]. Similarly, reports are mixed on the effect of seed weight on coleoptile length of wheat, with no effect reported for semi-dwarfs but a positive effect on some tall durum lines [25].

The pleiotropic effects associated with GA insensitive mutants on early growth have been reported and coleoptile length was positively correlated with plant height at maturity [6,8,26]. The major phenotypic effect quantitative trait loci (QTL) for coleoptile length were mapped on the short arm of 4B and 4D chromosomes in the close proximity of dwarfing genes, thus accounting for the pleotropic effect of dwarfing genes on coleoptile length [27]. Moreover, small-effect QTLs with additive effects were identified on wheat chromosomes 1A, 2B, 2D, 3B, 3A, 5A, 6A [27,28]. Further, genetic background of the cultivar and environmental conditions influence the relation between coleoptile length with plant height. This suggests the possibility to select for a longer coleoptile in reduced plant height genotypes [6], but more detailed systematic study is needed to understand the mechanism behind the relationship of coleoptile length, emergence, and plant height.

Our study was undertaken to gain a better understanding of wheat seedling emergence from deep planting depths and its relation to coleoptile length along with plant height, seed weight, and size. Another objective of the study was to identify the fastest and best emerging wheat cultivar to serve as a donor to improve the emergence trait in new and existing PNW cultivars. We used a population of 662 wheat cultivars from the major wheat growing areas of the world to capture variation present among cultivars. The collection included both winter and spring type habit and represented all major market classes of wheat.

Materials and Methods

Ethics Statement

Field experiments were conducted at the Washington State University (WSU) Dryland Research Station located near Lind, WA. No special permissions were required to conduct the experiment. The field studies did not involve any endangered or protected species. The wheat germplasm used in the study was obtained from the Germplasm Resources Information Network (GRIN), CIMMYT, and from WSU historical collection. These are open access seed sources available for research and education purposes.

Germplasm

We collected 662 wheat cultivars grown around the world; 360 with winter and 302 with spring growth habit (Table S1). The collection included 96 cultivars from the PNW, 125 from a historical US wheat collection including cultivars released since 1871, and 80 from the International Maize and Wheat Improvement Center (CIMMYT), Mexico representing the major wheat growing regions of the world. The remaining 361 entries were US cultivars collected from major wheat growing regions (outside the PNW) that have been widely grown since 1950. The most popular cultivars during each 5-year interval since 1950 from all major breeding programs in the US were included. Our intention was to capture as much variation among cultivated wheat as possible without making the population size unmanageably large. The world collection also represented different market classes depending upon hardness, color, and kernel shape. The collection consisted of 112 soft white winter (SWW), 90 soft white spring (SWS), 101 soft red winter (SRW), 35 soft red spring (SRS), 97 hard red winter (HRW), 127 hard red spring (HRS), 9 hard white winter (HWW), 57 hard white spring (HWS), and 34 club wheat cultivars. Due to concerns about seed purity and vigor, a single seed of each entry was grown in the greenhouse to obtain single-plant seeds. The harvested single-plant seed was then multiplied in the greenhouse and used for all the experiments.

Coleoptile Length Measurement

For coleoptile length measurement, 15 uniform-sized seeds of each cultivar with no physical damage were placed in the middle of a moist germination paper (Heavy Germination paper #SD 7615L), about one centimeter apart with germ end down. The germination paper was then folded vertically in half with the seed placed in the crease, the folded half was again folded horizontally four times and placed in a plastic tray with holes at the base to drain excess water. The plastic trays were then placed inside a completely darkened box and kept in a growth chamber at a constant temperature of 22°C. After 10 days, the average coleoptile length of 10 randomly-selected seedlings was recorded to the nearest millimeter measuring from the base of the seed to

the coleoptile tip. Germination percentage of all the lines was recorded.

Field Test for Emergence from Deep Planting

Emergence from deep planting was conducted at the WSU Dryland Research Station near Lind, WA during the 2009–2010 crop season. The experiment was planted on land that had been fallowed for 13 months (i.e., since the last wheat harvest). Soil type was Shano silt loam (coarse-silty, mixed, superactive, mesic Xeric Haplocambids) with a soil textural size distribution of 10% clay, 51% silt, and 39% fine sand. Slope was <2%. Soil volumetric water content in the seed zone was determined gravimetrically at time of planting on September 3. These measurements were obtained in 2-cm increments avoiding wheel tracks, to a depth of 24 cm using an incremental soil sampler. Mean water content for each increment was determined from four soil cores.

For each entry, 50 uniform-sized seeds with no physical damage were planted using a four-opener deep-furrow drill in 3-m-long rows with a 0.38-m spacing between rows. The experimental design was randomized complete block with four replicates per entry. To account for any differences among openers, each entry was planted once with each of the four openers.

Seeds were planted 150 mm below the soil surface into a seed zone having a water potential of −0.50 MPa (11.5% water by volume). The depth of the soil covering the seeds was 125 mm. Wheat seedling emergence was determined by counting individual seedlings at 24-hour intervals. Seedling emergence counts were obtained on 6, 7, 8, 9, 10, 15, and 21 DAP. No rainfall occurred at the site from time of planting through 21 DAP.

Plant Height, Seed Weight and Seed Size

Plant height of each entry was recorded at maturity by measuring the average distance from the soil surface to the top of the spike (excluding awns) of the three tallest plants. Several entries (mainly spring types) were winterkilled and, therefore, plant height could obviously not be obtained. Thousand kernel weight (TKW) for each entry was determined using electronic counter (Oldmill Company, Model 850.2) and then weighing on a digital scale. Seed size and grain hardness was determined using 200 seeds of each entry with the Single-Kernel Characterization System (Perten Instruments, Springfield, IL, Model 4100).

Statistical Analysis

Data analysis was conducted using the PROC GLM procedure of Statistical Analysis System (SAS; SAS Inst. 2000). Means and standard deviations were determined using PROC MEANS in SAS. Analysis of variance was performed using PROC MIXED (SAS Institute, 2000). Pearson's linear correlation coefficients were calculated by PROC CORR. Significance level was tested at $P<0.05$. Graphical representation of data was conducted using JMP genomics (JMP, Version 10. SAS Institute Inc., Cary, NC, 1989–2007).

Results

Coleoptile Length Variation

Mean coleoptile length among entries ranged from 34 mm to 114 mm (Figure 1A–B). The vast majority of entries had a mean coleoptile length between 41 to 90 mm. Only 10 of the 662 lines had coleoptiles shorter than 41 mm and 40 lines had coleoptiles longer than 90 mm (Figure 1A). The median coleoptile length was 62.4 mm. Among the nine coleoptile length intervals made in 10 mm increments, the interval 51 to 60 mm accounted for the largest number of entries (168). Coleoptile length of spring wheat lines ranged from 37 to 114 mm with a median of 62 compared to the winter lines with a range of 34 to 103 mm and a median of 63. Among various grain market classes, mean coleoptile length was the highest (71) for club wheat and the lowest for white classes (HWW, HWS, SWW), having a mean coleoptile length of 60 mm. The mean coleoptile length was 68 mm for SWS, 66 mm for both SRW and HRW, 65 mm for SRS and 64 mm for the HRS classes.

Variation in Emergence from Deep Planting Depth

The fastest emergence was recorded in four cultivars (Spinkcota, Regent, Pacific Bluestem, ARS95-452) that had an average of at least two emerged seedlings on 7 DAP. By 8 DAP, 43% of the entries had an average of at least one emerged seedling, and this was 96% on 9 DAP. No differences were observed for EDP between spring and winter growth habit. By 21 DAP, 658 lines (99.4%) had some emergence while four entries (0.6%) did not emerge at all. Of the entries that emerged, the extent of emergence of total seeds planted ranged from 4% to 66% by 21 DAP. Emergence in some lines was fast and uniform (e.g., Spinkcota, Indian, White Federation, and Barbee) and was completed within 2–3 days after initial emergence, whereas emergence of other lines (e.g., Golden Cross, Triumph 64 and 90451ARS) was gradual and extended over many days. Eleven lines showed >60% emergence by 21 DAP. The four entries with no emergence at all had coleoptile lengths ranging from 53 to 85 mm (Table 1).

The percent emergence at 10 DAP among coleoptile length classes ranged from 14% to 38%. There was a gradual increase in emergence percentage with increasing coleoptile length up to 90 mm, beyond which there were no statistical differences (Figure 2A). In general, emergence decreased for entries with coleoptile length >110 mm, although this interpretation is based only on two entries. The best early emergers on 7 DAP had coleoptiles ranging from 50–110 mm with no significant improvement in emergence above 60 mm (data not shown). The emergence trend was very similar at 15 and 21 DAP with no significant improvement in emergence of entries with coleoptiles longer than 90 mm (Figure 2B–C). Mean emergence percentage among entries on 10 and 15 DAP ranged from 2 to 62% and on 21 DAP from 4% to 66%. Overall, five of the nine coleoptile

Table 1. Coleoptile length (CL), thousand kernel weight (TKW) and emergence percentage of selected cultivars in the world wheat collection having maximum and minimum emergence from deep planting on 15 and 21 days after planting.

Entry Name	CL (mm)	TKW (gm)	% Emergence	
			DAY15	DAY21
Klein Dragon	51.0	26	52	66
Fortuna	90.8	33.8	56	64
Canadian Red	99.1	37.2	58	64
Luft	82.3	37.2	58	64
Sonora	106.3	29.8	60	62
Okla. 61STW8637	60.5	33.3	50	62
J961051	96.7	33.5	46	62
Bounty 309	46.8	39.1	4	8
WA 64250	49.0	32.0	2	4
Chinook	74.0	39.7	0	0
Blizzard	85.4	38.7	0	0
Newton	53.4	39.0	0	0

length classes differed significantly in their mean percent emergence (Figure 2A–C).

Within each of coleoptile length class, the range of emergence percentage was wide and variable (Figure 2D–F). The entry with the greatest emergence (66%) had a coleoptile length of only 52 mm. The minimum emergence in this (51–60 mm) coleoptile length class was 12%. The emergence percentage of entries with coleoptiles >100 mm (average coleoptile length of 112 mm) was 46% and ranged from 36% to 62%. The within-coleoptile class range was wider at 10 DAP compared to that at 21 DAP, suggesting that final emergence is better correlated with coleoptile

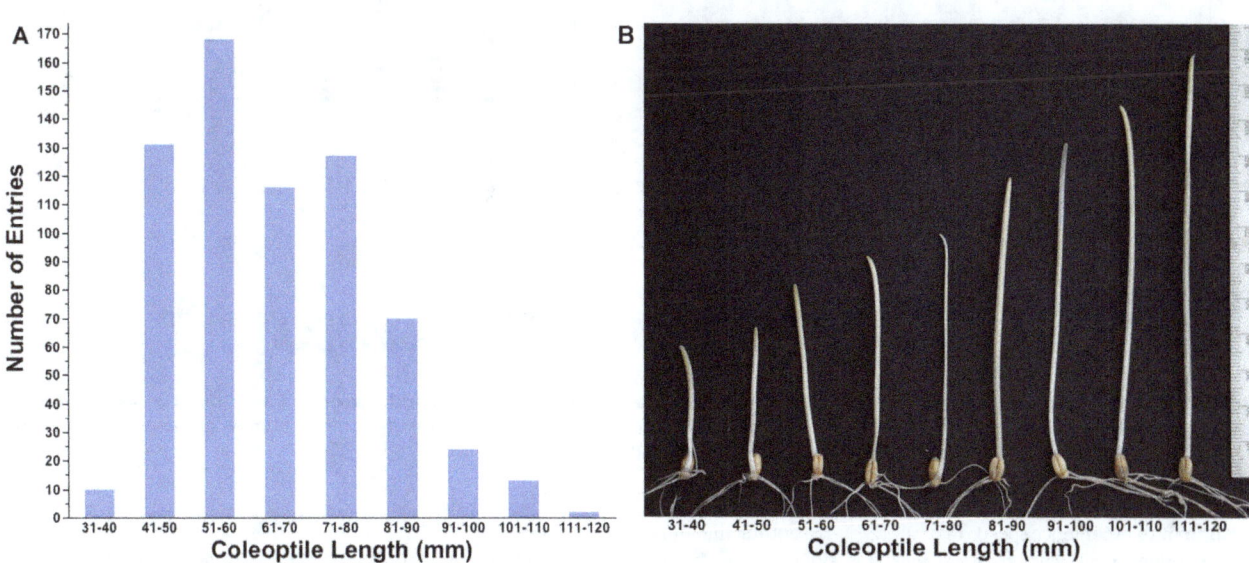

Figure 1. Coleoptile length distribution among 662 entries of the world wheat collection in 10-mm intervals (A). Representative picture of coleoptiles in each 10-mm increment class interval (B).

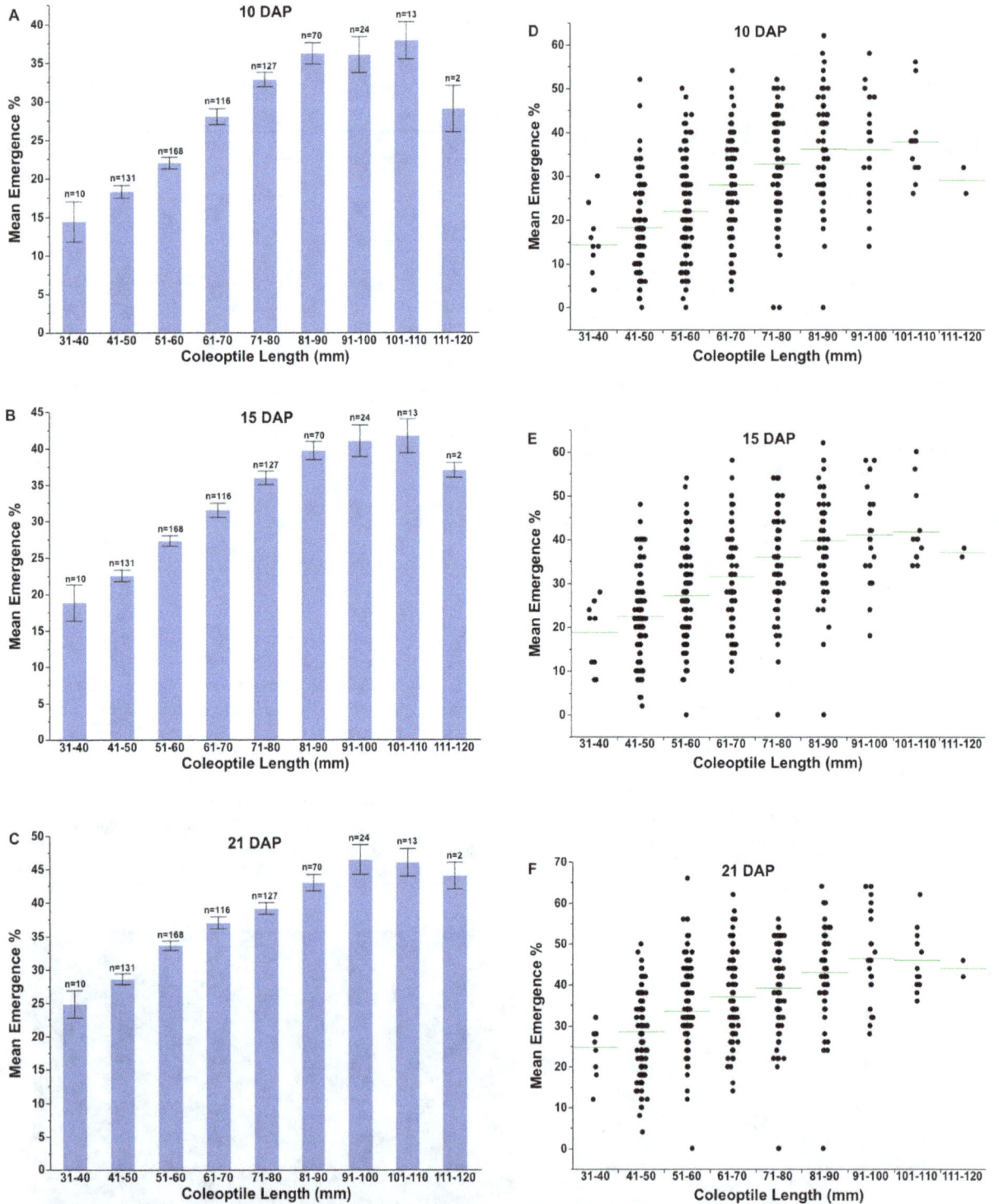

Figure 2. Mean emergence percentage among various coleoptile length classes (A–C) and, range of variation in coleoptile length classes on 10, 15, and 21 days after planting (D–F). The error bar represents the standard error of the mean. Number of lines in each coleoptile length class (n) is given above each bar. The horizontal line represents the mean in each coleoptile length class (D–F).

length than to emergence speed (Figure 2E–F). Except for the four non-emerging entries that spanned four coleoptile length classes with the highest being 80 to 90 mm, minimum emergence in each coleoptile length class gradually increased with the increasing coleoptile length (Figure 2D–F). The greatest relative emergence

improvements occurred for the three shortest coleoptile classes (31–40, 41–50, and 51–60 mm) and then leveled off for the next four classes (Figure 2D–F). Maximum emergence for the two longest coleoptile length classes gradually decreased from 64% in 91–100 mm class to 46% in the 111–120 mm class (Figure 2D–F).

Correlation Between Emergence and Coleoptile Length

There was an overall positive correlation between coleoptile length and emergence from deep planting (Figure 3A–B). On 8 DAP, correlation between emergence and coleoptile length was weak ($r^2 = 0.12$, p<0.0001), but somewhat improved on 9 DAP ($r^2 = 0.23$, p<0.0001). Coefficients of determination (r^2) for the relation of coleoptile length with emergence on 10, 15, and 21 DAP were 0.28, 0.29, and 0.23 respectively (p<0.0001) (Figure 3A, data for 10 and 21 DAP not shown).

Coefficients of determination of emergence with coleoptile length were similar among market classes and, except for the HWW class where there was no significant correlation (p = 0.1) on any DAP. Club wheat showed the highest coefficients of determination (p<0.001) on 10, 15 and 21 DAP ($r^2 = 0.43$, 0.43, 0.34, respectively).

Kernel Weight Versus Seedling Emergence

Thousand kernel weight (TKW) of the world wheat collection cultivars ranged from 15.0 to 47.4 g and averaged 32.0 g. Mean TKW for winter and spring cultivars was 33.2 and 30.8 g, respectively. There was no overall correlation between coleoptile length and TKW for spring and winter type and all the market classes except for HWS (Figure 4 A–B). The TKW of coleoptile class 31–40 and 110–120 mm entries differed significantly compared to other coleoptile classes (Figure 5A). The widest variability in TKW occurred in coleoptile classes between 41–80 mm (Figure 5B). Although, the coefficient of determination for TKW and seed size was 0.27 (p<0.001), seed size had negligible correlation with coleoptile length ($r^2 = 0.01$, p<0.01). The weak correlation held true for both winter and spring types and all market classes except for HWS, which showed the most positive correlation with coleoptile length ($r^2 = 0.22$, p<0.001). Thousand kernel weight did not affect emergence on any DAP nor was it associated with plant height at maturity (data not shown).

Plant Height

Plant height at maturity of 357 winter and two spring entries ranged from 46 to 116 cm with an average height of 82 cm. In general, plant height was positively correlated with coleoptile length ($r^2 = 0.25$, p<.001, Figure 6). There were no differences in plant height among entries with coleoptiles longer than 80 mm (Figure 7A). The tallest entry had a height of 116 cm with a coleoptile length of 74 mm. The shortest entry had a plant height of 46 cm with a coleoptile length of 54 mm. There was wide variability in plant height within each coleoptile length class (Figure 7B).

Discussion

Variation among wheat cultivars for coleoptile length and its correlation with seedling emergence has been reported from several authors [13,29,30] but the number of cultivars evaluated in those studies was relatively small. Here, we present a comprehensive analysis of variation among 662 wheat cultivars from major wheat growing regions around the world and provide a detailed analysis of the correlation between coleoptile length and seedling emergence.

Our wheat collection was carefully selected to capture the variation present in cultivated wheat while minimizing redundancy. We observed unprecedented variation in the population for the three traits studied. With a range from 34 mm to 114 mm, our collection had more than a three-fold difference in coleoptile lengths. Similarly, variation for seedling emergence from deep planting was high with some lines showing some emergence as early as 7 DAP and four lines with no emergence even on 21 DAP. Variation for seedling emergence was related to speed, extent, and uniformity of final plant stand.

It has previously been reported that the commonly used GA-insensitive semi-dwarf gene mutations (RhtB1b and RhtD1b) reduce coleoptile length by 30–40% compared to standard-height cultivars [15]. The GA insensitivity reduces cell wall extensibility, thus reduces plant height and also negatively affects growth of the coleoptile and first leaf [10,31]. Coleoptile length variation among the CIMMYT lines in our collection was very narrow with the majority having a coleoptile length between 40–60 mm. This narrow variation is probably due to the semi-dwarf nature of all the CIMMYT lines and/or due to a common genetic background. The CIMMYT lines had an average 32% (ranging from 12–54%) of seedlings emerged on 21 DAP. In addition to genetics, other factors including soil temperature, texture and moisture may also influence coleoptile length [20,25,32,33].

Emergence from deep planting is particularly important in the low-precipitation dryland regions of the world, such as the inland PNW, where wheat is planted as deep as 200 mm with deep-furrow drills and seedlings must emerge through 150 mm or more

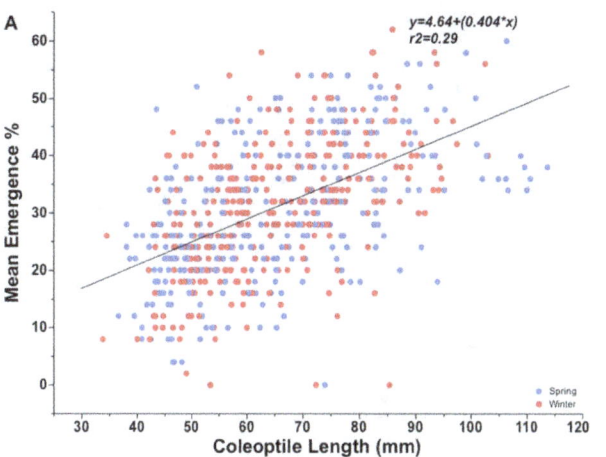
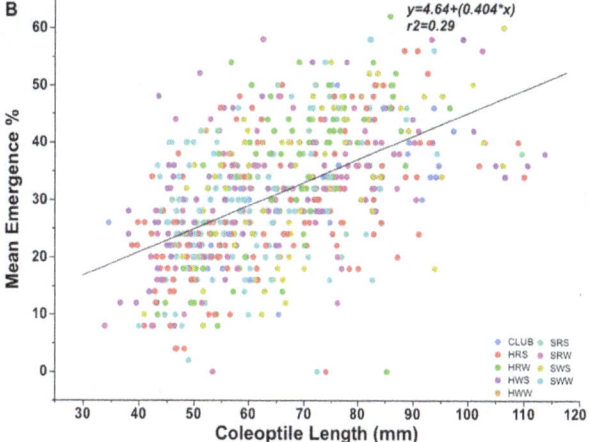

Figure 3. Mean emergence percentage of the (A) spring and winter type and (B) different market class of wheat world collection entries plotted against coleoptile length on 15 days after planting.

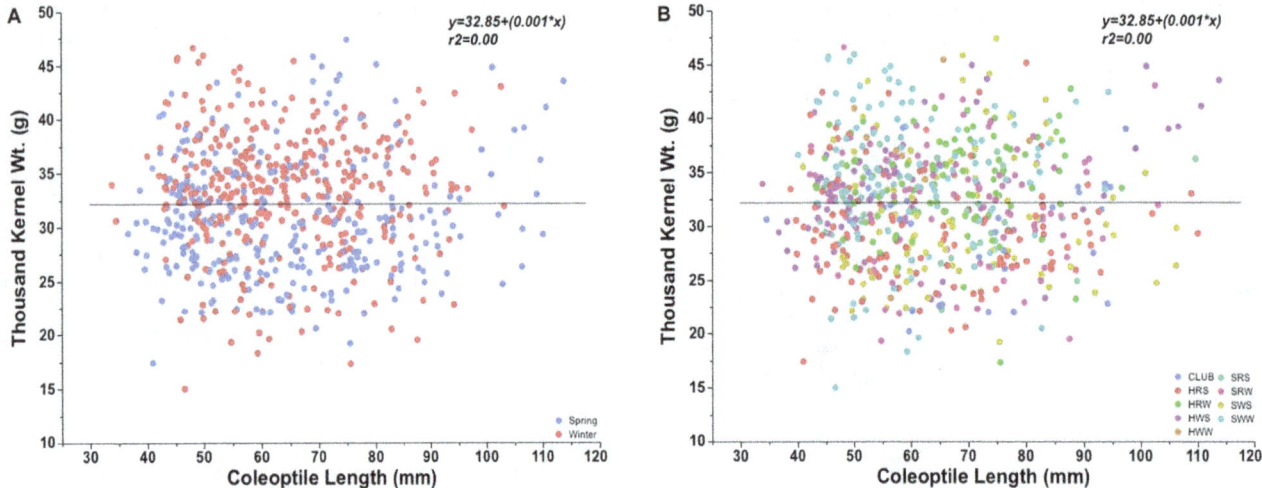

Figure 4. Thousand kernel weight of (A) spring and winter type wheat and (B) different market classes of world wheat collection entries plotted against coleoptile length.

soil cover. The coleoptile protects the emerging leaf (i.e., until the first leaf protrudes through the tip of the coleoptile while still below the soil surface) and also provides the mechanical support required for seedling emergence. Coleoptile length has previously been reported to be important for seedling emergence. In the present study, coleoptile length was significantly correlated with seedling emergence but accounted for only 28% of the variability for emergence on 10 and 15 DAP (Figure 2A and B). The emergence percentage of entries in coleoptile class 31–40 mm (25%) on 21 DAP was significantly lower than any other class (Figure 3C) with a gradual increase in mean emergence percentage up to a coleoptile length of 90 mm. Comparing various coleoptile length classes, emergence increased more dramatically with increased coleoptile length at 10 DAP compared to that at 21 DAP (Figure 3A and C). Our linear regression coefficients of determination for the relationship of coleoptile length to emergence were considerably lower compared to previous reports

[6,8,13,29]. Careful evaluation of present data suggests that coleoptile length is not as critical for seedling emergence as previously thought. Our five best emerging cultivars had coleoptile length range from 51–106 mm. Further, a wide range of emergence was observed within each of the coleoptile length classes. The lower value of emergence decreased with decreasing coleoptile length, suggesting an underlying effect of coleoptile length on emergence, but the maximum emergence values did not increase with coleotiptiles longer than 90 mm, suggesting that coleoptile lengths longer than 90 mm may not be beneficial for emergence.

Several entries in our study with relatively short coleoptiles emerged well from deep planting while other entries with longer coleoptiles had poor emergence. This difference is not due to uneven germination as during the coleoptile length measurement experiment all lines germinated well. Except for the shortest class, each coleoptile length class showed a wide range of seedling

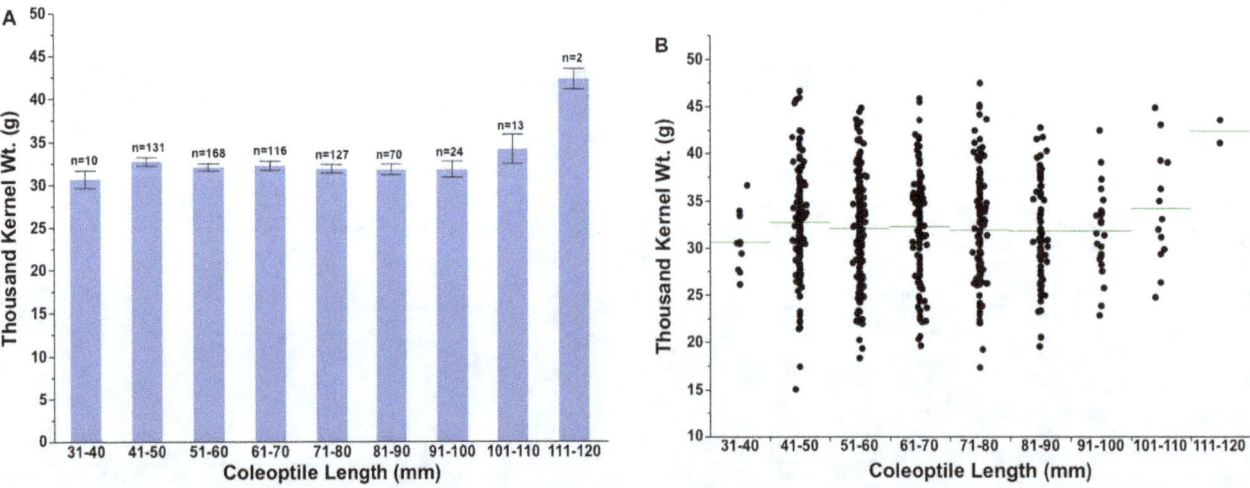

Figure 5. Thousand kernel weight distribution (A) and range variation (B) among different coleoptile length classes. Bars represent the standard error of the mean. The sample size (n) of each coleoptile class length is provided above the data bars. The horizontal line represents the mean in each coleoptile class (B).

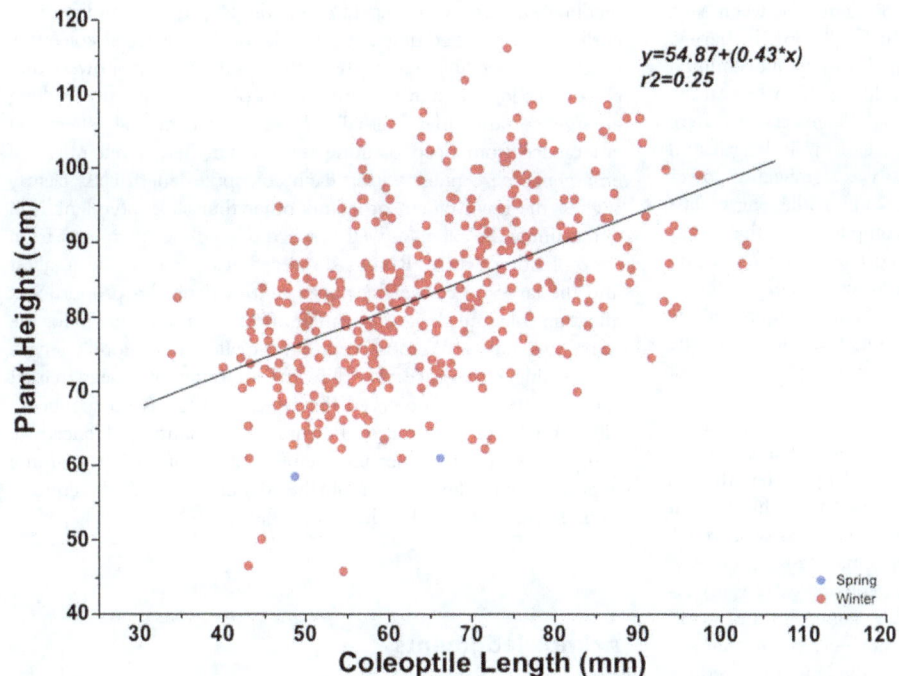

$y=54.87+(0.43*x)$
$r2=0.25$

Figure 6. Mean plant height at maturity of spring and winter type cultivars in the world wheat collection entries plotted against coleoptile length.

emergence, suggesting that coleoptile length is not as strongly correlated to seedling emergence as previously reported, thus making emergence a complex trait. Similar observations were made in barley (*Hordeum vulgare L.*) where lines with widely different coleoptile lengths showed similar seedling emergence [34]. Another reason for range of coleoptile length variation within the coleoptile class might be genotype parentage which affects coleoptile length [35]. For example, club wheat had higher mean coleoptile length among the different market classes studied.

Lower dependence of emergence on coleoptile length in comparison to previous reports may partly be due to our much larger and highly diverse material providing increased precision to study this correlation. Another explanation for the weaker correlation could be that previous studies used wheat lines with limited geographic diversity, whereas our wheat lines collected from around the world may respond differently under different environmental conditions.

In general, seed weight had no effect on coleoptile length or seedling emergence. This is in agreement with most other studies,

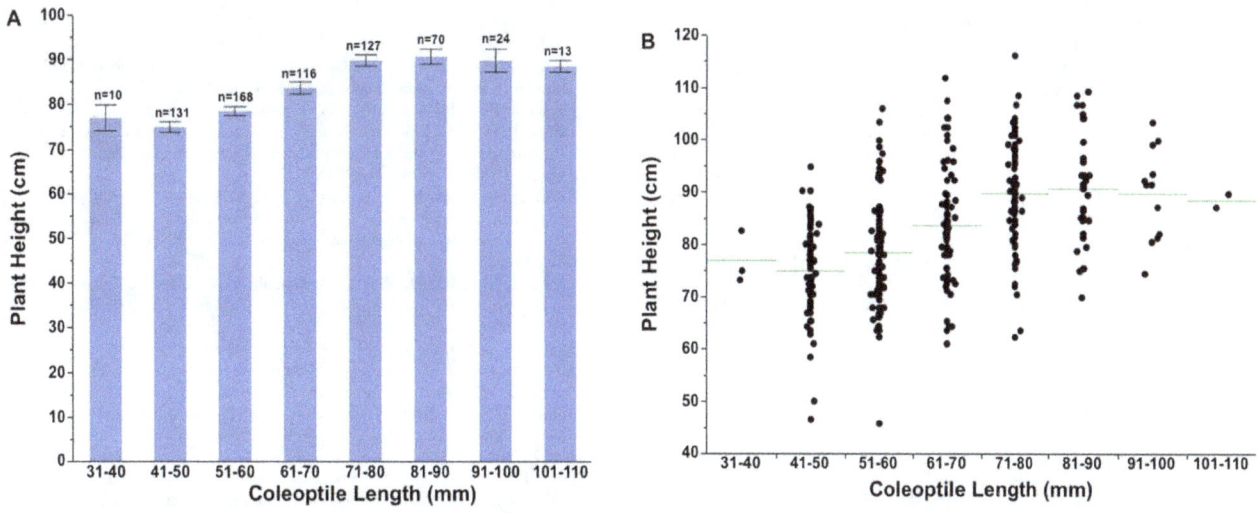

Figure 7. Mean (A) and range variation (B) in plant height at maturity in different coleoptile length classes of the world wheat collection. The bars are the standard error of the mean. The sample size (n) of each coleoptile class is provided above the data bars. The horizontal line represents the mean in each coleoptile class (B).

but there are some reports of a strong correlation between seed weight/size and coleoptile length in wheat [22,35,36]. In barley, seed weight had no effect on coleoptile length though it accounted for 31–53% of variability in coleoptile width [37]. In our study, except for the lightest and the heaviest seed weight classes, there were essentially no emergence differences. Coleoptile length and emergence of the lightest seed weight class was significantly lower than the other classes. Similarly, the highest coleoptile length class had significantly higher seed weight compared to the other coleoptile classes. In previous studies reported in the literature, essentially no effect of seed weight was reported on emergence or yield [18,20,24]. In our study the diverse genetic nature of the entries in each coleoptile length class might account for the differences in the coleoptile length and emergence within a particular seed weight range.

Coleoptile length showed a positive overall correlation with final plant height in our study (Figure 6). These results are in accordance with the earlier reports [38]. Coleoptile length and plant height are polygenic traits and many QTLs have been reported that influence traits [26,28,39–41]. We also observed a full range of plant heights in particular coleoptile class e.g. cultivar with short plant height (29 cm) with a long coleoptile (92 mm). As we do not know the nature of dwarfing genes in the world collection, these could possibly be GA sensitive genotypes that are reported to have no adverse effect on coleoptile length but reduce plant height, or possibly have favorable alleles for coleoptile length in semi-dwarf background [41]. In general, plant height was positively correlated with emergence at 10, 15, and 21 DAP. There was a general trend of increased plant height up to 90 mm coleoptile length, but no differences in the mean plant height for entries with coleoptiles longer than 90 mm.

Conclusions

A collection of 662 carefully selected wheat cultivars from around the world representing a full range of variation for the studied traits allowed us to gain improved understanding of the relation between coleoptile length and emergence from deep planting depths. Refuting the previous claims of coleoptile length explaining 60% or more of the variation for seedling emergence, our extensive study explains only 28% of the variability for

seedling emergence related to coleoptile length. Contradictory to earlier reports suggesting a linear relationship between coleoptile length and seedling emergence, our study clearly showed that coleoptiles longer than 90 mm has no positive affect on seedling emergence and may actually have a detrimental effect on emergence from deep planting depths. Further, a full range of emergence percentage within each coleoptile length class clearly suggest the involvement of factors other than coleoptile length in determining the extent as well as speed of seedling emergence from deep planting depths. Results showed that except for the smallest and the largest seed weight classes that evidenced a proportional affect on coleoptile length, a similar affect was not observed for the remaining classes. We observed that the effect of coleoptile length on seedling emergence was different for different wheat market classes. Further, we observed that seed weight has no significant affect on seedling emergence from deep planting. Hence, we conclude that wheat seedling emergence from deep planting depths is a complex trait and further detailed studies are required to understand the underlying mechanisms.

Acknowledgments

The authors gratefully acknowledge the excellent technical support provided by Timothy Smith and Steven Schofstoll at the WSU Dryland Research Station at Lind and by Patrick Reisenauer in collecting coleoptile data. We also thank WSU graduate students Amandeep Dhaliwal and Ramanjot Kaur for their help with the field emergence experiment at Lind and Maninder Chahal with statistical analysis.

Author Contributions

Conceived and designed the experiments: KSG WFS AM. Performed the experiments: AM KSG WFS. Analyzed the data: AM KSG. Contributed reagents/materials/analysis tools: KSG WFS. Wrote the paper: AM KSG WFS.

References

1. Schillinger WF, Papendick RI (2008) Then and now: 125 years of dryland wheat farming in the Inland Pacific Northwest. Agronomy Journal 100: S–166–S–182. doi:10.2134/agronj2007.0027c.
2. Peng J, Richards DE, Hartley NM, Murphy GP, Devos KM, et al. (1999) "Green revolution" genes encode mutant gibberellin response modulators. Nature 400: 256–261. doi:10.1038/22307.
3. Brooking I, Kirby E (1981) Interrelationships between stem and ear development in winter wheat: the effects of a Norin 10 dwarfing gene, Gai/Rht2. Journal of Agricultural Science 97: 373–381.
4. Evans L (1993) Crop evolution, adaptation, and yield. Cambridge; New York: Cambridge University Press.
5. Miralles DJ, Slafer GA (1995) Yield, biomass and yield components in dwarf, semi-dwarf and tall isogenic lines of spring wheat under recommended and late sowing dates. Plant Breeding 114: 392–396. doi:10.1111/j.1439-0523.1995.tb00818.x.
6. Sunderman DW (1964) Seedling emergence of winter wheats and its association with depth of sowing, coleoptile length, and plant height. Agronomy Journal 56: 23–25.
7. Feather JT, Qualset CO, Vogt HE (1968) Planting depth critical for short-statured wheat varieties. California Agriculture: 12–14.
8. Fick G, Qualset C (1976) Seedling emergence, coleoptile length, and plant height relationships in crosses of dwarf and standard-height wheats. Euphytica 25: 679–684.
9. Ellis M, Rebetzke G, Chandler P, Bonnett D (2004) The effect of different height reducing genes on the early growth of wheat. Functional Plant 31: 583–589.
10. Keyes G, Sorrells M, Setter T (1990) Gibberellic acid regulates cell wall extensibility in wheat (Triticum aestivum L.). Plant Physiology 92: 242–245.

11. Hoogendoorn J, Rickson JM, Gale MD (1990) Differences in leaf and stem anatomy related to plant height of tall and dwarf wheat (Triticum aestivum L.). Journal of Plant Physiology 136: 72–77. doi:10.1016/S0176-1617(11)81618-4.
12. Matsui T, Inanaga S, Shimotashiro T (2002) Morphological characters related to varietal differences in tolerance to deep sowing in wheat. Plant Production Science 5: 169–174.
13. Rebetzke GJ, Richards RA, Fettell NA, Long M, Condon AG (2007) Genotypic increases in coleoptile length improves stand establishment, vigour and grain yield of deep-sown wheat. Field Crop Research 100: 10–23. doi:10.1016/j.fcr.2006.05.001.
14. Rebetzke GJ, Richards RA (1999) Genetic improvement of early vigour in wheat. Australian Journal of Agricultural Research 50: 291. doi:10.1071/A98125.
15. Allan R, Vogel O, Burleigh J (1962) Length and estimated number of coleoptile parenchyma cells of six wheat selections grown at two temperatures. Crop Science 2: 522–524.
16. Nebreda IM, Parodi PC (1977) Effect of seed type on coleoptile length and weight in triticale, X Triticosecale Wittmack. Cereal Research Communications 5: 387–398.
17. Jessop RS, Stewart LW (1983) Effects of crop residues, soil type and temperature on emergence and early growth of wheat. Plant and Soil 74: 101–109.
18. Chastian T, Ward K, Wysocki D (1995) Stand establishment response of soft white winter wheat to seedbed residue and seed size. Crop science 35: 213–218.
19. Rebetzke G, Richards R (1999) Breeding long coleoptile, reduced height wheats. Euphytica 106: 159–168.

20. Botwright TL, Rebetzke GJ, Condon AG, Richards RA (2001) Influence of variety, seed position and seed source on screening for coleoptile length in bread wheat (*Triticum aestivum* L.). Euphytica 119: 349–356.

21. Richards R, Lukacs Z (2002) Seedling vigour in wheat: sources of variation for genetic and agronomic improvement. Australian Journal of Agricultural Research 53: 41–50.

22. Nik MM, Babaeian M, Tavassoli A (2011) Effect of seed size and genotype on germination characteristic and seed nutrient content of wheat. Scientific Research and Essays 6: 2019–2025.

23. Lafond G, Baker R (1986) Effects of genotype and seed size on speed of emergence and seedling vigor in nine spring wheat cultivars. Crop Science 26: 341–346.

24. Mian AR, Nafziger ED (1992) Seed size effects on emergence, head number, and grain yield of winter wheat. Journal of Production Agriculture 5: 265–268.

25. Trethowan R, Singh R, Huerta-Espino J, Crossa J, Van Ginkel M (2001) Coleoptile length variation of near-isogenic Rht lines of modern CIMMYT bread and durum wheats. Field Crops Research 70: 167–176.

26. Liatukas Z, Ruzgas V (2011) Coleoptile length and plant height of modern tall and semi-dwarf European winter wheat varieties. Acta Societatis Botanicorum Poloniae 80: 197–203. doi:10.5586/asbp.2011.018.

27. Rebetzke G, Ellis M (2007) Molecular mapping of genes for coleoptile growth in bread wheat (*Triticum aestivum* L.). Theoretical and Applied Genetics 114: 1173–1183. doi:10.1007/s00122-007-0509-1.

28. Spielmeyer W, Hyles J, Joaquim P, Azanza F, Bonnett D, et al. (2007) A QTL on chromosome 6A in bread wheat (*Triticum aestivum*) is associated with longer coleoptiles, greater seedling vigour and final plant height. Theoretical and Applied Genetics 115: 59–66. doi:10.1007/s00122-007-0540-2.

29. Schillinger WF, Donaldson E, Allan RE, Jones SS (1998) Winter Wheat Seedling Emergence from Deep Sowing Depths. Agronomy Journal 90: 582–586. doi:10.2134/agronj1998.00021962009000050002x.

30. Rebetzke GJ, Bruce SE, Kirkegaard JA (2005) Longer coleoptiles improve emergence through crop residues to increase seedling number and biomass in wheat (*Triticum aestivum* L.). Plant and Soil 272: 87–100. doi:10.1007/s11104-004-4040-8.

31. Tonkinson CL, Lyndon RF, Arnold GM, Lenton JR (1995) Effect of the *Rht3* dwarfing gene on dynamics of cell extension in wheat leaves, and its modification by gibberellic acid and paclobutrazol. Journal of Experimental Botany 46: 1085–1092. doi:10.1093/jxb/46.9.1085.

32. Pereira MJ, Pfahler PL, Barnett RD, Blount AR, Wofford DS, et al. (2002) Coleoptile length of dwarf wheat isolines: Gibberellic acid, temperature, and cultivar interactions. Crop Science 42: 1483–1487.

33. Radford B, Key A (1993) Temperature affects germination, mesocotyl length and coleoptile length of oats genotypes. Australian Journal of Agricultural Research 44: 677–688.

34. Kaufmann M (1968) Coleoptile length and emergence in varieties of barley, oats, and wheat. Canadian Journal of Plant Science 361: 357–361.

35. Addae P, Pearson C (1992) Variability in seedling elongation of wheat, and some factors associated with it. Australian Journal of Experimental Agriculture 32: 377–382. doi:10.1071/EA9920377.

36. Cornish P, Hindmarsh S (1988) Seed size influences the coleoptile length of wheat. Australian Journal of Experimental Agriculture 28: 521–524.

37. Ceccarelli S, Pegiati M (1980) Effect of seed weight on coleoptile dimensions in barley. Canadian Journal of Plant Science 60: 221–225.

38. Landjeva S, Karceva T, Korzun V, Ganeva G (2011) Seedling growth under osmotic stress and agronomic traits in Bulgarian semi-dwarf wheat: comparison of genotypes with *Rht8* and/or *Rht-B1* genes. Crop Pasture Science 62: 1017–1025.

39. Yu J-B, Bai G-H (2010) Mapping Quantitative Trait Loci for long coleoptile in chinese wheat landrace Wangshuibai. Crop Science 50: 43–50. doi:10.2135/cropsci2009.02.0065.

40. Rebetzke GJ, Appels R, Morrison AD, Richards RA, McDonald G, et al. (2001) Quantitative trait loci on chromosome 4B for coleoptile length and early vigour in wheat (*Triticum aestivum* L.). Crop and Pasture Science 52: 1221–1234. doi:10.1071/AR01042.

41. Wang J, Chapman SC, Bonnett DG, Rebetzke GJ (2009) Simultaneous selection of major and minor genes: use of QTL to increase selection efficiency of coleoptile length of wheat (*Triticum aestivum* L.). Theoretical and Applied Genetics 119: 65–74. doi:10.1007/s00122-009-1017-2.

8

Changes in Soil Carbon and Nitrogen following Land Abandonment of Farmland on the Loess Plateau, China

Lei Deng[1], Zhou-Ping Shangguan[1]*, Sandra Sweeney[2]

1 State Key Laboratory of Soil Erosion and Dryland Farming on the Loess Plateau, Northwest A&F University, Yangling, Shaanxi, China, 2 Institute of Environmental Sciences, University of the Bosphorus, Istanbul, Turkey

Abstract

The revegetation of abandoned farmland significantly influences soil organic C (SOC) and total N (TN). However, the dynamics of both soil OC and N storage following the abandonment of farmland are not well understood. To learn more about soil C and N storages dynamics 30 years after the conversion of farmland to grassland, we measured SOC and TN content in paired grassland and farmland sites in the Zhifanggou watershed on the Loess Plateau, China. The grassland sites were established on farmland abandoned for 1, 7, 13, 20, and 30 years. Top soil OC and TN were higher in older grassland, especially in the 0–5 cm soil depths; deeper soil OC and TN was lower in younger grasslands (<20 yr), and higher in older grasslands (30 yr). Soil OC and N storage (0–100 cm) was significantly lower in the younger grasslands (<20 yr), had increased in the older grasslands (30 yr), and at 30 years SOC had increased to pre-abandonment levels. For a thirty year period following abandonment the soil C/N value remained at 10. Our results indicate that soil C and TN were significantly and positively correlated, indicating that studies on the storage of soil OC and TN needs to focus on deeper soil and not be restricted to the uppermost (0–30 cm) soil levels.

Editor: Ben Bond-Lamberty, DOE Pacific Northwest National Laboratory, United States of America

Funding: The study was funded by the Strategic Priority Research Program of the Chinese Academy of Sciences (XDA05060403) and the Scholarship Award for Excellent Doctoral Student granted by Ministry of Education. The funders had no role in study design, data collection and analysis, decision to publish, or preparation of the manuscript.

Competing Interests: The authors have declared that no competing interests exist.

* E-mail: shangguan@ms.iswc.ac.cn

Introduction

Changes in land use have important effects on regional ecological processes and global climate change [1–2]. During the past two decades, many studies have focused on the effects of land-use change on soil organic carbon (SOC) and total nitrogen (TN) in terrestrial ecosystems [3–5]. The large differences in climatic conditions [6], soil properties [7], and type of land use change [8] are three factors whose effects are not yet well understood.

The revegetation of degraded land is one of the principal strategies for the control of both soil erosion and ecosystem recovery in fragile regions where either anthropogenic activities or severe environmental conditions can lead to disturbance [9]. Revegetation also greatly influences soil quality, C and N cycling, land management, as well as regional socioeconomic development [10–11]. The development of managed grassland and forest accelerates ecosystem restoration [12], affecting C and N cycles and C and N pools stored in soils [10,13]. Altered C and N cycles and C and N pools influence the production of biomass and ecosystem function [14]. Furthermore, C–N interactions are very important in determining whether the C sink in land ecosystems can be sustained over the long term [15–16]. Luo et al. [15] have proposed that N dynamics are a key factor in the regulation of long-term terrestrial C sequestration. N may become progressively more limiting as C accumulates in ecosystems under elevated CO_2 if the total N content in an ecosystem does not change [15]. Therefore, studying the dynamics of organic carbon (OC) and N in soils along a restoration succession gradient and analyzing the

relationships between C and N storage dynamics following restoration may be of importance in improving our knowledge of the sustainable management of land resources and predictions of future global C and N cycling.

Large-scale monocropping and over-grazing [17] have affected the semi-arid northern Loess Plateau in China. In addition, over the last century, the expanding human population, combined with a changing lifestyle, has accelerated ecosystem fragmentation and degradation [9]. To stabilize the fragile natural ecosystems characteristic of the Loess Plateau and to alleviate the degradation of land, the Chinese government has launched a series of nationwide conservation projects. One such project converts degraded farmland to either grassland or forest [18–19] to control soil erosion, increase storage of SOC and N, and prevent the occurrence of soil desiccation on the Loess Plateau [11]. Restoration succession may affect SOC and N decomposition. With regard to these ecosystem functions, the retention of C and N in soil is crucial [20]. This is of particular concern at sites with substantial N saturation, which are becoming increasingly widespread due to elevated atmospheric N deposition.

Previous studies report that the revegetation of degraded land can increase OC and N storages in soil [9,13]. Wang et al. [21] reported that both SOC and TN increase as a linear function of years abandoned. However, many previous studies have focused primarily on the topsoil (0–30 cm) [21–24]. Little is known about long-term changes to SOC and N in the deeper soil layers of the restoration succession on the Loess Plateau. This information can be useful in estimating the temporal distribution of storage of SOC and N and for evaluating OC and N dynamics throughout the conversion from managed to natural communities in semi-arid regions.

The objectives of this work were to investigate changes in SOC and TN concentration, soil OC and N storage and the relationship between SOC and TN with time since abandonment of farmland with depth in the soil profile.

Materials and Methods

Site description

This study was conducted in the Zhifanggou watershed in Ansai County, Shaanxi Province, NW China (36°46′28″–36°46′42″N, 109°13′03″–109°16′46″E; 1,010–1,431 m a.s.l., 8.27 km²) (Figure 1). The study area is characterized by a semi-arid climate and a deeply incised hilly-gully Loess landscape. Slopes vary between 0°–65°. The mean annual temperature range is 9.1±0.1°C from 1970 to 2010, and the annual mean temperature has increased over time (Figure 2a), in summer, the highest temperature is 35.3±2.3°C and in winter, the lowest temperature is −20.3±3.7°C from 1970 to 2010; the average frost-free period is 157 days. Mean annual precipitation is 503±15 mm (from 1970 to 2010) (Figure 2b), of which 70% falls between July and September. The loess-derived soils are fertile but extremely susceptible to erosion. The sand, silt and clay contents are 65%, 24% and 11%, respectively [21]. The main grassland species are *Stipa bungeana*, *Bothriochloa ischaemum*, *Artemisia sacrorum*, *Potentilla acaulis*, *Stipa grandis*, *Androsace erecta*, *Heteropappus altaicus*, *Lespedeza bicolor*, *Artemisia capillaries*, and *Artemisia frigid*, of which, *S. bungeana* is the most widely distributed. In addition, shrubland species such as *Rosa xanthina*, *Spiraea pubescens*, and *Hippophae rhamnoides* can be found in gullies. The primary planted trees in the study area are *Robinia pseudoacacia*, *Populus simonii*, *Caragana microphylla*, and *Platycladus orientalis*.

Experimental design and sampling

A common method used to study vegetation restoration is to monitor plants and soils under similar climatic conditions following the sequence of vegetation development [25]. This chronosequence method is widely adopted in applied ecosystem research [26–27] and is considered a "retrospective" research method because it compares existing conditions with original conditions and treatments [28]. The substitution of "space" for "time" is an effective way of studying change over time [28–29]. It makes the critical assumption that each site in the sequence differs only in age and that each site has traced the same history in both its abiotic and biotic components [27]. Before revegetation soil conditions are largely driven by geomorphological processes.

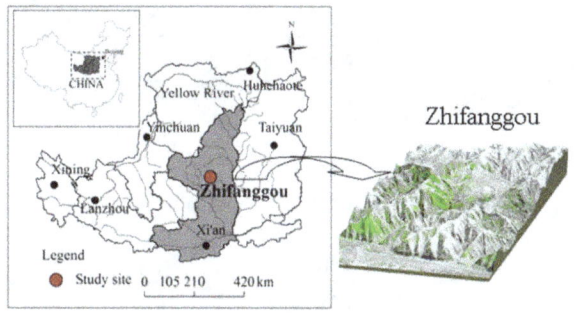

Figure 1. Location and DEM model of the Zhifanggou watershed (adapted from Wang et al., 2011).

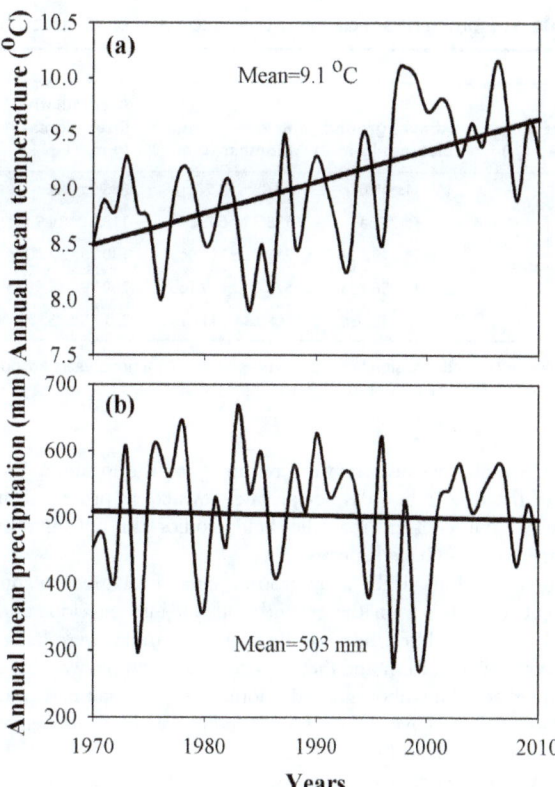

Figure 2. Annual mean temperature (a) and Annual mean precipitation (b) at the study site 1970–2010. Note: the solid lines in the figures a, b are the dynamic fitting curve for annual mean temperature, annual mean precipitation changes with time.

Thus, sites stabilized through revegetation for different periods of time offer an ideal opportunity to understand vegetation succession processes in extreme environments.

An abandoned farmland chronosequence in the watershed was selected for study after the history of the sites was determined through interviews with local farmers (Mr. Haibin Zhang, Soil and Water Conservation Experiment Station, Northwest A&F University, Ansai County, Shaanxi, NW China). Five age classes, 1, 7, 13, 20, and 30 years were selected. In August 2011, when the grassland community biomass peaked, five sites were established in each of the age classes, and the sites were separated by 0.5–1.5 km apart. At each site we set up a plot of 20 m×20 m. In each plot, five quadrats (1 m×1 m) were separately chosen in each of the four corners and center of the plot. In total, we surveyed five plots with twenty-five quadrats in each age class, and twenty-five plots with one hundred and twenty-five quadrats for our study. In each quadrat, the coverage and height, above- and belowground biomass, litter biomass, and soil samples in 0–100 cm soil cores were observed. The morphological traits of the herbage in each age group are listed in Table 1. The plots were all located near the top of loess mounds and there was little difference among the sites in altitude (1185–1341 m), aspect (half south-facing and half north-facing), gradient (18°–37°), or previous farming practices. In the study area, the soils were loess-derived. In addition, five sites on maize farmland (CK) were selected for comparison. Before the farmland was abandoned maize (*Zea mays*) had been widely seeded. The average amount of fertilizer applied was 225–300 kg ha⁻¹ of sheep manure as the base fertilizer in April and 300–450 kg ha⁻¹ of urea which was applied in June as topdressing. We state clearly

Table 1. Dominant species and their biomass, total cover, height at different number of years abandoned.

Years aban-doned (yr)	Above- ground biomass (g m^{-2})	Below- ground biomass (g m^{-2})	Accumulative litter biomass (g m^{-2})	Coverage (%)	Height (cm)	Dominant species
1	169.34±49.67a	78.92±13.57c	28.90±4.26c	8.8±0.58c	66±6.96ab	Artemisia scoparia
7	73.73±7.96a	138.73±6.72c	118.51±14.93bc	19.4±2.86bc	41.6±7.88a	Lespedeza bicolor+ Setaira viridis
13	100.26±19.01a	218.71±36.29c	130.78±33.00bc	24.4±5.61abc	68.2±6.01a	Agropyron cristatum+ Heteropappus altaicus
20	210.42±56.73a	575.21±129.16b	209.79±35.99ab	41±10.77ab	69±5.79a	Artemisia sacrorum+ Bothriochloa ischaemum
30	159.08±12.03a	1080.34±111.18a	281.86±25.77a	48±6.04a	52±2.07ab	Bothriochloa ischaemum+ Stipa bungeana

Different letters indicate significant differences at $P<0.05$ among years abandoned.
Values are mean ±SE of 5 sites.

that no specific permissions were required for the location. We confirm that the location is not privately-owned or protected in any way. We confirm that the field studies do not involve endangered or protected species.

In each quadrat, all the aboveground parts of the green plants were cut, collected from the ground and put into envelops and tagged, as was all litter. Because the biomass samples were large, they were weighed fresh and then a part of each sample was dried and weighed. The aboveground biomass of the samples was calculated by multiplying the ratio of the dry weight/fresh weight ratio by the fresh weight.

Soil samples were taken at five points in the quadrats of each plot. These were the four corners and center of the biomass sampling sites as described above. Litter horizons were removed before soil sampling. Soil sampling, using a soil drilling sampler (9 cm inner diameter), was done in seven soil layers, 0–5, 5–10, 10–20, 20–30, 30–50, 50–70, and 70–100 cm. We then mixed the

same layers together to make one sample. All samples were sieved through a 2 mm screen, and roots and other debris were removed. Each sample was air-dried and stored at room temperature for the determination of soil physical and chemical properties. The soil bulk density (g cm^{-3}) of the different soil layers was measured using a soil bulk sampler with a 5 cm diameter and 5 cm high stainless steel cutting ring (5 replicates) at points adjacent to the soil sampling quadrats. The original volume of each soil core and its dry mass after oven-drying at 105°C were measured.

To measure belowground biomass, soil sampling was done three times in seven soil layers, 0–5, 5–10, 10–20, 20–30, 30–50, 50–70, and 70–100 cm at a depth of 0–100 cm in each quadrat using a 9 cm diameter root auger. The majority of the roots were found in the soil samples thus obtained and then isolated using a 2 mm sieve. The remaining fine roots taken from the soil samples were isolated by spreading the samples in shallow trays, overfilling the trays with water and allowing the outflow from the trays to pass

Table 2. One-way analysis of variance (ANOVA) for soil properties in the seven soil layers among abandoned farmland in different years.

Soil layer (cm)	Df	SOC		TN	
		F	sig. (P)	F	sig. (P)
0–5	5	10.367	0.0001**	14.557	0.0001**
5–10	5	2.604	0.0668	3.651	0.0217*
10–20	5	0.88	0.4936	0.397	0.8081
20–30	5	5.322	0.0044**	0.132	0.9688
30–50	5	6.892	0.0012**	3.348	0.0298*
50–70	5	5.787	0.0029**	6.963	0.0011**
70–100	5	2.567	0.0697	1.825	0.1636
Soil depth (cm)	Df	C storages		N storages	
		F	sig. (P)	F	sig. (P)
0–5	5	5.71	0.0031**	5.809	0.0029**
0–10	5	4.053	0.0145*	4.527	0.0091**
0–20	5	0.783	0.5496	0.732	0.5811
0–30	5	1.404	0.2688	0.139	0.966
0–50	5	4.483	0.0095**	1.014	0.4237
0–70	5	7.042	0.0010**	3.089	0.0392*
0–100	5	7.905	0.0005**	3.783	0.0190*

*indicates significant at $P<0.05$ and **indicates significant at $P<0.01$.

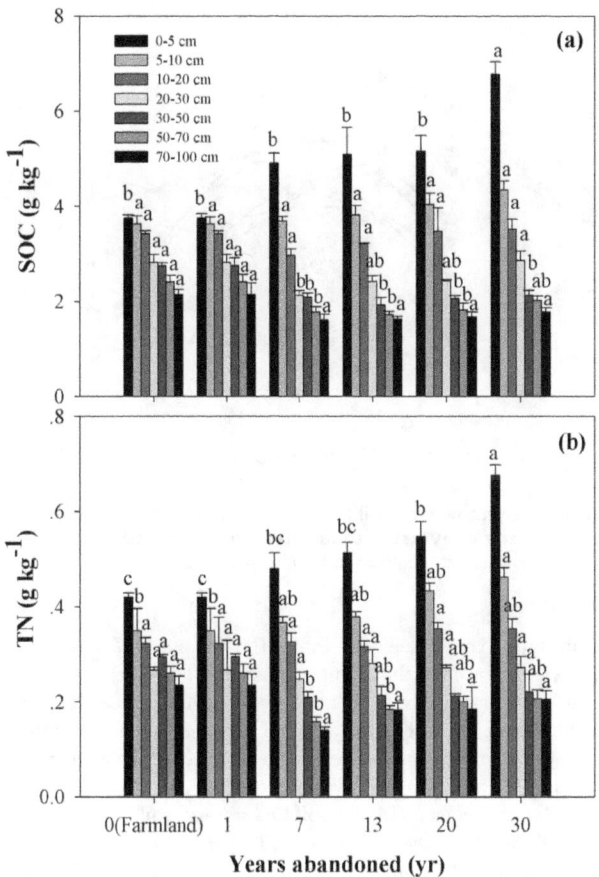

Figure 3. SOC and TN dynamics for seven soil layers of 0–100 cm following the conversion of farmland to grassland. Values are in the form of Mean ± SE and the sample size n = 5. Different lower-case letters above the bars mean significant differences in the same soil layers among abandoned land in different years (P<0.05).

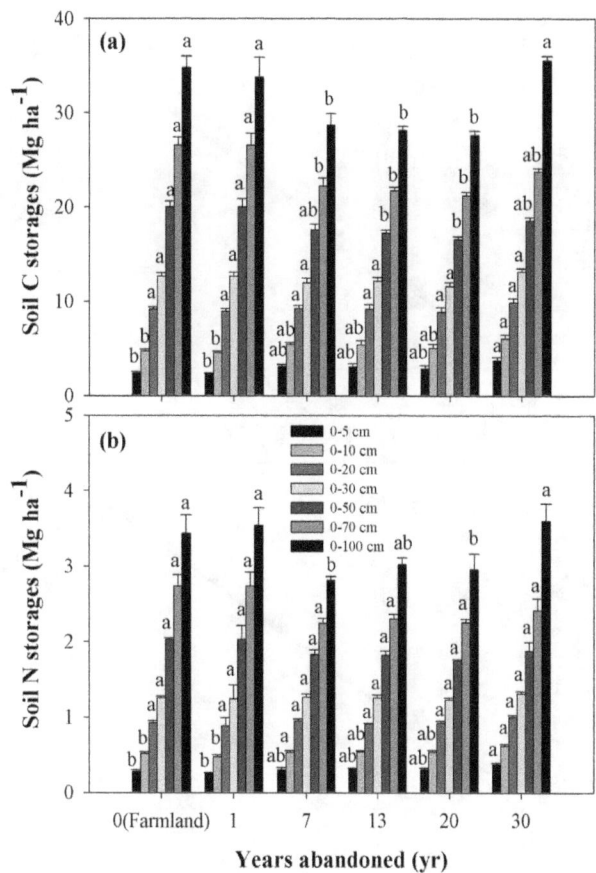

Figure 4. Dynamics of soil OC and N storages in 0–100 cm soil depth following the conversion of farmland to grassland. Values are in the form of Mean ± SE and the sample size n = 5. Different lower-case letters above the bars mean significant differences in the same soil layers among abandoned land in different years (P<0.05).

through a 0.5 mm mesh sieve. No attempts were made to distinguish between living and dead roots. All the roots thus isolated were oven-dried at 65°C and weighed to within 0.01 g.

Physical and chemical analysis

Soil bulk density (BD) was calculated depending on the inner diameter of the core sampler, sampling depth and the oven dried weight of the composite soil samples [30]. Soil OC content was assayed by dichromate oxidation [31] and soil TN content was assayed using the Kjeldahl method [32].

Calculation of soil C and N storages

Our sample soils did not have any coarse fraction (>2 mm). Therefore, the study used the following equation to calculate soil organic carbon storage (Cs) [22]:

$$Cs = BD \times SOC \times D/10 \qquad (1)$$

in which, Cs is SOC storages (Mg ha^{-1}); BD is soil bulk density (g cm^{-3}); SOC is soil organic carbon concentration (g kg^{-1}); and D is soil thickness (cm).

The following equation was used to calculate soil N storage (Ns) [33]:

$$Ns = BD \times TN \times D/10 \qquad (2)$$

in which, Ns is soil N storage (Mg ha^{-1}); BD is soil bulk density (g cm^{-3}); TN is soil TN concentration (g kg^{-1}); and D is soil thickness (cm).

Statistical analysis

One-way ANOVA was used to analyze the means of the same soil layers among the different abandoned years. Differences were evaluated at the 0.05 significance level. When significance was observed at the P<0.05 level, Tukey's post hoc test was used to carry out the multiple comparisons. All statistical analyses were performed using the software program SPSS, ver. 17.0 (SPSS Inc., Chicago, IL, USA).

Results

Dynamics of SOC and TN

SOC in 0–5 cm soil was higher in older grassland following abandonment (Table 2, Figure 3a). In the first 20 years, 0–5 cm SOC showed no significant changes; after 20 years, it had significantly increased (Figure 3a). The 5–20 cm SOC had not significantly increased (Table 2). The 20–70 cm SOC was significantly lower in the younger grasslands (<20 yr) and higher

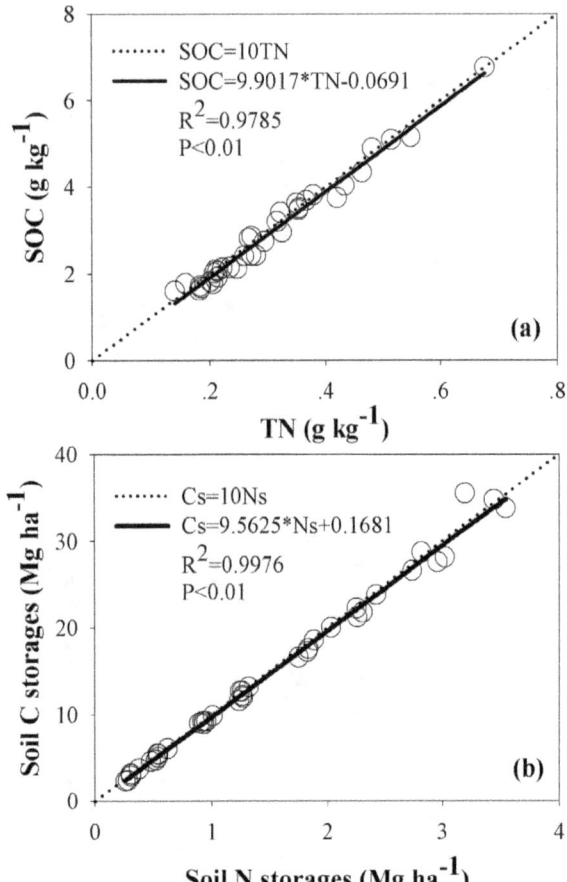

Figure 5. Relationship between SOC and TN, soil OC storages and soil N storages, and accumulative soil C storages and soil N storage.

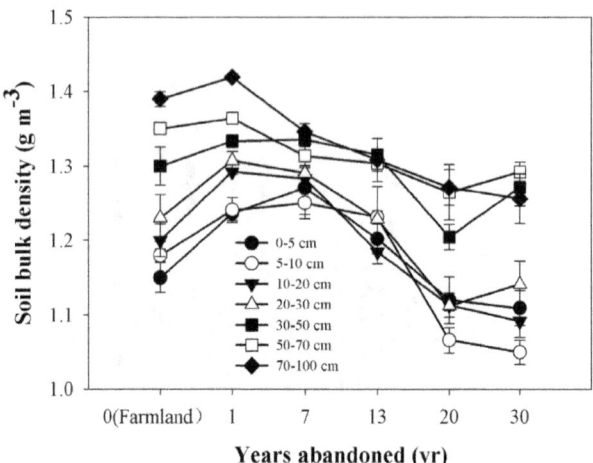

Figure 6. Dynamics of soil bulk density in 0–100 cm soil depth following the conversion of farmland to grassland. Values are in the form of Mean ± SE and the sample size n = 5.

in the older grasslands (30 yr) (Figure 3a). When abandoned for 30 years, 20–70 cm SOC had increased to the level prior to abandonment (maize) (Figure 3a). The 70–100 cm SOC showed no significant changes since having been abandoned, but also showed a tendency to be low in younger grasslands and higher in the older grasslands (Table 2, Figure 3a).

Soil TN in 0–5 cm soil was also higher in older grassland following abandonment (Table 2, Figure 3b). In the first 13 years after having been abandoned, TN in the 0–5 cm soil layer showed no significant changes; after 13 years, it had significantly increased (Figure 3b). The TN in the 5–10 cm soil layer showed no significant changes in the first 13 years, whereas after 13 years, it had significantly increased (Figure 3b). The 10–30 cm soil TN did not increase significantly (Table 2, Figure 3b). The 30–70 cm soil TN first decreased significantly and then increased to the level prior to abandonment (maize) (Figure 3b). Similar to the SOC, 70–100 cm soil TN had showed no significant changes since abandonment, but also showed a tendency to be lower in younger grasslands and higher in older grasslands (Table 2, Figure 3b).

C and N storages dynamics

Soil OC storage in the 0–10 cm soil was also higher in older grassland since abandonment (Table 2, Figure 4a). After having been abandoned for 30 years, the 0–10 cm soil OC storage had increased significantly. 0–30 cm soil OC storage had not increased

significantly (Table 2, Figure 4a). 0–100 cm soil OC storage was significantly lower in the younger grasslands (<20 yr) and higher in the older grasslands (30 yr); at 30 years it had increased to pre-abandonment levels (Figure 4a). Soil C storage changed mainly in the top 10 cm of soil following the conversion of farmland to grassland for a period of 30 years.

Similar to the accumulative soil OC storage, soil N storage had changed mainly in the top 10 cm soil layer 30 years after having been abandoned. The conversion of farmland into grassland significantly increased the soil N storage in the 0–10 cm soil (Table 2, Figure 4b). After having been abandoned for 30 years, 0–10 cm soil N storage had significantly increased. 0–50 cm soil N storage had not significantly increased (Table 2, Figure 4b). 0–100 cm soil N storage was also significantly lower in younger grasslands (<20 yr) and higher in older grasslands (30 yr), until after thirty years it had increased to pre-abandonment levels (maize) (Figure 4b).

Relationship between Soil C and N

Soil C and N showed significant positive correlations (Figure 5). The relationship between SOC and TN, soil C storage and soil N storage were significant ($P<0.01$). In the process of the revegetation, soil OC and TN, soil C storage and N storage approximately represents SOC = 10TN (Figure 5a) and Cs = 10Ns (Figure 5b).

Discussion

The soil C and N results supported the hypothesis that soil organic C and N conditions in both the top soil and in the deeper soil layer are significantly affected by land use change on the northern Loess Plateau. In our study, top soil OC and TN were higher in older post-abandonment grassland, especially in the 0–5 cm soil depth (figure 3), indicating the accumulation of soil OC and TN by revegetation [23–24]. These results agree with those of Wang et al. [21], who studied changes to the physico-chemical properties of top soil during natural succession on abandoned farmland in the Zhifanggou watershed. The evident increase may be partly attributed to a lower fraction of non-soluble materials in more readily decomposed plant residues. In the farmland, cultivation breaks up soil aggregates, decreases total soil porosity, and accelerates composition and mineralization of soil organic

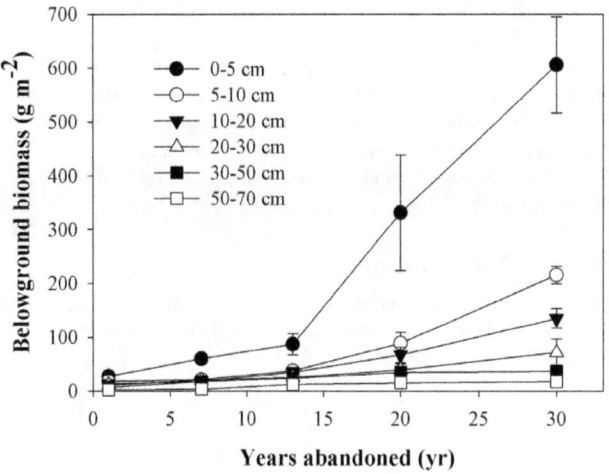

Figure 7. Dynamics of belowground biomass in 0–100 cm soil depth following the conversion of farmland to grassland. Values are in the form of Mean ± SE and the sample size n = 5. It didn't have roots in the 70–100 cm soil layers.

matter (SOM) due to exposure of previously accessible SOM to microbial attack [34]. This results in a reduction in the amounts of intra-aggregate light fraction organic carbon (LFOC) and some organomineral SOC [35]. In addition, the reduction of crop residue return to soil may also be a factor as farmers take away straw with grain harvesting each year. This speculation can be supported by the results of Wu et al. [36] in this region where plant residue in the top soil layer was reduced considerably after native grasslands were cultivated, contributing to the decrease in LFOC and SOC. Conversely, the conversion of farmland into grassland increases SOC and its fractions [37], and increases total soil porosity, thus resulting in a reduction in soil BD (Figure 6, [21]).

Aboveground vegetation plays an important role in regulating the biogeochemistry of ecosystems by fixing C and nutrients and preventing the loss of nutrients under disturbance, such as plant decline, acid-rain and climate change [38]. It is also clear from these results that revegetated grassland has a great impact on the storage of soil OC and N. In our study, both soil OC and N storage in the 0–10 cm soil depth was higher in older grassland that had been abandoned for thirty years (Figure 4). This indicates that during the 30 year period soil OC and N storage mainly had changed in the top 10 cm of soil since abandoned. There may be a range of potential mechanisms through which soil OC and N in the top soil increased with revegetation. A prime candidate is the return of both C and N from increased aboveground biomass and litter. As soil OC and N input are mainly derived from the decomposition of litter [21], primary productivity is the main driver of soil carbon sequestration [39], which resulted primarily in soil OC and N storage increasing in the top soil. Secondly, belowground biomass (dead roots, mycorrhizae, and exudates) is an important element of soil carbon sequestration [40]. Below-ground biomass increased in the time after the farmland had been abandoned (Table 1, Figure 7). Thirdly, changes in vegetation composition and the dominant plant functional group could affect the sequestration of C and N in the soil [39,41]. Plant functional composition strongly influences the chemical and physical composition of litter inputs, and thereby their decomposability, carbon loss through soil respiration and leaching, and carbon immobilization in hummified plant residues [39], and the increase of C3 plants can increase soil C and N accumulation in meadow

soils [41]. Revegetation had a direct effect on the dominant vegetation species, vegetation cover, height, above- and below-ground biomass (Table 1).

SOC and TN at deeper soil levels (>30 cm) were lower in younger grasslands and higher in older grasslands (Figure 3), this is probably due to long-term natural organic fertilizer and inorganic fertilizer inputted into the soil resulting in higher SOC and N in the farmland stage. So, this initial loss of soil OC and N following the conversion of farmland has been commonly attributed to the net effect of decreased organic matter inputs and losses through decomposition [42]. During the course of grasslands development, we observed an increase in the soil SOC and N storage, after 30 years of revegetation, where deeper soil C and N had returned to pre-abandonment levels (Figure 3); the return of belowground biomass to the deeper soil layer is another reason (Figure 7). Previous studies of soil C dynamics have emphasized the role of physical protection from different particle fractions (sand, silt and clay) [43], microbes and enzymes within aggregates [44], microaggregates and macroaggregates [45], and bacterial and fungal [46], therefore, the mechanism in the soil C and N dynamics in the deeper soil during the process of conversion from farmland into grassland probably relate to those factors. However, our lack of a full understanding of this process calls for more attention to be paid to soil C and N dynamics in the deeper soil layers. Wang et al. [21] found that SOC and TN had a negative relationship with soil bulk density, and Singh et al. [23–24] reported that SOC and TN were significantly greater while BD was significantly lower during the restoration of degraded lands, our results agreed with them (Figures 3 and 6). So, we can infer the trend of SOC and TN according to the trend of soil bulk density.

We observed that soil C and N were significantly and positively correlated (Figure 5). In the restoration process, soil OC and TN, soil C storage and N storage approximately represents SOC = 10TN (Figure 5a) and Cs = 10Ns (Figure 5b). So, we can conclude that soil C/N value was 10 in the process of 30 years of conversion from farmland to grassland in the Loess Plateau, a value greater that reported by Liu et al. [35] (C/N = 7.62). Ammonium-N and the sum of NO_3^- -N and NO_2^- -N are the readily available forms of soil N for root uptake. Liu et al. [35] reported that ammonium-N and the sum of NO_3^- -N and NO_2^- -N were lower in grassland than in cropland with the addition of chemical fertilizer N, because the conversion of farmland into grassland increased the C/N ratio and reduced soil mineral N by enhancing soil N immobilization.

Grasslands play an important role in the global C and N cycles [3–4,10]. Grassland in good condition should be in balance in terms of C and N input and output or in the state where C and N input is greater than their output [41]. At least one study has shown that carbon input is greater than carbon output in enclosed grassland [47]. In our study, 0–100 cm soil OC and N storage was significantly lower in the younger grasslands (<20 yr) and higher in the older grasslands (30 yr), at 30 years they had increased to pre-abandonment levels, and that the values in farmland were higher compared to grasslands abandoned for 20 years. Our results indicate that the study of the storage of soil OC and N need to include the deeper soil layer and not focus solely on the top soil. Li et al. [48] also found that SOC density in the deeper soil layer was significantly higher than that of the farmland in the Zhifanggou watershed, Loess Plateau. Li et al. [49] found that soil C and N storage in the deeper layers (mineral layer) show significant difference in top soil layers (organic layer) with time after afforestation at the global scale. Therefore, estimating soil OC and N input or output requires that research consider not only soil depth but also time since abandonment.

Conclusions

The results of this study indicate that plant succession after land has been abandoned resulted in a significant improvement in the physico-chemical properties of soil. Thirty years following abandonment, soil OC and N storage had increased primarily in the top 10 cm of the soil depth. After 30 years of restoration, deeper soil C and N storage had increased to pre-abandonment levels. This finding indicates that deeper soil has a higher potential to fix both C and N in the future (>30 yr). Thus, in the semi-arid environment of the Loess Plateau, vegetation recovery following abandonment is slow and the improvement of soil properties is likely to require a considerably long period of time (>30 yr). Therefore, the findings are important for assessing the resilience of these degraded ecosystems and developing a more effective strategy of vegetation restoration for the management of degraded grassland from a long-term perspective. More research, for example, on soil physico-chemical properties, soil enzyme activities, soil microbial, animal and plant function and composition, is required to better understand the mechanism behind how the soil fixes C and N in the deeper soil profile of sub-arid regions, for example, the Loess Plateau, China.

Author Contributions

Conceived and designed the experiments: LD ZS. Analyzed the data: LD. Contributed reagents/materials/analysis tools: LD. Wrote the paper: LD ZS SS.

References

1. Kalnay E, Cai M (2003) Impact of urbanization and land-use change on climate. Nature 423: 528–531.
2. Ficetola GF, Maiorano L, Falcucci A, Dendoncker N, Boitani L, et al. (2010) Knowing the past to predict the future: land-use change and the distribution of invasive bullfrogs. Global Change Biol 16: 528–537.
3. Russell AE, Laird DA, Parkin TB, Mallarino AP (2005) Impact of nitrogen fertilization and cropping system on carbon sequestration in Midwestern Mollisols. Soil Sci Soc Am J 69: 413–422.
4. Brown J, Angerer J, Salley SW, Blaisdell R, Stuth JW (2010) Improving estimates of rangeland carbon sequestration potential in the US Southwest. Rangeland Ecol Manag 63: 147–154.
5. Qiu SJ, Ju XT, Ingwersen J, Qin ZC, Li L, et al. (2010) Changes in soil carbon and nitrogen pools after shifting from conventional cereal to greenhouse vegetable production. Soil Till Res 107: 80–87.
6. Jobbágy EG, Jackson RB (2000) The Vertical Distribution of Soil Organic Carbon and Its Relation to Climate and Vegetation. Ecol Appl 10: 423–436.
7. Piao SL, Fang JY, Ciais P, Peylin P, Huang Y, et al. (2009) The Carbon balance of terrestrial ecosystems in China. Nature 458: 1009–1013.
8. Arora VK, Boer GJ (2010) Uncertainties in the 20th century carbon budget associated with land use change. Global Change Biol 16: 3327–3348.
9. Jia XX, Wei XR, Shao MA, Li XZ (2012) Distribution of soil carbon and nitrogen along a revegetational succession on the Loess Plateau of China. Catena 95: 160–168.
10. Eaton JM, McGoff NM, Byrne KA, Leahy P, Kiely G (2008) Land cover change and soil organic carbon stocks in the Republic of Ireland 1851–2000. Climate Change 91: 317–334.
11. Fu XL, Shao MA, Wei XR, Horton R (2010) Soil organic carbon and total nitrogen as affected by vegetation types in Northern Loess Plateau of China. Geoderma 155: 31–35.
12. She DL, Shao MA, Timm LC, Reichardt K (2009) Temporal changes of an alfalfa succession and related soil physical properties on the Loess Plateau, China. Pesquisa Agropecuária Brasileira 44: 189–196.
13. Wei XR, Shao MA, Fu XL, Horton R (2010) Changes in soil organic carbon and total nitrogen after 28 years of grassland afforestation: effects of tree species, slope position, and soil order. Plant Soil 331: 165–179.
14. Foster D, Swanson F, Aber J, Burke I, Brokaw N (2003) The importance of land-use legacies to ecology and conservation. BioScience 53: 77–88.
15. Luo YQ, Field CB, Jackson RB (2006) Does nitrogen constrain carbon cycling, or does carbon input stimulate nitrogen cycling? Ecology 87: 3–4.
16. Luo YQ, Su B, Currie WS, Dukes JS, Finzi A, et al. (2004) Progressive nitrogen limitation of ecosystem responses to rising atmospheric carbon dioxide. BioScience 54: 731–739.
17. Fu BJ, Chen LD, Ma KM, Zhou HF, Wang J (2000) The relationships between land use and soil conditions in the hilly area of the Loess Plateau in northern Shaanxi, China. Catena 39: 69–78.
18. Deng L, Shangguan ZP (2011) Food security and farmers' income: impacts of the Grain for Green Programme on rural households in China. J Food Agric Environ 9: 826–831.
19. Deng L, Shangguan ZP, Li R (2012) Effects of the grain-for-green program on soil erosion in China. Int J Sediment Res 27: 120–127.
20. Prietzel J, Bachmaan S (2012) Changes in soil organic C and N stocks after forest transformation from Norway spruce and Scots pine into Douglas fir, Douglas fir/spruce, or European beech stands at different sites in Southern Germany. Forest Ecol Manag 269: 134–148.
21. Wang B, Liu GB, Xue S, Zhu BB (2011) Changes in soil physico-chemical and microbiological properties during natural succession on abandoned farmland in the Loess Plateau. Environ Earth Sci 62: 915–925.
22. Guo LB, Gifford RM (2002) Soil carbon storage and land use change: a meta analysis. Global Change Biol 8: 345–360.
23. Singh K, Pandey VC, Singh B, Singh RR (2012) Ecological restoration of degraded sodic lands through afforestation and cropping. Ecol Eng 43: 70–80.
24. Singh K, Singh B, Singh RR (2012) Changes in physico-chemical, microbial and enzymatic activities during restoration of degraded sodic land: Ecological suitability of mixed forest over monoculture plantation. Catena 96: 57–67.
25. Bhojvaid PP, Timmer VR (1998) Soil dynamics in an age sequence of Prosopis Juliflora planted for sodic soil restoration in India. Forest Ecol Manag 106: 81–193.
26. Fang W, Peng SL (1997) Development of species diversity in the restoration process of establishing a tropical man-made forest ecosystem in China. Forest Ecol Manag 99: 185–196.
27. Johnson EA, Miyanishi K (2008) Testing the assumptions of chronosequences in succession. Ecol Let 11: 419–431.
28. Li XR, Kong DS, Tan HJ, Wang XP (2007) Changes in soil and vegetation following stabilization of dunes in the southeastern fringe of the Tengger Desert, China. Plant Soil 300: 221–231.
29. Sparling GP, Schipper LA, Bettjeman W, Hill R (2003) Soil quality monitoring in New Zealand: practical lessons from a 6-year trial. Agr Ecosyst Environ 104: 523–534.
30. Jia GM, Cao J, Wang CY, Wang G (2005) Microbial biomass and nutrients in soil at the different stages of secondary forest succession in Ziwulin, northwest China. Forest Ecol Manag 217: 117–125.
31. Kalembasa SJ, Jenkinson DS (1973) A comparative study of titrimetric and gravimetric methods for the determination of organic carbon in soil. J Sci Food Agr 24: 1085–1090.
32. Bremner JM (1996) Nitrogen-total. In: Sparks, D.L. (Ed.), Methods of Soil Analysis, Part 3. America Society of Agronomy, Madison, pp. 1085–1121 (SSSA Book Series: 5).
33. Rytter RM (2012) Stone and gravel contents of arable soils influence estimates of C and N stocks. Catena 95: 153–159.
34. Shepherd TG, Saggar S, Newman RH, Ross CW, Dando JL (2001) Tillage-induced changes to soil structure and organic carbon fractions in New Zealand soils. Aust J Soil Res 39: 465–489.
35. Liu X, Li FM, Liu DQ, Sun GJ (2010) Soil Organic Carbon, Carbon Fractions and Nutrients as affected by Land Use in Semi-Arid Region of Loess Plateau of China. Pedosphere 20: 146–152.
36. Wu T, Schoenau JJ, Li F, Qian P, Malhi SS, et al. (2004) Influence of cultivation and fertilization on total organic carbon and carbon fractions in soils from the Loess Plateau of China. Soil Till Res 77: 59–68.
37. Zeng ZX, Liu XL, Jia Y, Li FM (2007) The effect of conversion of cropland to forage legumes on soil quality in a semiarid agroecosystem. J Sustain Agr 32: 335–353.
38. Bormann BT, Sidle RC (1990) Changes in productivity and distribution of nutrients in a chronosequence at Glacier Bay National Park, Alaska. Journal of Ecology 78: 561–578.
39. De Deyn GB, Cornelissen JHC, Bardgett RD (2008) Plant functional traits and soil carbon sequestration in contrasting biomes. Ecol Lett 11: 516–531.
40. Langley JA, Hungate BA (2003) Mycorrhizal controls on belowground litter quality. Ecology 84: 2302–2312.
41. Wu GL, Liu ZH, Zhang L, Chen JM, Hu TM (2010) Long-term fencing improved soil properties and soil organic carbon storage in an alpine swamp meadow of western China. Plant Soil 332: 331–337.
42. Zhao WZ, Xiao HL, Liu ZM, Li J (2005) Soil degradation and restoration as affected by land use change in the semiarid Bashang area, Northern China. Catena 59: 173–186.
43. He NP, Wu L, Wang YS, Han XG (2009) Changes in carbon and nitrogen in soil particle-size fractions along a grassland restoration chronosequence in northern China. Geoderma 150: 302–308.
44. Udawatta RP, Kremer RJ, Adamson BW, Anderson SH (2008) Variation in soil aggregate stability and enzyme activities in a temperate agroforestry practice. Appl Soil Ecol 39: 153–160.
45. Chen FS, Zeng DH, Fahey TJ, Liao PF (2010) Organic carbon in soil physical fractions under different-aged plantations of Mongolian pine in semi-arid region of Northeast China. Appl Soil Ecol 44: 42–48.

46. Six J, Frey SD, Thiet RK, Batten KM (2006) Bacterial and fungal contributions to carbon sequestration in agroecosystems. Soil Sci Soc Am J 70: 555–569.

47. Li YQ, Zhao HL, Zhao XY, Zhang TH, Chen YP (2006) Soil respiration, carbon balance and carbon storage of sandy grassland under post-grazing natural restoration. Acta Prataculturae Sinica 15: 25–31. (in Chinese with English abstract).

48. Li MM, Zhang XC, Pang GW, Han FP (2013) The estimation of soil organic carbon distribution and storage in a small catchment area of the Loess Plateau. Catena 101: 11–16.

49. Li DJ, Niu SL, Luo YQ (2012) Global patterns of the dynamics of soil carbon and nitrogen stocks following afforestation: a meta-analysis. New Phytol 195: 172–181.

9

Potential of Global Cropland Phytolith Carbon Sink from Optimization of Cropping System and Fertilization

Zhaoliang Song[1,2,3]*, Jeffrey F. Parr[4], Fengshan Guo[2]

1 Zhejiang Provincial Key Laboratory of C Cycling in Forest Ecosystems and C Sequestration, Zhejiang Agricultural and Forestry University, Lin'an, Zhejiang, China, **2** School of Environment and Resources, Zhejiang Agricultural and Forestry University, Lin'an, Zhejiang, China, **3** State Key Laboratory of Environmental Geochemistry, Institute of Geochemistry, Chinese Academy of Sciences, Guiyang, Guizhou, China, **4** Southern Cross GeoScience, Southern Cross University, Lismore, New South Wales, Australia

Abstract

The occlusion of carbon (C) by phytoliths, the recalcitrant silicified structures deposited within plant tissues, is an important persistent C sink mechanism for croplands and other grass-dominated ecosystems. By constructing a silica content-phytolith content transfer function and calculating the magnitude of phytolith C sink in global croplands with relevant crop production data, this study investigated the present and potential of phytolith C sinks in global croplands and its contribution to the cropland C balance to understand the cropland C cycle and enhance long-term C sequestration in croplands. Our results indicate that the phytolith sink annually sequesters 26.35 ± 10.22 Tg of carbon dioxide (CO_2) and may contribute $40 \pm 18\%$ of the global net cropland soil C sink for 1961–2100. Rice (25%), wheat (19%) and maize (23%) are the dominant contributing crop species to this phytolith C sink. Continentally, the main contributors are Asia (49%), North America (17%) and Europe (16%). The sink has tripled since 1961, mainly due to fertilizer application and irrigation. Cropland phytolith C sinks may be further enhanced by adopting cropland management practices such as optimization of cropping system and fertilization.

Editor: Senjie Lin, University of Connecticut, United States of America

Funding: The authors are grateful for support from the National Natural Science Foundation of China (Grant No. 41103042), the Zhejiang Province Key Science and Technology Innovation Team (No. 2010R50030), and the Opening Project of State Key Laboratory of Environmental Geochemistry (SKLEG9011). The funders had no role in study design, data collection and analysis, decision to publish, or preparation of the manuscript.

Competing Interests: The authors have declared that no competing interests exist.

* E-mail: songzhaoliang78@163.com

Introduction

Present understanding of the global carbon (C) cycle and climate feedbacks is limited by uncertainty over terrestrial C balance [1–5]. As one of the largest terrestrial ecosystems deeply influenced by human activities, the croplands cover an area of 15.33×10^8 hm^2 globally and may play a significant role in terrestrial C balance [3,6]. Although croplands were traditionally considered to be the largest biospheric source of C lost to the atmosphere in most areas of the world [7–12], they may also be significant C sinks under proper management [3,6,13–15].

Phytolith-occluded C (PhytOC), where C is entrapped within recalcitrant silicified structures when they are deposited within plant tissues [16–18], is particularly prolific in many crops such as rice [19], wheat [20], millet [21] and sugarcane [22]. PhytOC is highly resistant against decomposition [18,23–25] and may accumulate in soil for several thousands of years after plant decomposition [18], demonstrating the potential of phytoliths in the long-term biogeochemical sequestration of atmospheric carbon dioxide (CO_2) [5,26]. Soil PhytOC accumulation is an important persistent C sink mechanism for croplands [18–22] and other grass-dominated ecosystems [26,27]. Moreover, Jansson et. al. [28] suggest that the production of PhytOC in croplands could be greatly enhanced through crop breeding. However, the present and potential of global cropland phytolith C sink have not been revealed.

In the present study, we quantifed the present and potential of phytolith carbon sink and its contribution to the global cropland C balance by constructing a silica content- phytolith content transfer function and calculating the magnitude of the phytolith C sink in global croplands with relevant crop data including the PhytOC and silica content, farm crop output, the Si-rich organ ratio (mass ratios of the Si-rich organ: crop output) and the PhytOC stability factor. The purposes of the study are to guide the management of cropland ecosystems to maximize phytolith C sequestration and mitigate climate change.

Materials and Methods

Ethics Statements

No specific permits were required for the described field studies, because the experimental field is owned by Zhejiang Agricultural and Forestry University, and the School of Environment and Resources performs the management. No specific permits were required for these locations/activities, because the location is not privately-owned or protected in any way, the field studies did not involve endangered or protected species, and each sample consisted of no more than 500 grams (fresh weight).

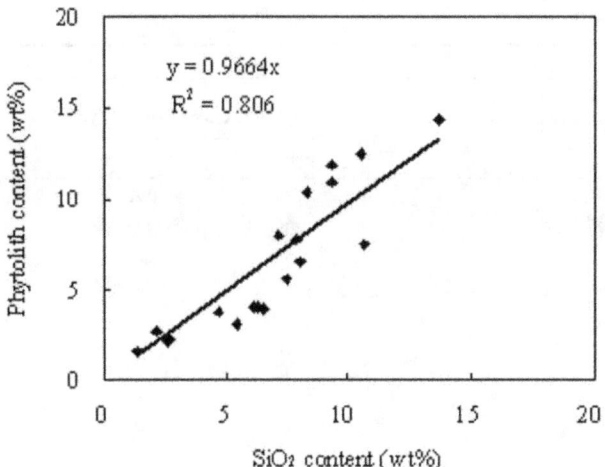

Figure 1. The correlation of phytolith content (%) to SiO$_2$ (%) content in different crop species (p<0.01).

Table 1. General information and PhytOC content of the dominant arable crops.

Farm crops	Area (10^6 hm^2)[a]	Plant Si-rich organs	PhytOC (%)[b]	
			mean	SE
Crops (total)	1532.6		0.13	0.05
Cereals (total)	697.7	Stem, sheath and leaf	0.19	0.07
Rice	164.1	Stem, sheath and leaf	0.25	0.07
Wheat	220.4	Stem, sheath and leaf	0.16	0.08
Maize	170.4	Stem, sheath and leaf	0.16	0.05
Soybeans	103	Stem and leaf	0.02	0.01
Roots and Tubers	54.3	Stem and leaf	0.02	0.01
Oil-bearing crops	62	Stem and leaf	0.08	0.08
Seed cotton	35.2	Stem and leaf	0.02	0.01
Sugar cane	25.4	Sheath and leaf	0.25	0.07

[a] values from FAO (2012).
[b] estimated from phytolith and silica content data (This study; ref. [19–21,30–33]) using equation (1) and occluded-C content in phytolith of 3±1% (This study; ref. [19–21,27]).

Constructing the Transfer Function for the Phytolith:Silica Content

Plant phytolith content may be estimated from plant silica content data using the transfer function for the phytolith:silica content [26]. To construct the silica content- phytolith content transfer function, mature crop organ samples were collected– each sample consisted of approximately 500 g of composite plant material.

Plant samples were oven-dried at 65°C to a constant mass and cut into small pieces (<5 mm). They were ashed at 500°C to remove organic matter, fused with lithium metaborate, dissolved in dilute nitric acid and analyzed for silica content using inductively coupled plasma-optical emission spectroscopy (ICP-OES; Optima 7000 DV, Perkin Elmer, Massachusetts, USA). Plant phytoliths were isolated using a microwave digestion process followed by a Walkley–Black type digestion to ensure the removal of extraneous organic material [19,27]. The isolated phytoliths were dried to a constant mass at 75°C for 24 h in a fan-forced oven and weighed to determine the plant phytolith content. The occluded-C content within phytoliths was also determined [19,27]. The error was <5% in phytolith and silica measurements and <10% in PhytOC measurements using plant standards (GSV-1) and triplicate analyses.

The plant silica content- phytolith content transfer function was constructed using regression analysis based on the phytolith and silica contents determined for the samples (Figure 1). Silica content was converted to phytolith content using the following equation ($R^2 = 0.806$, p<0.01):

Phytolith content (wt %) = 0.9664 × silica content (wt %)   (1)

Data Collection, Phytolith and PhytOC Content Estimation

Farm productivity data was obtained from Food and Agriculture Organization of the United Nations (FAO) Statistics [29]. Silica content data was obtained from published monographs [30,31], papers [19–21,32,33] and also determined in the present study. Silica content of crop species was used to estimate phytolith content using a conversion factor of 0.9664; see equation (1). The

PhytOC content in plant organs was estimated from phytolith content data using an occluded-C content in phytolith of 2~4% (average 3%) according to the present study and references [19–21,27].

Estimating PhytOC Production and the Phytolith C Sink

The production of PhytOC is primarily affected by plant PhytOC concentration and aboveground net primary productivity (ANPP) of Si-rich organs [34], where plant PhytOC is mainly determined by PhytOC content in phytoliths [26] and plant Si content [30,33]. This allowed the crop PhytOC production rate to be estimated from the PhytOC content and total ANPP of the Si-rich organs of an area as:

$$\text{PhytOC production rate} = \text{PhytOC content} \times \text{ANPP of Si-rich organs} \times 44/12 \quad (2)$$

where PhytOC production rate is the PhytOC production by a particular crop's Si-rich organs per year (Tg CO$_2$ yr^{-1}), PhytOC content is the concentration of PhytOC in a crop's Si-rich organs (wt %) and ANPP is the total aboveground net primary productivity of Si-rich crop organs (Tg yr^{-1}) of an area estimated from Si-rich organ factor [35,36] and crop output [29].

As the PhytOC sequestration rate is controlled by the PhytOC production rate in plants and the stability of phytolith in environments, the phytolith C sink rate can be estimated from data of PhytOC production rate and phytolith stability factor as:

$$\text{Phytolith C sink rate} = \text{PhytOC production rate} \times \text{phytolith stability factor} \quad (3)$$

where PhytOC production rate may be estimated from equation (2) and the phytolith stability factor is assumed to be 0.9±0.05 as most phytoliths have been proved stable for thousands of years though some small phytolith particles containing little carbon may be partly dissolved depending on formation sites and chemical

Table 2. Estimated phytolith C sink produced by global farm crops in 2011.

Farm crops	Crop output (Tg yr^{-1})[a]	Si-rich organ factor[b]	ANPP[c] of Si-rich organs (Tg yr^{-1})[c]	Phytolith C sink (Tg CO$_2$ yr^{-1})[d]	
				Mean	SE
Crops (total)		1.43	8091	26.35	10.22
Cereals (total)	2587	1.37	3557	22.39	8.41
Rice	723	1.1	795	6.60	1.99
Wheat	704	1.29	906	4.93	2.30
Maize	883	1.35	1194	6.14	2.46
Soybeans	261	1.5	391	0.27	0.14
Roots and Tubers	807	0.58	468	0.3	0.17
Oil-bearing crops	105	2.2	231	0.59	0.70
Seed cotton	77	2.91	225	0.18	0.07
Sugar cane[e]	1794	0.18	323	2.63	0.74

[a]values from FAO [29];
[b]mass ratios of the Si-rich organ: crop output from Huang et al. [35] and Zhu et al. [36];
[c]ANPP: above-ground net primary productivity;
[d]estimated from the crop output and Si-rich organ factor and PhytOC content in Table 1 using equations (2, 3);
[e]The crop output of sugar cane is fresh cane weight.

composition of phytoliths in plant organs, and deposition environments of phytoliths after plant decay [18,25].

Results

Distribution of PhytOC in Dominant Arable Crops

The global area of croplands is 1532.6 10^6 hm^2, about half of which is covered by cereals (Table 1). The PhytOC content varies greatly among different crops (0.02–0.25%, with an average of 0.13%) (Table 1). Generally, sugar cane and cereals have higher PhytOC contents in dry biomass (0.16–0.25%) than than other crops (0.02–0.08%). Within cereals, rice has higher PhytOC content in dry biomass (0.25±0.07%) than other cereal crops such as wheat (0.16±0.08%) and maize (0.16±0.05%).

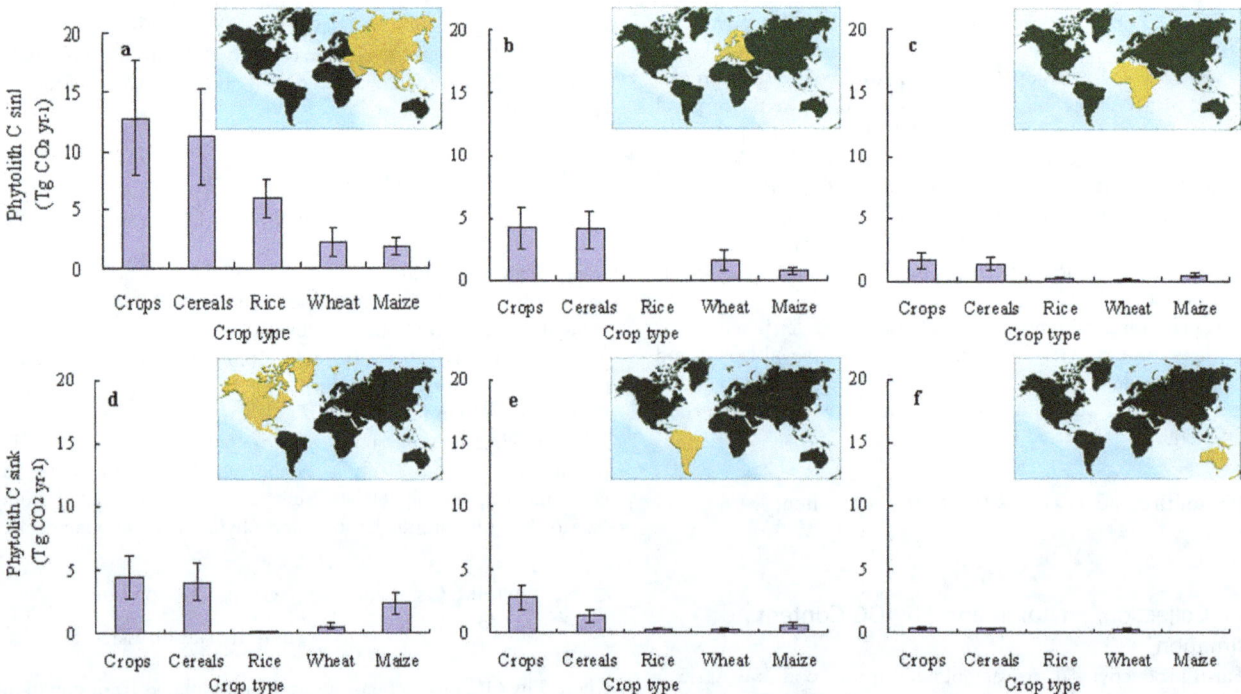

Figure 2. Phytolith carbon sink production by farm crops from different continents in 2011. Where **A:** Asia, **B:** Europe, **C:** Africa, **D:** North America, **E:** South America, **F:** Oceania. 'Crops' represents the sum of all farm crops and "Cereals" represents the sum of all cereal crops including rice, wheat, maize etc.

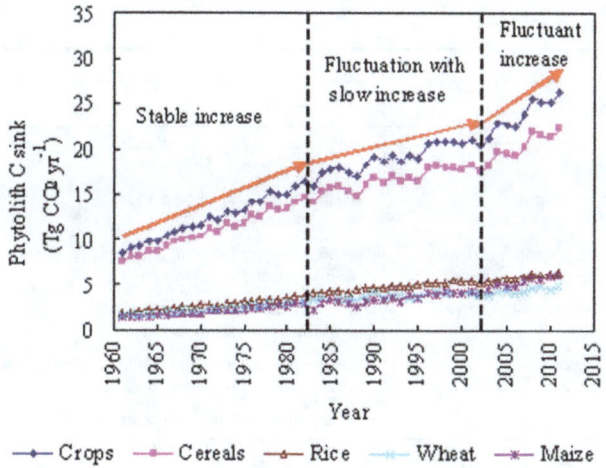

Figure 3. Phytolith carbon sink produced by global farm crops from 1961 to 2011. Crops include all farm crops in Table 1. Cereals include rice, wheat, maize, millet, barley and sorghum.

Phytolith Carbon Sink of Global Croplands

The phytolith C sink varies greatly among different crops (Table 2). The phytolith C sinks generated by rice, wheat and maize (6.60 ± 1.99, 4.93 ± 2.30 and 6.14 ± 2.46 Tg CO_2 yr^{-1}, respectively) are much higher than other crops. The total phytolith C sink produced by global farm crops is around 26.35 ± 10.22 Tg CO_2 yr^{-1}, 85% of which is contributed from cereals, including rice (25%), wheat (19%) and maize (23%).

Figure 2 displays the relative land area of the major continents and the phytolith C sink produced by farm crops in each in 2011. The largest phytolith C sinks occur in Asia (12.80 ± 4.90 Tg CO_2 yr^{-1}), North America (4.50 ± 1.74 Tg CO_2 yr^{-1}) and Europe (4.21 ± 1.66 Tg CO_2 yr^{-1}), which account for 49, 17 and 16% of the total global croplands, respectively.

The total phytolith C sink of global croplands has tripled since 1961 (Figure 3). In general, the evolution of the phytolith C sink since 1961 may be divided into three stages:

(1) 1961–1982: the total phytolith C sink increased steadily from 8.61 to 16.57 Tg CO_2 yr^{-1}.

(2) 1983–2002: the total phytolith C sink fluctuated, with a slow increase from 15.99 to 20.51 Tg CO_2 yr^{-1}.

(3) 2003–2011: the total phytolith C sink increased, with some fluctuation, from 21.14 to 26.35 Tg CO_2 yr^{-1}.

Discussion

Contribution of Phytolith C Sink to Global Cropland C Balance

By comparing phytolith C sink data and the global cropland soil C balance during 1961–2100, the contribution of the phytolith C sink to the net global cropland C balance was estimated (Table 3). Ruddiman [37] estimated that the emission from land-use conversion during the postindustrial era (i.e. 200 years), using 0.8 Gt C yr^{-1} (or 2.93 Pg CO_2 yr^{-1}), at 160 Gt C (or 587 Pg CO_2). Using an average soil C sink rate of -2.93 Pg CO_2 yr^{-1} during 1961–2015, the total soil C sink is about -161.2 Pg CO_2. Lal [6,14] estimated a high and attainable soil C sequestration potential of 0.55 Gt C yr^{-1} (or 2.02 Pg CO_2 yr^{-1}) for global croplands assuming judicious land use and recommended management practices (RMPs) were taken world-wide. Taking an average soil C sink rate of 2.02 Pg CO_2 yr^{-1} during 2016–2100, the total soil C sink is about 171.7 Pg CO_2. The total net soil C sink of global croplands during 1961–2100 is about 10.6 Pg CO_2. Taking an average phytolith C sink rate of 0.03 Pg CO_2 yr^{-1}, the total phytolith C sink of global croplands during 1961 and 2100 is 4.2 ± 1.9 Pg CO_2 yr^{-1}, 40 ± 18% of the total net soil C sink (Table 3).

Enhancing Phytolith Carbon Sink through the Optimization of Cropping System and Fertilization

Carbon sink trading has been carried out in many fields [38]. If phytolith carbon sink can be measured exactly, we believe that the sink may also be traded to increase the income of farmers. Therefore, in the future, farmers will be optimising carbon sequestration besides increasing yields.

Regional analysis of crop structures and farm productivity (FAO, 2012) suggests that the high phytolith C sinks in Asia, North America and Europe are due to the relatively wide production of rice, maize and wheat, respectively. The rapid increase of total phytolith C sink since 1961 has been due to cropland expansion and increase in the cereal yield per unit area as a result of fertilizer application and irrigation.

Although the global cropland area is difficult to increase significantly in the near future, the findings of the study suggest that the present global cropland phytolith carbon sink could be further enhanced through the optimization of cropping system and fertilization (Table 4).

Cropping system optimization measures include enhancement of cereal area percentage in croplands and enhancement of multi-cropping index (Table 4). For example, Parr and Sullivan [20] and Li et al. [19] revealed that the enhancement of rice and wheat area percentage in croplands might significantly increase the total phytolith C sink in croplands because of their higher phytolith contents than other crops with low costs. Enhancement of multi-

Table 3. Contribution of the phytolith C sink to the global cropland C balance for 1961–2100.

	Phytolith C	Soil C			Phytolith C contribution (%)
Period	1961–2100	1961–2015[a]	2016–2100[b]	1961–2100	1961–2100
Sink rate (Pg CO_2 yr^{-1})	0.03 ± 0.01	-2.93	2.02	0.08	40 ± 18
Total sink (Pg CO_2)	4.2 ± 1.9	-161.2	171.7	10.6	40 ± 18

Sinks are positive values and sources are negative values.
[a] the average soil C sink rate data of 1961–2015 are after Ruddiman [37].
[b] the average soil C sink rate data of 2016–2100 are after Lal [6,14] assuming judicious land use and recommended management practices (RMPs) are applied worldwide during 2016–2100.

Table 4. Potential measures to enhance global cropland phytolith carbon sink.

Types	Measures	Mechanisms	Comments
Optimization of cropping system	Enhancement of cereal percentage in croplands	Enhancing crop output and phytolith content	High efficiency in all croplands with low costs
	Enhancement of multi- cropping index	Enhancing crop output	High efficiency in all croplands with low costs
Fertilization	Silicon fertilizer application	Enhancing crop phytolith content	High efficiency in cereal croplands and sugarcane with high costs
	Rock powder amendment	Enhancing crop phytolith content	High efficiency in cereal croplands and sugarcane with low costs
	Organic mulching	Enhancing crop output and phytolith content	High efficiency in cereal croplands and sugarcane with low costs
	Traditional fertilization	Enhancing crop output	High efficiency with high costs

cropping index may significantly increase the total cropland phytolith C sink by enhancing crop output with low costs.

Fertilization measures include silicon fertilizer application, rock powder amendment, organic mulching, and traditional fertilization (Table 4). Silicon fertilizer application, rock powder amendment and organic mulching will increase soil bioavailable silicon input, plant silicon uptake and phytolith content for cereals and sugarcane [5,19]. Traditional fertilization (N, P, K fertilizer application) may also increase total phytolith C sink in croplands by enhancing crop output.

Although the potential measures proposed for promoting cropland phytolith C sink based on the study are meritable, more data is required. The exact efficiency and costs of the proposed measures need further assessment before practical measures may be implemented to sequester globally-significant amounts of atmospheric CO_2.

Conclusions

Relative to the liable biomass C sink, the phytolith C sink in croplands is certain and stable, and can be sustained for several hundreds or thousands of years in most regions of the world. The phytolith sink of global croplands is a stable net sink of

26.35 ± 10.22 Tg CO_2 yr^{-1}, and may play a significant role in global cropland C balance for 1961–2100. The high phytolith sinks in Asia, North America and Europe can be attributed to the relatively high production of rice, maize, and wheat, respectively. The total phytolith C sink of global croplands has tripled since 1961 mainly due to fertilization, irrigation and cropland expansion. Taking an average phytolith C sink rate of 0.03 Pg CO_2 yr^{-1}, the total phytolith C sink of global croplands during 1961 and 2100 is 4.2 ± 1.9 Pg CO_2 yr^{-1}, $40 \pm 18\%$ of the total net soil C sink. Our data suggest that the cropland phytolith C sinks may be further enhanced by adopting cropland management practices such as optimization of cropping system and fertilization.

Acknowledgments

We thank Mr. David Cushley of ISE for language editing.

Author Contributions

Conceived and designed the experiments: ZLS JFP. Performed the experiments: ZLS FSG. Analyzed the data: ZLS. Contributed reagents/materials/analysis tools: ZLS. Wrote the paper: ZLS JFP FSG.

References

1. Cao M, Woodward FI (1998) Dynamic responses of terrestrial ecosystem carbon cycling to global climate change. Nature 393: 249–252.
2. Heimann M, Reichstein M (2008) Terrestrial ecosystem carbon dynamics and climate feedbacks. Nature 451: 289–292.
3. Piao S, Fang J, Ciais P, Peylin P, Huang Y, et al. (2009) The carbon balance of terrestrial ecosystems in China. Nature 458: 1008–1013.
4. Pan Y, Birdsey RA, Fang J, Houghton R, Kauppi PE, et al. (2011) A large and persistent carbon sink in the world's forests. Science 333: 988–993.
5. Song ZL, Wang HL, Strong PJ, Li ZM, Jiang PK (2012a) Plant impact on the coupled terrestrial biogeochemical cycles of silicon and carbon: Implications for biogeochemical carbon sequestration. Earth-Sci Rev 155, 319–331.
6. Lal R (2004a) Soil carbon sequestration impacts on global climate change and food security. Science 304: 1623–1627.
7. Houghton RA (1995) Changes in the storage of terrestrial carbon since 1850. In: Lal R, Kimble JM, Levine E, Stewart BA (Eds.), Soils and Global Change. CRC/Lewis, Boca Raton, FL, pp, 45–65.
8. Houghton RA (1999) The annual net flux of carbon to the atmosphere from changes in land use 1850 to 1990. Tellus 50B: 298–313.
9. Houghton RA, Hackler JL, Lawrence KT (1999) The U.S. carbon budget: contributions from land-use change. Science 285: 574–578.
10. Schimel DS (1995) Terrestrial ecosystems and the carbon cycle. Global Change Biol 1: 77–91.
11. Intergovernmental Panel on Climate Change (IPCC) (2001) Climate Change: The Scientific Basis. Cambridge Univ Press, Cambridge, UK.
12. Smith PC (2004) Carbon sequestration in croplands: the potential in Europe and the global context. Eur J Agron 20: 229–236.
13. Lal R (2001) Managing world soils for food security and environmental quality. Adv Agron 71: 155–192.
14. Lal R (2004b) Soil carbon sequestration to mitigate climate change. Geoderma 123: 1–22.
15. Six J, Ogle SM, Jay breidt F, Conant RT, Mosier AR, et al. (2004) The potential to mitigate global warming with no-tillage management is only realized when practised in the long term. Global Change Biol 10: 155–160.
16. Siever R, Scott OL (1963) Organic geochemistry of silica, in: Berger IA (Ed.), Organic Geochemistry. Pergammon Press, Elmsford, NY, pp, 579–595.
17. Piperno DR (1988) Phytolith Analysis: An Archaeological and Geological Perspective. Academic Press, London.
18. Parr JF, Sullivan LA (2005) Soil carbon sequestration in phytoliths. Soil Biol Biochem 37: 117–124.
19. Li ZM, Song ZL, Parr JF, Wang HL (2013) Occluded C in rice phytoliths: implications to biogeochemical carbon sequestration. Plant Soil. Published online first. DOI: 10.1007s11104-013-1661-9.
20. Parr JF, Sullivan LA (2011) Phytolith occluded carbon and silica variability in wheat cultivars. Plant Soil 342: 165–171.
21. Zuo X, Lü H (2011) Carbon sequestration within millet phytoliths from dry-farming of crops in China. Chinese Sci Bull 56: 3451–3456.
22. Parr JF, Sullivan LA, Quirk R (2009) Sugarcane phytoliths: Encapsulation and sequestration of a long-lived carbon fraction. Sugar Tech 11: 17–21.
23. Wilding LP (1967) Radiocarbon dating of biogenetic opal. Science 156: 66–67.
24. Mulholland SC, Prior C (1992) Processing of phytoliths for radiocarbon dating by AMS. Phytolitharien Newsletter 7: 7–9.

25. Meunier JD, Colin F, Alarcon C (1999) Biogenic silica storage in soils. Geology 27: 835–838.
26. Song ZL, Liu HY, Si Y, Yin Y (2012b) The production of phytoliths in the grasslands of China: implications to biogeochemical sequestration of atmospheric CO_2. Global Change Biol 18: 3647–3653.
27. Parr JF, Sullivan LA, Chen B, Ye G, Zheng W (2010) Carbon bio-sequestration within the phytoliths of economic bamboo species. Global Change Biol 16: 2661–2667.
28. Jansson C, Wullschleger SD, Kalluri UC, Tuskan GA (2010) Phytosequestration: Carbon biosequestration by plants and the prospects of genetic engineering. BioScience 60: 685–696.
29. Food and Agriculture Organization of the United Nations. FAO (2012): Statistics. Available: http://faostat.fao.org/site/567/default.aspx#ancor, from 1961 to 2011.
30. Hou X (1982) Vegetation Geography of China and Chemical Composition of its Dominant Plants, Science Press, Beijing, pp, 188–243.
31. Xu X, Yang L, Dong Y (1998) Rice field ecosystem in China. China Agriculture Press.
32. Zhou Q, Li Y, Lin J, Ye Y, Yang L (2005) Effects of irrigation on the vascular bundle structures and contents of silicon, magnesium and zinc in leaves of two sugarcane varieties. Sugar Crops of China (in Chinese with English abstract) 2: 1–4.
33. Ding TP, Tian SH, Sun L, Wu LH, Zhou JX et al. (2008) Silicon isotope fractionation between rice plants and nutrient solution and its significance to the study of the silicon cycle. Geochim Cosmochim Ac 72: 5600–5615.
34. Blecker SW, McCulley RL, Chadwick OA, Kelly EF (2006) Biologic cycling of silica across a grassland bioclimosequence. Global Biogeochem Cy 20: 1–11.
35. Huang Y, Zhang W, Sun WJ, Zheng XH (2007) Net primary production of Chinese croplands from 1950 to 1999. Ecol Appl 17: 692–701.
36. Zhu J, Li R, Yang X (2012) Spatial and temporal distribution of crop straw resources in 30 years in China. Journal of Northwest A & F University (in Chinese with English abstract) 40: 139–145.
37. Ruddiman WF (2003) The anthropogenic greenhouse era began thousands of years ago. Climate Change 61: 261–293.
38. Fang J, Chen A, Peng C, Zhao S, Ci L (2001) Changes in forest biomass carbon storage in China between 1949 and 1998. Science 292: 2320–2323.

10

Effects of Controlled-Release Fertiliser on Nitrogen Use Efficiency in Summer Maize

Bin Zhao, Shuting Dong*, Jiwang Zhang, Peng Liu

State Key Laboratory of Crop Biology, College of Agronomy, Shandong Agricultural University, Tai'an, China

Abstract

Nitrogen (N) is a nutrient element necessary for plant growth and development. However, excessive inputs of N will lead to inefficient use and large N losses to the environment, which can adversely affect air and water quality, biodiversity and human health. To examine the effects of controlled-release fertilisers (CRF) on yield, we measured ammonia volatilisation, N use efficiency (NUE) and photosynthetic rate after anthesis in summer maize hybrid cultivar Zhengdan958. Maize was grown using common compound fertiliser (CCF), the same amount of resin-coated controlled release fertiliser (CRFIII), the same amount of sulphur-coated controlled release fertiliser (SCFIII) as CCF, 75% CRF (CRFII) and SCF (SCFII), 50% CRF (CRFI) and SCF (SCFI), and no fertiliser. We found that treatments CRFIII, SCFIII, CRFII and SCFII produced grain yields that were 13.15%, 14.15%, 9.69% and 10.04% higher than CCF. There were no significant differences in grain yield among CRFI, SCFI and CCF. We also found that the ammonia volatilisation rates of CRF were significantly lower than those of CCF. The CRF treatments reduced the emission of ammonia by 51.34% to 91.34% compared to CCF. In addition, after treatment with CRF, maize exhibited a higher net photosynthetic rate than CCF after anthesis. Agronomic NUE and apparent N recovery were higher in the CRF treatment than in the CCF treatment. The N uptake and physiological NUE of the four yield-enhanced CRF treatments were higher than those of CCF. These results suggest that the increase in NUE in the CRF treatments was generally attributable to the higher photosynthetic rate and lower ammonia volatilisation compared to CCF-treated maize.

Editor: Randall P. Niedz, United States Department of Agriculture, United States of America

Funding: This research was supported by the National Natural Science Foundation of China (31171497, 31071358), System of Maize Modern Industrial Technologies (nyhyzx07-003), Ministry of Agriculture System of Maize Industrial Technologies (CARS-02), The Major State Basic Research Development Program of China (973 Program) (2011CB100105), Special Fund for Agro-scientific Research in the Public Interest (20120306, HY1203096), "Five-twelfth" National Science and Technology Support Program (2011BAD16B14, 2011BAD11B01, 2011BAD11B02). The funders had no role in study design, data collection and analysis, decision to publish, or preparation of the manuscript.

Competing Interests: The authors have declared that no competing interests exist.

* E-mail: stdong@sdau.edu.cn

Introduction

Nitrogen (N) is a critical element for plant growth, and adding N to crops is a valuable agronomic practice. During the past decade, China has made considerable progress in terms of grain yield (GY) and feeding its growing population; however, this increase in agricultural yield has partly resulted from excessive application of N fertilisers [1]. Excessive application can result in inefficiencies and large losses of excess N to the environment, which can impact air and water quality, biodiversity and human health [2]. The overuse of fertilisers contributes to NO_3-N contamination of both surface water and soil water, and high profile NO_3-N accumulation can reduce N use efficiency (NUE) [1,3]. Releases of nitrous oxide (mainly via the application of N fertiliser) can degrade stratospheric ozone and contribute to global warming [4]. Ammonia (NH_3) volatilisation from soil and plants can also aggravate environmental contamination and contribute to acid deposition [5]. Therefore, interventions to increase NUE and reduce N inputs are important not only for reducing environmental risk but also for lowering agricultural production costs [6].

Controlled-release fertiliser (CRF) is a possible alternative to common compound fertiliser (CCF) to increase N uptake efficiency and minimise N losses to the environment. However, current grower acceptance is limited due to a lack of experience with CRF performance and its high relative cost [7]. As one kind of enhanced-efficiency fertiliser, CRF has several advantages compared to CCF. Some of the advantages and disadvantages are listed in Table 1. The greatest benefits of switching from CCF to CRF include increased profitability and reductions in the environmental impact of crop production.

In sandy nursery soils, CRF was shown to be effective for seedling production, due to the increased residence time of CRF in the soil relative to conventional fertilisation [8,9]. Oliet et al. [10] found that CRF promoted suitable morphological values and nutritional status in *Pinus halepensis* planting stock, suggesting that the CRF types used in their study were suitable for the nursery production of *P. halepensis*. Tang et al. [11] reported major increases in rice yield following a single basal application of CRF, that was attributed to increased soil availability of N, superior development of the root systems, better nutrient absorption capacity, delayed senescence and enhanced lodging resistance.

To improve CRF, studies need to describe nutrient release characteristics. Du et al. [12] revealed that the thickness of the coating membrane was the most important parameter for controlling nitrate release, followed by temperature, granule radius and the saturated concentration of nitrate. However, the global use of CRF has so far been limited due to the higher cost (at least 2 or more times the price) compared to CCF.

Table 1. Advantages and disadvantages of CRF over CCF.

Advantages	Disadvantages
Slower release rate – plants are able to take up most of the fertilisers	Very high costs
Reduced fertiliser loss – slower leaching and run-off	
Reduced labour capital – less frequent application is required	Lower consumption
Lower salt index – reduced plant damage from high concentrations of salts	
Fertiliser burn is not a problem with CRF even at high rates of application	Limited to nursery stock

Note(s): CRF, controlled-release fertiliser; CCF, common compound fertiliser.

To date, few investigations have been carried out in the field on the performance of crops grown with CRF. Even if CRF use becomes economical, the widespread acceptance by growers will likely be limited as a result of grower concern about field performance [7]. It has mainly been applied to nursery stock in foreign countries. Until now, reports on CRF in crops have focused mainly on domestic rice and little information is available about the effects of CRF on maize. Consequently, it is valuable to clarify the mechanisms of the impacts of CRF on maize. Therefore, in the present study, we investigated the following: how to select the right application rate of CRF in maize, the photosynthetic traits and physiological mechanisms associated with yields and the NUE of CRF in maize. Our results will assist in the successful application of CRF to maize fields.

Materials and Methods

Experimental Design

Summer maize hybrid Zhengdan958 (released in 2000) was planted on a farm at Shandong Agricultural University, Shandong Province, China (36°10′19′′N, 117°9′03′′E). The soil, classified as a silt loam, is considered to be highly suitable for crop production. The soil pH was 6.1. The average organic matter content in the tillage layer was 19.7 g kg^{-1} and the available N, phosphorus (P) and potassium (K) were 124.38 mg kg^{-1}, 45.23 mg kg^{-1} and 81.78 mg kg^{-1}, respectively. Methods of soil analysis referenced from "Agricultural Soil Analysis" written by Bao S D [13]. Two kinds of CRF, a resin-coated CRF (hereafter, CRF) and a sulphur-coated CRF (hereafter, SCF) offered by Shandong Kingenta Ecological Engineering Co., Ltd. located in No. 19 Xingda West Street, Linshu, Shandong Province were used in our experiment and common compound fertiliser (CCF) was used as a control. The content of N, P$_2$O$_5$ and K$_2$O in each CCF, CRF and SCF was 24%, 8% and 16%; 21%, 7% and 14%; and 18%, 6% and 12%, respectively. The experiment was conducted as a random complete block design with three replications. There were eight treatments: CCF applied at 1250 kg ha^{-1} (the local average commercial fertiliser N application rate; hereafter, CCF); CRF applied at 714.29 kg ha^{-1} (CRFI, 50% CCF), 1071.43 kg ha^{-1} (CRFII, 75% CCF) and 1428.57 kg ha^{-1} (CRFIII, 100% CCF); SCF applied at 833.33 kg ha^{-1} (SCFI, 50% CCF), 1250 kg ha^{-1} (SCFII, 75% CCF) and 1666.67 kg ha^{-1} (SCFIII, 100% CCF); and control plots without fertiliser application (CK). All fertilisers were applied at a basal dose.

Measurement of Net Photosynthetic Rate

Net photosynthetic rate (P_N) was measured with a portable, open-flow portable photosynthetic system (LI-COR, LI-6400 System, UK) in 2006. The photosynthetic photon flux density (PPFD) was 1400 μmolm^{-2}s^{-1} provided by the internal light source of the leaf chamber. The measurements were taken at approximately 10-day intervals following pollination on cloudless days. All of the leaves (22 or 23) were fully expanded at the time of measurement. Only ear leaves were used for P_N measurements, because these structures are metabolically active for the longest period of time and their relative contribution to total photosynthetic assimilates is high [14]. The leaves for experiments were all fully exposed and oriented to normal irradiation during measurements. Five plants per treatment were randomly selected for measurements.

Plant Sampling and N Content Determination

To measure aboveground N uptake, five plants were collected per treatment in about 10-day intervals 30 days after sowing. At the mature stage, five plants were manually harvested per treatment. Rows per ear (RE), kernels per row (KR) and kernels per ear (KE) were counted. Aboveground dry matter (DM) was determined by oven-drying the samples at 80°C until a constant weight was achieved. Subsequently, samples were manually separated into the vegetative and grain portions. Then the GY and thousand-kernel weight (TKW) were determined.

The grains and straw were ground using a cyclone sample mill with a mesh size of 0.5 mm. Then the grain N concentration (GNCT) and the straw N concentration (SNCT) were measured using the micro-Kjeldahl method (CN61M/KDY-9820, Beijing).

The following parameters were calculated:

$$Plant\ N\ uptake = GNCT\,(the\ grain\ N\ concentration)$$
$$\times GW\,(the\ grains\ weight)$$
$$+ SNCT\,(the\ straw\ N\ concentration)$$
$$\times DMSW\,(the\ dry\ matter\ of\ straw\ weight)$$

$$Agronomic\ NUE\,(ANUE)$$
$$= (grain\ weight[fertiliser] - grain\ weight[no\ fertiliser])/N$$
$$fertiliser\ applied$$

$$Physiological\ NUE\ (PNUE)$$
$$= (grain\ weight[fertiliser] - grain\ weight[no\ fertiliser])$$
$$/[plant\ N\,(fertiliser) - plant\ N\,(nofertiliser)]$$

$$Apparent\ N\ recovery\,(AR)$$
$$= (plant\ N[fertiliser] - plant\ N[no\ fertiliser])/N$$
$$fertiliser\ applied$$

The plant N included N in dry matter of straw and grain.

Ammonia (NH₃) Volatilisation Measuring Device and Procedure

Devices. The vented chamber (Figure 1) was made of grey round polyvinyl chloride (PVC) tubing (15 cm internal diameter and 10 cm high), as described by Liao [15]. Two pieces of round sponge (16 cm in diameter and 2 cm in thickness) were moistened with a 15 mL phosphate-glycerol solution (50 mL analytical phosphate and 40 mL glycerol diluted to 1000 mL with pure water) and then inserted into each chamber. Because the volume of the solution only accounted for 3.7% of the sponge's volume, the sponge was still ventilative after being moistened. The sponge inside the chamber absorbed NH₃ volatilised from the soil, and the top sponge absorbed NH₃ from the ambient air. Glycerol in the sponges absorbed moisture from the air inside or outside the chamber and prevented the sponges from drying [16].

Detection of Ammonia Volatilization in Field

On June 12, 2006, when N fertiliser was applied to summer maize at a basal dose, five sets of both vented chambers were evenly placed at different locations in each plot, with the bottom edge pushed 2 cm into the soil. The two pieces of sponge in the vented chamber and the boric acid solution in the closed chamber were replaced every day during the first week and every 2 to 3 days during the second and third week. Simultaneously, each chamber was moved to a new location in the plot. Ammonia in the phosphate solution in each sponge inside the vented chamber was extracted with 300 mL of 1 M KCl after 60 min of oscillation. Ammonium quantities in the KCl extract solution were measured using the micro-Kjeldahl method (CN61M/KDY-9820, Beijing). NH₃ volatilisation from the soil was estimated by the following formula:

$$NH_3 - N\left(kgNha^{-1}d^{-1}\right) = M/(A \times D) \times 10^{-2}$$

where M is the NH₃ (mg N) captured by the vented chamber during each sampling, A is the cross-section area (m²) of the round chamber, D (days) is the duration of each sampling, and 10^{-2} is 10,000 m² ha⁻¹ × 10⁻⁶ kg mg⁻¹.

In the field, five sets of vented-chamber devices were evenly placed at different locations in each plot in the manner described above. During the summer maize-growing season, NH₃ emission was measured from June 12 to August 3, 2006, after basal N fertilisation. Measurements continued for 53 days.

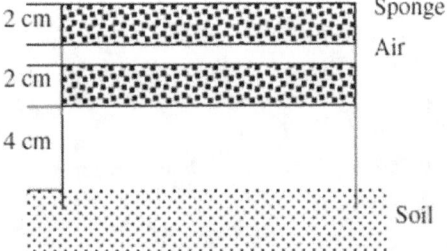

Figure 1. Vented-chamber methods used in field experiments to capture NH₃ emitted from the soil.

Statistical Analysis

Statistical analyses were performed using the analysis of variance (ANOVA) in the General Linear Model procedure of SPSS (Ver. 11, SPSS, Chicago, IL, USA). Results are presented as means of the 2 years of experimentation, because the trends of these parameters were consistent between years. The least significant differences (LSDs) between the means were estimated at the 95% confidence level. Unless indicated otherwise, significant differences among different plants are given at P<0.05. LSD was used to compare adjacent means arranged in order of magnitude. Calculations and linear regressions were performed using a *SigmaPlot 10.0* program.

Results

GY and GY Components

The application of fertilisers increased GY significantly compared to that of no fertiliser (Table 2), and the effect of CRF was much more pronounced than that of CCF. Furthermore, CRFIII, SCFIII, CRFII and SCFII were 13.15%, 14.15%, 9.69% and 10.04% higher in GY than CCF. No significant difference in GY was found between CRFI, SCFI and CCF, and there was no significant difference in GY between the two CRFs. The average economic efficiency of CRFIII/SCFIII was 1190.50 yuan hm⁻² more than CCF; CRFII/SCFII was 1753.75 yuan hm⁻² more than CCF; CRFI/SCFI was 758.75 yuan hm⁻² more than CCF.

Net Photosynthetic Rate (Post-anthesis Changes in the Light-saturated Photosynthesis Rate)

There was no significant difference in net P_N among treatments (Figure 2). All P_N of the ear leaves decreased after flowering, and P_N values for CRFIII, CRFII, SCFII and SCFII decreased more slowly than CCF. The P_{Ns} of CRFIII, CRFII, SCFIII and SCFII on the 10th day after flowering were 24.4%, 21.6%, 23.0% and 24.5%, higher, respectively, than CCF (P<0.05); on the 30th day after flowering were 13.6%, 14.9%, 15.3% and 17.8%, higher, respectively, than those of CCF (P<0.05); on the 50th day after flowering were 27.3%, 20.8%, 23.4% and 26.0%, higher, respectively, than those of CCF (P<0.05). No significant difference in P_N was observed among CRFI, SCFI and CCF. These results suggest that the yield increases afforded by the CRF treatments were generally attributable to their higher photosynthetic rates.

NH₃ Volatilisation

The application of N fertilisers in the field increased the volatilisation of NH₃ (Figure 3). The maximum flux of NH₃ increased to 3.36 kg N ha⁻¹ d⁻¹ 2 days after the application of CCF, and then rapidly decreased to approximately 1.18 kg N ha⁻¹ d⁻¹. However, the flux of NH₃ from CRF treatments was significantly lower than that of the CCF treatment. NH₃ volatilisation fluxes from CRF treatments peaked later than those of CCF. The treatments CRFIII and SCFIII reached their peak NH₃ volatilisation fluxes of 1.87 kg N ha⁻¹ d⁻¹ and 1.06 kg N ha⁻¹ d⁻¹, respectively, 9 days following fertiliser application (Figure 3 A B).

Cumulative rates of NH₃ volatilisation generally displayed similar patterns of increase and temporal characteristics up to 53 days following treatment (Figure 3 C D), after which parameters remained relatively constant. Cumulative fluxes of NH₃ emitted from the field were 10.56 kg N ha⁻¹ d⁻¹, 2.23 kg N ha⁻¹ d⁻¹, 3.16 kg N ha⁻¹ d⁻¹, 5.65 kg N ha⁻¹ d⁻¹, 3.79 kg N ha⁻¹ d⁻¹, 4.06 kg N ha⁻¹ d⁻¹ and 5.88 kg N ha⁻¹ d⁻¹ in the CCF, CRFI, CRFII, CRFIII, SCFI, SCFII and SCFIII

Table 2. Effect of controlled-release fertiliser on yield and its component of summer maize.

Treatments	Rows per ear	Kernels per row	Kernels per ear	Wt. per 1000-kernel(g)	Grain yield (kg hm^{-2})		Benefit increase than CK (yuan hm^{-2})	
					2005	2006	2005	2006
CK	14.44	39.02	563.65	281.17e	9383.8d D	8896.1c D	–	–
CCF	14.95	40.74	609.02	293.72d	11380.4c C	11046.1b BC	1056e	1349e
CRFI	14.53	39.93	580.36	298.20cd	11558.6bc BC	10826.8b C	2257c	1790d
CRFII	15.20	40.83	620.67	300.71c	12382.7ab AB	11990.4a A	2882a	3064a
CRFIII	14.60	41.70	608.82	316.29a	12716.9a A	12108.0a A	2571b	2339b
SCFI	14.67	40.57	595.04	296.58cd	11331.8c C	10986.4b BC	1763d	2035c
SCFII	14.80	42.07	622.59	299.77cd	12523.5a AB	11909.5a AB	3060a	2819a
SCFIII	15.07	41.16	620.08	308.53b	12810.5a A	12011.0a A	2629b	2033c

CCF, common compound fertiliser; CRF, a resin-coated CRF; SCF, a sulphur-coated CRF.
CCF, applied at 1250 kg ha^{-1} (the local average commercial fertiliser N application rate); CRFI, CRF applied at 714.29 kg ha^{-1} (50% CCF), CRFII,1071.43 kg ha^{-1} (75% CCF), CRFIII, 1428.57 kg ha^{-1} (100% CCF); SCFI, SCF applied at 833.33 kg ha^{-1} (50% CCF), SCFII, 1250 kg ha^{-1} (75% CCF), SCFIII, 1666.67 kg ha^{-1} (100% CCF); CK, control plots without N application.
Yield component of summer maize includes rows per ear, kernels per row, kernel No. per ear and Wt. per 1000-kernel.
According to the average market price at present, that is 1911 yuan t^{-1} for maize, 2190 yuan t^{-1} for CCF, 2660 yuan t^{-1} for CRF, 2350 yuan t^{-1} for SCF.
All data are means of 3 replications.
Means values marked with different capital letters indicate significant differences at $P=0.01$ level; different small letters indicate significant differences at $P=0.05$ level.

treatments, respectively. Volatilisation rates of NH_3 were 78.8%, 70.0%, 46.5%, 64.1%, 61.5% and 44.3% lower than those of CCF for CRFI, CRFII, CRFIII, SCFI, SCFII and SCFIII, respectively ($P<0.05$). These results suggest that the application of CRF considerably decreased NH_3 volatilisation rates.

N Uptake and NUE

Following the field application of CCF, N uptake increased rapidly during the first phase (i.e., the stage prior to flowering) and then increased slowly after flowering (Figure 4). Phenotypic phenomena such as vigorous growth before flowering and premature senescence after flowering were observed (data not shown). However, N uptake increased relatively constantly with

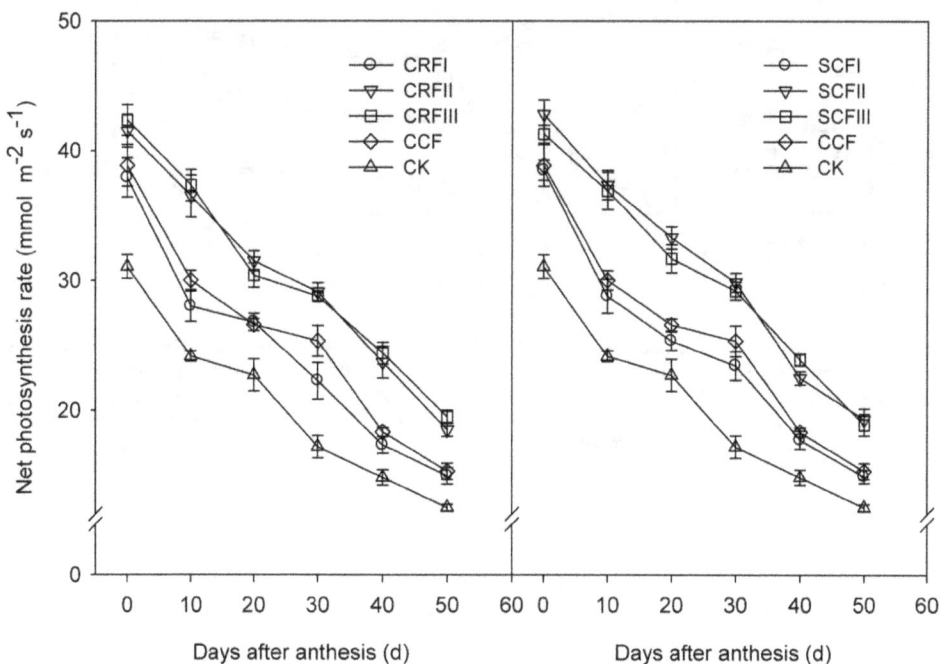

Figure 2. Effects of controlled-release fertiliser on net photosynthetic rate in ear leaves of summer maize. CCF, common compound fertiliser; CRF, a resin-coated CRF; SCF, a sulphur-coated CRF; CCF, applied at 1250 kg ha^{-1} (the local average commercial fertiliser N application rate); CRFI, CRF applied at 714.29 kg ha^{-1} (50% CCF), CRFII,1071.43 kg ha^{-1} (75% CCF), CRFIII, 1428.57 kg ha^{-1} (100% CCF); SCFI, SCF applied at 833.33 kg ha^{-1} (50% CCF), SCFII, 1250 kg ha^{-1} (75% CCF), SCFIII, 1666.67 kg ha^{-1} (100% CCF); CK, control plots without N application. Error bars are SE (n = 5).

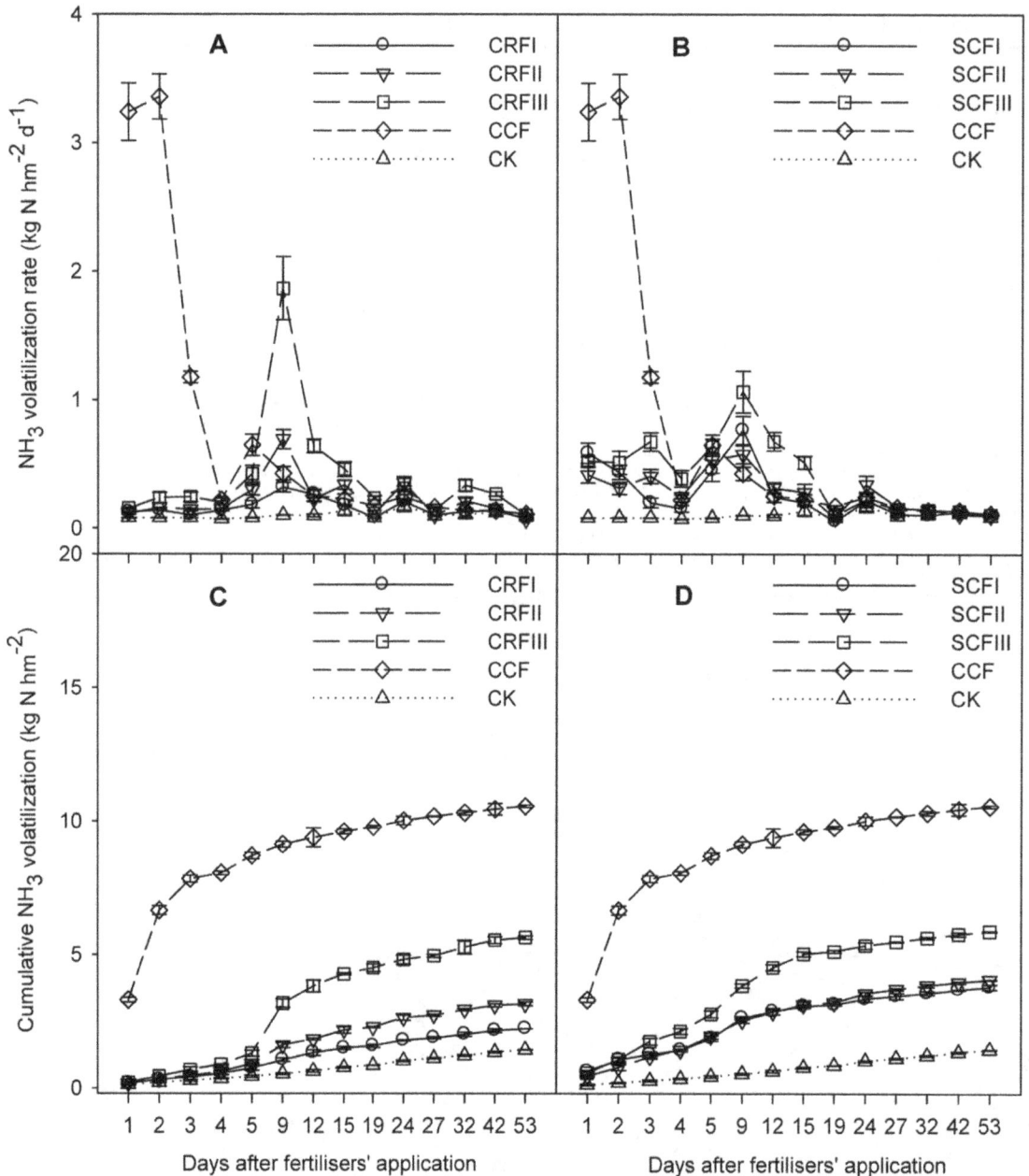

Figure 3. Soil NH₃ volatilisation rates (A and B) and changes in cumulative NH₃ volatilisation (C and D) following basal fertilisation.
CCF, common compound fertiliser; CRF, a resin-coated CRF; SCF, a sulphur-coated CRF; CCF, applied at 1250 kg ha⁻¹ (the local average commercial fertiliser N application rate); CRFI, CRF applied at 714.29 kg ha⁻¹ (50% CCF), CRFII,1071.43 kg ha⁻¹ (75% CCF), CRFIII, 1428.57 kg ha⁻¹ (100% CCF); SCFI, SCF applied at 833.33 kg ha⁻¹ (50% CCF), SCFII, 1250 kg ha⁻¹ (75% CCF), SCFIII, 1666.67 kg ha⁻¹ (100% CCF); CK, control plots without N application. Error bars are SE (n = 15; some SE bars are smaller than the symbols).

the increased application of CRF throughout the growth period (Figure 4), and caused obvious delays in leaf senescence. After flowering, the rank of shoot N uptake among all treatments was CRFIII>CRFII>CRFI>CCF>CK and SCFIII>SCFII>CCF >SCFI>CK.

ANUE and NR were significantly higher for CRF than for CCF (91.7% and 89.8%, respectively; $P<0.05$) (Table 3). Although no significant differences in PNUE were found between CRF and CCF, PNUE was slightly higher for CRFIII, CRFII, SCFIII and SCFII than for CCF.

Discussion

Suitable Application Rates for CRF in the Field

Fertiliser use efficiency has become a critical measure of sustainable agriculture. Efforts are underway to improve crop production and enhance N use efficiency in two main ways: by breeding new varieties of maize with high NUE and by improving fertiliser application management [17,18]. Among the improved N management practices, the use of enhanced-efficiency fertilisers such as SRF and CRF, nitrification inhibitors (NI) and urease inhibitors (UI) are being studied extensively under a variety of

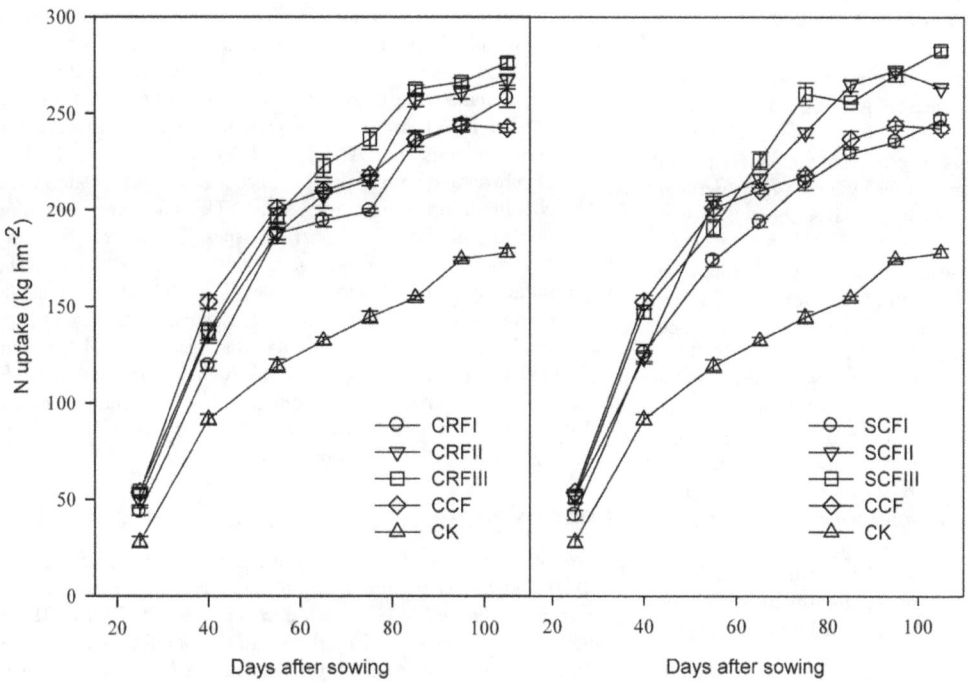

Figure 4. Dynamics of N uptake by aboveground parts after fertilisation. CCF, common compound fertiliser; CRF, a resin-coated CRF; SCF, a sulphur-coated CRF; CCF, applied at 1250 kg ha^{-1} (the local average commercial fertiliser N application rate); CRFI, CRF applied at 714.29 kg ha^{-1} (50% CCF), CRFII,1071.43 kg ha^{-1} (75% CCF), CRFIII, 1428.57 kg ha^{-1} (100% CCF); SCFI, SCF applied at 833.33 kg ha^{-1} (50% CCF), SCFII, 1250 kg ha^{-1} (75% CCF), SCFIII, 1666.67 kg ha^{-1} (100% CCF); CK, control plots without N application. Error bars are SE (n = 3; some SE bars are smaller than the symbols).

environmental conditions and agricultural systems to determine their effectiveness for increasing agricultural production and reducing environmental N losses [19]. CRFs are the fertilisers of the future, especially for open-field crops. Therefore, choosing the appropriate application rate is critical for the successful field application of CRF. In the present study, values of GY were not significantly different between CRFI/SCFI and CCF, CRFIII/ SCFIII or CRFII/SCFII. However, values of GY for CRFIII, CRFII, SCFIII and SCFII were significantly higher than for CCF.

Furthermore, agronomic NUE, apparent N recovery and economic efficiency of fertiliser for CRFII/SCFII (equivalent to 75% of CRF/SCF) were higher significantly than CCF. These results suggest that both of the two CRF types used in this study were effective for the agricultural production of maize, which also indicates that the CRFII/SCFII treatments corresponded to the optimum application rate of CRF for the maize fields studied in the North China Plain.

Table 3. Effects of different controlled-release fertiliser treatments on NUE of maize.

Treatments	Grain yield (t hm^{-2})	Total N uptake (kg N hm^{-2})	Agronomic N use efficiency (kg$_{grain}$ kg^{-1}N)	Apparent N recovery (%)	Physiological N use efficiency (kg$_{grain}$ kg^{-1}N)
CK	9.38d	177.83	–	–	–
CCF	11.38c	242.3	6.66c	21.49f	30.97ab
CRFI	11.56bc	257.82	14.50a	53.33a	27.19b
CRFII	12.38ab	267.28	12.56ab	39.76c	31.59ab
CRFIII	12.72a	276.11	11.11b	32.76e	33.92ab
SCFI	11.33c	246.82	12.99ab	45.99b	28.24b
SCFII	12.52a	263.28	13.95ab	37.98cd	36.74a
SCFIII	12.81a	282.48	11.42ab	34.88de	32.74ab

CCF, common compound fertiliser; CRF, a resin-coated CRF; SCF, a sulphur-coated CRF; CCF, applied at 1250 kg ha^{-1} (the local average commercial fertiliser N application rate); CRFI, CRF applied at 714.29 kg ha^{-1} (50% CCF), CRFII,1071.43 kg ha^{-1} (75% CCF), CRFIII, 1428.57 kg ha^{-1} (100% CCF); SCFI, SCF applied at 833.33 kg ha^{-1} (50% CCF), SCFII, 1250 kg ha^{-1} (75% CCF), SCFIII, 1666.67 kg ha^{-1} (100% CCF); CK, control plots without N application.
All data are means of 3 replications.
Means values marked with different letters indicate significant differences at $P = 0.05$ level.

Photosynthesis Rates and Physiological Mechanisms of NUE

Active photosynthesis has always been considered a desirable characteristic during the growing season [20]. Active photosynthesis in maize is primarily associated with the plants' ability to produce grain [20,21]. It had been suggested that yield increases in maize may be at least partly accounted for by increases in net leaf photosynthesis [22]. Leaf photosynthesis has been studied extensively as a plant trait in relation to NUE [23].

In the present study, after flowering, P_N in the yield-enhanced treatments (CRFIII, SCFIII, CRFII and SCFII) was significantly higher than that in the CCF treatments. No significant difference was found in P_N and yield among CRFI, SCFI and CCF. This suggests that the yield increase in the four CRF treatments may be attributable to the higher net photosynthetic rate. Wang et al. [14] also suggested that a yield increase in two cross-pollination treatments was generally due to a higher photosynthetic rate and related photosynthetic traits. Delaying or slowing down senescence may improve yield by increasing photosynthetic leaf area, which increases total photosynthate transported to sink tissue [24]. Indeed, in the present study, phenotypic delayed leaf senescence was obvious in the fourth yield-enhanced treatments (CRFIII, SCFIII, CRFII and SCFII), in accordance with Iain et al. [24].

Reduced NH3 Volatilisation

Ammonium ions (NH_4^+) in the soil exist in equilibrium with NH_3. If this conversion occurs at the soil surface and is accompanied by warm sunny days, NH_3 is subject to gaseous losses to the atmosphere. NH_3 is the most prolific atmospheric reactive N species emitted [25], and agricultural NH_3 emissions are predicted to increase significantly in Asia from 13.8 Tg N year^{-1} in 2000 to 18.8 Tg N year^{-1} in 2030 [26]. NH_3 emissions may result in N deposition to neighbouring ecosystems, which can damage vegetation [27]. In addition, some of the NH_3 may be oxidised and converted into nitric acid, which together with sulphuric acid, make up acid rain. This acidic deposition also damages vegetation, and can acidify both soil and surface water, inducing aluminium toxicity in terrestrial and aquatic organisms [28]. NUE are GY are complex traits that depend on interactions among several component traits [18]. Both appear to be most affected by the NH_3 volatilisation losses of N fertilisers [19].

NH_3 volatilisation in the present study was highest for the CCF treatment, and the majority of N losses occurred within the first 2–

12 days after CCF application. However, the majority of N losses occurred within the first 9–20 days after CRF application. The measured N uptake rates in the CRFIII, SCFIII, CRFII and SCFII treatments were higher than those of the CCF treatment, and ANUE and NR were significantly higher for CRF than for CCF. This suggests that ANUE and NR are significantly and positively correlated with N uptake, and negatively correlated with NH_3 volatilisation. Improvements in NUE and environmental protection are increasingly important issues. The use of CRF can partially resolve both of these issues. The reduced NH_3 volatilisation of CRF will help improve NUE while also decreasing the environmental contamination associated with excess N leaching. In addition, we found that the residual N of CRF in 0–100 cm soil profile was significantly higher than that of CCF (data not shown). Therefore, we conclude that CRF can be used to help conserve both air and water quality by maximising NUE and reducing N losses to the environment.

Conclusion

We found that GY was significantly higher for CRFIII, CRFII, SCFIII and SCFII treatments than for CCF treatment, while no significant difference in GY was found between CRFIII/SCFIII or CRFII/SCFII. These results indicate that 75% CRF/SCF was the optimum application rate for CRF in maize fields of the North China Plain. Further research is necessary to determine the effects, if any, of the type, frequency and timing of CRF applications on maize. In addition, the yield increases afforded by CRF were partly due to higher rates of net photosynthesis and lower rates of NH_3 volatilisation. Finally, we need to develop integrative approaches for enhancing societal *acceptance* of CRF, and for promoting the global application of this technology in economically sound and environmentally friendly agricultural systems.

Acknowledgments

Thanks are due to Dr Ruifang Wang, Qingen Xie, MS, Congfeng Li, MS, Hao Dong, MS, Dejun Gao, MS, for their help in the field and laboratory. We thank the anonymous reviewers for constructive suggestions.

Author Contributions

Conceived and designed the experiments: STD BZ. Performed the experiments: JWZ BZ. Analyzed the data: PL BZ. Contributed reagents/materials/analysis tools: STD BZ. Wrote the paper: BZ.

References

1. Zhu ZL, Chen DL (2002) Nitrogen fertilizer use in China contributions to food production, impacts on the environment and best management strategies. Nutrient Cycling in Agroecosystems 63: 117–127.
2. Goulding K, Jarvis S, Whitmore A (2008) Optimizing nutrient management for farm systems. Philosophical Transactions of the Royal Society B: Biological Sciences 363: 667–680.
3. Ju X, Liu X, Zhang F, Roelcke M (2004) Nitrogen fertilization, soil nitrate accumulation, and policy recommendations in several agricultural regions of China. AMBIO: A Journal of the Human Environment 33: 300–305.
4. Milich L (1999) The role of methane in global warming: where might mitigation strategies be focused? Global Environmental Change, Part A: Human and Policy Dimensions 9: 179–201.
5. Harper L, Sharpe R (1995) Nitrogen dynamics in irrigated corn: soil-plant nitrogen and atmospheric ammonia transport. Agronomy Journal 87: 669–675.
6. Wang RF, An DG, Hu CS, Li LH, Zhang YM, et al. (2011) Relationship between nitrogen uptake and use efficiency of winter wheat grown in the North China Plain. Crop & Pasture Science 62: 1–11.
7. Medina CL, Obreza TA, Sartain JB, Rouse RE (2008) Nitrogen release patterns of a mixed controlled-release fertilizer and its components. Hort Technology 18: 475–480.
8. Dobrahner J, Lowery B, Iyer JG (2007) Slow-release fertilization reduces nitrate leaching in bareroot production of Pinus strobes seedlings. Soil Science 172: 242–255.

9. Zotarelli L, Scholberg JM, Dukes MD, Munoz-Carpena R (2008) Fertilizer residence time affects nitrogen uptake efficiency and growth of sweet corn. Journal of Environmental Quality 37: 1271–1278.
10. Oliet J, Planelles R, Segura ML, Artero F, Jacobs DF (2004) Mineral nutrition and growth of containerized Pinus halepensis seedlings under controlled-release fertilization. Sci Hortic (Amsterdam) 103: 113–129.
11. Tang SH, Yang SH, Chen JS, Xu PZ, Zhang FB, et al. (2007) Studies on the mechanism of single basal application of controlled-release fertilizers for increasing yield of rice (Oryza sativa L.). Agricultural Sciences in China 6: 586–596.
12. Du C, Tang D, Zhou J, Wang H, Shaviv A (2008) Prediction of nitrate release from polymer-coated fertilizers using an artificial neural network model. Biosyst. Eng. 99: 478–486.
13. Bao SD (2005) Agricultural Soil Analysis [M]. Beijing:China Agriculture Press.
14. Wang RF, An DG, Xie QE, Wang KJ, Jiang GM (2009) Leaf photosynthesis is enhanced in normal oil maize pollinated by high oil maize hybrids. Industrial Crops and Products 29: 182–188.
15. Liao XL (1983) The methods of research of gaseous loss of nitrogen fertilizer. Progress Soil Sci. 11: 49–55.
16. Wang CH, Liu XJ, Ju XT, Zhang FS, Malhi SS (2004) Ammonia Volatilization Loss from Surface-Broadcast Urea: Comparison of Vented- and Closed-Chamber Methods and Loss in Winter Wheat–Summer Maize Rotation in North China Plain. Communications in Soil Science and Plant Analysis 35: 2917–2939.

17. Baligar V, Fageria N, He Z (2001) Nutrient use efficiency in plants. Communications in Soil Science and Plant Analysis 32: 921–950.
18. Dawson J, Huggins D, Jones S (2008) Characterizing nitrogen use efficiency in natural and agricultural ecosystems to improve the performance of cereal crops in low-input and organic agricultural systems. Field Crops Research 107: 89–101.
19. Motavalli PP, Goyne KW, Udawatta R (2008) Environmental impacts of enhanced efficiency nitrogen fertilizers. Online. Crop Management. doi:10.1094/CM-2008-0730-02-RV.
20. Ding L, Wang KJ, Jiang GM, Liu MZ, Niu SL, et al. (2005) Post-anthesis changes in photosynthetic traits of maize hybrids released in different years. Field Crop Research 93: 108–115.
21. Araya T, Noguchi K, Terashima I (2006) Effects of carbohydrate accumulation on photosynthesis differ between sink and source leaves of Phaseolus vulgaris L. Plant & Cell Physiology 47: 644–652.
22. Dwyer LM, Tollenaar M (1989) Genetic improvement in photosynthetic response of hybrid maize cultivars 1959 to 1988. Can.J. Plant Sci. 69: 81–91.
23. Foulkes M, Hawkesford M, Barraclough P, Holdsworth M, Kerr S, et al.(2009) Identifying traits to improve the nitrogen economy of wheat: Recent advances and future prospects. Field Crops Research 114: 329–342.
24. Iain SD, Alan PG, Howard T, Keith JE, David E, et al. (2007) Modification of nitrogen remobilization, grain fill and leaf senescence in maize (Zea mays) by transposon insertional mutagenesis in a protease gene. New Phytology 173: 481–494.
25. Galloway JN, Cowling EB (2002) Reactive nitrogen and the world: 200 years of change. Ambio 31: 64–71.
26. Zheng XH, Fu CB, Xu XK, Yan XD, Huang Y, et al. (2002) The Asian nitrogen cycle case study. Ambio 31: 79–87.
27. Newbould P (1989) The use of fertiliser in agriculture. Where do we go practically and ecologically? Plant Soil 115: 297–311.
28. Reuss JO, Johnson DW(1986) Acid deposition and acidification of soil and waters. Ecol Studies No. 59. Springer-Verlag, NY.

11

Distinct Soil Bacterial Communities Revealed under a Diversely Managed Agroecosystem

Raymon S. Shange[1]*, Ramble O. Ankumah[1], Abasiofiok M. Ibekwe[2], Robert Zabawa[3], Scot E. Dowd[4]

1 Department of Agricultural and Environmental Science, Tuskegee University, Tuskegee, Alabama, United States of America, 2 United States Department of Agriculture-Agricultural Research Service-United States Salinity Lab, Riverside, California, United States of America, 3 George Washington Carver Agricultural Experiment Station, Tuskegee University Tuskegee, Alabama, United States of America, 4 Molecular Research LP, Shallowater, Texas, United States of America

Abstract

Land-use change and management practices are normally enacted to manipulate environments to improve conditions that relate to production, remediation, and accommodation. However, their effect on the soil microbial community and their subsequent influence on soil function is still difficult to quantify. Recent applications of molecular techniques to soil biology, especially the use of 16S rRNA, are helping to bridge this gap. In this study, the influence of three land-use systems within a demonstration farm were evaluated with a view to further understand how these practices may impact observed soil bacterial communities. Replicate soil samples collected from the three land-use systems (grazed pine forest, cultivated crop, and grazed pasture) on a single soil type. High throughput 16S rRNA gene pyrosequencing was used to generate sequence datasets. The different land use systems showed distinction in the structure of their bacterial communities with respect to the differences detected in cluster analysis as well as diversity indices. Specific taxa, particularly Actinobacteria, Acidobacteria, and classes of Proteobacteria, showed significant shifts across the land-use strata. Families belonging to these taxa broke with notions of copio- and oligotrphy at the class level, as many of the less abundant groups of families of Actinobacteria showed a propensity for soil environments with reduced carbon/nutrient availability. Orders Actinomycetales and Solirubrobacterales showed their highest abundance in the heavily disturbed cultivated system despite the lowest soil organic carbon (SOC) values across the site. Selected soil properties ([SOC], total nitrogen [TN], soil texture, phosphodiesterase [PD], alkaline phosphatase [APA], acid phosphatase [ACP] activity, and pH) also differed significantly across land-use regimes, with SOM, PD, and pH showing variation consistent with shifts in community structure and composition. These results suggest that use of pyrosequencing along with traditional analysis of soil physiochemical properties may provide insight into the ecology of descending taxonomic groups in bacterial communities.

Editor: David L. Kirchman, University of Delaware, United States of America

Funding: The work for this publication was funded by a United States Department of Agriculture-McIntyre-Stennis Forest Service Grant (ALX-MSFSW 0219038; http://www.csrees.usda.gov/about/offices/legis/mcintirestennis.html), and a United States Department of Agriculture-National Institute of Food and Agriculture Grant (ALX-SWQ 0211874; http://www.csrees.usda.gov/). The funders had no role in study design, data collection and analysis, decision to publish, or preparation of the manuscript.

Competing Interests: The authors have declared the following interest. Scot E. Dowd is employed by Molecular Research LP. There are no patents, products in development or marketed products to declare.

* E-mail: rshange2946@mytu.tuskegee.edu

Introduction

Land use change and management in the conversion of forested, pastured, and cropped land have ecosystem-scale impacts such as: soil cycling of organic compounds [1–3], biodiversity [4–5], and soil nutrient dynamics [6]. The sensitivity of microbial communities to changes in management as well as their importance to nutrient cycling is a reason why they have been considered as early indicators of change in the quality of the soil ecosystem [7]. Although changes in bacterial communities have been reported in prior research [8,9], advances in sequencing technology have provided researchers with the ability to assess bacterial diversity at lower costs, and quicker turnaround than prior 16S rRNA and sequencing methods. These advances have allowed researchers to enhance work in the much needed area of bacterial community structure at varying scales [10–12].

Physical disturbance of the soil has been reported as being a crucial factor in determining soil biotic characteristics in agroecosystems [13]. The loss/disturbance of a stratified soil microhabitat has been attributed to a decrease in the density of species that inhabit agroecosystems. It has been proposed that such soil biodiversity reductions may be negative because the recycling of nutrients and proper balance among organic matter, soil organisms and plant diversity are necessary components of a productive and ecologically balanced soil environment [14]. For example, Acosta-Martinez et al. [15] evaluated the physical disturbance to bacterial communities with respect to tillage and reported that tillage reduced the bacterial diversity due to the interruption to physical diversity of the soil environment. Torsvik et al. [16] conducted a study comparing biodiversity of bacterial communities in perturbed soils versus relatively undisturbed, pristine soils, and found that human-induced pollution can lead to profound changes in microbial community structure by greatly reducing bacterial diversity. Other studies that have introduced organic amendments to soils have showed changes in microbial community structure in addition to biochemical changes [17,18].

Table 1. Sequences Recovered, Observed and Predicted OTUs for each sample.

Sample	Total Sequences Recovered	OTUs	ACE	Chao1
Cultivated 1	10,860	2065	7709	4699
Cultivated 2	5,318	589	1326	1009
Cultivated 3	11,494	1272	3150	2253
Forested 1	11,812	990	3094	2088
Forested 2	15,724	914	2104	1626
Forested 3	6,402	359	743	596
Pastured 1	12,382	1470	5491	3461
Pastured 2	19,981	1488	5174	3461
Pastured 3	18,678	1767	6816	4171

16S rRNA gene sequences recovered from soil DNA for each sample and the corresponding OTUs identified for each richness estimate.

Three primary areas of land use in the southeastern United States include pine plantation, annual cropping, and livestock pastures. The incorporation of forestry (particularly pine plantation) into agroecosystems has become an important economic, social [19], and environmental [20] issue in ecosystem management of the southeastern USA. Grazing of pine forests to control understory growth is one of the management systems which have been proposed to further diversify agroecosystems, and manage fire hazards [21,22]. All three of these land use types have the potential to alter soil chemical and physical disturbance, and ultimately to influence changes in soil bacterial community structure and composition. The ecological impact of management practices has been demonstrated as being a consistent source of disturbance to soil ecosystems [23]. Participating landowners and farmers are often engaged in multiple management practices within the same agroecosystem, creating a level of ecological complexity that is difficult to replicate in controlled settings. Hobbs and Huenneke [23] concluded that such disturbance has the potential to alter plant communities, but this process is poorly understood in the case of microbial species. It is therefore the objective of this study to utilize 16S rRNA sequencing techniques to elucidate the bacterial community structure and composition that exists in a single soil type exhibiting the three land use types discussed above. DNA was extracted from soils under each type of land use introduced above, amplified, and subjected to pyrosequencing. We therefore hypothesized that the specified land use types across the landscape will exhibit distinct soil bacterial communities.

Results and Discussion

Richness and Diversity Estimates

The maximum operational taxonomic units (OTUs) detected across the study site according to the observed clusters (sobs) at 3% dissimilarity was 2065 (Table 1) in the Cultivated sample 1, despite there being an obvious dominance of sequence recovery from the pastured samples. The maximum amount of OTUs observed is also reflected in the Chao and ACE values predicted for that sample. Though the highest number of OTUs was observed and predicted for this particular sample, the richness values were quite variable. All four of the estimators followed a distinct trend (pine forested<cultivated<grazed pasture) (Figures 1a and 1b). That trend exemplified an agroecosystem in which the lowest number of OTUs was found in soils sampled under the forest land use system

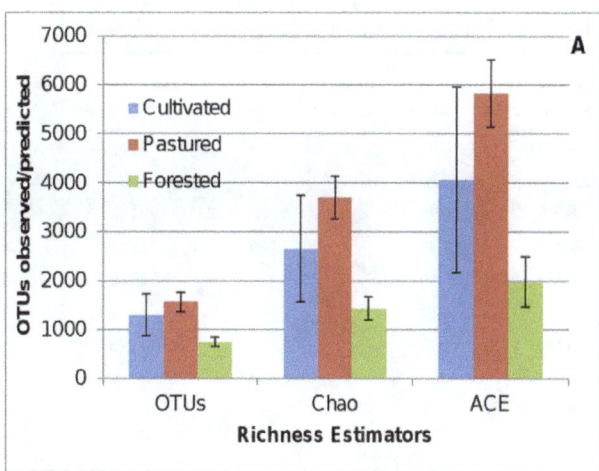

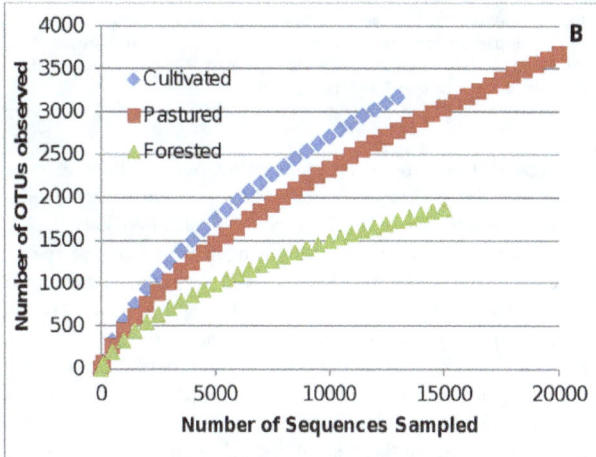

Figure 1. Diversity estimates. Richness/diversity estimators (**a**) and rarefaction curves (**b**) are presented as calculated by MOTHUR at a level of 3% dissimilarity.

and the highest under the grazed pasture soils. The indices reflected the same trend when calculated at 5% dissimilarity; however no differences were found between the land management systems for any of the diversity estimators.

In the richness/diversity data presented in Figures 1a and 1b, the grazed pasture system further establishes this point as its richness trends higher than the pine forest for all three of the estimators and trended higher than the community under the cultivated system. Increased richness in semi-stable pastures can be expected, but the added effects of grazing has been reported by other researchers as serving to enhance microbial communities attributing to three major factors: organism and substrate diversification from fecal and urine deposition, stimulation rhizosphere activity as a result of mowing, and the mixing and dispersal of microbial communities through trampling [24–26]. This pattern is not only noticeable in richness, but is also reflected in the highest SOC, TN, PD, and APA values (Table 2). Though the richness values of the cultivated community are not as high as those of the grazed pasture, they do not show a significant difference. In prior studies [27,28] no significant differences were detected between agricultural and pastured soils. It may be assumed that bacterial community richness under cultivated management would differ due to habitat difference introduced

Table 2. Selected soil properties.

Soil Properties	Cultivated	Forested	Pastured
pH (H$_2$O)	6.08a	4.90b	6.06a
SOC (g kg^{-1} soil)	2.70a	3.69b	4.41c
TN (g kg^{-1} soil)	0.26a	0.28a	0.39b
APA†	1.43a	2.10b	3.21b
ACP†	3.60a	3.30a	3.73a
PD†	1.62a	1.10b	1.96a
Sand‡	0.17a	0.31b	0.32b
Silt‡	0.34a	0.30b	0.27b
Clay‡	0.49a	0.39b	0.40ab

Means of selected soil properties and their means amongst different land use strata. Different letters denote significant differences between stratified land uses at $P \leq 0.05$ (n = 45).
†Values for enzyme activity are in units of μmol p-nitrophenol g soil^{-1} hr^{-1}.
‡Values for particle size are expressed as a fraction of total soil particles (1.00).
APA = acid phosphatase ACP = alkaline phosphatase PD = phosphodiesterase.
SOC = soil organic carbon TN = total nitrogen.

by inorganic amendments and tillage, but Acosta-Martinez et al. [11] suggest diverse crop residues and root systems may be cause for increased richness in diversely planted cultivated soils.

The forested system is the most distinguishable land use regime according to the acidic pH (Table 2) and the reduced amount of OTUs detected in the communities. In figures 1a & 1b, the forested soil bacterial community consistently shows the lowest values for richness. The forested soil presents harsh conditions to resident organisms, specifically in the acidic pH and polyphenolic compounds found in loblolly pine litter [29]. Soil pH has been found in numerous studies that utilize 16S sequencing techniques, as the driving abiotic factor shaping community structure in soils [30,9,12,31]. Though not directly measured in the study, polyphenolic compounds resulting from leaf litter have been reported as having a negative impact on soil bacterial communities [32,33] as well. This chemical disturbance creates a community with distinct abiotic and biotic characteristics, which impacts the growth and stability of the soil bacterial community.

Relative Abundance of Bacterial Phyla and Classes

Bacterial community compositions of the soils were examined at descending levels of biological classification to determine the effect of diversified land use on community membership. Detailed phylogenetic analyses grouped the soil associated bacterial sequences into 26 phyla (including unknown). The relative abundances of the 10 most abundant phyla are presented in Figure 2. The phylum distribution showed that Proteobacteria was the most dominant phyla with mean relative abundance values ranging from 39.9 % in cultivated soil, to 41.0 % in forested soil and to 45.5% in pastured soil. Other dominant phyla of note were Actinobacteria (representing 20.1–34.3% of the bacterial sequences in the samples) and Acidobacteria (which represented 4.4–20.6% of the bacterial sequences in each sample). Results from running GLM substantiate that the land uses are significantly different when considering the composition of the bacterial communities. Specific differences were observed between the cultivated and forested land use types for the phyla Acidobacteria ($P < 0.01$), while Actinobacteria showed significant distinction ($P < 0.05$) under the pine forested land use system.

The remaining phyla accounted for fewer than 25% of the relative abundance observed and were designated as minor (reference to abundance only). Of these groups, Gemmatimonadetes showed significantly distinguishable populations in all three land use types (forested/cultivated, $P < 0.001$; forested/pastured, $P \leq 0.001$; cultivated/pastured, $P < 0.05$), while Chloroflexi (forested/cultivated, $P < 0.05$; forested/pastured, $P < 0.05$) relative abundance is distinct in the forested system, and Verrucomicrobia (pastured/cultivated, $P < 0.05$; forested/cultivated, $P < 0.05$), BRC1 (pastured/cultivated, $P < 0.001$; forested/cultivated, $P < 0.001$) and Nitrospirae (pastured/cultivated, $P < 0.001$; forested/cultivated, $P < 0.001$) showed significant distinction in the cultivated area. Another significant difference observed the pastured and cultivated land use was in the phyla Cyanobacteria ($P \leq 0.005$). Shannon-Wiener indices for evenness calculated at 3% dissimilarity showed that the cultivated system had the highest value of evenness (0.93) compared to the pastured (0.89) and forested (0.87) with significant differences ($P < 0.05$) occurring between the cultivated and forested systems. When considering the difference of the composition of the bacterial community under the cultivated land use compared to the other management systems, it has been suggested [34] and supported [35] that the cultivation of a field, no matter how long ago, leaves an indelible imprint on the soil bacterial community. Through tillage, organic matter is incorporated throughout the plough layer of the soil and benefits unique microbial communities as the community reverts to an earlier (and more unstable) stage of ecological succession [36]. These communities are described as having quick responses to conditions of feast or famine that would be created by growing season-fallow pattern in cultivated systems. Figure 2 exhibits this phenomenon as the minor phyla in the cultivated system accounted for almost 20% of the sequences detected, and showed a more even distribution of abundance. With the values of evenness presented above, the data provides further evidence of a more even community in the cultivated system disturbed by tillage.

At the class level, α-proteobacteria, Acidobacteria (class), and Actinobacteria (class) were the dominant bacterial classes found across the site. Of the classes of Proteobacteria, α-proteobacteria appeared to be the most dominant among them. The remaining classes of Proteobacteria (with exception to γ-proteobacteria) showed higher relative abundance compared to that of any other taxa at the class level (excluding class Actinobacteria). Based on GLM of transformed data, relative abundance was significantly higher for α-proteobacteria and Acidobacteria in the forested than the cultivated system, while β-proteobacteria and δ-proteobacteria were significantly higher in the cultivated system (N = 3; $P < 0.05$). Seemingly, the pastured area was transitionary in its bacterial class abundance, as it differs only from the forested system only including an increase in Acidobacteria (N = 3; $P < 0.01$) and a decrease in β-proteobacteria and δ-proteobacteria (N = 3; $P < 0.05$). The observed shifts in taxonomic groups may suggest that microbial community composition changes in response to land management or environmental perturbation.

It has recently been reported that Actinobacteria and β-proteobacteria follow copiotrophic lifestyles [37,38], while Acidobacteria can be classified as oligotrophic. The results of the other soil properties present the land use regimes as environments that could support both types of lifestyles. The copiotrophic environment could be considered as the pastured system which has significantly higher SOC, TN, and nutrient cycling due to the activity of alkaline phosphatase and phosphodiesterase. The oligotrophic environment could be then considered as the forested system which is lacking in SOM and enzymatic activity as compared to the pastured system. The trend with respect to the

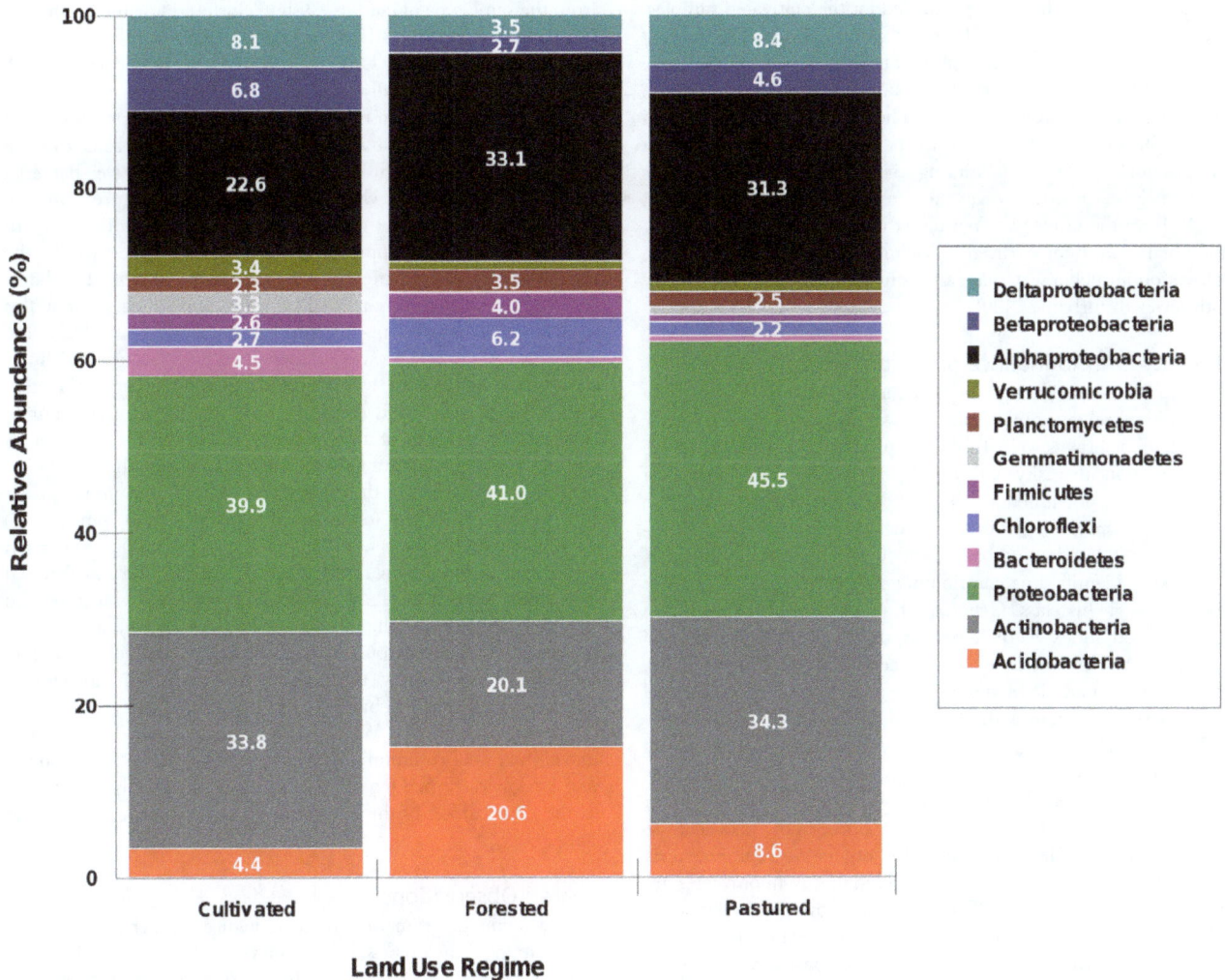

Figure 2. Relative abundance of major taxonomic groups across land use systems. Phyla included in this figure had relative abundance values consistently greater than 1%, as well as the abundant classes of the phylum Proteobacteria. Values presented are the mean percent. CLT = Cultivated; GRP = Grazed Pasture; PNF = Pine Plantation.

two trophic groups can be seen in Figure 2, in which the class Actinobacteria and the classes of Proteobacteria (with the exception of α-proteobacteria) trend down in the forested system with the opposite being observed for Acidobacteria. The instability of the cultivated community may also explain the lowest values for key taxa in the study. Even though α-proteobacteria was not explicitly designated as an oligotroph, in this study, as well as in other studies [27,31], its members have shown the ability to outcompete most (if not all other) bacterial classes in acidic soils. What this data suggests is that the chemical ecology of the forested system may provide the foundation for significant differences in community membership as well as growth. Studies have also shown that shifts from forest to grassland soil [31] as well as cultivated to pasture [11] result in changes to bacterial community composition. The assertion of trophic lifestyles was stated as a general guideline, so a deeper look into descending taxonomic groups was observed for these particular classes to determine if this assertion holds at these levels.

A heatmap (with hierarchal clustering) (Figure 3) of the bacterial families found in the samples was generated with respect to the

classes previously mentioned as following copiotrophic/oligotrophic lifestyles. The heatmap demonstrates that the top three families found in the study belong to the most abundant classes Actinobacteria and α-proteobacteria. The forested system appears to have the most distinction and was verified with ANOVA. The forested system differed from the other two systems significantly (P<0.05; n = 3) for the families: Acidobacteriaceae, Solibactereaceae, Solirubrobacterales (unclassified), Solirubrobactereaceae, Patullibactereaceae, Micromonosporaceae, Streptomycetaceae, Nocardioidaceae, Sphingomonodaceae, Alcaligenaceae, Burholderiaceae, Haliangiaceae, Polyangiaceae, Pseudomonadaceae. Out of the 42 Families graphed from these groups, less than half show abundance patterns that would support this assertion. As these are generally the most abundant families, prior assertions may be biased to those groups who are most identifiable using current methods. The evenness of the bacterial community under the cultivated system that was observed in Figure 2 and the reported Shannon evenness values is again exemplified by abundance spread further across the specified families than in any of the other systems. This is shown by the coverage of purple to orange colors in the columns labeled for the cultivated samples. The opposite is true for the Forested system, in which there is a large and visible gap in class abundance (exhibited by the large amount of black cells) in the columns designated as those belonging to the forested system. The samples of the Pastured and Cultivated systems seem to be more closely related in composition, as their clustering in the X-direction shows divergence at ~1.05, while the Forested system diverges at ~1.40 from the other two. Wardle et al. have also suggested that microbial community diversity/richness may not respond to cultivation in such a way that can distinguish it from a community with fewer disturbances [39].

Diversity and Abundance of Actinomycetes

To further demonstrate the differences in bacterial community composition, relative abundance was assessed at the level of Order for Actinobacteria as well. Table 3 displays the relative abundance and Shannon diversity indices of the most salient orders of Actinobacteria identified in the soils across the three land use types. The orders Solirubrobacterales, Actinomycetales, 0319-7L14, MC47, Acidomicrobiales, and Unclassified Actinobacteria (class) were identified as contributing substantially to the relative abundance of this class. Generally, both relative abundance and the Shannon index were lowest in the forested system for all Orders of Actinobacteria with the exception of Acidomicrobiales, which actually had the highest relative abundance and was second to the pastured system with respect to diversity.

These results suggest that pH shapes the Acidobacterial community structure more than the other environmental variables measured. Although the chemical environment seems to play a major role in the patterns of abundance, some of the taxa seem to respond to other factors. From the orders of Actinobacteria detected across the study site it is observed that Actinomycetes are the most abundant order (Table 3), and show a trend in their relative abundance and diversity across the site that reflects the amount of disturbance (pine forested<grazed pasture≤cultivated). Actinomycetes have been shown to have a preference for animal and human activity [40], and may be a possible biomarker in determining effects of land use change. Prior studies utilizing different primers from various geographic locations have suggested that Actinomycetes show higher abundance in agricultural and pasture soils, [41,42,9]. Although Solirubrobacterales has not been extensively studied, recent studies have shown its members to be adaptive in their ability to colonize different ecosystems: fungal growing ant colonies [43], spinach phyllosphere [44], desert and Antarctic soil [45,46]. In this study, the order seems to favor physical disturbance as its highest relative abundance is in the cultivated system, and its highest diversity is in the pastured system. In this way, Solirubrobacterales performs similar to Actinomycetales with respect to disturbance.

Principle Coordinates Analysis

Unifrac metrics were used to assess community similarity between two or more samples according to their structure (weighted/quantitative) and membership (unweighted/qualitative). In the 2-dimensional plot visualized from the Unifrac weighted distance matrix principle coordinates analysis (3% dissimilarity), the samples of each system distinctively responded to the majority of the variation detected in the samples across two axes (Figure 4a). Axis 1 accounted for 34.6% of the variation, and Axis 2 accounted for 14.8% of variation. In Figure 4b, the same 2-dimensional plot was shown for the unweighted method which shows that in consideration of community membership, samples

from the same type of land use system clustered together, although less distinctive (Axis 1 = 18.5%, Axis 2 = 14.1%).

The results from the Unifrac weighted and unweighted PCoA plots demonstrate the distinction that the bacterial communities under different land use management in both structure and composition. This is perhaps the most persuasive evidence that the three types of land use cultivate three distinctive bacterial communities. In the weighted PCoA plot (Figure 4a) the communities are clearly distinguished from one another quantitatively. There does appear to be some divergence within the samples from the cultivated system, as sample Cultivated 1, differs in its location on the y-axis. The divergence suggests that this sample exhibits different structure than the other two samples. This same phenomenon was evident when calculating the richness of the cultivated system, which explains the high error displayed in Figure 1a (as richness is considered a component of community structure). Regardless of its divergence, this community still seems distinct as clear separation is still detectable from the other land use types, primarily in the x-direction. The communities in the unweighted PCoA plot (constructed with respect to composition) show clear distinction as well (Figure 4b), as the samples from each of the land use systems cluster within the system. The pine forested and cultivated systems show the most dissimilarity with respect to structure and membership when considering the amount of space which separates them on the plot. Although the cultivated and pastured systems show the most similarity in physiochemical characteristics and richness/diversity values, they did show distinction in structure. Wardle [39] suggested that the difference in soil microbial communities of lightly disturbed (reduced/no till) and heavily disturbed (conventional tillage) systems is exemplified in the composition of the communities as seen in Figures 4a and 4b.

Shared Observations

The Venn diagram presented in Figure S1 demonstrates the distribution of phylotypes at the 3% level of dissimilarity. The total number of phylotypes found in the entire agroecosystem is 7771. All three land use systems shared 2.2% of these phylotypes, while the most were shared between the pastured and cultivated systems (8.9%). The cultivated and forested systems share the least amount of phylotypes (2.5%), and the forested and pastured system share 4.5%.

Clustering results showed three discrete communities present under the agroecosystem. Though the grazed pasture and forested systems show the greatest dissimilarity in the data, it may be suggested that the grazed pasture and cultivated systems may show more similarity. The Venn diagram further establishes this point. Of the 11.2% of the phylotypes that the grazed pasture system shares, 8.9% is shared with the cultivated community, as compared to the 4.5% that it shares with the forested community. Even though these similarities are observed, the data also distinguishes the grazed pasture and cultivated systems in the Unifrac PCoA plots (Figures 4a & 4b). This distinction suggests that regardless of the phylotype sharing there are significant differences in the structure and membership of the communities. With the unique membership and structure elucidated by the three systems studied on a single soil type, the development of these communities with respect to management raises further questions as to the influence of environmental history and dispersal within the soil ecosystem.

In conclusion, components of this study further establish the distinction of soil bacterial communities with respect to differing land management strategies. As has been demonstrated in other studies using both 16S rRNA and PLFA techniques we notice

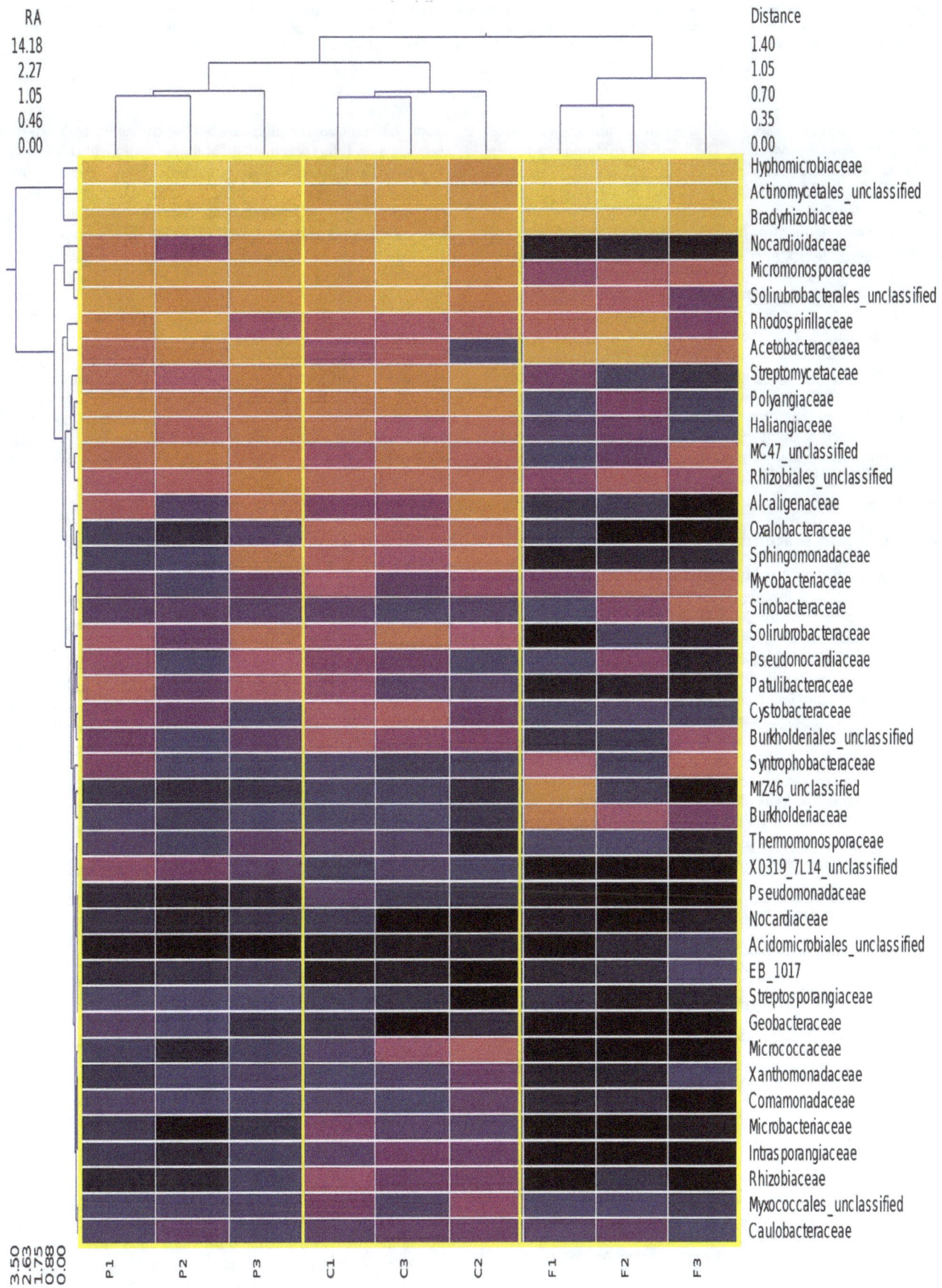

Figure 3. Hierarchical cluster analysis presented as a double dendrogram. A double cluster dendrogram that demonstrates the relative abundance of Families across the 9 samples across the three land use systems. Clustering in the Y-direction is indicative of abundance, not phylogenetic similarity. RA = Relative Abundance; CLT = Cultivated; PST = Grazed Pasture; FST = Grazed Pine Plantation.

distinction of forested soils with respect to community composition, structure, and phylogeny. The bacterial communities under in the cultivated and grazed pasture soils were similar in richness and composition, furthering the point that in a community with moderate disturbance, new individuals and groups could be introduced in a manner that promotes competition and diversity of the community, thus establishing a more stable community [47]. Though community richness and composition observed under the cultivated and grazed pasture regimes were not distinguishable, metrics assessing phylogenetic differences showed distinct communities under all three land use types. Further analysis sheds more light on the behavior of soil bacterial Orders and Families under differing land management strategies. As has been noted in other studies, phyla and class taxa behave according to predetermined guidelines regarding the chemical ecology of microbial communities, but at the lower levels of taxa a majority of groups depart from these generalizations. Further study will be needed to actually elucidate and validate the divergence in ecology of the descending taxa.

Materials and Methods

Sites and Sampling for Physical and Biochemical Analyses

The study site was located on the Sundown Ranch demonstration farm located at 32° 26′ latitude and −87° 27′ longitude on 17 hectares of land in Perry County, Alabama, USA. The soil series of the site was Kipling clay loam (Fine, smectitic, thermic Vertic Paleudalfs). For the past 10 years this land has been used as a demonstration farm for a consortium of Alabama Land Grant Universities.

A preliminary geostatistical study of soil biochemical characteristics provided the initial evidence that soil biochemical and biological factors spatially vary with respect to land use type on this site. The preliminary study allowed for the reasonable designation of three major sampling areas that coincided with land management considering the spatial variability of the soil surface. The sampling methodology of choice was a stratified random sampling

Table 3. Abundance and diversity of orders from class actinobacteria.

Orders of Actinobacteria	Cultivated		Forested		Pastured	
	RA[a]	SI[b]	RA[a]	SI[b]	RA[a]	SI[b]
Solirubrobacterales	6.21	3.15	1.67	2.23	2.93	4.00
MC47	1.86	2.52	1.03	2.38	2.37	2.88
0319-7L14	0.39	1.12	0	N/A	0.98	2.07
Acidomicrobiales	0.05	1.40	0.34	1.56	0.15	1.69
Actinomycetales	24.04	4.97	16.88	3.65	23.36	3.83
Unclassified	0.27	2.04	0.22	0.97	0.31	1.92

The mean relative abundance and Shannon index values as calculated for the Orders of the Class Actinobacteria.
[a]Relative abundance (%) of taxonomic group with respect to total OTUs observed for community.
[b]Shannon diversity index.

design that utilized sampling areas that coincided with specific land use types. Effort was made to obtain as statistically useful data set as possible that was well distributed throughout the site, while achieving a cost effective sampling method. The three sampling areas were constructed of sizes which were reflective of the amount of area each covered of the farm and overlaid by sampling grids. Sample numbers were assigned to each vertex in the grids, and the actual sample locations were randomly selected from these grid points. The amount of samples collected were proportional to the land area each land use system covered (loblolly pine forest = 8 ha; grazed pasture = 5 ha; cultivated = 3 ha).

The grazed pastured area is where pasture poultry are grazed twice yearly, and goats graze moderately throughout the year. This has been the management practice for the past 5 years. The loblolly pine plantation, which accounted for approximately half of the total area of the site, was designated as the forested system. It is characterized by a pine plantation at about 15 years of growth, and is lightly grazed by a goat herd. The crop-growing area was designated as the cultivated system, was used annually as a mixed cropping system (corn, beans, tomatoes, squash, and okra). This area has been under differential cropping management during the 10 years with the last 5 years being plastic mulch, inorganic fertilizer, and no herbicide.

Sampling was conducted on 14 October 2008 following the summer growing season. Soil samples of approximately 120-g field-moist weight were systematically collected from the upper 15 cm of soil at 45 sampling points in the landscape using a soil auger. In between sampling, the auger was sterilized with ethanol. Samples were preserved on ice during transport to the laboratory. Upon arrival soils were shortly stored at field moist conditions at 4°C. Prior to biochemical and physical assays soils were air-dried for 48 h, plant residues were removed by hand and the soil was sieved using a 2 mm mesh and mixed thoroughly thereafter. Five grams of soil were separated and kept at 4°C for molecular analysis.

Soil Enzyme Activity

Enzyme analysis was performed according to the methods of Tabatabai [48], with slight modification. The artificial substrate, p-nitrophenyl (1 mL, 0.05 M), and a pH buffer (pH values were 11 for Acid Phosphatase [APA], 6.5 for Acid Phosphatase [ACP], and 8 for Phosphodiesterase [PD]) were incubated in 25 mL glass flasks and capped at 37°C for 1 h with 1 g of soil. At the end of incubation, enzyme activity was stopped by addition of 4 mL of 0.5 M NaOH for phosphomonoesterases and 4 mL of 0.5 THAM-NaOH for phosphodiesterase followed by extraction with 1 mL of 0.5 M $CaCl_2$. The mixture was then filtered (Whatman No. 2) and the extract analyzed using a Genesys 10 VIS spectrophotometer at (Thermo Fisher Scientific Inc., Waltham MA, USA) at 420 nm. Enzyme activity in filtrates was determined from a standard curve developed using p-nitrophenol standards. To account for non-enzymatic hydrolysis, values for controls were subtracted from sample readings. Toluene was not used in accordance with Bandick and Dick [49] and Elsgaard et al. [50], who showed that with incubation periods fewer than two hours, the absence of toluene was inconsequential to measured enzyme activity. All enzyme activities reported are expressed on a moisture-free basis.

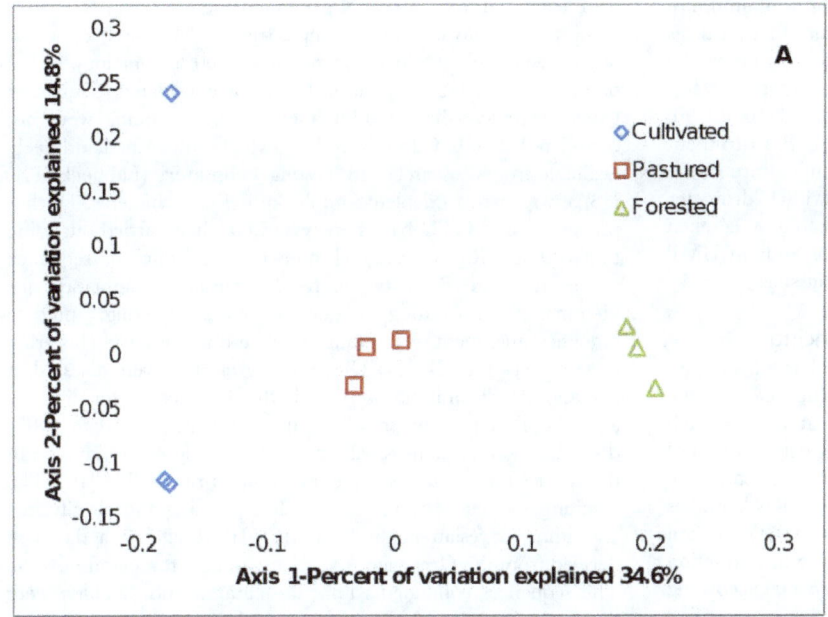

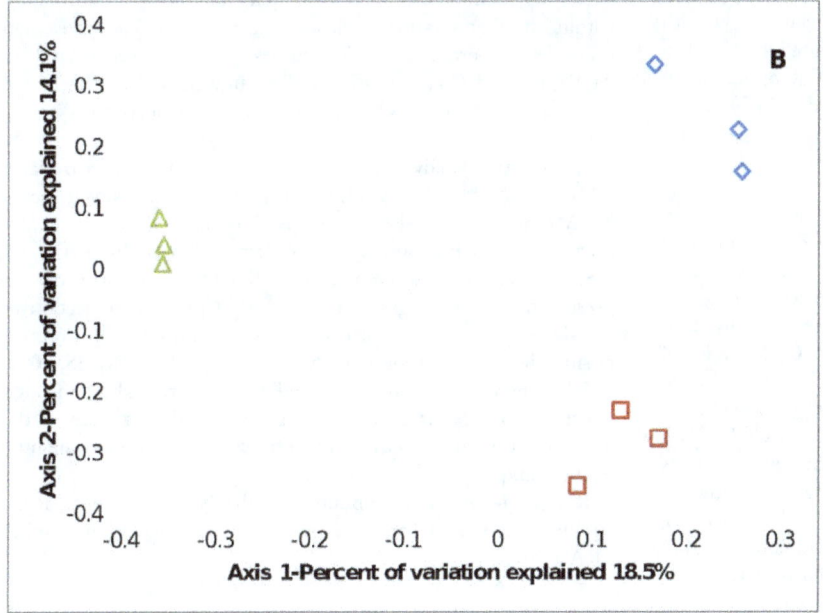

Figure 4. Ordination of unifrac metrics. PCoA plots are presented of the first two axes based on (a) weighted and (b) unweighted Unifrac distance matrices showing the quantitative and qualitative clustering of samples. The pastured system is represented by the red; the cultivated by the yellow; and the forested by the green.

Soil pH and SOM

Soil pH was analyzed using a 1:2 soil/water ratio according to those methods described in McLean [51]. The samples were then analyzed with an S500 pH Meter (following calibration at a pH of 4 and a pH of 10) [A. Daigger & Co., Vernon Hills, Illinois USA], measured to the nearest 0.01. For soil organic carbon (SOC) and total nitrogen (TN), air-dried soils were transferred to Auburn University Soil Testing Laboratory for analysis by the dry combustion method using an Elementar Vario Macro Combustion Analyzer (Elementar Americas, Inc., Mt. Laurel, NJ).

Soil Textural Analysis

Soil textural analysis was determined using the Buyocos hydrometer method for 40 g duplicate sample and <2 mm diameter. To enhance dispersion of the soil particles chemically, 100 mL of 0.05 g mL–1 Sodium Hexametaphosphate (HMP) solution was added to each sample and shaken (for mechanical dispersion) for 12 h [52]. After shaking, the HMP solution and soil were transferred into a 1000 mL sedimentation cylinder. Deionized water was added to bring the total volume in the sedimentation cylinder to 1000 mL. To begin sedimentation process, the sedimentation cylinder containing the sample was agitated by shaking the cylinder back and forth for a minimum of 30 s. During the agitation, care was taken to ensure that particles were not stuck on the sedimentation cylinder. After agitation the cylinder was placed on the countertop, signifying time zero. Readings were then taken at elapsed times of 0.667, 3, 10, 30, 90, 120, and 720 min using a 152 H hydrometer (H-B Instrument Company, Collegeville, Pennsylvania, USA). Temperature of the suspension liquid was recorded simultaneously with hydrometer readings. To calibrate the hydrometer, a blank reading was taken in a solution containing 0.05 g mL–1 100 mL of Sodium HMP solution and 900 mL deionized water, but without soil.

DNA Extraction, Amplification, and Sequencing

For molecular analysis of community DNA, nine soil samples were randomly selected from the soil samples within each of the three strata. DNA was extracted from these soil samples for each of the three strata and composited following extraction and validation (providing three representative samples per land use system). As previously demonstrated in other 16S rRNA studies [11,31,35], DNA was extracted from approximately 0.25 g of soil (oven dry basis of field-moist soil) using the Power Soil Extraction Kit (MO BIO Laboratories, Soloana Beach, California) according to the included protocol. Extracted DNA (2 μL) was quantified using Nanodrop ND-1000 spectrophotometer (Nanodrop Technologies, Wilmington DE), and run on 0.8% agarose gel with 0.5 M TBE buffer. The samples were then submitted to Research and Testing Laboratories (Lubbock, TX) for PCR optimization and pyrosequencing analysis. Bacterial tag-encoded FLX amplicon pyrosequencing PCR, massively parallel pyrosequencing and tag design were carried out according to procedure described previously by Dowd et al. [53,54].

Samples were evaluated using Tag-encoded FLX amplicon pyrosequencing (bTEFAP), which has had prior description and utilization by Dowd et al. in characterizing bacterial populations in a variety of studies [28,54,55]. All DNA samples were diluted to 20 ng/μl. A 20 ng (1 μl) aliquot of each sample DNA was used for a 25 μl PCR reaction with 5 min denature at 95°C, 30 cycles of 94°C 30 sec –52°C 40 sec –70°C 40 sec with a final extension of 70°C for 5 minutes. The 16S universal Eubacterial primers 28 F (5'- GGC GVA CGG GTG AGT AA) and 530 R (5'-CCG CNG CNG CTG GCA CS) Amplicons were mixed in equal volumes and purified using Agencourt Ampure beads (Agencourt Bioscience Corporation, MA, USA). In preparation for FLX sequencing (Roche, Nutley, NJ), the DNA fragments size and concentration were measured by using DNA chips under a Bio-Rad Experion Automated Electrophoresis Station (Bio-Rad Laboratories, Hercules, CA) and a TBS-380 Fluorometer (Promega Corporation, Madison, WI). A 9.6×10^6 sample of double-stranded DNA molecules/μl with an average size of 625 bp were combined with 9.6 million DNA capture beads, and then amplified by emulsion PCR. After bead recovery and bead enrichment, the bead-attached DNAs were denatured with NaOH, and sequencing primers were annealed. The 454 Titanium sequencing run was performed on a 70×75 GS PicoTiterPlate by using a Genome Sequencer FLX System (Roche, Nutley, NJ).

Bioinformatics and Statistical Analysis

Quality trimmed sequences were provided with the sequencing services by Research and Testing Laboratories (Lubbock, TX) following the process described in Acosta-Martinez et al. [11]. As described in the aforementioned studies, each sequence was trimmed to utilize only high quality sequence information; tags

were extracted from the FLX generated multi-FASTA file, while being parsed into individual sample specific files based upon the tag sequence. Tags which did not have 100% homology to the original sample tag designation were not considered. Sequences which were less than 250 bp after quality trimming were not considered. The B2C2 software [56], which is described and freely available from Research and Testing Laboratory (Lubbock, TX, USA), was used to deplete samples of definite chimeras. Further processing and OTU based analyses were then carried out using the MOTHUR v.1.19.4 [57] suite of algorithms for sequence processing and diversity analysis, including commands for identifying/consolidating unique sequences, filtering, multiple sequence alignment, generating distance matrices, and clustering of sequences into OTUs. The resulting clusters were assessed at 3% and 5% dissimilarity to provide the data needed for diversity analysis. Based upon the literature we can expect that 0% dissimilarity in sequences will provide dramatic overestimation of the species present in a sample, based upon rarefaction [10]. The resulting sequences were then evaluated using the classify.seqs algorithm (Bayesian method) in MOTHUR against a database derived from the Greengenes set using a bootstrap cutoff of 65%. The sequences contained within the curated 16S database were those considered of high quality based upon Greengenes [58] standards and which had complete taxonomic information within their annotations. Clusters at 3% and 5% were then utilized to generate rarefaction curves and the (diversity) indices ACE [59] and Chao1 [60] as well as unweighted and weighted UniFrac for Principle Coordinate Analysis (PCoA) plots and a venn diagram.

All statistical analysis was performed using the SPSS package (SPSS Inc, v 17.0, Chicago, Illinois). The generalized linear model (GLM) was used to assess the means of soil physical, chemical and microbial properties among the systems followed by a Tukey's HSD test for pairwise comparisons. Relative abundance data is presented as percentages/proportions, but prior to subjection to GLM, they were transformed using the arcsine function for normal distribution prior to analysis. NCSS package (NCSS, 2007, v 7.1.2, Kaysville, Utah) was used for cluster analysis through which double dendrograms were generated through use of the Manhattan distance method with no scaling, and the unweighted pair technique.

Raw sequences were submitted to the NCBI Sequence Read Archive (SRA) and can be found under the accession number SRA007616.

Supporting Information

Figure S1 Venn diagram of shared OTUs. A venn diagram of the phylotype richness among the three land use systems at 3% dissimilarity. The size of the spheres is not consistent with the amount of phylotypes present.

Acknowledgments

The authors would like to thank Drs. Conrad Bonsi, Desmond Mortley, and Berhanu Tameru for their suggestions and edits to the manuscript.

Author Contributions

Conceived and designed the experiments: RS ROA RZ. Performed the experiments: RS SED. Analyzed the data: RS. Contributed reagents/materials/analysis tools: ROA SED. Wrote the paper: RS AMI.

References

1. Parfitt RL, Scott NA, Ross DJ, Salt GJ, Tate KR (2003) Landuse change effects on soil C and N transformations in soils of high N status: comparisons under indigenous forest, pasture and pine plantation, Biogeochemistry 66: 203–221.

2. Rillig MC, Ramsey PW, Morris S, Paul EA (2003) Glomalin, an arbuscular-mycorrhizal fungal soil protein, responds to land-use change. Plant Soil 253: 293–299.

3. Houghton RA, Goodale C (2004) Effects of land-use change on the carbon balance of terrestrial ecosystems. In: DeFries R, Asner G, & Houghton RA, editors. Ecosystems and Land Use Change. Geophysical Monograph Series. Vol. 153. Washington, DC: American Geophysical Union. 85–98.

4. Dupouey JL, Dambrine E, Laffite JD, Moares C (2002) Irreversible impact of past land use on forest soils and biodiversity. Ecology 83: 2978–84.

5. Honnay O, Piessens K, Van Landuyt W, Hermy M, Gulinck H (2003) Satellite based land use and landscape complexity indices as predictors for regional plant species diversity. Landscape Urban Plan 63: 241–250.

6. Fu B, Chen L, Ma K, Zhou H, Wang J (2000) The relationships between land use and soil conditions in the hilly area of the loess plateau in northern Shaanxi, China. Catena 39: 69–78.

7. Kennedy AC, Stubbs TL (2006) Soil microbial communities as indicators of soil health. Ann Arid Zone 45: 287–308.

8. Ibekwe AM, Kennedy AC, Frohne PS, Papiernik SK, Yang CH, et al. (2002) Microbial diversity along a transect of agronomic zones. FEMS Microbiol Ecol 39: 183–191.

9. Lauber CL, Strickland MS, Bradford MA, Fierer N (2008) The influence of soil properties on the structure of bacterial and fungal communities across land-use types. Soil Biol Biochem 40: 2407–2415.

10. Roesch LF, Fulthrope RR, Riva A, Casella G, Hadwin AKM, et al. (2007) Pyrosequencing enumerates and contrasts soil microbial diversity. ISME J 1: 283–290.

11. Acosta-Martinez V, Dowd S, Sun Y, Allen V (2008) Tag-encoded pyrosequencing analysis of bacterial diversity in a single soil type as affected by management and land use. Soil Biol Biochem 40: 2762–2770.

12. Lauber CL, Hamady M, Knight R, Fierer N (2009) Pyrosequencing-based assessment of soil pH as a predictor of soil bacterial community structure at the continental scale. Appl Environ Microbiol 75: 5111–5120.

13. Brye KR, Slaton NA, Savin MC, Norman RJ, Miller DM (2003) Short-term effects of land leveling on soil physical properties and microbial biomass. Soil Sci Soc Am J 67: 1405–1417.

14. Hendrix PH, Crossley Jr DA, Coleman DC (1990) Soil biota as components of sustainable agro-ecosystem. In: Edwards CA, Lal R, Madden P, Miller RH, & House G, editors. Sustainable Agricultural Systems. Iowa: Soil Water Conservation Society. 637–654.

15. Acosta-Martinez V, Dowd SE, Bell CW, Lascano R, Booker JD, et al. (2010a) Microbial Community Composition as Affected by Dryland Cropping Systems and Tillage in a Semiarid Sandy Soil. Diversity 2: 910–931.

16. Torsvik V, Daae FL, Sandaa RA, Ovreas L (1998) Novel techniques for analyzing microbial diversity in natural and perturbed environments. J Biotechnol 64: 53–62.

17. Parham JA, Deng SP, Da HN, Sun HY, Raun WR (2003) Long-term cattle manure application in soil. II. Effect of soil microbial populations and community structure. Biol Fert Soils 38: 209–215.

18. Acosta-Martinez V, Harmel RD (2006) Soil microbial communities and enzyme activities under various poultry litter application rates. J Environ Qual 35: 1309–1318.

19. Jacobson MG (2002) Ecosystem management in the southeast United States: Interest of forest landowners in joint management across ownerships. Small-scale Forest Economics, Management and Policy 1: 71–92.

20. Lugo AE (1997) The apparent paradox of reestablishing species richness on degraded lands with tree monocultures. Forest Ecol Manag 99: 9–19.

21. Tsiouvaras CN, Havlik NA, Bartolome JW (1989) Effects of Goats on Understory Vegetation and Fire Hazard Reduction in a Coastal Forest in California. Forest Sci 35: 1125–1131.

22. Hart SP (2001) Recent perspectives in using goats for vegetation management in the USA. J Dairy Sci 84(E. Suppl.): E170–E176.

23. Hobbs RJ, Hueneke LF (1992) Disturbance, diversity, and invasion: Implications for conservation. Conserv Biol 6: 324–337.

24. Kohler F, Hamelin J, Gillet F, Gobat J-M, Buttler A (2005) Soil microbial community changes in wooded mountain pastures due to simulated effects of cattle grazing. Plant Soil 278: 327–40.

25. Patra AK, Abbadie L, Clays-Josserand A, et al. (2005) Effects of grazing on microbial functional groups involved in soil N dynamics. Ecol Monogr 75: 65–80.

26. Sorensen LI, Mikola J, Kytoviita M-M, Olofsson J (2009) Trampling and spatial heterogeneity explain decomposer abundances in a sib-arctic grassland subjected to simulated reindeer grazing. Ecosystems 12: 830–842.

27. Jangid K, Williams MA, Franzluebbers AJ, Sanderlin JS, Reeves JH, et al. (2008) Relative impacts of land-use, management intensity and fertilization upon soil microbial community structure in agricultural systems. Soil Biol Biochem 40: 2843–2853.

28. Acosta-Martinez V, Dowd SE, Sun Y, Webster D, Allen VG (2010b) Pyrosequencing analysis for characterization of soil bacterial populations as affected by an integrated livestock-cotton production system. Appl Soil Ecol 45: 13–25.

29. Karchesy JJ, Hemingway RW (1980) Loblolly pine bark flavonoids. J Agr Food Chem 28: 222–228.

30. Fierer N, Jackson RB (2006) The diversity and biogeography of soil bacterial communities. P Natl Acad Sci USA 103: 626–631.

31. Nacke H, Thurmer A, Wollherr A, Will C, Hodac L, et al. (2011) Pyrosequencing-based assessment of bacterial community structure along different management types in German forest and grassland soils. PLoS ONE 6(2): e17000. doi:10.1371/journal.pone.0017000.

32. Kraus TEC, Dahlgren RA, Zasoski RJ (2003) Tannins in nutrient dynamics of forest ecosystems – a review. Plant Soil 256: 41–66.

33. Selvakumar G, Saha S, Kundu S (2007) Inhibitory activity of pine needle tannin extracts on some agriculturally resourceful microbes. Indian J Microbiol 47: 267–270.

34. Buckley DH, Schmidt TM (2001) The structure of microbial communities in soil and the lasting impact of cultivation. Microb Ecol 42: 11–21.

35. Osborne CA, Zwart AB, Broadhurst LM, Young AG, Richardson AE (2011) The influence of sampling strategies and spatial variation on the detected soil bacterial communities under three different land use types. FEMS Microbiol Ecol 78: 70–79.

36. Neher DA (1999) Soil community composition and ecosystem processes: Comparing agricultural ecosystems with natural ecosystems. Agroforest Syst 45: 159–185.

37. Fierer N, Bradford MA, Jackson RB (2007) Toward an ecological classification of soil bacteria. Ecology 88: 1354–1364.

38. Singh BK, Bardgett RD, Smith P, Reay DS (2010) Microorganisms and climate change: terrestrial feedbacks and mitigation options. Nat Rev Microbiol 8: 779–790.

39. Wardle DA, Yeates GW, Watson RN, Nicholson KS (1995) The detritus food-web and the diversity of soil fauna as indicators of disturbance regimes in agro-ecosystems. Plant Soil 170: 35–43.

40. Hill P, Kristufek V, Dijfkhuizen L, Boddy C, Kroetsch D, et al. (2011) Land use intensity controls actinobacterial community structure. Microb Ecol 61: 286–302.

41. Burke RA, Molina M, Cox JE, Osher LJ, Piccolo MC (2003) Stable carbon isotope ratio and composition of microbial fatty acids in tropical soils. J Environ Qual 32: 198–206.

42. Waldrop MP, Balser TC, Firestone MK (2000) Linking microbial community composition to function in a tropical soil. Soil Biol Biochem 32: 1837–1846.

43. Ishak HD, Miller JL, Sen R, Dowd SE, Meyer E, et al. (2011) Microbiomes of ant castes implicate new microbial roles in the fungus-growing ant Trachymyrmex septentrionalis. Sci Rep 1: 204. doi: 10.1038/srep00204.

44. Lopez-Velasco G, Welbaum GE, Boyer RR, Mane SP, Ponder MA (2011) Changes in spinach phylloepiphytic bacteria communities following minimal processing and refrigerated storage described using pyrosequencing of 16S rRNA amplicons. J Appl Microbiol 110: 1203–1214.

45. Saul-Tcherkas V, Steinberger Y (2011) Soil Microbial Diversity in the Vicinity of a Negev Desert Shrub—*Reaumuria negevensis*. Microb Ecol 61: 64–81.

46. Chong CW, Convey P, Pearce DA, Tan IKP (2012) Assessment of soil bacterial communities on Alexander Island (in the maritime and continental Antarctic transitional zone. Polar Biol 35: 387–399.

47. Connell JH (1978) Diversity in tropical rain forests and coral reefs. Science 199: 1302–1310.

48. Tabatabai MA (1994) Enzymes, In: Weaver RW, Augle S, Bottomly PJ, Bezdicek D, Smith S, Tabatabai MA, Wollum A, editors. Methods of Soil Analysis, Part 2: Microbiological and Biochemical properties. Madison: Soil Science Society of America. 775–833.

49. Bandick AK, Dick RP (1999) Field management effects on soil enzyme activities. Soil Biol Biochem 31: 1471–1479.

50. Elsgaard L, Anderson GH, Eriksen J (2002) Measurement of arylsulphatase activity in agricultural soils using a simplified assay. Soil Biol Biochem 34: 79–82.

51. Mclean EO (1982) Soil pH and lime requirements. In: Page AL, Miller RH, Keeney, DR, editors. Methods of Soil Analysis, Part 2. Chemical and Microbiological Properties. Madison: Soil Science Society of America. 199–224.

52. Keller JM, Gee GW (2006) Comparison of American Society of Testing Materials and Soil Science Society of America hydrometer methods for particle-size analysis. Soil Sci Soc Am J 70: 1094–1100.

53. Dowd SE, Sun Y, Secor PR, Rhoads DD, Wolcott BM, et al. (2008a) Survey of bacterial diversity in chronic wounds using pyrosequencing, DGGE, and full ribosome shotgun sequencing. BMC Microbiol 8: 43.

54. Dowd SE, Sun Y, Wolcott RD, Carroll JA (2008b) Bacterial tag-encoded FLX amplicon pyrosequencing (bTEFAP) for microbiome studies: bacterial diversity in the ileum of newly weaned Salmonella-infected pigs. Foodborne Pathog Dis 5: 459–472.

55. Feingold SM, Dowd SE, Gontcharova V, Liu C, Henley KE, et al. (2010) Pyrosequencing study of fecal microflora of autistic and control children. Anaerobe 16: 444–453.

56. Gontcharova V, Youn E, Wolcott RD, Hollister EB, Gentry TJ, et al. (2010) Black Box Chimera Check (B2C2): a windows-based software for batch

depletion of chimeras from bacterial 16S rRNA gene datasets. Open Microbiol J. 4: 47–52.

57. Schloss PD, Westcott SL, Ryabin T, Hall JR, Hartmann M, et al. (2009) Introducing mothur: open source, platform-independent, community-supported software for describing and comparing microbial communities. Appl Environ Microbiol 75: 7537–7541.

58. DeSantis TZ, Hugenholtz P, Larsen N, Rojas M, Brodie EL, et al. (2006) Greengenes, a Chimera-Checked 16S rRNA Gene Database and Workbench Compatible with ARB. Appl Environ Microbiol 72: 5069–72.

59. Chao A, Ma MC, Yang MCK (1993) Stopping rules and estimation for recapture debugging with unequal failure rates. Biometrika 80: 193–201.

60. Chao A (1984) Nonparametric estimation of the number of classes in a population. Scand. J Stat 11: 265–270.

12

Chinese Tallow Trees (*Triadica sebifera*) from the Invasive Range Outperform Those from the Native Range with an Active Soil Community or Phosphorus Fertilization

Ling Zhang[1,2], Yaojun Zhang[1], Hong Wang[1], Jianwen Zou[1]*, Evan Siemann[1,2]

1 College of Resources & Environmental Sciences, Nanjing Agricultural University, Nanjing, China, **2** Department of Ecology & Evolutionary Biology, Rice University, Houston, Texas, United States of America

Abstract

Two mechanisms that have been proposed to explain success of invasive plants are unusual biotic interactions, such as enemy release or enhanced mutualisms, and increased resource availability. However, while these mechanisms are usually considered separately, both may be involved in successful invasions. Biotic interactions may be positive or negative and may interact with nutritional resources in determining invasion success. In addition, the effects of different nutrients on invasions may vary. Finally, genetic variation in traits between populations located in introduced versus native ranges may be important for biotic interactions and/or resource use. Here, we investigated the roles of soil biota, resource availability, and plant genetic variation using seedlings of *Triadica sebifera* in an experiment in the native range (China). We manipulated nitrogen (control or 4 g/m^2), phosphorus (control or 0.5 g/m^2), soil biota (untreated or sterilized field soil), and plant origin (4 populations from the invasive range, 4 populations from the native range) in a full factorial experiment. Phosphorus addition increased root, stem, and leaf masses. Leaf mass and height growth depended on population origin and soil sterilization. Invasive populations had higher leaf mass and growth rates than native populations did in fresh soil but they had lower, comparable leaf mass and growth rates in sterilized soil. Invasive populations had higher growth rates with phosphorus addition but native ones did not. Soil sterilization decreased specific leaf area in both native and exotic populations. Negative effects of soil sterilization suggest that soil pathogens may not be as important as soil mutualists for *T. sebifera* performance. Moreover, interactive effects of sterilization and origin suggest that invasive *T. sebifera* may have evolved more beneficial relationships with the soil biota. Overall, seedlings from the invasive range outperformed those from the native range, however, an absence of soil biota or low phosphorus removed this advantage.

Editor: Harald Auge, Helmholtz Centre for Environmental Research - UFZ, Germany

Funding: This work was supported by the National Natural Science Foundation of China (NSFC-41225003), the PADA (the Priority Academic Program Development of Jiangsu Province), the Ministry of Education 111 project (B12009), and US-NSF (DEB 0820560). The funders had no role in study design, data collection and analysis, decision to publish, or preparation of the manuscript.

Competing Interests: The authors have declared that no competing interests exist.

* E-mail: jwzou21@njau.edu.cn

Introduction

Exotic plant invasions threaten ecosystem functions and stability [1–3]. Identifying the mechanisms underlying successful plant invasions will help guide effective invasive plant control and aid in ecosystem restoration. Two mechanisms that have been proposed to explain successful plant invasions are: 1) that exotic plants benefit from greater resource availability (the increased resource availability hypothesis or "IRAH"; [4,5]) and 2) exotic plants benefit from weak effects of natural enemies (the enemy release hypothesis or "ERH"; [6]) and/or strong effects of mutualists (the enhanced mutualists hypothesis or EMH, [7]).

The IRAH posits that the opportunities for invasions increase as resource availability increases in a community [4]. This increased resource availability does not necessarily reflect higher nutrient input because resource availability reflects the balance of resource supply and uptake by resident plants [5]. While most, but not all, exotic invaders may be better adapted to high nutrient conditions

than native species ("ruderals" [8]), pre-adaptation or post-introduction adaptation of exotic plants to high nutrient conditions may confer an advantage to exotic plants compared to less well-adapted native plants. For instance, invasive plants may be favored by increased soil resources (*e.g.* nitrogen [N], phosphorus [P]) that favor plants with low root to shoot ratios [9]. Similarly, plants with high N dependent maximal growth rates will be favored over those with high N use efficiencies when that N availability is high. Because plants with high N demand may not also have high P demand, for instance because of different symbiotic relationships (*e.g.* rhizobial or mycorrhizal) or allocation to high N (proteins) or P (nucleic acids) compounds, soil resources may vary in their impacts on invasions [2,10]. Moreover, nutrient assimilation by invasive plant species may vary due to positive and/or negative biotic interactions with more positive or less negative interactions facilitating nutrient uptake of the host plant.

The ERH posits that exotic plants benefit from introduction to a new range without specialist enemies in combination with not being preferred by generalist enemies [6]. Recent studies suggested that escape from soil pathogens may be at least as important as escape from aboveground specialist insect herbivores in their contribution to successful plant invasions [11–13]. Since soil communities include pathogens, parasites, and herbivores as well as beneficial groups (*e.g.* mycorrhizae, rhizobia) [14,15], the overall impact of soil biota on plant performance will reflect the net effect of both negative and positive interactions [16]. Strong negative impacts of soil microbial communities on invasive plants have mostly been observed in natural population of these plants growing in their native ranges [11,17] indicating that negative interactions are relatively stronger than beneficial ones [16]. This could reflect stronger negative effects or weaker positive effects on plant performance [7,18,19].

Differences in biotic or abiotic factors between the native and invasive ranges of plants can lead to genetic differences in morphological or physiological traits between populations in the native and introduced ranges [20–23]. One example of a shift in morphological traits is a lower root to shoot ratio [24,25]. In general, a lower root to shoot ratio provides an advantage in competition for aboveground resources and a disadvantage in competition for belowground resources [9]. In addition, escape from natural enemies, in particular specialists, in the invasive range may lead to a reallocation from defense to growth [26–28]. Moreover, more beneficial soil mutualisms in the invasive range [11] may lead to genetic differences in plant traits relevant to these interactions. However, resource requirements and biotic interactions are not independent [29,30]. In addition, shifts in traits of invasive plants may lead to altered soil microbial communities [25,31], which may in turn impact soil N and P use [32–35]. However, the dependence of invasive plant performance on genetic variation in plant traits, interactions with the soil biota, and availability of N and P is poorly understood.

Here, we examined effects of interactions between soil nutrients (N and P), soil microbial communities (active or sterilized), and population origin (native or invasive range) using Chinese tallow tree (*Triadica sebifera* (L.) Small, henceforth *T. sebifera*) as a model plant. *T. sebifera* is native to China and was first introduced into the USA in 1772 to Savannah, GA then subsequently to several sites along the Gulf Coast and is now invasive in grasslands, forests, and disturbed habitats throughout the southeastern USA, converting them to monospecific forests [36–39]. Previous studies have demonstrated that invasive *T. sebifera* had unusually positive interactions with the soil biota relative to native tree species in the introduced range [40]. Conducting studies in the native range with populations from the native and introduced ranges provides additional insights into how genetic differences in *T. sebifera* populations may influence the net effects of the soil biota on *T. sebifera* performance. In an experiment conducted in the native range, we addressed the following questions: (1) Do *T. sebifera* seedlings perform better with N and/or P addition? (2) What are the net effects of the soil biota in the native range? (3) Do *T. sebifera* seedling responses to nutrient additions and soil biota manipulations differ between population origins?

Materials and Methods

Focal Species

T. sebifera is native to China where it has been cultivated for 14 centuries and is now an aggressive invader in the southeastern USA [36,41]. Studies demonstrated *T. sebifera* in the invasive range (invasive populations) are faster-growing relative to native conspecifics (native populations) or non-invasive co-occurring plant species [25,27,42]. Invasive *T. sebifera* rapidly accumulates soil pathogens in the invasive range relative to co-occurring native resident species which decreases the performance of *T. sebifera* seedlings under conspecifics [43,44]. However, *T. sebifera* has also been shown to be more mycorrhizal dependent in its invasive range compared with native trees [40,44]. In addition, *T. sebifera* seedlings from the invasive range have stronger responses to N addition than ones from the native range perhaps partly due to facilitation of N mineralization [34].

Seeds and Seedlings

In November 2009, we hand collected seeds of naturalized *T. sebifera* in China and the USA (Table 1). All seed collections were from public areas where no permission was required for collection. *T. sebifera* is not an endangered or protected species in either country. All seeds were collected from at least five haphazardly selected trees. Seeds used for planting were weighed by populations to evaluate the potential impacts of seed provisioning on seedling performance. Results of an ANOVA showed that seed masses of populations were independent of population origin ($F_{1,6} = 3.99$, $P = 0.09$). Also, seedling height ($F_{1,6} = 0.25$, $P = 0.64$) and number of leaves ($F_{1,6} = 2.59$, $P = 0.16$) at the time of transplanting were independent of population origin. Together these results suggest that there were no strong maternal effects due to differences in seed provisioning. In January, we treated seeds in a 10% bleach rinse and then soaked seeds in water with lab detergent to remove the waxy seed coat [43]. All seeds were then surface sterilized by 0.5% potassium permanganate and planted in 100 ml Conetainers (Stuewe & Sons, Corvallis, OR, USA) filled with sterilized field soil (see below). Seeds germinated in early April, 2010. After seedlings had secondary leaves, seedlings of similar heights were transplanted into pots (1.5 L). Pots received three soil treatments in a full-factorial design (N = 256, 2 population origins×4 populations×2 soil sterilization×2 N×2 P×4 replicates). To coincide with the growing season of *T. sebifera* in this area seeds were grown for 4 months in a non-heated greenhouse from June 2010 to November 2010 at Nanjing Agricultural University, Nanjing, China.

Soil Treatments

Soil was collected from the top 20 cm in a fallow agricultural field. *T. sebifera* trees were at least 200 m away from where soil was collected to reduce the potential buildup of specific soil organisms [16]. Soils characteristics were: carbon % = 2.32±0.11; nitrogen % = 0.22±0.007; C:N = 10.53±1.65 (means ±1 se). Previous

Table 1. Native (China) and invasive (USA) *T. sebifera* populations used in this experiment.

Source population	Latitude	Longitude
China		
Hefei, Anhui	31°38~39′N	117°50~51′E
Bengbu, Anhui	32°57~58′N	117°20~21′E
Nanjing, Jiangsu	32°02~03′N	118°50~51′E
Shanghai	31°31~32′N	121°52~53′E
USA		
Limehouse, SC	32°09~10′N	81°05~07′W
Hutchinson Island, GA	31°23~24′N	81°15~16′W
Houston, TX	29°41~42′N	95°25~26′W
Gainesville, FL	29°34~35′N	82°21~22′W

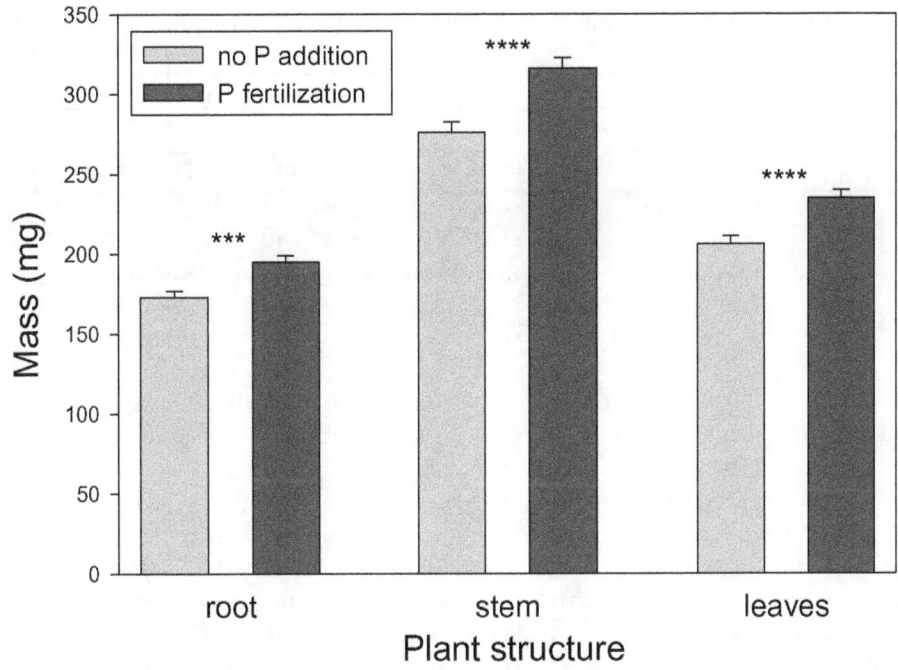

Figure 1. The dependence of root, stem and leaf masses of *T. sebifera* **seedlings on P addition.** Means+1 SE. ***: $P<0.001$; ****: $P<0.0001$.

studies focused on home- and away-soil effects indicate buildup of negative soil organisms in conspecific (home) soils in both the native and introduced ranges [43,44]. The soils used here are suited for making inferences about the effects of soil nutrients and the soil biota during the process of colonization in the native range and spread in the introduced range. Half of the soil was autoclaved at 121°C for 40 minutes ("sterilized soil") and the other half was left untreated ("fresh soil").

Pots that were in the N fertilizer treatment received 4 g m^{-2} of N as KNO_3 (equivalent to 15.1 mg/L of soil). Plants in the control (no addition) N treatment received an equivalent volume of deionized water. Pots in the P fertilizer treatments received P at a

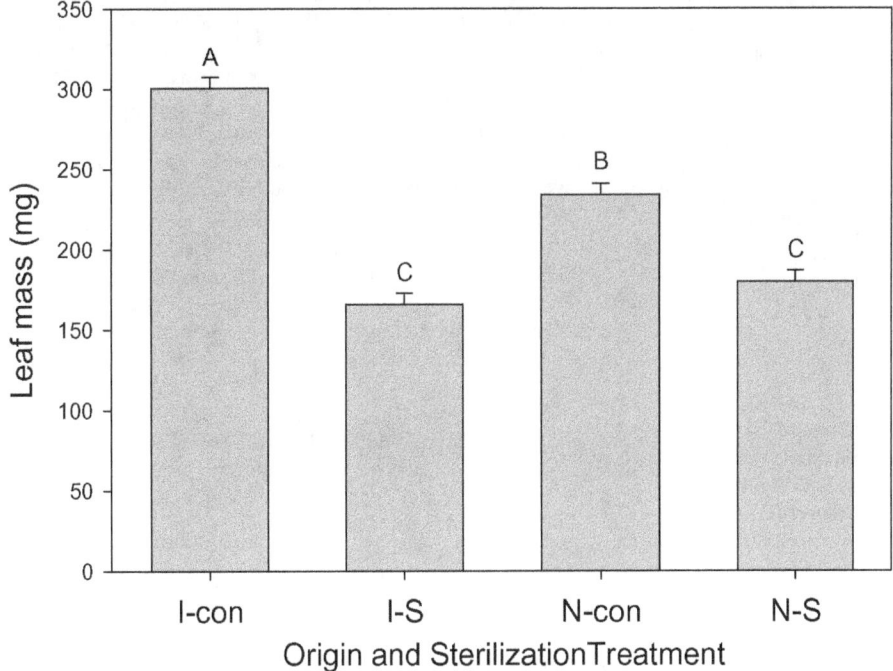

Figure 2. The dependence of leaf mass on population origin ("I" is invasive, "N" is native) and soil sterilization treatments ("con" is control, "S" is sterilization). Means+1 SE. Means with the same letter were not significantly different in post-hoc tests ($P<0.05$).

Table 2. The dependence of root, stem, leaf biomass on origin, N addition, P addition and soil sterilization and their interactions in a MANOVA and follow-up ANOVAs.

Effect	MANOVA			Root			Stem			Leaf		
	DF	F	P	DF	F	P	DF	F	P	DF	F	P
Origin	3,4	2.52	0.1971	1,6	1.09	0.3366	1,6	0.12	0.7446	1,6	5.35	0.0600
N	3,189	1.01	0.3911	1,191	2.40	0.1227	1,191	0.25	0.6167	1,191	0.18	0.6723
P	**3,189**	**7.15**	**<0.0001**	**1,191**	**12.20**	**0.0006**	**1,191**	**16.79**	**<0.0001**	**1,191**	**16.21**	**<0.0001**
Sterilization	**3,189**	**83.84**	**<0.0001**	1,191	0.23	0.6339	1,191	0.02	0.8918	**1,191**	**171.29**	**<0.0001**
Origin×N	3,4	0.48	0.7147	1,6	0.44	0.5323	1,6	0.42	0.5402	1,6	0.01	0.9915
Origin×P	3,4	3.95	0.1087	1,6	1.46	0.2717	1,6	0.09	0.7773	1,6	0.99	0.3589
Origin×Sterilization	**3,4**	**7.16**	**0.0437**	1,6	0.69	0.4376	1,6	0.48	0.5142	**1,6**	**14.52**	**0.0089**
N×P	3,189	0.59	0.6228	1,191	0.06	0.8117	1,191	0.25	0.6205	1,191	0.53	0.4684
N×Sterilization	3,189	1.75	0.1574	1,191	0.07	0.7855	1,191	0.32	0.5716	1,191	2.26	0.1348
P×Sterilization	3,189	1.26	0.2910	1,191	2.30	0.1311	1,191	3.56	0.0609	1,191	0.57	0.4509
Origin×N×P	3,4	0.22	0.8789	1,6	0.43	0.5385	1,6	0.42	0.5420	1,6	0.80	0.4056
Origin×N×Sterilization	3,4	0.39	0.7676	1,6	1.34	0.2914	1,6	0.88	0.3850	1,6	0.05	0.8314
Origin×P×Sterilization	3,4	0.81	0.5514	1,6	0.15	0.7111	1,6	0.06	0.8211	1,6	1.36	0.2872
N×P×Sterilization	3,189	0.57	0.6342	1,191	0.64	0.4243	1,191	0.01	0.9176	1,191	0.59	0.4445
Origin×N×P×Sterilization	**3,4**	**8.45**	**0.0332**	1,6	0.43	0.5385	1,6	1.39	0.2834	1,6	3.85	0.0973

Significant results shown in bold.

rate of 0.5 g m^{-2} as KH_2PO_4 (equivalent to 1.9 mg/L of soil) and control (no addition) P pots received an equivalent volume of deionized water. Fertilizer additions were made one month after seedlings were transplanted.

Data Collection

We measured stem height of each seedling from ground surface to terminal bud at both the beginning and the end of the experiment. We thoroughly cleaned equipment between measurements. At the end of the experiment (4 months), seedlings were clipped at ground level (then separated into leaves and stems) and roots were gently washed from the soil. Total leaf area (cm^2) was obtained by scanning fresh leaves and analyzing them with SCNIMAGE (Scion Corporation; www.scioncorp.com). Seedling roots, stems, and leaves were then dried at $60°C$ to constant mass and weighed. We calculated height growth rates (HGR, mm cm^{-1} day^{-1}) as: HGR = ln (harvest stem height/initial stem height at transplanting)/days. Specific leaf area (SLA, leaf area per unit leaf dry mass, cm^2 g^{-1}) was calculated dividing leaf area by leaf biomass.

Statistical Analyses

We first conducted a MANOVA to examine the effects of seedling origin, N treatment, P treatment, soil treatment, and their interactions on *T. sebifera* root mass, stem mass, and leaf mass. We used variation among populations to test for differences between population origins (and corresponding interaction terms with population to examine interactive effects with origin). Because there were significant MANOVA results, we then conducted ANOVAs for each of the biomass variables. We also conducted ANOVAs to examine the dependence of height growth rate and specific leaf area on our treatments. We used partial difference adjusted means contrast tests to examine differences among treatment means for significant interactive effects. Data did not need to be transformed to meet the assumptions of ANOVA.

Differences at $\alpha = 0.05$ level are reported as significant. All statistical analyses were carried out in SAS (SAS Institute, Cary, NC, USA).

Results

Plant Biomass

In the MANOVA, P addition, sterilization, origin×sterilization, and origin×N×P×sterilization all had significant effects on root, stem and leaf biomass (Table 2). In follow-up ANOVAs, P addition significantly increased biomass of roots, stems and leaves (Table 2; Fig. 1). In addition, leaf biomass depended on sterilization and origin×sterilization with greater increases in leaf biomass in fresh soil compared to sterilized soil for seedlings from invasive populations versus native populations (Fig. 2).

Plant Growth Rate and Specific Leaf Area

Height growth rate depended on origin, P addition, soil sterilization, origin×P, and origin×sterilization (Table 3). Seedlings from invasive populations had significantly higher growth rates with P addition but ones from native populations did not (Fig. 3A).

In addition, the height growth increases in fresh soil compared to sterilized soil were significantly larger for invasive populations (Fig. 3B). Specific leaf area was significantly higher in fresh soil (Fig. 4A) and SLA also depended on N addition, P addition, and origin×N×P (Table 3). This interactive effect reflected significantly higher SLA for seedlings from invasive populations but significantly lower SLA for those from native populations when both N and P were added (Fig. 4B).

Discussion

Root, stem and leaf biomass of both origins were increased with P addition. In previous studies of plant invasions and soil P, most reported increased P availability in invaded areas [32,33,45–47]

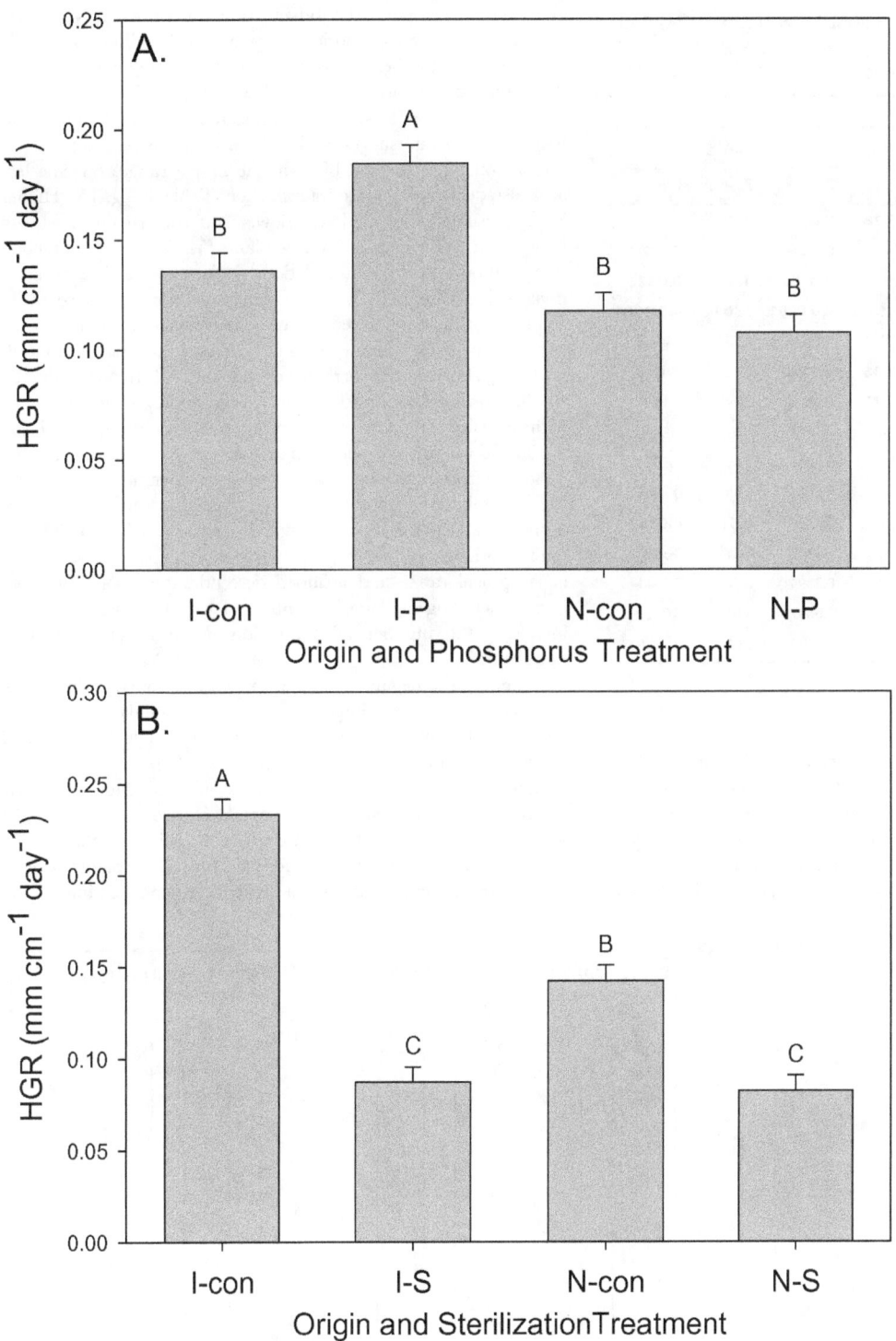

Figure 3. The dependence of height growth (HGR) on A) population origin ("I" is invasive, "N" is native) and P treatment ("con" is control, "P" is addition) and B) population origin and soil sterilization treatments ("con" is control, "S" is sterilization). Means+1 SE. Means with the same letter were not significantly different in post-hoc tests (P<0.05).

suggesting that invasive species may have evolved to mineralize soil P at a higher efficiency relative to native ones. Additional studies have demonstrated the importance of P availability for competitive ability and range expansion for invasive plant species [48,49]. Our results indicated that seedlings from both native and invasive origins were P limited since each responded positively to P addition, but had no response to N addition. However, N is another important soil nutrient that may limit plant growth and range expansion of *T. sebifera*. Zou *et al.* [34] found higher soil organic N mineralization in soils associated with *T. sebifera* of

Table 3. The dependence of height growth rate (HGR) and specific leaf area (SLA) on origin, N addition, P addition and soil sterilization and their interactions in ANOVAs.

		HGR		SLA	
Effect	DF	F	P	F	P
Origin	1,6	**33.78**	**0.0011**	3.70	0.1026
N	1,233	0.02	0.8999	**5.67**	**0.0181**
P	1,233	**5.94**	**0.0156**	4.65	**0.0322**
Sterilization	1,233	**146.88**	**<0.0001**	**150.86**	**<0.0001**
Origin×N	1,6	2.91	0.1388	0.52	0.4979
Origin×P	1,6	**12.36**	**0.0126**	0.74	0.4218
Origin×Sterilization	1,6	**27.72**	**0.0019**	2.44	0.1691
N×P	1,233	0.34	0.5618	0.02	0.8884
N×Sterilization	1,233	1.11	0.2940	1.05	0.3074
P×Sterilization	1,233	0.82	0.3664	1.55	0.2145
Origin×N×P	1,6	0.09	0.7728	**8.51**	**0.0267**
Origin×N×Sterilization	1,6	0.35	0.5749	1.01	0.3528
Origin×P×Sterilization	1,6	1.03	0.3494	0.02	0.8811
N×P×Sterilization	1,233	0.18	0.6687	3.03	0.0832
Origin×N×P×Sterilization	1,6	0.04	0.8530	0.01	0.9739

Significant results shown in bold.

invasive origin, which might lead to increased soil N availability. In addition, invasive *T. sebifera* plants have been shown to have a stronger positive response to inorganic N levels relative to those from native populations [34]. However, growth of *T. sebifera* seedlings from invasive populations invading coastal prairies in the introduced range responded significantly to N and K addition

alone but only responded positively to P addition when N was also added [48]. The strong positive response to P addition but not N addition we found here may reflect the extremely high levels of N deposition in the native range of *T. sebifera* [50].

The negative effects of soil sterilization on leaf biomass and height growth rate suggested *T. sebifera* seedlings had net positive interactions with the soil biota in the native range. Specific leaf area also decreased with soil sterilization (Table 3; Fig. 4A). Higher SLA is usually associated with lower leaf construction cost and higher N use efficiency in invasive plants [51–53]. One interaction that is important for P assimilation by plant species is arbuscular mycorrhizal fungi [54]. The higher arbuscular mycorrhizal colonization level observed for invasive *T. sebifera* relative to the native tree species in the introduced range is evidence that *T. sebifera* is arbuscular mycorrhizae dependent [40]. In our study, soil sterilization interacted with seedling origin to impact leaf biomass, with invasive origin seedlings more strongly inhibited by soil sterilization relative to ones of native origin. Thus, it appears that *T. sebifera* from both origins have overall positive interactions with the soil microbial communities but that those interactions are more beneficial for those of invasive origin relative to those of native origin. Although our populations spanned a broad geographical range and included descendants of both introductions, including a larger number of populations may have increased the number of population origin effects that were significant.

Assuming the negative effect of soil sterilization was simply the removal of mutualists important in P or N uptake [10,55], the negative effect of soil sterilization on growth might be weakened when N and/or P were added. However, there was not such a significant interactive effect on the mass of leaves, stems, or roots or on height growth rate. Perhaps uptake was so poor in sterilized soils that additional nutrients were not available to plants. The greater decline in leaf biomass for invasive origin plants relative those of native origin indicated a greater net beneficial interaction

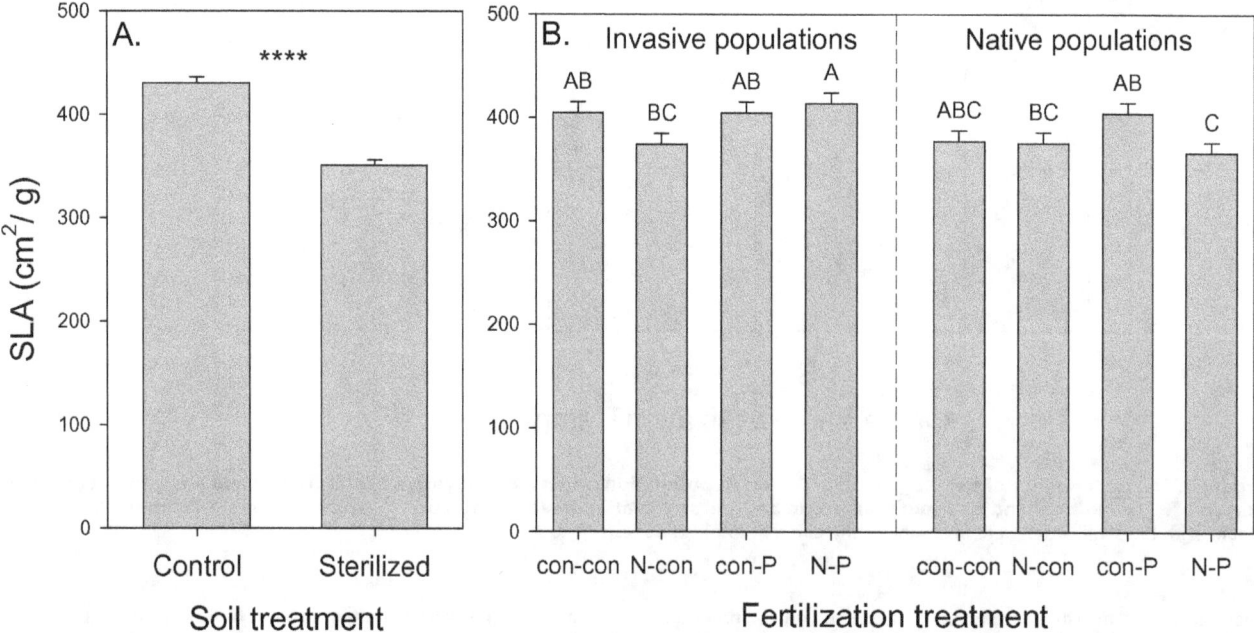

Figure 4. The dependence of specific leaf area (SLA) on A) soil treatment and B) population origin ("I" is invasive, "N" is native) and fertilization treatment ("con-con" is no fertilization, "N-con" is N addition, "con-P" is P addition, and "N-P" is N and P addition). Means+1 SE. Means with the same letter were not significantly different in post-hoc tests (P<0.05). ****: *P*<0.0001.

with the soil biota [40]. This could reflect greater positive interactions or weaker negative interactions but these possibilities cannot be evaluated in this study. If a similar pattern exists in the introduced range, it might be a mechanism contributing to its successful invasion.

Height growth rate of seedlings from the invasive range significantly increased with P addition but those from the native range did not respond to P addition (Table 3; Fig. 3). Generally, in a high resource, low stress environment, plants with a higher growth rate would be more successful when competing for light [56,57]. There was a significant interactive effect of origin, N addition, and P addition in which seedlings from invasive populations had especially high SLA and seedlings from native populations had especially low SLA (Fig. 4B). This is consistent with seedlings from invasive populations being more responsive to increased resources. Overall, the strong P response of seedlings from the invasive range together with comparable performance of seedlings without P addition suggests that seedlings from invasive populations may only have a competitive advantage in high P conditions [58].

It should be noted that this study focused on interactions with generalists in the native range since we collected soil more than 200 m away from any *T. sebifera* trees [59]. It is possible that we would have observed overall more negative effects of interactions with the soil biota had we used soil collected near conspecifics [60]. The interactions of *T. sebifera* seedlings of different origins might also differ if the soil community included more specialists [21]. If *T. sebifera* interacts with few specialists in the introduced range, the results of this study may help to understand the role of plant-soil interactions and soil resources in invasions. Research conducted on *Robinia pseudoacacia* by collecting soil from native, expanded (naturalized), and invasive ranges indicated that invasive plants are successful due to acquiring mutualisms and meanwhile, escaping from pathogens to gain a net positive effect of soil biota [11]. Further studies conducted in areas where *T. sebifera* is naturalized but not invasive [41,61] would increase our knowledge of the role soil communities play in range expansion of *T. sebifera*.

Acknowledgments

We thank: Chunxiao Wang, Hao Wang, Mengyue Wang and Rui Qu for their assistance in collecting and processing soil; Nannan Chen and Xiaomei Yang for their help in data collection; colleagues at Rice University, Wuhan Botanical Garden and Fudan University for assistance in collecting seeds. Further thanks should be addressed to the anonymous reviewers for their considerate comments that have greatly improved the earlier version of this manuscript.

Author Contributions

Conceived and designed the experiments: LZ JWZ. Performed the experiments: LZ YJZ HW. Analyzed the data: LZ ES. Wrote the paper: LZ JWZ ES.

References

1. Mack RN, Simberloff D, Lonsdale WM, Evans H, Clout M, et al. (2000) Biotic invasions: Causes, epidemiology, global consequences, and control. Ecol Appl 10: 689–710.

2. Ehrenfeld JG (2010) Ecosystem consequences of biological invasions. Annu Rev Ecol Evol Syst 41: 59–80.

3. Vitousek PM, DAntonio CM, Loope LL, Rejmanek M, Westbrooks R (1997) Introduced species: A significant component of human-caused global change. New Zeal J Ecol 21: 1–16.

4. Davis MA, Grime JP, Thompson K (2000) Fluctuating resources in plant communities: a general theory of invasibility. J Ecol 88: 528–534.

5. Shea K, Chesson P (2002) Community ecology theory as a framework for biological invasions. Trends Ecol Evol 17: 170–176.

6. Keane RM, Crawley MJ (2002) Exotic plant invasions and the enemy release hypothesis. Trends Ecol Evol 17: 164–170.

7. Reinhart KO, Callaway RM (2004) Soil biota facilitate exotic *Acer* invasions in Europe and North America. Ecol Appl 14: 1737–1745.

8. Grime JP (2001) Plant strategies, vegetation processes, and ecosystem properties. Chichester: Wiley. 417 p.

9. Reynolds HL, Pacala SW (1993) An analytical treatment of root-to-shoot ratio and plant competition for soil nutrient and light. Am Nat 141: 51–70.

10. Richardson DM, Allsopp N, D'Antonio CM, Milton SJ, Rejmanek M (2000) Plant invasions - the role of mutualisms. Biol Rev 75: 65–93.

11. Callaway RM, Bedmar EJ, Reinhart KO, Silvan CG, Klironomos J (2011) Effects of soil biota from different ranges on *Robinia* invasion: acquiring mutualists and escaping pathogens. Ecology 92: 1027–1035.

12. Wardle DA, Bardgett RD, Klironomos JN, Setala H, van der Putten WH, et al. (2004) Ecological linkages between aboveground and belowground biota. Science 304: 1629–1633.

13. van der Putten WH, Klironomos JN, Wardle DA (2007) Microbial ecology of biological invasions. ISME J 1: 28–37.

14. Kempel A, Schmidt AK, Brandl R, Schadler M (2010) Support from the underground: Induced plant resistance depends on arbuscular mycorrhizal fungi. Funct Ecol 24: 293–300.

15. Reinhart KO, Johnson D, Clay K (2012) Conspecific plant-soil feedbacks of temperate tree species in the southern Appalachians, USA. PloS ONE 7: e40680.

16. Reinhart KO, Callaway RM (2006) Soil biota and invasive plants. New Phytol 170: 445–457.

17. Inderjit, van der Putten WH (2010) Impacts of soil microbial communities on exotic plant invasions. Trends Ecol Evol 25: 512–519.

18. Parker MA, Wurtz AK, Paynter Q (2007) Nodule symbiosis of invasive *Mimosa pigra* in Australia and in ancestral habitats: a comparative analysis. Biol Invasions 9: 127–138.

19. Callaway RM, Newingham B, Zabinski CA, Mahall BE (2001) Compensatory growth and competitive ability of an invasive weed are enhanced by soil fungi and native neighbours. Ecol Lett 4: 429–433.

20. Bossdorf O, Auge H, Lafuma L, Rogers WE, Siemann E, et al. (2005) Phenotypic and genetic differentiation between native and introduced plant populations. Oecologia 144: 1–11.

21. Joshi J, Vrieling K (2005) The enemy release and EICA hypothesis revisited: incorporating the fundamental difference between specialist and generalist herbivores. Ecol Lett 8: 704–714.

22. Orians CM, Ward D (2010) Evolution of plant defenses in nonindigenous environments. Annu Rev Entomol 55: 439–459.

23. Fukano Y, Yahara T (2012) Changes in defense of an alien plant *Ambrosia artemisiifolia* before and after the invasion of a native specialist enemy *Ophraella communa*. PLoS ONE 7: e49114.

24. Allred BW, Fuhlendorf SD, Monaco TA, Will RE (2010) Morphological and physiological traits in the success of the invasive plant *Lespedeza cuneata*. Biol Invasions 12: 739–749.

25. Zou J, Rogers WE, Siemann E (2007) Differences in morphological and physiological traits between native and invasive populations of *Sapium sebiferum*. Funct Ecol 21: 721–730.

26. Huang W, Siemann E, Wheeler GS, Zou JW, Carrillo J, et al. (2010) Resource allocation to defence and growth are driven by different responses to generalist and specialist herbivory in an invasive plant. J Ecol 98: 1157–1167.

27. Zou JW, Siemann E, Rogers WE, DeWalt SJ (2008) Decreased resistance and increased tolerance to native herbivores of the invasive plant *Sapium sebiferum*. Ecography 31: 663–671.

28. Blossey B, Notzold R (1995) Evolution of increased competitive ability in invasive nonindigenous plants - a hypothesis. J Ecol 83: 887–889.

29. Blumenthal D, Mitchell CE, Pysek P, Jarosik V (2009) Synergy between pathogen release and resource availability in plant invasion. P Natl Acad Sci USA 106: 7899–7904.

30. Bozzolo F, Lipson D (2013) Differential responses of native and exotic coastal sage scrub plant species to N additions and the soil microbial community. Plant Soil doi:10.1007/s11104-013-1668-2.

31. Norton JB, Monaco TA, Norton U (2007) Mediterranean annual grasses in western North America: kids in a candy store. Plant Soil 298: 1–5.

32. Vanderhoeven S, Dassonville N, Chapuis-Lardy L, Hayez M, Meerts P (2006) Impact of the invasive alien plant *Solidago gigantea* on primary productivity, plant nutrient content and soil mineral nutrient concentrations. Plant Soil 286: 259–268.

33. Thorpe AS, Archer V, DeLuca TH (2006) The invasive forb, *Centaurea maculosa*, increases phosphorus availability in Montana grasslands. Appl Soil Ecol 32: 118–122.

34. Zou JW, Rogers WE, DeWalt SJ, Siemann E (2006) The effect of Chinese tallow tree (*Sapium sebiferum*) ecotype on soil-plant system carbon and nitrogen processes. Oecologia 150: 272–281.

35. Hawkes CV, Wren IF, Herman DJ, Firestone MK (2005) Plant invasion alters nitrogen cycling by modifying the soil nitrifying community. Ecol Lett 8: 976–985.

36. Bruce KA, Cameron GN, Harcombe PA, Jubinsky G (1997) Introduction, impact on native habitats, and management of a woody invader, the Chinese tallow tree, *Sapium sebiferum* (L) Roxb. Nat Area J 17: 255–260.

37. Dewalt SJ, Siemann E, Rogers WE (2006) Microsatellite markers for an invasive tetraploid tree, Chinese tallow (*Triadica sebifera*). Mol Ecol Notes 6: 505–507.

38. DeWalt SJ, Siemann E, Rogers WE (2011) Geographic distribution of genetic variation among native and introduced populations of Chinese tallow tree, *Triadica sebifera* (Euphorbiaceae). Ann Bot 98: 1128–1138.

39. Siemann E, Rogers WE (2006) Recruitment limitation, seedling performance and persistence of exotic tree monocultures. Biol Invasions 8: 979–991.

40. Nijjer S, Rogers WE, Lee CTA, Siemann E (2008) The effects of soil biota and fertilization on the success of *Sapium sebiferum*. Appl Soil Ecol 38: 1–11.

41. Wang HH, Grant WE, Gan JB, Rogers WE, Swannack TM, et al. (2012) Integrating spread dynamics and economics of timber production to manage Chinese tallow invasions in southern US forestlands. PloS ONE 7: e33877.

42. Siemann E, Rogers WE (2003) Herbivory, disease, recruitment limitation, and success of alien and native tree species. Ecology 84: 1489–1505.

43. Nijjer S, Rogers WE, Siemann E (2007) Negative plant-soil feedbacks may limit persistence of an invasive tree due to rapid accumulation of soil pathogens. Proc Biol Sci 274: 2621–2627.

44. Yang Q, Carrillo J, Jin HY, Shang L, Hovick SM, et al. (2013) Plant–soil biota interactions of an invasive species in its native and introduced ranges: Implications for invasion success. Soil Biol Biochem 65: 78–85.

45. Chapuis-Lardy L, Vanderhoeven S, Dassonville N, Koutika LS, Meerts P (2006) Effect of the exotic invasive plant *Solidago gigantea* on soil phosphorus status. Biol Fert Soils 42: 481–489.

46. Herr C, Chapuis-Lardy L, Dassonville N, Vanderhoeven S, Meerts P (2007) Seasonal effect of the exotic invasive plant *Solidago gigantea* on soil pH and P fractions. J Plant Nutr Soil Sc 170: 729–738.

47. Kueffer C, Klingler G, Zirfass K, Schumacher E, Edwards PJ, et al. (2008) Invasive trees show only weak potential to impact nutrient dynamics in phosphorus-poor tropical forests in the Seychelles. Funct Ecol 22: 359–366.

48. Siemann E, Rogers WE (2007) The role of soil resources in an exotic tree invasion in Texas coastal prairie. J Ecol 95: 689–697.

49. Suding KN, LeJeune KD, Seastedt TR (2004) Competitive impacts and responses of an invasive weed: dependencies on nitrogen and phosphorus availability. Oecologia 141: 526–535.

50. Liu XJ, Zhang Y, Han WX, Tang AH, Shen JL, et al. (2013) Enhanced nitrogen deposition over China. Nature 494: 459–462.

51. Baruch Z, Goldstein G (1999) Leaf construction cost, nutrient concentration, and net CO_2 assimilation of native and invasive species in Hawaii. Oecologia 121: 183–192.

52. Grotkopp E, Rejmanek M (2007) High seedling relative growth rate and specific leaf area are traits of invasive species: Phylogenetically independent contrasts of woody angiospermis. Ann Bot 94: 526–532.

53. Feng YL, Fu GL, Zheng YL (2008) Specific leaf area relates to the differences in leaf construction cost, photosynthesis, nitrogen allocation, and use efficiencies between invasive and noninvasive alien congeners. Planta 228: 383–390.

54. Schweiger PF, Thingstrup I, Jakobsen I (1999) Comparison of two test systems for measuring plant phosphorus uptake via arbuscular mycorrhizal fungi. Mycorrhiza 8: 207–213.

55. de Groot CC, Marcelis LFM, van den Boogaard R, Kaiser WM, Lambers H (2003) Interaction of nitrogen and phosphorus nutrition in determining growth. Plant Soil 248: 257–268.

56. Poorter H, Niklas KJ, Reich PB, Oleksyn J, Poot P, et al. (2012) Biomass allocation to leaves, stems and roots: meta-analyses of interspecific variation and environmental control. New Phytol 193: 30–50.

57. Valladares F, Wright SJ, Lasso E, Kitajima K, Pearcy RW (2000) Plastic phenotypic response to light of 16 congeneric shrubs from a Panamanian rainforest. Ecology 81: 1925–1936.

58. te Beest M, Stevens N, Olff H, van der Putten WH (2009) Plant-soil feedback induces shifts in biomass allocation in the invasive plant *Chromolaena odorata*. J Ecol 97: 1281–1290.

59. Johnson DJ, Beaulieu WT, Bever JD, Clay K (2012) Conspecific negative density dependence and forest diversity. Science 336: 904–907.

60. McCarthy-Neumann S, Kobe RK (2010) Conspecific plant-soil feedbacks reduce survivorship and growth of tropical tree seedlings. J Ecol 98: 396–407.

61. Bower MJ, Aslan CE, Rejmanek M (2009) Invasion potential of Chinese tallow tree (*Triadica sebifera*) in California's central valley. Invasive Plant Sci Manag 2: 386–395.

13

Intercropping Competition between Apple Trees and Crops in Agroforestry Systems on the Loess Plateau of China

Lubo Gao[1], Huasen Xu[1], Huaxing Bi[1,2]*, Weimin Xi[3], Biao Bao[1], Xiaoyan Wang[1], Chao Bi[1], Yifang Chang[1]

1 College of Water and Soil Conservation, Beijing Forestry University, Beijing, P.R. China, 2 Key Laboratory of Soil and Water Conservation, Ministry of Education, Beijing, P.R. China, 3 Department of Biological and Health Sciences, Texas A&M University-Kingsville, Kingsville, Texas, United States of America

Abstract

Agroforestry has been widely practiced in the Loess Plateau region of China because of its prominent effects in reducing soil and water losses, improving land-use efficiency and increasing economic returns. However, the agroforestry practices may lead to competition between crops and trees for underground soil moisture and nutrients, and the trees on the canopy layer may also lead to shortage of light for crops. In order to minimize interspecific competition and maximize the benefits of tree-based intercropping systems, we studied photosynthesis, growth and yield of soybean (*Glycine max* L. Merr.) and peanut (*Arachis hypogaea* L.) by measuring photosynthetically active radiation, net photosynthetic rate, soil moisture and soil nutrients in a plantation of apple (*Malus pumila* M.) at a spacing of 4 m × 5 m on the Loess Plateau of China. The results showed that for both intercropping systems in the study region, soil moisture was the primary factor affecting the crop yields followed by light. Deficiency of the soil nutrients also had a significant impact on crop yields. Compared with soybean, peanut was more suitable for intercropping with apple trees to obtain economic benefits in the region. We concluded that apple-soybean and apple-peanut intercropping systems can be practical and beneficial in the region. However, the distance between crops and tree rows should be adjusted to minimize interspecies competition. Agronomic measures such as regular canopy pruning, root barriers, additional irrigation and fertilization also should be applied in the intercropping systems.

Editor: Randall P. Niedz, United States Department of Agriculture, United States of America

Funding: This paper was supported by National Scientific and Technology Program of China (No. 2011BAD38B02) and CFERN & GENE Award Funds on Ecological paper. The funders had no role in study design, data collection and analysis, decision to publish, or preparation of the manuscript.

Competing Interests: The authors have declared that no competing interests exist.

* E-mail: bhx@bjfu.edu.cn

Introduction

The Loess Plateau is the birthplace of China's primitive agriculture. However, because of unsound land use and destruction of forests, the Loess Plateau has suffered serious soil erosion. At the same time, rapid population growth has also brought greater pressure to the environment in the region. The ensuing ecological and environmental problems have slowed down the economic development and living standards of local people. These problems lead to further deterioration of ecological environment, forming a vicious cycle. The local government is facing dual pressures from both economy and ecology.

Agroforestry systems have been considered as an effective practice to alleviate the conflicts between the rapidly growing population and the limited arable land resources [1,2]. In recent years, agroforestry management has been widely applied in the Loess Plateau region for reducing soil erosion and water loss, restoring ecological balance, raising land utilization rate and increasing economic benefits [3,4]. However, in most agroforestry systems, competition for light, moisture and nutrients exists at the interface between trees and crops which can cause a reduction of crop yield [5]. It is a major constraint that has affected stability of the structure and the function of the agricultural ecosystems. The competition between woody tree species and understory crop species not only exists aboveground (competition for light) but also comes from belowground (competition for soil moisture and nutrients), leading to lower crop yield. According to Friday and Fownes, the competition between trees and crops is overwhelmingly for light which is the main reason for the reduction of maize in alley cropping system in Hawii, USA [6]. Similar results were reported by Peng et al. in loess area of Weibei in Shaanxi Province, China [7]. Elsewhere in southern Australia, studies showed that reduced crop yields are associated with the competition for water in windbreak and alley systems [8,9]. Kowalchuk and Jong found that, especially in drought years, competition for water is the principal factor affecting the yield of spring wheat intercropped with shelterbelts in Western Saskatchewan [10]. In some related studies, the results indicated that competition for nutrients does not exist in intercropping systems [11–13]. However, others reported that as one of the main reasons leading to the reduction of crop yield, the competition for soil nutrients does exist in the interface of trees and crops and has a negative impact [14,15]. It is very important to explore the competitive mechanism in intercropping systems, in order to provide optimum management strategies and technologies for managing intercropping system with high-yield, high-efficiency and stabilization.

Table 1. Characteristics of apple trees intercropped with soybean and peanut in the experimental sites in July 2011.

Measurement	Intercropped with soybean	Intercropped with peanut
Tree height (m)	2.4	2.5
DBH (cm)	4.1	4.2
Depth of live crown (m)	1.7	1.8
Mean radius of crown (m)	1.3	1.2

Apple-crop intercropping system is one of the most commonly applied agroforestry systems in the Loess Plateau region owing to its good ecological, social and economic benefits. However, only few studies focused on this intercropping system in the area. In order to explore the biological reasons of the competition in typical intercropping systems and to provide effective management techniques, we report on a study of two apple-crop intercropping systems (apple-soybean, apple-peanut) on the Loess Plateau region in the western portion of Shanxi Province. The objectives of our research were (1) to analyze the interspecies competition relationship between trees and crops; (2) to find the limiting factors in the development of intercropping systems in this area; (3) to offer possible solutions to minimize the interspecies competitions and maximize resource utilization; (4) to enrich the related study and to improve the management of the intercropping systems in this region.

Materials and Methods

Study site

The study site was located in the Baidong Village, Jixian County, Shanxi Province, China (36°06′ N, 110°35′ E, 1025 m a.s.l.). The area is a typical hill and gully region of the Loess Plateau. The annual mean rainfall is about 575 mm, and the mean annual temperature is 10°C (1991–2010). The precipitation is unevenly distributed seasonally, with an average rainfall of 463 mm from June to August (1991–2010), which contributed about 80% of annual precipitation. The parent material of the soil is loess, and the soil properties are uniform. The bulk density, pH, total porosity, $CaCO_3$ content, cation exchange capacity, organic C, total N and available P of the top soil layer (100 cm) were 1.32 $Mg•m^{-3}$, 8.24, 50.16%, 18.35%, 18.43 $cmol•kg^{-1}$, 6.27 $g•kg^{-1}$, 0.39 $g•kg^{-1}$ and 4.39 $mg•kg^{-1}$, respectively. The main intercropping tree species are Apple (*Malus pumila* M.), Apricot (*Prunus armeniaca* L.), Pear (*Pyrus bretschneideri* R.), Chinese arborvitae (*Platycladus orientalis* (L.) and Franco) and Black locust (*Robinia pseudoacacia* L.).

Ethics Statement

No specific permits were required for the described field studies. The sampling locations were not privately-owned or protected in any way and the field studies did not involve endangered or protected species.

Treatments and Crop Cultivation

Two typical intercropping systems of apple-soybean and apple-peanut were chosen for this study during the crop growing season of 2011 and 2012. The apple trees were planted in an East-West orientation in 2007. The characteristics of the apple trees intercropped with soybeans and peanuts in July 2011 are listed in Table 1. There were four treatments in this study: apple-soybean intercropping treatment (AS), soybean monoculture

served as control (CS), apple-peanut intercropping treatment (AP) and peanut monoculture served as control (CP). Each treatment had three replicates. Each replicate of intercropping treatment (AS and AP) was an 8 × 10 m plot that included 12 trees planted in three rows with 4 m between trees and 5 m between rows. Each replicate of control treatment (CS and CP) was the same size of 8 × 10 m. For all treatments, the crops were planted at a spacing of 0.4 m with in rows and 0.5 m between rows and received the same agricultural management practices. Soybean and peanut were grown 0.3 m from an adjacent tree row in the intercropping systems. All plots received 147 kg N ha^{-1}, 30 kg P ha^{-1} and 30 kg K ha^{-1} as basal fertilizer and no additional fertilizer or irrigation in the rest of the year.

Measurements of Plant Photosynthesis, Soil Moisture and Nutrients

For the sampling of plant photosynthesis, soil moisture and soil nutrients, six sampling locations at distances of 0.5 m, 1.5 m and 2.5 m, respectively, from both side of tree row were identified as sampling points in each intercropping plot (Figure 1). The sampling points were further divided into three equal groups and denoted as F0.5, F1.5 and F2.5 based on the distance (0.5 m, 1.5 m and 2.5 m) from the tree row. Measurement parameters of F0.5, F1.5 and F2.5 were used to represent the major locations of 0.5 m, 1.5 m and 2.5 m away from apple tree row. For each control plot, five selected points were established with an S-shaped sampling method.

Photosynthetically active radiation (PAR) and net photosynthetic rate (NPR) of crops were performed by two portable Li-6400 photosynthesis systems which had a 6 cm^2 clamp-on leaf chamber connected to the main engine (Li-6400, Li-Cor Inc., Lincoln, NE, USA) under ambient humidity, temperature and irradiance. One fully expanded leaf from the upper part of the crop canopy in each sampling point was selected and measured five times with 2 h intervals during daytime (0900–1700 h). During each measurement period, all sampling points of intercropping treatment and control treatment were visited. These treatments were measured in mid-August 2011 and again in late August 2012, the typical phenological phases of peanut and soybean. For all measurements, the flow velocity was set at 500 $μmol•s^{-1}$ and the airstream entering the chambers was kept at the growth CO_2 concentration (370 $μmol•mol^{-1}$) by a computer-controlled CO_2 injector system supplied with Li-6400. PAR and CO_2/H_2O exchanged by the leaf were measured concurrently with the quantum sensor and the infrared gas analyzer on LI-6400. The data were recorded and calculated automatically with the software in the photosynthesis system.

For soil moisture, the samples were taken at different phenological phases of soybean and peanut: 8 July, 23 August, and 23 September in 2011; and 4 July, 11 August, and 22 September in 2012. A drill was used to remove the soil from 0–100 cm in 20 cm intervals in soil profile. The soil moisture content

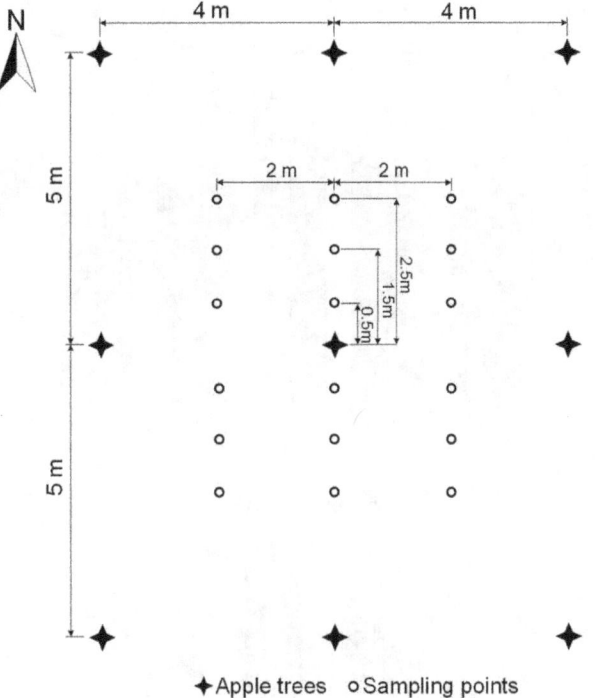

Figure 1. Sampling points of plant photosynthesis, soil moisture and nutrients in the intercropping study sites.

was determined gravimetrically in each layer. The mean soil moisture content of the five layers in all sampling time (2011 and 2012) was calculated and used as the final value of the sampling point.

For the sampling of soil nutrients, the soil samples were taken on 23 August, 2011 and 11 August, 2012, the typical phenological phases of peanut and soybean. The soil samples were collected from a depth of 0–100 cm in soil profile with a drill. Organic matter content was determined by H_2SO_4-$K_2Cr_2O_7$ pyrogenation. Total N was determined using the Kjeldahl method, with a KDY-9830 N Analyzer. Available P was determined by Olsen sodium-bicarbonate extraction. Available K was determined by flame photometer.

Measurements of Crop Growth and Yields

For the sampling of crop growth, we used the same sample locations as for soil moisture. A single crop plant was sampled at each sampling point on 24 August 2011 and 12 August 2012. A total of 69 soybean plants and 69 peanut plants were harvested during each measurement period. In the lab, plant height, hundred leaf dry weight and total above-ground biomass of all plants were measured and recorded.

At the end of the growing season, in each intercropping plot, peanuts and soybeans were harvested from both sides of the tree row in two rectangular areas. The rectangular area was 4.0 m long and 2.7 m wide. As the convenience of the study, the two rectangular areas were divided into three groups: (1) the area of 0.3–1.0 m away from tree row; (2) the area of 1.0–2.0 m away from tree row; and (3) the area of 2.0–3.0 m away from tree row. The yields of the three groups were used to represent the crop yield of F0.5, F1.5 and F2.5, respectively. In the control plots, 2 m × 2 m quadrates of soybean and peanut were harvested to get the grain production. The peanuts and soybeans were dried at 70 °C

and then weighed to obtain an average dry weight. Yield values were reported on a per hectare basis.

Data analysis

All parameters (PAR, NPR, soil moisture, soil nutrients content, crop growth and yields) measured for control treatments and three major locations (F0.5, F1.5 and F2.5) of intercropping treatments were described in terms of mean values followed by respective standard deviations. Simple regression analysis was used to examine the relationships between the data of PAR, NPR, soil moisture and the distance from the tree row. Differences among groups for each crop (soybean or peanut) were determined by one-way ANOVA, and the results of the multiple comparisons were performed with least significant difference (LSD) test at $P<0.05$. NPR, total above-ground biomass and yield values of soybean and peanut had a correlated analysis with environmental parameters to decide the effect of apple trees competition on crop growth and productivity via bivariate correlation (Pearson) analysis at $P<0.05$ and $P<0.01$. All the analyses were performed by using the software IBM SPSS Statistics 20.0 for Windows.

Results

Light Interception and Plant Photosynthesis

For both crops, diurnal variation of photosynthetically active radiation (PAR) in the intercropping systems and the monoculture configurations (control treatments) showed a single peak curve with time (Figure 2). The peak of PAR appeared at 13:00 pm and the minimum value appeared in 17:00 pm. Because of reflectance, absorbance and transmittance by the apple tree canopy, the PAR of crops in the intercropping systems were lower than that in the monoculture configurations during the same period. On the horizontal distribution, the general trend was that the closer the crops to the tree rows, the lower the PAR received. The same tendency was found in diurnal variation of net photosynthetic rate (Figure 3).

The daily mean values of PAR showed a clear linear relationship with distance from the apple tree row in both intercropping treatments (Figure 4A). The trend lines of PAR (Y, $\mu mol \cdot s^{-1} \cdot m^{-2}$) and distance from trees rows (X, m) were $Y = 78.5 \times +865.3$ ($R^2 = 0.999$) in apple-soybean intercropping treatment (AS) and $Y = 82.5 \times +881.9$ ($R^2 = 0.873$) in apple-peanut intercropping treatment (AP). The slopes of both regression lines suggested that the PAR in AP treatment had a higher growth than that in AS treatment as the distance from the tree increased. As shown in Figure 4A, PAR reaching the upper parts of the crop canopy in AP treatment also had higher values than that in AS treatment at the same distance away from the tree row. It indicated that peanut canopy could obtain more solar radiation in AP treatment than soybean canopy in AS treatment. At confidence level of 95%, the control treatment PAR mean fell within the confidence intervals of F2.5 in the corresponding intercropping systems. Compared with the corresponding control treatment, PAR at F0.5 and F1.5 showed a reduction of 17.9% and 10.4% in AS treatment, respectively, 17.8% and 5.4% in AP treatment. Similar linear relationships were also obtained through regression analysis of the relationship between NPR and distance from the apple tree row (Figure 4B). The trend lines of NPR (Y, $\mu mol \cdot s^{-1} \cdot m^{-2}$) and distance from trees rows (X, m) were $Y = 1.025 \times +12.003$ ($R^2 = 0.902$) in AS treatment, and $Y = 0.940 \times +10.983$ ($R^2 = 0.951$) in AP treatment. The NPR in AS treatment had higher values and growth than that in AP treatment as the distance from the tree increased which was different from the measurement of PAR. The control treatments

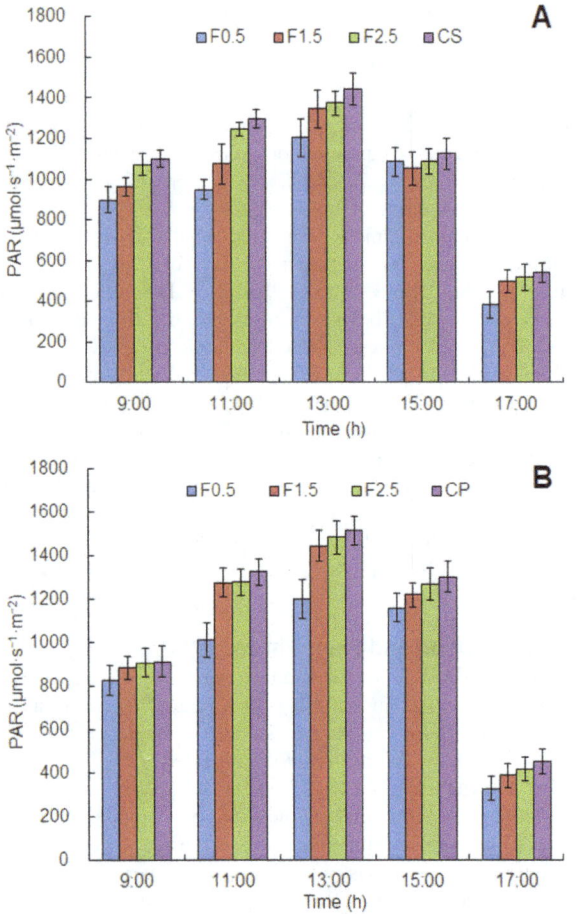

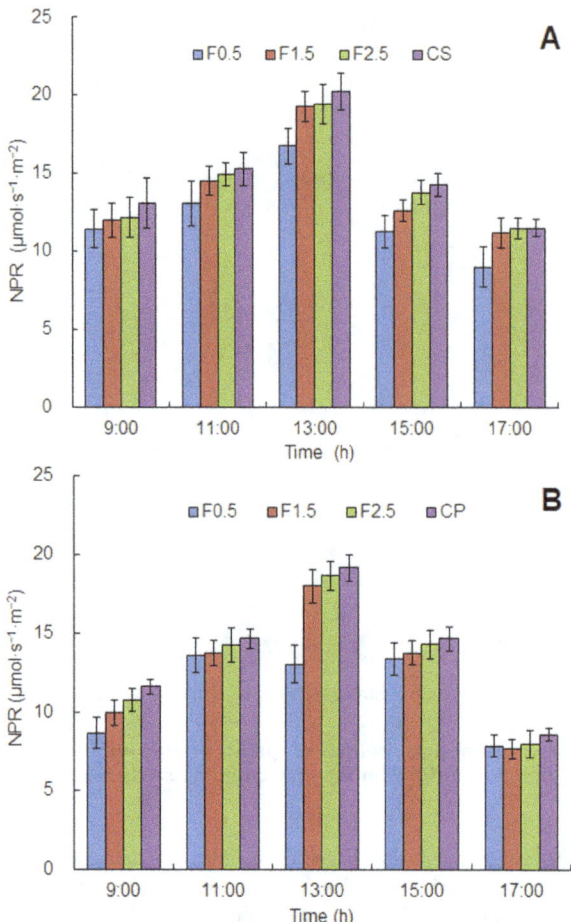

Figure 2. Diurnal variation of photosynthetically active radiation (PAR) for the intercropping systems and its control (A. apple–soybean and B. apple–peanut). F0.5, F1.5 and F2.5 were used to represent the sampling points which had different distance (0.5 m, 1.5 m and 2.5 m) from the tree row. Error bars indicate standard deviation.

Figure 3. Diurnal variation of net photosynthetic rate (NPR) for the intercropping systems and its control (A. apple–soybean and B. apple–peanut). F0.5, F1.5 and F2.5 were used to represent the sampling points which had different distance (0.5 m, 1.5 m and 2.5 m) from the tree row. Error bars indicate standard deviation.

mean fell within the confidence intervals of F2.5 in the corresponding intercropping systems at confidence level of 95%.

Spatial Distribution of Soil Moisture

Although the soil moisture content in the whole soil profile (0 to 100 cm in depth) in AS was different from AP, the trend of spatial distributions of soil moisture was similar (Figure 5). Soil moisture content in AS was related to distance from the apple tree row and showed a clear linear relationship ($Y = 0.465 \times + 11.602$, $R^2 = 0.999$), and AP showed the same trend ($Y = 0.590 \times + 11.002$, $R^2 = 0.900$). Compared with AP, AS had higher values at the same distance away from the tree row. However, with increasing distance from the tree row, soil moisture in AP had a higher growth than that in AS. The lowest soil moisture content was 11.83% in AS and 11.41% in AP, showed a decrease of 10.31% and 11.14% when compared with the corresponding control treatments. The soil moisture at F2.5 in both intercropping systems also had slightly lower values than that in monoculture configurations, however no difference was observed at significance level of 5%, since the control treatments mean fell within the confidence intervals of F2.5 in the corresponding intercropping

systems (confidence level 95%). Otherwise, the average soil moisture content in AP was lower than that in AS.

Spatial Distribution of Soil Nutrients

The soil nutrients content in the 0 to 100 cm interval was calculated (Table 2). It represented that organic matter, total N, available P and available K in AS had different degrees of reduction when compared with CS, and showed significant differences ($P<0.05$). Similar results were found between AP and CP, except that no significant difference was observed for total N and available K at the location of F2.5. The average content of organic matter, total N, available P and available K in AS decreased by 30.77%, 63.24%, 56.08% and 27.83% when compared with CS–the monoculture configuration. For AP and CP, the decreased percentages were 18.32%, 21.05%, 36.27% and 7.49% respectively. In addition, except available K, soil nutrients content in AP was higher than that in AS at the same spatial location. With the increasing distance from tree row, the distribution trend of soil nutrients was different from that of PAR, NPR or soil moisture in the same intercropping condition. The lowest content of organic matter, total N, available P and available K in AS were present at the location of F1.5. The similar

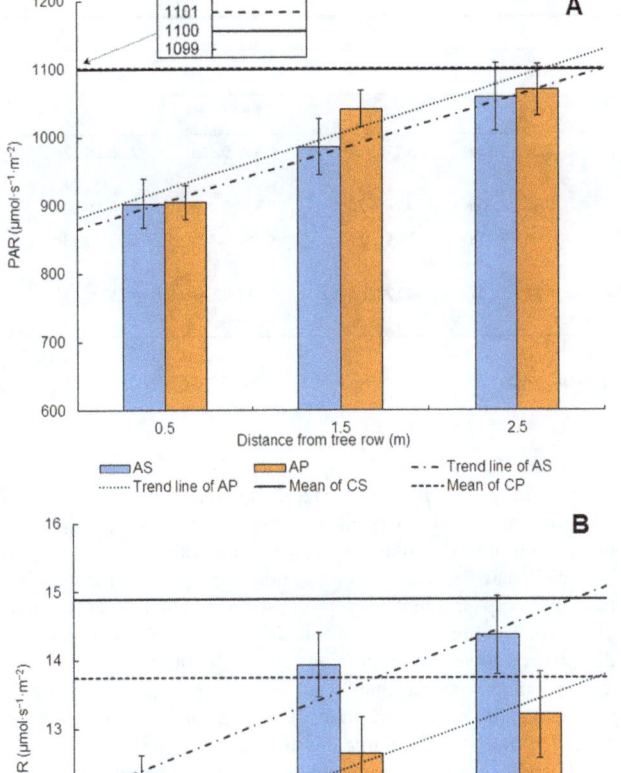

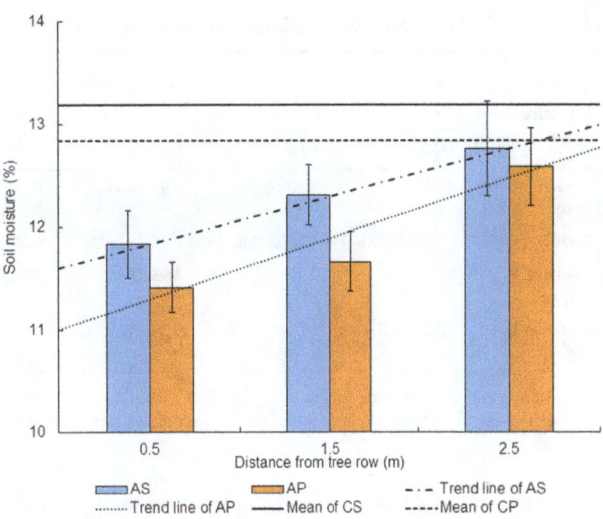

Figure 5. Soil moisture of 0 to 100 cm depth for the intercropping systems and its control. Vertical lines indicate confidence interval at 95% level.

Figure 4. Daily mean of photosynthetically active radiation (PAR) and net photosynthetic rate (NPR) for the intercropping systems and its control (A. PAR and B. NPR). Vertical lines indicate confidence interval at 95% level.

result was found in AP, except that the lowest value of available K was the location of F0.5.

Crop Growth and Yields

Plant height, hundred leaf dry weight and total above-ground biomass in both intercropping systems had lower values when compared with the monoculture configuration ($P<0.05$; Table 3). The locations of F0.5 and F1.5 in all these parameters showed significant differences with corresponding monoculture configuration ($P<0.05$); however, there were no difference observed in the location of F2.5.

The yield of soybean in AS was significantly related to the distance from the row of apple trees ($Y=0.180 \times +1.400$, $R^2 = 0.991$), and the yield of peanut in AP showed the same trend ($Y=0.095 \times +1.554$, $R^2 = 0.900$) which showed that yield of soybean had greater impacted by distance from the tree row. The yields at F0.5 and F1.5 in AS were lower than that in CS ($P<0.05$), with a reduction of 22.45% and 11.95%, and in AP the yields of reduction were 13.31% and 11.03% when compared with

CP. No differences were observed between the locations of F2.5 in both intercropping systems and the corresponding monoculture configuration ($P<0.05$).

Within plot differences in these parameters were significantly correlated (Table 4). NPR was highly correlated with PAR, and soil moisture. The total above-ground biomass of soybean was highly correlated with PAR, soil moisture and available P. The total above-ground biomass of peanut was correlated with PAR, soil moisture, total N and available P. Yield of soybean was highly correlated with PAR, soil moisture, available P and total N, with a trend of soil moisture >PAR>available P>total N. For peanut, the trend was soil moisture > PAR > total N > available P. It showed that, for both of the intercropping system in the study region, the primary factor affecting the yield is soil moisture, and the secondary factor is photosynthetically active radiation, and soil nutrient also have an impact on crop yield in some depth.

Discussion

Agroforestry system has been studied for a long time and has been widely used in the agricultural production practices in China [3,16,17]. However, there has been little research done on the agroforestry system in the Loess Plateau region. The main intercropping models which have been studied are always walnut-wheat and apple-wheat [18–20]. The types of fruit trees intercropping with economic crops such as soybean and peanut has not been well studied. In fact, compared with wheat, soybean and peanut could bring more economic income to farmers. At the same time, these two crops could be rotated with wheat in order to improving land-use efficiency, and re-establishing the economic viability of the Loess Plateau.

Our study observed a clearly positive linear relationship between distance from the apple tree rows and the daily mean values of PAR and NPR in the intercropping systems. For both apple-crop intercropping systems, the shading of the 4–5 years old apple trees had a significant negative effect on the crops in the range of 1.5 m away from the tree rows and further caused the reduction of crop yield. In other researches of temperate agroforestry systems, the similar results were reported by Reynolds et al. [21] about maize and soybean intercropped with poplar and

Table 2. Soil nutrients for the intercropping systems and control configurations.

Measurement	AS			CS	AP			CP
	F0.5	F1.5	F2.5		F0.5	F1.5	F2.5	
Organic matter (g·kg^{-1})	4.63±0.27a	3.40±0.29b	4.93±0.27a	6.24±0.31c	6.26±0.36a	5.21±0.26b	6.37±0.36a	7.28±0.27c
Total N (g·kg^{-1})	0.28±0.05a	0.22±0.04b	0.25±0.04ab	0.68±0.03c	0.31±0.06a	0.24±0.05b	0.35±0.07ac	0.38±0.04c
Available P (mg·kg^{-1})	2.82±0.56a	2.50±0.54a	3.93±0.31b	7.02±0.22c	5.06±0.61a	3.44±0.44b	4.75±0.42a	6.93±0.33c
Available K (mg·kg^{-1})	97.33±1.77a	84.02±2.09b	95.43±2.97a	127.84±2.75c	87.45±2.81a	90.73±2.45a	98.11±2.42b	99.55±2.19b

Data were given as the means ± SD.
Different lowercase letters within a row of each crop indicate significant differences (LSD, $P<0.05$).

silver maple in Canada and Peng et al. [22] about mungbean and pepper intercropped with walnut and plum in Weibei area, China. For total above-ground biomass and yield of both crops, PAR of soybean had higher correlations than that of peanut, which indicated that soybean is more adversely impacted by tree shading. Within tree-based intercropping systems, many factors such as tree species, tree height, crown shape, tree row orientation and distance between tree rows can influence tree shading of adjoining agricultural crops. Light reduction would depend on the extent and duration of the shade of trees [21]. Regular pruning of fruit trees could reduce light competition within the intercropping system, improving crop yields.

In semiarid and arid regions, it is still a focus of studies whether intercropping system has an overall negative or positive effect on soil moisture [23]. In some related studies, it was considered that the trees can improve soil moisture holistic conditions in intercropping systems [17,24]. In other studies, the opposite results were reported [11,12,25,26]. However, little research has been carried out in this aspect on the Loess Plateau. Our research confirmed that the competition of water between trees and crops do exist, and showed adverse effects in the study site. A clear linear relationship was observed between the distance from the tree row and soil moisture in both of the intercropping systems. The closer to the tree row, the more intense the competition. The lowest soil moisture content in apple-soybean intercropping system and apple-peanut intercropping system showed a reduction of 10.31% and 11.14%, respectively. Only considering competition of water, the mainly affected region of the apple trees was 1.5 m away from the tree rows under the current tree age.

Another key factor of crop growth is soil nutrients in the intercropping systems. Elsewhere, Thomas et al. [27] and Thevathasan et al. [13] have reported that competition for nutrients in intercropping systems does not exist. In our study, it identified that there were competition for soil nutrients between trees and crops in the intercropping systems. The average content of organic matter, total N, available P and available K showed different degrees of reduction in both of the apple-crops intercropping systems than that of the corresponding control treatments. In particular, total N and available P had higher reduction rate than organic matter and available K, and had significantly correlation with yield of crops. As leguminous plants, soybean and peanut could fix nitrogen from the air via a symbiotic relationship with rhizobium bacteria and increase the mineral soil nitrogen content [28,29]. However, the nitrogen coming from biologically fixed N_2 of symbiosis could not meet all the demand of crops growth, and any gaps between N supply by N_2 fixation and crop N demand must be met by N uptake from soil [30]. The deficiency of light and water in the intercropping systems reduced the physiological activity of the crop, and then affected the N fixation capacity, resulting more intense competition for nitrogen between trees and crops. Compared with soybean and peanut, the growth of other non-nitrogen-fixing crop species (i.e. wheat, maize and millet) would be more severely affected because of nitrogen deficiency in the intercropping systems. Different from understory light distribution and soil moisture, soil nutrients had a different variation pattern in both intercropping systems. The main reasons for this phenomenon might be: (1) the crops close to tree row were seriously affected by tree shading, soil moisture stress and human activities, resulting in low physiological activity and low absorption

Table 3. Crop growth, biomass and yield for soybean and peanut intercropped with apple trees and its control.

Measurement	AS			CS	AP			CP
	F0.5	F1.5	F2.5		F0.5	F1.5	F2.5	
Crop height (cm)	43.7±2.8a	44.4±2.3a	47.7±3.2b	50.2±2.6b	19.1±0.7a	21.9±1.0b	22.6±0.9bc	23.5±0.8c
Hundred leaf dry weight (g)	14.12±0.81a	14.14±0.51a	14.95±0.38b	15.37±0.57b	4.39±0.23a	5.79±0.22b	5.90±0.31bc	6.14±0.28c
Total above-ground biomass (g)	54.55±4.66a	66.06±3.81b	76.96±4.30c	79.76±4.22c	50.10±2.62a	51.57±2.45a	79.16±3.28b	79.65±2.46b
Yield (t/ha)	1.48±0.06a	1.69±0.04b	1.84±0.06c	1.91±0.04c	1.62±0.04a	1.66±0.04a	1.81±0.05b	1.86±0.04b

Data were given as the means ± SD.
Different lowercase letters within a row of each crop indicate significant differences (LSD, $P<0.05$).

Table 4. Correlations of soybean and peanut net photosynthetic rate, biomass, and yield with environmental or physiological parameters measured in Jixian, China.

Independent variable	NPR ($\mu mol \cdot s^{-1} \cdot m^{-2}$)	Total above-ground biomass (g)	Yield (kg/ha)
Soybean			
PAR ($\mu mol \cdot s^{-1} \cdot m^{-2}$)	0.973**	0.996**	0.952**
Soil moisture (%)	0.953**	0.977**	0.957**
Organic matter (g•kg^{-1})	0.441	0.575	0.566
Total N (g•kg^{-1})	0.537	0.555	0.601*
Available P (mg•kg^{-1})	0.697*	0.750**	0.763**
Available K (mg•kg^{-1})	0.469	0.538	0.565
Peanut			
PAR ($\mu mol \cdot s^{-1} \cdot m^{-2}$)	0.986**	0.773**	0.816**
Soil moisture (%)	0.926**	0.965**	0.959**
Organic matter (g•kg^{-1})	0.450	0.562	0.583
Total N (g•kg^{-1})	0.513	0.843**	0.770**
Available P (mg•kg^{-1})	0.424	0.628*	0.646*
Available K (mg•kg^{-1})	0.479	0.531	0.446

*Significant at 5% level.
**Significant at 1% level.

of soil nutrients; (2) the decomposition of tree litter leaded to high nutrients content in the area near the tree row; (3) the overlapping of apple tree roots and crop roots resulting in lower nutrient content at F1.5; (4) the tree roots reduced with the increase of the distance from the tree, therefore, the soil nutrients had relatively high content F2.5. Therefore, in the area of 1.5 m away from tree row, strengthen the application of fertilizer (especially nitrogen and phosphorus) would be helpful to alleviate interspecific competition for soil nutrients.

In the apple-crop intercropping systems, the competition of light, water and nutrients resulted in a greater negative impact on crop growth and yields. For the two apple-crop intercropping systems in our study, the primary factor affecting the crop yield was soil moisture, and the secondary factor was light, and deficiency of the soil nutrient also had a negative impact on crop yields. In the same study area, Yun et al. reported a similar research with a different conclusion: the light is the primary limiting factor leading to reduction of crops, followed by soil moisture [31]. In their research, the apple trees had greater crown width, canopy density and root depth due to elder age (9-year-old) and smaller tree spacing (3 m×4 m). Affected by the impact of canopy structure, the obvious microclimate effect inhibited evapotranspiration of soil moisture to some extent [32], in the same time, the low transmittance led to more intense light stress to crops. Furthermore, the effect of hydraulic lift by tree roots also alleviated the interspecific competition for soil water [24]. Combined with these reasons, different results were found. For different intercropping patterns and tree ages, the intensity of competition for resources would be different in the intercropping system. Therefore, a long-term observation should be carried out in this region to obtain more details about the mechanism of interspecific competition in the intercropping systems. In our study, under the current tree age and growth conditions, the influence scope of the apple trees was 1.5 m away from the tree rows. Compared with the corresponding monoculture configuration, the yield of peanut in the intercropping system had a lower reduction than that of soybean. With comprehensive consideration, peanut is more suitable for inter-cropping with apple trees in this region.

As we have demonstrated in this study, soil moisture, light and soil nutrients were the limiting factors of crop yield. In order to obtain more production, appropriate management measures were needed to minimize competition between trees and crops. Namirembe [33] and Friday [6] have suggested that the competition for light between trees and crops could be alleviated by pruning of trees crown and increasing the intercropping distance. In general, the aboveground competition could be intuitively observed and managed. However, the competition belowground is invisible and easily ignored by farmers or managers. To avoid these yield losses, root barrier in the intercropping interface is considered to be a useful agricultural management practice according to some related studies [12,34,35]. Combined with their research achievements and our experiment result, we offered several specific recommendations to reduce the competition exist in apple-crop intercropping systems: (1) the selection of crop varieties which is more suitable for apple-crop intercropping systems; (2) appropriate distance increase between the crops and apple tree rows; (3) regular pruning of fruit trees, in order to increase canopy light transmittance rate; (4) additional fertilization and irrigation in the key phenological phase of the crops; (5) differences of irrigation and fertilization based on the distance from the apple trees. Management measures such as plastic film and straw mulching have been widely used in agricultural production. Whether these measures have overall positive effects on intercropping system would be one of the focus of our future research work in this region.

Conclusions

As an effective method to increase the efficiency of land use and economic returns, tree-based intercropping systems are particularly important on Loess Plateau. We concluded that the competitions exist both above-ground and below-ground between apple trees and crops. The competition for soil moisture is the primary limiting factor for the crop productivity in this region. Furthermore, the tree shading and the competition for soil nutrients in the interface of trees and crops also have a negative impact on the understory crops. However, it could be minimized by better agricultural technology and management measures.

In summary, our study suggests that there is great potential for intercropping systems in the Loess Plateau. Therefore, in order to relieve the shortage of arable land and promote the sustainable development of natural resources, the intercropping systems would continue to be the hot spot for future research. Canopy structure, roots distribution of trees, the application of different agronomic measures and the role they play in the competition process in the intercropping systems will be the focus of our future research.

Acknowledgments

We are grateful for the support from the Shanxi Jixian Forest Ecosystem Research Station. We also would like to thank the three anonymous reviewers and the editors for their helpful comments.

Author Contributions

Conceived and designed the experiments: LG HX HB BB. Performed the experiments: LG HX HB BB. Analyzed the data: LG HX WX BB XW CB YC. Contributed reagents/materials/analysis tools: XW CB YC. Wrote the paper: LG HB WX.

References

1. Burel F (1996) Hedgerows and their role in agricultural landscapes. Critical Reviews in Plant Sciences 15: 169–190.
2. Gene Garrett HE, Buck L (1997) Agroforestry practice and policy in the United States of America. Forest Ecology and Management 91: 5–15.
3. Li W, Lai S (1994) Agroforestry in China. Beijing: Chinese Science Press. pp. 14–18.
4. Zhu Q, Zhu J (2003) Sustainable management technology for conversion of cropland to forest in loess area. Beijing:Chinese Forestry Press. pp. 160–165.
5. Ong CK, Huxley P (1996) Tree-crop interactions: a physiological approach. Wallingford:CAB International. pp. 386.
6. Friday JB, Fownes JH (2002) Competition for light between hedgerows and maize in an alley cropping system in Hawaii, USA. Agroforestry Systems 55: 125–137.
7. Peng X, Zhang Y, Cai J, Jiang Z, Zhang S (2009) Photosynthesis, growth and yield of soybean and maize in a tree-based agroforestry intercropping system on the Loess Plateau. Agroforestry Systems 76: 569–577.
8. Hall DJM, Sudmeyer RA, McLernon CK, Short RJ (2002) Characterisation of a windbreak system on the south coast of Western Australia. 3. Soil water and hydrology. Australian Journal of Experimental Agriculture 42: 729–738.
9. Unkovich M, Blott K, Knight A, Mock I, Rab A, et al. (2003) Water use, competition, and crop production in low rainfall, alley farming systems of south-eastern Australia. Australian Journal of Agricultural Research 54: 751–762.
10. Kowalchuk TE, Jong E (1995) Shelterbelts and their effect on crop yield. Canadian Journal of Soil Science 75: 543–550.
11. Jose S, Gillespie AR, Seifert JR, Biehle DJ (2000) Defining competition vectors in a temperate alley cropping system in the midwestern USA: 2. Competition for water. Agroforestry Systems 48: 41–59.
12. Miller AW, Pallardy SG (2001) Resource competition across the crop-tree interface in a maize-silver maple temperate alley cropping stand in Missouri. Agroforestry Systems 53: 247–259.
13. Thevathasan NV, Gordon AM, Simpson JA, Reynolds PE, Price G, et al. (2004) Biophysical and ecological interactions in a temperate tree-based intercropping system. Journal of Crop Improvement 12: 339–363.
14. Newman SM, Bennett K, Wu Y (1997) Performance of maize, beans and ginger as intercrops in Paulownia plantations in China. Agroforestry Systems 39: 23–30.
15. Yun L, Bi H, Gao L, Zhu Q, Ma W, et al. (2012) Soil moisture and soil nutrient content in walnut-crop intercropping systems in the Loess Plateau of China. Arid Land Research and Management 26: 285–296.
16. Meng P, Zhang J, Fan W (2003) Research on agroforestry in china. Beijing:Chinese Forestry Press. pp.235.
17. Meng P, Zhang J (2004) Effects of pear-wheat inter-cropping on water and land utilization efficiency. Forest Research 17: 167–171.
18. Zhang J, Meng P, Yin C (2002) Spatial distribution characteristics of apple tree roots in the apple-wheat intercropping. Scientia Silvae Sinicae 38: 30–33.
19. Zhang J, Meng P (2004) Model on wheat potential evapotranspiration in apple-wheat intercropping. Forest Research 17: 284–290.
20. Yun L, Bi H, Ren Y, Ma W, Tian X (2009) Soil moisture distribution at fruit-crop intercropping boundary in the Loess region of Western Shanxi. Journal of Northeast Forestry University 37: 70–78.
21. Reynolds PE, Simpson JA, Thevathasan NV, Gordon AM (2007) Effects of tree competition on corn and soybean photosynthesis, growth, and yield in a temperate tree-based agroforestry intercropping system in southern Ontario, Canada. Ecological Engineering 29: 362–371.
22. Peng X, Cai J, Jiang Z, Zhang Y, Zhang S (2008) Light competition and productivity of agroforestry system in loess area of Weibei in Shaanxi. Chinese Journal of Applied Ecology 19: 2414–2419.
23. Zhang J, Meng P, Yin C, Cui G (2003) Summary on the water ecological characteristics of agroforestry system. World Forestry Research 16: 10–14.
24. Hirota I, Sakuratani T, Sato T, Higuchi H, Nawata E (2004) A split-root apparatus for examining the effects of hydraulic lift by trees on the water status of neighbouring crops. Agroforestry Systems 60: 181–187.
25. Lott JE, Howard SB, Ong CK, Black CR (2000) Long-term productivity of a Grevillea robusta-based overstorey agroforestry system in semi-arid Kenya: II. Crop growth and system performance. Forest Ecology and Management 139: 187–201.
26. Lehmann J, Peter I, Steglich C, Gebauer G, Huwe B, et al. (1998) Below-ground interactions in dryland agroforestry. Forest Ecology and Management 111:157–169
27. Thomas J, Kumar BM, Wahid PA, Kamalam NV, Fisher RF (1998) Root competition for phosphorus between ginger and Ailanthus triphysa in Kerala, India. Agroforestry Systems 41: 293–305.
28. Cheng D (1994) Resource microbiology. Harbin:Northeast Forestry University Press. pp. 32.
29. Wani SP, Rupela OP, Lee KK (1995) Sustainable agriculture in the semi-arid tropics through biological nitrogen fixation in grain legumes. Plant Soil 174: 29–49.
30. Salvagiotti F, Cassman KG, Specht J E, Walters DT, Weiss A, et al. (2008). Nitrogen uptake, fixation and response to fertilizer N in soybeans: A review. Field Crops Research 108:1–13.
31. Yun L, Bi H, Tian X, Cui Z, Zhou H, et al. (2011) Main interspecific competition and land productivity of fruit-crop intercropping in Loess Region of West Shanxi. Chinese Journal of Applied Ecology 22:1225–1232
32. Zhang J, Meng P, Song Z, Gao J (2004) An overview on micro-climatic effects of agro-forestry systems in plain agricultural areas in China. Agricultural Meteorology 25:52–55
33. Namirembe S (1999) Tree management and resource utilization in agroforestry systems with Senna spectabilis in the drylands of Kenya. Bangor:University of Wales. pp. 206.
34. Singh RP, Saharan N, Ong CK (1989) Above and below ground interactions in alley-cropping in semi-arid India. Agroforestry Systems 9: 259–274.
35. Hou Q, Brandle J, Hubbard K, Schoeneberger M, Nieto C, et al. (2003) Alteration of soil water content consequent to root-pruning at a windbreak/crop interface in Nebraska, USA. Agroforestry Systems 57: 137–147.

14

Spatial and Temporal Variation of Archaeal, Bacterial and Fungal Communities in Agricultural Soils

Michele C. Pereira e Silva[1]*, **Armando Cavalcante Franco Dias**[2], **Jan Dirk van Elsas**[1], **Joana Falcão Salles**[1]

1 Department of Microbial Ecology, Centre for Life Sciences, University of Groningen, Groningen, The Netherlands, **2** Department of Soil Science, "Luiz de Queiroz" College of Agriculture, University of São Paulo, Piracicaba, São Paulo, Brazil

Abstract

Background: Soil microbial communities are in constant change at many different temporal and spatial scales. However, the importance of these changes to the turnover of the soil microbial communities has been rarely studied simultaneously in space and time.

Methodology/Principal Findings: In this study, we explored the temporal and spatial responses of soil bacterial, archaeal and fungal β-diversities to abiotic parameters. Taking into account data from a 3-year sampling period, we analyzed the abundances and community structures of *Archaea*, *Bacteria* and *Fungi* along with key soil chemical parameters. We questioned how these abiotic variables influence the turnover of bacterial, archaeal and fungal communities and how they impact the long-term patterns of changes of the aforementioned soil communities. Interestingly, we found that the bacterial and fungal β-diversities are quite stable over time, whereas archaeal diversity showed significantly higher fluctuations. These fluctuations were reflected in temporal turnover caused by soil management through addition of N-fertilizers.

Conclusions: Our study showed that management practices applied to agricultural soils might not significantly affect the bacterial and fungal communities, but cause slow and long-term changes in the abundance and structure of the archaeal community. Moreover, the results suggest that, to different extents, abiotic and biotic factors determine the community assembly of archaeal, bacterial and fungal communities.

Editor: A. Mark Ibekwe, U. S. Salinity Lab, United States of America

Funding: This work was supported by the NWO-ERGO Programme and was part of a collaborative project with Utrecht University, Utrecht, The Netherlands. The funders had no role in study design, data collection and analysis, decision to publish, or preparation of the manuscript.

Competing Interests: The authors have declared that no competing interests exist.

* E-mail: m.silva@rug.nl

Introduction

Understanding temporal and spatial patterns in the abundance and distribution of communities has been a fundamental quest in ecology. Such an understanding is crucial to allow an anticipation of responses of ecosystems such as soil to global changes [1]. Because local conditions are never constant, small disturbances that affect the soil microbial communities might occur [2–3] at different temporal and spatial scales. The assessment of microbial communities at a particular locality may result in patterns that vary greatly both within and between years, and these communities may be subjected to changes over longer time scales as a result of processes such as succession and evolutionary change [4]. One approach to investigate temporal (and spatial) variability in complex systems is to explore patterns of β-diversity. Whereas alpha (α-) diversity represents a measure of the total diversity of a given site, β-diversity is the variation of species composition (turnover) across space or time between paired sites. High β-diversity indicates large differences in community composition among different sites. Such high diversity can result from local as well as regional factors, e.g. changes in the local environmental conditions or limitation of dispersal between sites [5].

Temporal variation of conditions is a very common feature of ecosystems. Ecologists have long been interested in how such variation structures natural communities [6,7]. It can presumably affect the rate of microbial turnover, as microorganisms can process resources and adapt to changes in natural environments on a much faster time scale than macroorganisms [8]. Moreover, many functional microbial groups can show dramatic seasonal changes in soils [9].

The number of studies employing the concept of β-diversity to understand how microbial communities respond to biotic and abiotic parameters has increased substantially in soil ecology. Martiny and co-workers [10] studied the mechanisms driving ammonia-oxidizing bacterial (AOB) communities in salt marsh sediments. They found no evolutionary diversification when comparing the AOB community composition between three continents; although a negative relationship was observed between geographic distance and community similarity. Furthermore, in an attempt to determine to which extent a bacterial metacommunity that consisted of 17 rock pools was structured by different assembly mechanisms [11], the authors studied changes in β-diversity across different environmental gradients over time, including phosphorus concentration, temperature and salinity. They found that there were temporal differences in how the communities responded to abiotic factors. β-diversity allows not only the understanding of temporal but of spatial variations as well. For instance, in a survey of bacterial communities across more than 1000 soil cores in Great Britain [12], no spatial patterns were observed, but instead variations in β-diversity according to soil pH were found, which revealed that β-diversity (between sample variance in α-diversity) was higher in acidic soils (pH 4–5) than in more alkaline soils (pH 7–9) [12]. In the former soils, environmental heterogeneity was highest, calculated as the variance in environmental conditions [12]. In another study, different patterns of bacterial β-diversity were observed between different layers in sediment cores, which could be attributed to historical variation and geochemical stratification [13].

Of the soil microbial groups, bacteria have been mostly studied, as they exhibit an estimated species diversity of about 10^3 to up to 10^6 per g soil [14–16]. However, archaea and fungi are also important microorganisms found in soil. Previous studies have shown the ubiquity of archaea in soil, especially the crenarchaeota [17–19]. Fungal abundances in the order of 10^4 fungal propagules per g of dry soil were observed in Antarctic soils [20] and 10^7 per g of soil in soil crusts [21]. Fundamental differences in the physiology and ecology of members of such communities would suggest that their patterns of spatial and temporal variation are controlled by distinct edaphic factors.

In this study, we explored the temporal and spatial fluctuations of soil microbial communities and their relation to local environmental conditions. In order to do so, we investigated the spatiotemporal dynamics of the soil microbiota by analyzing the patterns of α- and β-diversity of archaea, bacteria and fungi in eight agricultural soils across the Netherlands. We sampled the soils eleven times, from 2009 to 2011. Furthermore, to complement the analyses, we applied TLA (time-lag analysis) [22], a distance-based approach to study the temporal dynamics of communities by measuring community dissimilarity over increasing time lags. TLA provides measures of model fit and statistical significance, allowing the quantification of the strength of temporal community change in a numerical framework [23]. We thus interrogated how the relationship between microbial abundance, species composition and the surrounding environment varies in space and time and how this relates to long-term compositional changes.

Materials and Methods

Study area and field sampling

The eight soil sites sampled are located in the Netherlands. Their characteristics and geographical coordinates are found in Table 1 and in Table S1. Sampling points were selected to reflect temporal differences in abiotic parameters. For each soil four replicates were taken. Each replicate consisted of 10 subsamples (15–20 cm deep) collected between plots with a spade, away from the plant roots. Soil samples were collected four times over an annual cycle in 2009 (April, June, September and November), three times in 2010 (April, June and October), and four times in 2011 (February, April, July and September). Each sample was placed in a plastic bag and thoroughly homogenized before analysis. A 100-g subsample was kept at 4°C and used for chemical analyses, whereas the remaining soil was kept at −20°C for subsequent DNA extraction and molecular analysis of community composition and total abundance (see below).

Soil chemical analysis

The environmental variables measured included pH, concentrations of nitrate (N-NO_3^- in mg/kg of soil), ammonium (N-NH_4^+ in mg/kg of soil), organic matter (OM in %) and clay content (in %). The pH was measured in $CaCl_2$ suspension 1:4.5 (g/v) (Hanna Instruments BV, IJsselstein, The Netherlands). Organic matter (OM) content is calculated after 4 hours at 550°C. Nitrate (N-NO_3^-) and ammonium (N-NH_4^+) were determined with a colorimetric method using the commercial kits Nanocolor Nitrat50 (detection limit, 0.3 mg N kg-1 dry weight, Macherey-Nagel, Germany) and Ammonium3 (detection limit, 0.04 mg N kg-1 dry weight; Macherey-Nagel, Germany) according to manufacturer's protocol.

Nucleic acid extraction

DNA was extracted from 0.5 g of soil using Power Soil MoBio kit (Mo Bio Laboratories Inc., NY), according to the manufacturer's instructions, after the addition of glass beads (diameter 0.1 mm; 0.25 g) to the soil slurries. The cells were disrupted by bead beating (mini-bead beater; BioSpec Products, United States) three times for 60 s. Following extraction, the DNA preparations were electrophoresed over agarose gels in order to assess DNA purity, quality (average size) and quantity. The quantity of extracted DNA was estimated on gel by comparison to a 1-kb DNA ladder (Promega, Leiden, Netherlands) and quality was determined based on the degree of DNA shearing (average molecular size) as well as the amounts of coextracted compounds.

Real-time PCR quantification (qPCR)

Absolute quantification was carried out in four replicates on the ABI Prism 7300 Cycler (Applied Biosystems, Germany). The 16S rRNA gene was amplified by qPCR using diluted extracted DNA as template. Specific primers for archaea (group 1 crenarchaeota) 771F/957R [24] and for V5–V6 region of bacteria 16SFP/16SRP [25] were used. We have chosen to focus on Crenarchaeota, as this group is often more common in soil environments than Euryarchaeota [26]. For Fungal community primers 5,8S/ITS1f [27] were chosen. Cycling programs and primer sequences are detailed in Table S2. The specificity of the amplification products was confirmed by melting-curve analysis and on 1.5% agarose gels. Standard curves were obtained using serial dilutions of plasmid DNA containing the cloned 16S rRNA gene obtained from *Burkholderia terrae* BS001 or ITS region of *Rhizoctonia solani* AG3. Dilutions ranged from 10^7 to 10^2 gene copy numbers/μl. The archaeal standard curve was obtained by serial dilution of PCR product generated from *Cenarchaeum symbiosum* with the aforementioned archaeal specific primers [24].

PCR for DGGE analysis

For DGGE analysis, bacterial 16S rRNA genes were PCR amplified using the forward primer F968 [28] with a GC-clamp attached to 5′ and the universal R1401.1b [29]. Archaeal 16S rRNA genes were amplified with the A2F/U1406R primer pair [30], following amplification using the *Archaea*-specific forward primer at position 344 with a 40-bp GC clamp [31] added to the 5′ end, and a universal reverse primer at position 517. The fungal ITS region was amplified with EF4 [32]/ITS4 [33], followed by a second amplification with primers ITS1f-GC [34]/ITS2 [33]. PCR mixtures, primer sequences and cycling conditions are described in Table S3. About 200 ng of amplicons were loaded onto a 6% (w/v) polyacrylamide gel in the Ingeny Phor-U system (Ingeny International, Goes, The Netherlands), with a 20–50%, 45–65% and 40–60% denaturant gradient for the fungal, bacterial

Table 1. List of soils included in this study.

Sampling Site	Soil type	Land use	Crops			North coordinate	East coordinate
			2009	2010	2011		
Buinen (B)	Sandy loam	agriculture	barley	potato	potato	52°55′386″	006°49′217″
Valthermond (V)	Sandy loam	agriculture	barley	potato	potato	52°50′535″	006°55′239″
Droevendaal (D)	Sandy loam	agriculture	triticale	barley	barley	51°59′551″	005°39′608″
Wildekamp (K)	Sandy loam	grassland	grass	grass	grass	51°59′771″	005°40′157″
Kollumerwaard (K)	Clayey	agriculture	potato	grass	sugar beet	53°19′507″	006°16′351″
Steenharst (S)	Silt loam	agriculture	grass	potato	grass	53°15′428″	006°10′189″
Grebedijk (G)	Clayey	agriculture	potato	wheat	wheat	51°57′349″	005°38′086″
Lelystad (L)	Clayey	agriculture	potato	grass	corn	52°32′349″	005°33′601″

and archaeal community, respectively (100% denaturant corresponded to 7 M urea and 40% (v/v) deionized formamide). Electrophoresis was performed at a constant voltage of 100 V for 16 h at 60°C. The gels were stained for 60 min in 0,5× TAE buffer with SYBR Gold (final concentration 0,5 μg/liter; Invitrogen, Breda, The Netherlands). Images of the gels were obtained with Imagemaster VDS (Amersham Biosciences, Buckinghamshire, United Kingdom). Genetic fingerprints were analyzed using GelCompar software (Applied Maths, Sint-Martens Latem, Belgium) [35,36].

Data analyses

The diversity of each of the soil bacterial, archaeal and fungal communities was determined on the basis of the PCR-DGGE profiles. Total diversity (α) of the dominant community members was estimated from these data using the Shannon index, as recommended by Hill et al. [37], as well as the number of DGGE bands (species richness). We calculated the temporal β-diversity of archaeal, bacterial and fungal communities as the mean of all pairwise Bray-Curtis dissimilarities based on the relative abundance of DGGE bands, as previously described [38,39,11]. To support results from the calculated β-diversity and to test the statistical significance and strength of community dynamics we used time-lag analysis (TLA) [22] by plotting Hellinger-transformed [40] distance values against the square root of the time lag for all lags. The time-lag analytical approach can produce a number of general theoretical patterns with time-series data [22]. The square root transformation reduces the probability that a smaller number of points at larger time lags will bias the analysis [41]. The Bray-Curtis matrices as well as Hellinger-transformed distances were determined in PRIMER-E (version 6, PRIMER-E Ltd, Plymouth, UK; [42]).

To test how α-diversity, β-diversity and microbial abundance varied in relation to environmental variables, parametric Pearson correlation coefficients were calculated between α and β diversities, soil pH, organic matter, nitrate, ammonium, clay content and soil moisture, as well as between total abundances and TLA slopes using SPSS v18.0.3 (SPSS Inc., Chicago, IL, USA). All variables except pH were transformed (Log(x+1)) prior to all analyses. Moreover, we applied variance partitioning to evaluate the relative contribution of the drivers of the microbial assemblages. Forward selection was used on CCA (Canonical Correspondence Analysis) to select a combination of environmental variables that explained most of the variation observed in the species matrices. For that, a series of constrained CCA permutations was performed in Canoco

(version 4.0 for Windows, PRI Wageningen, The Netherlands) to determine which variables best explained the assemblage variation, using automatic forward selection and Monte Carlo permutation tests (permutations = 999). The length of the corresponding arrows indicated the relative importance of the chemical factor explaining variation in the microbial communities.

Results

Variability of environmental parameters

Soil pH, nitrate, ammonium and organic matter levels were determined in triplicate across all soil samples. Soil pH was significantly higher (P<0.05) in soils K, G and L (7.32±0.06, n = 57) than in soils B, V, D, W and S (4.88±0.04, n = 99) during the whole experimental period and no significant variation over time was observed. In all soils, significant changes were observed in the levels of nitrate, with lower values at the end of the growing season for most of the soils (September 2009: 32.78 mg/kg±7.77; October 2010: 24.15 mg/kg±3.62; September 2011: 2.45 mg/kg±0.41) and higher at the beginning (April 2009: 75.6 mg/kg±12.5; April 2010: 56.4 mg/kg±5.63; April 2011:100.1 mg/kg±16.5). Levels of ammonium also varied over the whole period, with higher values being observed at the beginning of the season (April 2009: 13.3 mg/kg±1.14; April 2010: 16.0 mg/kg±1.19; April 2011: 12.1 mg/kg±2.72) and lower values at the end (September 2009: 1.93 mg/kg±0.16; October 2010: 8.86 mg/kg±1.22) (Table S1).

Considering each soil individually, they had characteristically different values, with higher levels of nitrate and ammonium found in soils B, V, D and S than in soils W, K, G and L (Table S1). In 2009 and 2010, variations in organic matter (OM) content were observed from September (5.63%±1.20) to November (7.34%±1.45) 2009 and from April (6.28%±0.85) to June (5.04%±0.89) 2010. Small but insignificant variations in OM were observed in 2011. On average, the OM content of all soils was in the range around 4%, except for soil V, which had on average 17% OM.

Temporal variations in the abundance of archaeal, bacterial and fungal communities and their responses to abiotic variables

We studied the variations in the abundances of archaeal, bacterial and fungal communities over time and across all samples in three years. The total bacterial abundance showed significant temporal variation during the whole period, ranging between

8.12±0.23 (mean ± standard error) (September 2011) and 10.93±0.06 (June 2010) log copy numbers per g dry soil and showing comparable copy numbers in sandy (9.65±0.13) and clayey soils (9.64±0.16). The archaeal abundance (crenarchaeota) ranged between 6.96±0.14 (April 2009) and 8.78±0.07 (April 2011) log copy numbers per g dry soil, and showed significant differences between sandy and clayey soils across almost all sampling times, with lower numbers in the sandy soils (7.77±0.13) than in the clayey soils (8.22±0.13). Fungal abundance varied between 8.76±0.16 (February 2011) and 10.00±0.09 (April 2011), and significantly higher abundance was observed in the sandy soils depending on the sampling time (Fig. 1). Overall and on average, the abundance of bacteria was higher than that of the fungi, except in September 2009 and during 2011.

We used Pearson's correlation to examine how soil parameters influenced the abundances of bacterial, archaeal and fungal communities. Whereas the archaeal abundances were positively correlated with soil pH (r = +0.883, P<0.001), they were negatively influenced by nitrate (r = −0.764, P<0.05). A positive relationship was observed between fungal abundance and soil organic matter (r = +0.722, P<0.05), and a negative relationship was observed between fungal abundance and archaeal abundance (r = −0.484, P<0.05). Relationships between the abundance of bacteria and fungi, as well as between bacteria and archaea, were not significant. Interestingly, none of the soil parameters measured influenced bacterial abundance significantly.

Patterns of α-diversity and response to abiotic variables

Understanding how species are distributed in space and time may yield a first avenue towards their assembly rules [43]. We used two ecological measures, i.e. the Shannon index (H′) and species richness, as proxies to study the variations in the α-diversities of the archaeal, bacterial and fungal communities. Differential patterns of archaeal, bacterial and fungal α-diversities were observed, as measured by H′ (Fig. 2). The H′ values of the archaeal communities ranged from 1.68±0.04 in June 2009 to 2.40±0.05 in February 2011, and they were consistently lower than the corresponding bacterial and fungal values. The bacterial H′ values varied from 2.52±0.04 in October 2010 to 3.85±0.04

in April 2009, whereas those of the fungal communities varied from 3.2±0.16 in April 2009 to 4.09±0.04 in April 2010 (Fig. 2). In general, the differences observed between sandy and clayey soils for the bacterial and fungal diversities (Shannon index) were time point-dependent. For archaea, a higher Shannon index was noticed in the sandy soils compared to the clayey ones in 2009 and 2010 but not in 2011.

Concerning correlations with edaphic factors, a positive effect of OM content was observed on the archaeal α-diversity (r = +0.691, P<0.05) (Table 2). When using the number of DGGE bands as a measure of α-diversity (species richness), a significant and strong positive correlation was found between archaeal α-diversity and nitrate levels (r = +0.962 , P<0.001) (Table 2). None of the soil parameters measured correlated significantly with bacterial or fungal α-diversity.

Patterns of temporal β-diversity and responses to abiotic variables

The patterns of temporal β-diversity of the archaeal, bacterial and fungal communities (taking into account the variations in community composition of each microbial group in individual soils over time) showed small but significant variations across soils (Fig. 3A). Bacterial β-diversities were in general higher than fungal ones across soils, except for soil V. There were slight but significant differences (P<0.05) between sandy and clayey soils regarding the temporal β-diversity of archaeal and bacterial but not of fungal communities (Fig. 3B).

Although chemical parameters might show variability over time, significant correlations could still be observed. The patterns in the archaeal temporal β-diversities observed were mainly due to positive correlations with nitrate (r = +0.874, P<0.05) (Table 2). None of the soil parameters measured were correlated with bacterial and fungal temporal β-diversities. Canonical correspondence analysis was used to test the significance of the influence of soil parameters on the community parameters. We used variance partitioning to control for the effect of each individual parameter, while all others are defined as covariables in the constrained analyses [44]. Considering the whole data set, soil parameters explained 45%, 6.6% and 6.9% of the temporal variability in

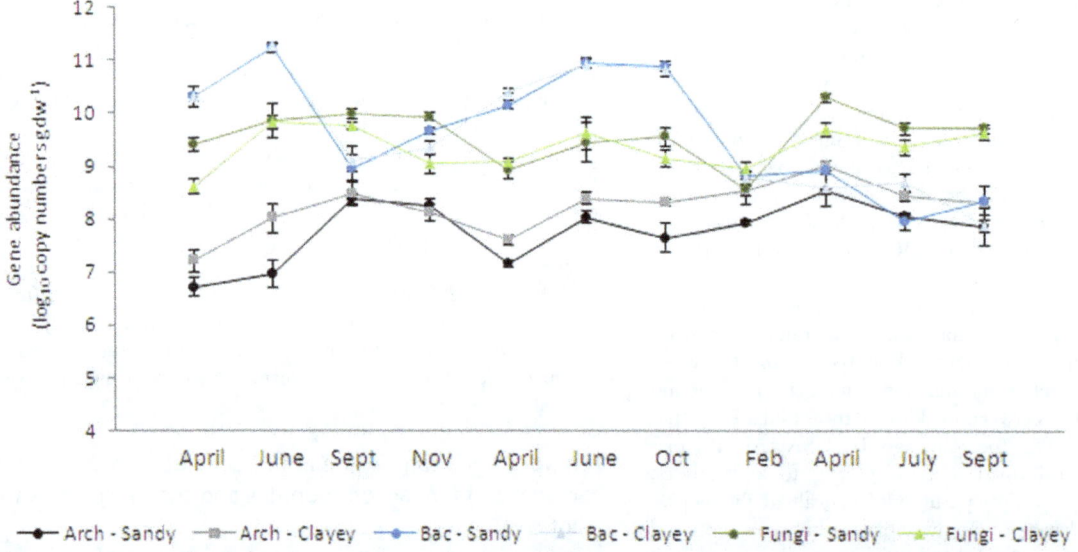

Figure 1. Changes in abundance of archaeal, bacterial and fungal communities. The copy number in each gram of dry soil was estimated by real-time PCR in the eight agricultural soils as an average of sandy and clayey soils at different sampling times. Bars are standard errors (n = 4).

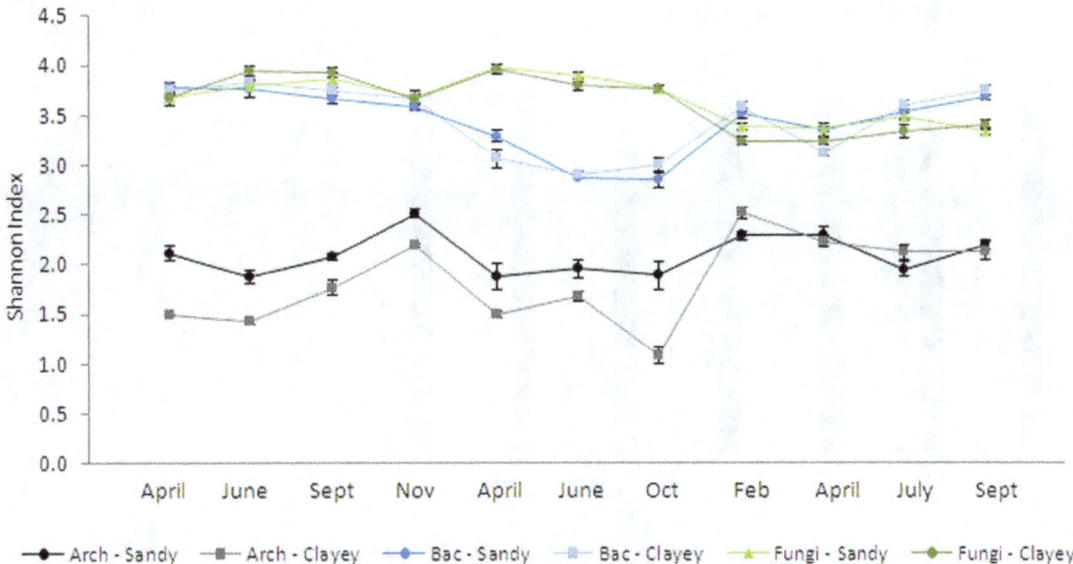

Figure 2. Total α-diversity of archaeal, bacterial and fungal communities. Alpha diversity was calculated as the average of the Shannon Index (H′) per soil type (sandy×clayey), from April 2009 to September 2011 (mean ± s.d.).

archaeal, bacterial and fungal community structures, respectively. The archaeal community was mostly affected by changes in OM (11.9%) and nitrogen specimens (nitrate + ammonium; 7.8% each), whereas the bacterial and fungal community variations were mostly related to ammonium (2.1% and 2.2% for bacteria and fungi, respectively) (Fig. S1).

Significant relationships were observed between variation in β-diversity and H′ for bacterial (r = +0.602, P<0.05) and fungal communities (r = −0.481, P<0.05) but not for archaeal communities.

Table 2. Pearson's correlation coefficient between soil chemical parameters, biotic parameters (total abundance, alpha diversity, beta diversity and slopes from TLA analysis), calculated from the eight soils over time.

	pH	N-NH$_4^+$ (mg^{-1}kg)	N-NO$_3^-$ (mg^{-1}kg)	OM (%)	Clay (%)
Total abundance					
Total archaeal community	**0.883*****	−0.498^{ns}	**−0.764***	0.030^{ns}	**−0.795***
Total bacterial community	−0.636^{ns}	0.379^{ns}	0.236^{ns}	−0.624^{ns}	0.417^{ns}
Fungi	0.363^{ns}	−0.476^{ns}	−0.356^{ns}	**−0.722***$	0.387^{ns}
Alpha Diversity (Shannon)					
Total archaeal community	0.284^{ns}	0.230^{ns}	0.137^{ns}	**0.691***	0.599^{ns}
Total bacterial community	−0.174^{ns}	−0.033^{ns}	0.470^{ns}	−0.158^{ns}	0.442^{ns}
Fungi	0.095^{ns}	−0.149^{ns}	0.175^{ns}	−0.370^{ns}	0.241^{ns}
Alpha Diversity (N° bands)					
Total archaeal community	−0.408^{ns}	−0.056^{ns}	**0.962*****	0.482^{ns}	0.442^{ns}
Total bacterial community	0.441^{ns}	−0.355^{ns}	−0.485^{ns}	−0.497^{ns}	−0.335^{ns}
Fungi	−0.154^{ns}	0.150^{ns}	−0.579^{ns}	−0.416^{ns}	−0.469^{ns}
Temporal Beta diversity					
Total archaeal community	−0.194^{ns}	−0.249^{ns}	**0.874***	0.541^{ns}	−0.415^{ns}
Total bacterial community	0.028^{ns}	−0.313^{ns}	−0.123^{ns}	−0.502^{ns}	0.074^{ns}
Fungi	−0.380^{ns}	0.167^{ns}	−0.456^{ns}	−0.232^{ns}	−0.035^{ns}

Notes: Values in boldface type indicate significant correlations with P values indicated in superscript.
*P<0.05;
**P<0.01;
***P<0.001.
^{ns}not significant at P<0.05.

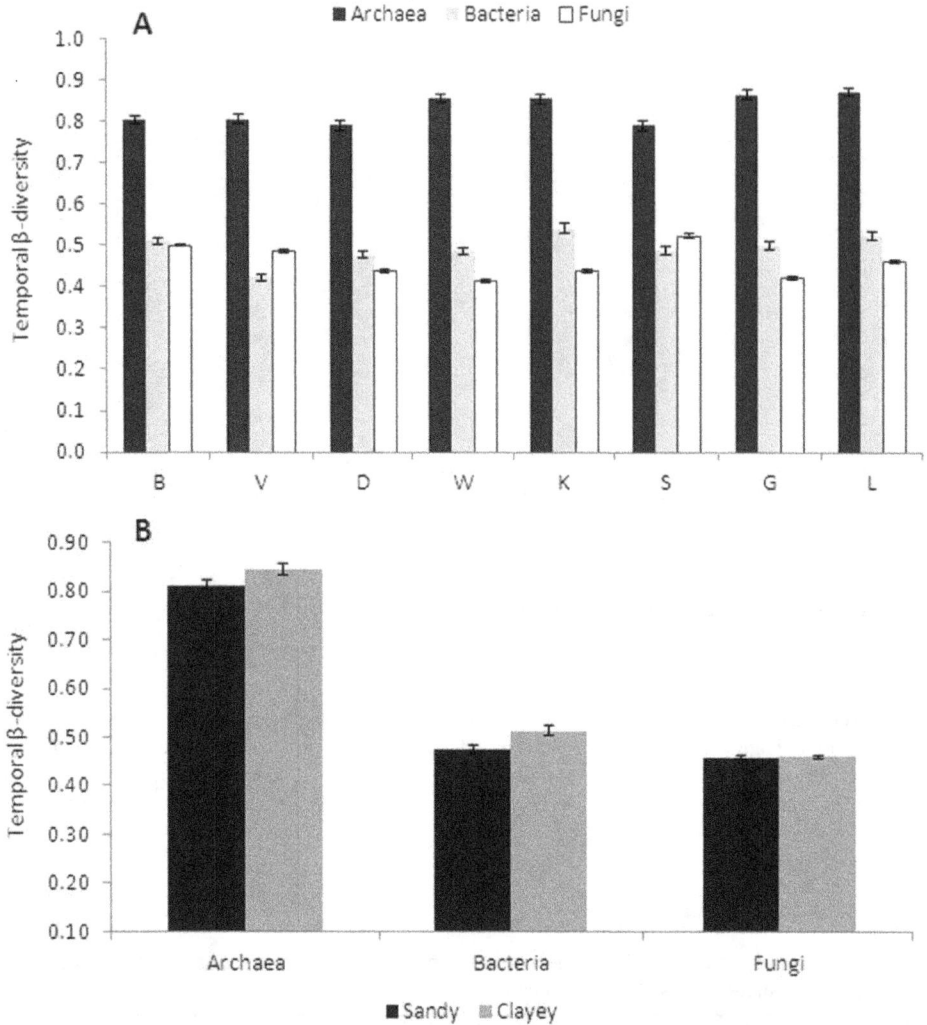

Figure 3. Temporal β-diversity of archaeal, bacterial and fungal communities. Temporal β-diversity, which takes into account temporal changes of each individual soil, was calculated across the different sampling points (A) and separated per soil type (B) (mean ± s.e.).

Quantifying temporal changes of archaeal, bacterial and fungal communities

The temporal changes of the microbial guilds were quantified and statistically tested via TLA. TLA analyses were performed separately per year and also considering all three years. For both analyses, the results and conclusions were similar. Therefore, we decided to include only the whole three year dataset. A statistically significant regression line (P<0.05) was observed for the archaeal community, with an overall slope of 1.835 (Fig. 4A and Table 3). Moreover, all eight soils showed indications of directional changes in community composition, yielding regression lines that were statistically different from zero (P<0.05) with the exception of the G soil (Table 3). Although the slopes were small (Table 3), they were mainly reflected in the positive Pearson correlations with nitrate levels (r = +0.814, P<0.05). The bacterial communities showed a similar trend as observed for the archaeal ones, with significant regression lines and a slope of 0.785 (Fig. 4B and Table 3). Although analyses of the fungal communities in the eight soils showed that only three soils were undergoing directional changes (B, V an L soils), the overall result based on the simultaneous analysis of all soils yielded statistically significant

regression lines (slope of 0.638, P<0.05, Fig. 4C). None of the soil parameters measured had significant effects on the rates of change of bacterial and fungal communities. Significant and contrasting relationships were observed between the TLA slopes and the H′ values of archaeal (r = +0.629, P<0.05) and bacterial (r = −0.523, P<0.05) communities but not of fungal communities.

Discussion

Temporal variation in the abundance of soil microbial communities

In our study, population sizes of archaea, bacteria and fungi, estimated using quantitative PCR, were found to be within the range observed in other soil systems [24,45]. Quantitative PCR of soil DNA, as any PCR based approach, has its inherent limitations, whether it be the biases of soil DNA extraction, PCR, or the core genes targeted. However, the method is highly reproducible and sensitive, enabling the quantification of microbial abundance changes across temporal and spatial scales. Moreover, this study performed multiple qPCR runs in order to ensure results were statistically significant. In our calculations, we also took into account the efficiency and amount of extracted

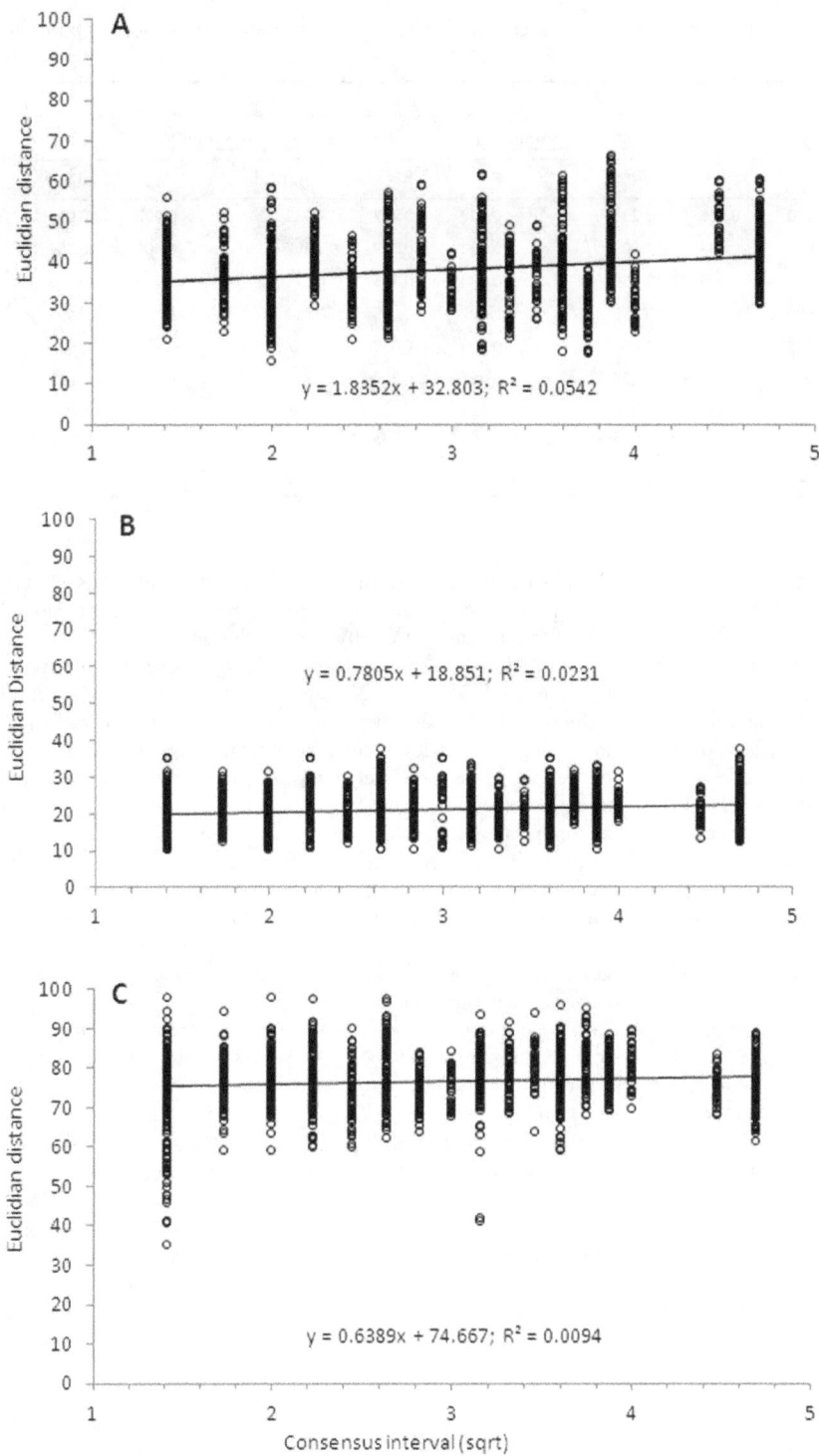

Figure 4. Quantification of archaeal, bacterial and fungal dynamics. Patterns of change (regression of square root of time-lag on Euclidian distance) of archaeal (A, slope 1.835), bacterial (B, slope 0.785) and fungal community (C, slope 0.638) in eight soils. The best-fit line is shown.

DNA from the soil samples. Therefore, we argue that our results are representative of the fluctuations observed between different times rather than pure noise.

A high abundance of crenarchaeota in soils has been previously observed [24,46], possibly indicating a crucial functional role for such organisms in agricultural soils. Furthermore, the bacterial abundance was often higher across soils than the fungal abundance (except at the end of 2009 and the end of 2011), supporting the finding that bacterial:fungal (B:F) ratios are quite high in agricultural or grassland soils as compared to, say, forest soils [45,47–49]. These comparisons are very important in the context of whether soils are thought of a being fungal (more

Table 3. Results of the time-lag analyses (TLA) performed for bacterial, archaeal and fungal communities for all soils separately, and an overall result considering all soils.

Sampling Site	Archaeal community			Bacterial community			Fungal community		
	Slope	P	R^2	Slope	P	R^2	Slope	P	R^2
Buinen (B)	1.749	0.000	0.116	3.298	0.000	0.227	0.989	0.007	0.031
Valthermond (V)	1.876	0.000	0.064	1.903	0.000	0.139	1.297	0.001	0.059
Droevendaal (D)	2.063	0.000	0.098	1.864	0.000	0.135	0.671	NS	0.013
Wildekamp (W)	1.765	0.000	0.105	1.059	NS	0.036	0.414	NS	0.004
Kollumerwaard (K)	1.378	0.028	0.021	1.860	0.009	0.079	−0.078	NS	0.000
Steenharst (S)	2.038	0.000	0.080	1.697	0.015	0.069	0.817	NS	0.012
Grebedijk (G)	0.853	NS	0.022	2.300	0.000	0.156	−0.029	NS	0.000
Lelystad (L)	1.282	0.016	0.026	0.552	NS	0.009	0.981	0.011	0.034
Overall	1.835	0.000	0.054	0.785	0.000	0.023	0.638	0.000	0.009

"natural") or bacterial (more highly cultivated) dominated. Indeed, such elevated B:F ratios may also reflect anthropogenic disturbances due to agricultural practices.

The variations in microbial abundances could be explained by several parameters, depending on the target group. Soil pH and nitrate explained more than 75% of the variation in archaeal abundance. Previous studies have reported negative effects of pH on group 1.1c Crenarchaeota [50] in acid forest soils and negative relationships between nitrate and archaeal abundance [21]. The positive correlations between archaeal abundance and soil pH observed here suggest that our soils might be dominated by crenarchaeal species that are adapted to conditions of higher soil pH (7.0–7.5) [51], which may be linked to the long agricultural history of the plots studied here.

Interestingly, the bacterial abundances didn't respond to soil pH or any other measured abiotic parameter, although several studies have reported pH as the main determinant of bacterial community composition [52–57]. It has been shown that some specific bacterial taxa abundances decrease or increase with a changing pH, for instance members of the *Acidobacteria* and *Actinobacteria* [56]. Although pH may have driven changes in the relative abundance of some bacterial classes, the abundance of total bacteria remained quite constant in the different pH ranges, indicating that the carrying capacity of the soil was not strongly affected by pH. Fungal abundance was also not affected by pH, but this was expected since the pH range in our soils was within the (wide) pH optimum for this group, often covering 5–9 pH units without significant inhibition of growth [58,59]. We also observed that when conditions apparently favored increases in fungal abundance, archaeal abundance decreased, suggesting that fungi and archaea might compete for similar niches. Nonetheless, fungal abundance was positively affected by OM content, which is consistent with the saprophytic status of most fungi [60].

Temporal variation in α-diversity

None of the soil parameters measured in this study were able to explain the patterns of α-diversity observed for bacteria and fungi. It might be that the taxonomic scale was too broad and a deeper analysis would allow a better understanding of the observed patterns, as observed by Rasche et al. [61]. With PCR-DGGE, only the most abundant taxa, comprising no more than 0.1–1% of the community, can be detected. In other words, only the most abundant organisms are detected with PCR-DGGE. Because of

these caveats, the parameters calculated from PCR-DGGE fingerprints and correlations based thereon should be interpreted as indications and not as absolute conclusions.

Archaeal α-diversity, on the other hand, was shown to respond to nitrate and OM levels. Nitrate had opposite effects on archaeal richness and diversity, depicting a community that responds to increasing nitrate with an increase of richness but a great decrease of evenness. This most likely indicates the outgrowth of previously undetectable OTUs. Thus, in addition to the strong negative effect on archaeal abundance observed by qPCR, nitrate availability seems to be a crucial factor determining archaeal community structure. The positive correlation between archaeal diversity and soil OM content indicate that the OM provides a substantial fraction of carbon to the local archaeal communities. Recently, genomic analyses of *Crenarchaeum symbiosum* and *Nitrosopumilus maritimus* suggested that these organisms are capable of mixotrophy [62,63] and that group 1.1c Crenarchaeota are able to grown on methanol and methane [64]. This suggests that archaea might not be solely sustained by ammonia oxidation [65,66].

Variation in β-diversity over time (species turnover)

To assess how dynamic each soil microbial group was over time, we calculated the temporal β-diversity (average Bray Curtis dissimilarity) for each soil over time and for each microbial group. We observed a higher temporal β-diversity for archaea than for bacteria and fungi across all soils. This indicates that the archaeal communities are much more dynamic than the bacterial or fungal ones along a time gradient. These differences are probably due to the differential physiologies and sensitivities to environmental perturbations of these microorganisms. It has been shown that changes in temperature, moisture [61,67] and resource availability due to seasonal variation [61] can affect soil archaeal as well as bacterial communities. Moreover, a clear pattern was observed for bacterial β-diversity in the metacommunity of 17 rock pools, with higher variations during summer and lower during autumn [11]. The temporal variations of archaeal and bacterial communities were also higher in the clayey soils than in sandy ones, suggesting that the latter harbors more dynamic communities.

One main finding of this study is that, although the β-diversity patterns of the three microbial domains investigated are related with the same set of abiotic factors, the total percentage of variation able to explain those patterns was much higher for archaeal (45.0%) than for bacterial (6.6%) or fungal (6.9%)

communities. This suggests that the archaeal communities might be much more sensitive to environmental changes than the bacterial or fungal ones. Based on these results, we hypothesize that the archaeal communities of agricultural soils with a long history of N-fertilization are more sensitive to disturbances than the corresponding bacterial or fungal communities.

Quantification of β-diversity

To be able to quantify community dynamics, allowing comparisons and providing a general overview of long-term trends in the complex soil system, we used the slopes obtained from TLA. TLA has been intensively used to identify directional changes and to quantify temporal dynamics of macroorganisms [69–73]but very few studies have focused on microorganisms. Although the TLA slopes for archaea and bacteria were small, they were significantly different from neutral. Clearly, even small changes can be part of a long-term trend. On the contrary, changes in the fungal communities were non-significant, suggesting stochastic species dynamics.

Changes in environmental variables within soil sites determine how time affects turnover (β-diversity), as different microbial interactions are favored if prevailing conditions change [74]. Only archaeal communities responded to changes in environmental parameters, being strongly correlated with nitrogen availability and with the degree of temporal variation quantified by TLA. This might suggest that at some level strongly deterministic processes are acting on the archaeal but not on the bacterial and fungal communities in these soils. Another explanation is that archaea are much more limited in their ecoversatility, whereas bacteria and fungi are highly functionally redundant. The observed relation between bacterial richness and TLA slopes, e.g. high turnover at low richness, was also noticed in a study of the distribution of British birds [75]. The authors discuss that low species richness areas tend to have relatively more random mixtures of species than high species richness areas. The relation observed between archaeal turnover and species richness suggests a less random distribution of species, caused mainly by nitrate contents.

In this study we demonstrate that changes in the community composition of bacteria and fungi could be linked to both environmental and biotic factors (e.g. species-species interactions), as their (α-) diversity co-varied significantly with their β-diversity over the time period of the study. Conversely, archaea showed no significant correlation between α- and β-diversity, and the community shifts were mainly driven by the surrounding environment, mostly by the effects of soil pH and nitrate concentrations. This might indicate that changes in archaeal community are mostly driven by environmental factors, as

previously observed by Zinger et al. [68] in a study on the patterns of archaeal, bacterial and fungal communities in an alpine landscape. Furthermore, we propose that different environmental and biological mechanisms act on each microbial niche. A more comprehensive understanding of the rules governing these important soil microorganisms will require additional field work as well as microcosm experiments to identify the key environmental and biotic factors driving the assemblage of these communities.

Ethic statement. No specific permits were required for the described field studies. The locations are not protected. The field studies did not involve endangered or protected species.

Supporting Information

Figure S1 Biplots of canonical correspondence analysis (CCA) of Archaeal, Bacterial and Fungal similarity matrices and vector fitting of the environmental variables. Similarity matrices from DGGE data were obtained from eight soils over three years (2009, 2010 and 2011). Physicochemical data, soil moisture (Humidity), soil nitrate (NO3), soil ammonium (NH4), organic matter (OM), clay content (clay + silt) and soil pH (pH) are presented with black arrows.

Table S1 Soil chemical parameters measured in this study.

Table S2 PCR mixtures for real time quantification of Archaeal 16S rDNA, Bacterial 16S rDNA and Fungal ITS region.

Table S3 PCR mixtures for DGGE analysis of Archaeal 16S rDNA, Bacterial 16S rDNA and Fungal ITS region.

Acknowledgments

We would like to thank our colleagues Alexander V Semenov, Jolanda Brons and our Utrecht partners, Heike Schmitt and Agnieszka Szturc-Koetsier for their help with sampling. We also thank Cyrus A. Mallon for revision and helpful comments on the manuscript.

Author Contributions

Performed the experiments: MCPS. Analyzed the data: MCPS ACFD. Contributed reagents/materials/analysis tools: JDvE JFS. Wrote the paper: MCPS .

References

1. Singh BK, Bardgett RD, Smith P, Reay DS (2010) Microorganisms and climate change: terrestrial feedbacks and mitigation options. Nature Microbiol Rev 8:779–790.
2. Hooper DU, Vitousek PM (1997) The effects of plant composition and diversity on ecosystem processes. Science 277:1302–1305.
3. Tilman D, Knops J, Wedin D, Reich P, Richie M, et al. (1997) The influence of functional diversity and composition on ecosystem processes. Science 277:1300–1302.
4. Bardgett RD, Yeates G, Anderson J (2005) Patterns and determinants of soil biological diversity. In Biological diversity and function in soils. pp. 100–118.
5. Lindström ES, Langenheder S (2011) Local and regional factors influencing bacterial community assembly. Environm Microbiol Rep 4:1–9.
6. Andrewartha HG, Birch LC (1953) The Lotka-Volterra theory of interspecific competition. Aust J Zoology 1:174–177. DOI: 10.1071/ZO9530174.
7. Lewontin R, Cohen D (1969) On population growth in a randomly varying environment. Proc Natl Acad Sci USA 62:1056–1060.
8. Schmidt SK, Costello EK, Nemergut DR, Cleveland CC, Reed SC, et al. (2007) Biogeochemical consequences of rapid microbial turnover and seasonal succession in soil. Ecology 88:1379–1385.
9. Lipson DA, Schadt CW, Schmidt SK (2002) Changes in soil microbial community structure and function in an alpine dry meadow following spring snow melt. Microbial Ecol 43:307–314.
10. Martiny JBH, Eisen JA, Penn K, Allison SD, Horner-Devine MC (2011) Drivers of bacterial β-diversity depend on spatial scale. Proc Natl Acad Sci USA 10: 7850–4.
11. Langenheder S, Berga M, Örjan Ö, Székely AJ (2012) Temporal variation of β-diversity and assembly mechanisms in a bacterial metacommunity. ISME J 6:1107–1114.
12. Griffiths RI, Thomson BC, James P, Bell T, Bailey M, et al. (2011) The bacterial biogeography of British soils. Environm Microbiol 13:1642–1654.
13. Wang J, Wu Y, Jiang H, Li C, Dong H, et al. (2008) High beta diversity of bacteria in the shallow terrestrial subsurface. Environm Microbiol 10:2537–2549.
14. Curtis T P, Sloan WT, Scannell JW (2002) Estimating prokaryotic diversity and its limits. Proc Natl Acad Sci USA 99:10494–10499.
15. Gans J, Wolinsky M, Dunbar J (2005) Computational improvements reveal great bacterial diversity and high metal toxicity in soil. Science 309:1387–1390.

16. Torsvik V, Ovreas L, Thingstad TF (2002) Prokaryotic diversity: magnitude, dynamics, and controlling factors. Science 296:1064–1066.

17. Buckley DH, Graber JR, Schmidt TM (1998) Phylogenetic analysis of nontermophilic members of the kingdom Crenarchaeota and their diversity and abundance in soils. Appl Environm Microbiol 64:4333–4339.

18. Jurgens G, Lindström K, Saano A (1997) Novel group within kingdom Crenarcheota from borest forest soil. Appl Environm Microbiol 63:803–805.

19. Ueda T, Suga Y, Matsuguchi T (1995) Molecular phylogenetic analysis of a soil microbial community in a soybean field. Eur J Soil Sci 46:415–421.

20. Jung J, Yeom J, Kim J, Han J, Lim HS, et al. (2011) Change in gene abundance in the nitrogen biogeochemical cycle with temperature and nitrogen addition in Antarctic soils. Res Microbiol 168:1028–1026.

21. Bates ST, Berg-Lyons D, Caporaso JG, Walters WA, Knight R, et al. (2011) Examining the global distribution of dominant archaeal populations in soil. ISME J 5:511–517.

22. Collins SL, Micheli F, Hartt L (2000) A method to determine rates and patterns of variability in ecological communities. Oikos 91:285–293.

23. Angeler DG, Viedma O, Moreno JM (2009) Statistical performance and information content of time lag analysis and redundancy analysis in time series modeling. Ecology 90:3245–3257.

24. Ochsenreiter T, Selezi D, Quaiser A, Bonch-Osmolovskaya L, Schleper C (2003) Diversity and abundance of Crenarchaeota in terrestrial habitats studied by 16S RNA surveys and real time PCR. Environm Microbiol 5:787–797.

25. Bach H-J, Tomanova J, Schloter M, Munch JC (2002) Enumeration of total bacteria and bacteria with genes for proteolytic activity in pure cultures and in environmental samples by quantitative PCR mediated amplification. J Microbiol Met 49:235–45.

26. Nicol GW, Webster G, Glover LA, Prosser JI (2004) Differential response of archaeal and bacterial communities to nitrogen inputs and pH changes in upland pasture rhizosphere soil. Environm Microbiol 6:861–867.

27. Fierer N, Jackson JA, Vilgalys R, Jackson RB (2005) Assessment of soil microbial community structure by use of taxon-specific quantitative PCR assays. Appl Environ Microbiol 71: 4117–4120.

28. Gomes NCM, Heuer H, Schonfeld J, Costa R, Hagler-Mendonça L, et al. (2001) Bacterial diversity of the rhizosphere of maize (Zea mays) grown in tropical soil studied by temperature gradient gel electrophoresis. Plant and Soil 232:167–180.

29. Brons JK, van Elsas JD (2008) Analysis of bacterial communities in soil by use of denaturing gradient gel electrophoresis and clone libraries, as influenced by different reverse primers. Appl Environm Microbiol 74:2717–2727.

30. Bano N, Ruffin S, Ransom B, Hollibaugh T (2004) Phylogenetic composition of arctic ocean archaeal assemblages and comparison with Antarctic assemblages Appl. Environm Microbiol 70:781–789.

31. Myers RM, Fischer SG, Lerman LS, Maniatis T (1985). Nearly all single base substitutions in DNA fragments joined to a GC-clamp can be detected by denaturing gradient gel electrophoresis. Nucleic Acids Res 13:3131–3145.

32. Smit E, Leeflang P, Glandorf B, van Elsas JD, Wernars K (1999) Analysis of fungal diversity in the wheat rhizosphere by sequencing of cloned PCR-Amplified genes encoding 18S rRNA and temperature gradient gel rlectrophoresis. Appl Environm Microbiol 65:2614–2621.

33. White TJ, Bruns T, Lee S, Taylor JW (1990) Amplification and direct sequencing of fungal ribosomal RNA genes for phylogenetics. In: Innis MA, Gelfand DH, Sninsky JJ and White TJ, editors. PCR Protocols: A Guide to Methods and Applications. Academic Press, New York , pp. 315–322.

34. Gardes M, Bruns TD (1993) ITS primers with enhanced specificity for basidiomycetes - application to the identification of mycorrhizae and rusts. Mol Ecol 2: 113–118.

35. Kropf S (2004) Nonparametric multiple test procedures with data-driven order of hypotheses and with weighted hypotheses. J Statis Plann Inf 125:31–47.

36. Rademaker J L, de Bruijn AF (1999) Molecular microbial ecology manual. In van Elsas, JD, Akkermans ADL and de Bruijn AF, Eds. Molecular Microbial Ecology Manual. Dordrecht, The Netherlands: Kluwer Academic Publishers, pp. 1–33.

37. Hill TCJ, Walsh KA, Harris JA, Moffett BF (2003) Using ecological diversity measures with bacterial communities. FEMS Microbiol Ecol 43:1–11.

38. Legendre P, Borcard D, Peres-Neto PR (2005) Analyzing β-diversity: partitioning the spatial variation of community composition data. Ecological Monographs 75:435–450.

39. Peres-Neto PR, Legendre P, Dray S, Borcard D (2006) Variation partitioning of species data matrices: estimation and comparison of fractions. Ecology 87:2614–25.

40. Legendre P, Gallagher ED (2001) Ecologically meaningful transformations for ordination of species data. Oecologia 129:271–280.

41. Kampichler C, Geissen V (2005) Temporal predictability of soil microarthropod communities in temperate forests. Pedobiologia 49:41–50.

42. Clarke KR, Gorley RN (2006) PRIMER v6: User manual/tutorial, PRIMER-E, Plymouth UK, (192pp).

43. Magurran AE, Dornelas M (2010) Biological diversity in a changing world. Phil Trans R Soc B 365:3593–3597.

44. Leps J, Smilauer P (2003) Multivariate analysis of ecological data using CANOCO. Cambridge:CambridgeUniversityPress.

45. Bailey VL, Smith JL, Bolton H (2002) Fungal-to-bacterial ratios in soils investigated for enhanced C sequestration. Soil Biol Biochem 34:997–1007.

46. Kemnitz D, Kolb S, Conrad R (2007) High abundance of Crenarchaeota in a temperate acidic forest soil. FEMS Microbiol Ecol 60:442–448.

47. Bossuyt H, Denef K, Six J, Frey SD, Merckx R, et al. (2001) Influence of microbial populations and residue quality on aggregate stability. Appl Soil Ecol 16:195–208.

48. Högberg MN, Högberg P, Myrold DD (2007) Is microbial community composition in boreal forest soils determined by pH, C-to-N ratio, the trees, or all three? Oecologia 150:590–601.

49. Treseder KK (2004) A meta-analysis of mycorrhizal responses to nitrogen, phosphorus, and atmospheric CO2 in field studies. New Phytologist 164:347–355.

50. Lethovirta LE, Prosser JI, Nicol GW (2009) Soil pH regulates the abundance and diversity of groups 1.1c Crenarchaeota. FEMS Microbiol Ecol 70:367–376.

51. Bengtson P, Sterngren AE, Rousk J (2012) Archaeal abundance across a pH gradient in an arable soil and its relationship with bacterial and fungal growth rates. Appl Environ Microbiol doi:10.1128/AEM.01476-12.

52. Fierer N and Jackson RB (2006) The diversity and biogeography of soil bacterial communities. Procl Natl Acad Sci USA 103:623–631.

53. Lauber CL, Strickland MS, Bradford MA, Fierer N (2008) The influence of soil properties on the structure of bacterial and fungal communities across land-use types. Soil Biol Biochem 40:2407–2415.

54. Baker BJ, Comolli LR, Dick GJ, Hauser LJ, Hyatt D, et al. (2010) Enigmatic, ultrasmall, uncultivated Archaea. Proc Natl Acad Sci USA 107:8806–8811.

55. King AJ, Freeman KR, McCormick KF, Lynch RC, Lozupone C, et al. (2010) Biogeography and habitat modeling of high-alpine bacteria. Nature Comm 1:53. DOI: 10.1038/ncomms1055.

56. Lauber CL, Hamady M, Knight R, Fierer N (2009) Pyrosequencing-based assessment of soil pH as a predictor of soil bacterial community structure at the continental scale. Appl Environm Microbiol 75:5111–5120.

57. Rousk J, Baath E, Brookes PC, Lauver CL, Lozupone C, et al. (2010) Soil bacterial and fungal communities across a pH gradient in an arable soil. ISME J 4:1340–1351.

58. Wheeler KA, Hurdman BF, Pitt JI. (1991) Influence of pH on the growth of some oxigenic species of Aspergillus, Penicillium and Fusarium. Int J Food Microbiol 12: 141–150.

59. Nevarez L, Vasseur V, Le Madec L, Le Bras L, Coroller L, et al. (2009) Physiological traits of Penicillium glabrum strain LCP 08.5568, a filamentous fungus isolated from bottled aromatised mineral water. Int J Food Microbiol 130: 166–171.

60. de Boer W, Folman LB, Summerbell RC, Boddy L (2005) Living in a fungal world: impact of fungi on soil bacterial niche development. FEMS Microbiol Rev 29:795–811.

61. Rasche F, Knapp D, Kaiser C, Koranda M, Kitzler B, et al. (2011) Seasonality and resource availability control bacterial and archaeal communities in soils of a temperate beech forest. ISME J 5:389–402.

62. Hallam SJ, Konstantinidis KT, Putnam N, Schleper C, Watanabe Y-i, et al. (2006) Genomic analysis of the uncultivated marine crenarchaeote Cenarchaeum symbiosum. Proc Natl Acad Sci USA 103: 18296–18301.

63. Walker CB, de la Torre JR, Klotz MG, Urakawa H, Pinel N, et al. (2010) Nitrosopumilus maritimus genome reveals unique mechanisms for nitrification and autotrophy in globally distributed marine crenarchaea. Proc Natl Acad Sci USA 107:8818–8823.

64. Bomberg M, Timonen S (2007) Distribution of cren- and euryarchaeota in scots pine mycorrhizosphere and boreal forest humus. Microb Ecol 54:406–416.

65. Ouverney CC, Fuhrman JA (2000) Marine planktonic archaea take up amino acids. Appl Environ Microbiol 66:4829–4833.

66. Jia Z, Conrad R (2009) Bacteria rather than Archaea dominate microbial ammonia oxidation in an agricultural soil. Environm Microbiol 11:1658–1671.

67. Tourna M, Freitag TE, Nicol GW, Prosser JI (2008) Growth, activity and temperature responses of ammonia-oxidizing archaea and bacteria in soil microcosms. Environ Microbiol 10:1357–1364.

68. Zinger L, Lejon DPH, Baptist F, Bouasria A, Aubert S, et al. (2011) Contrasting diversity patterns of crenarchaeal, bacterial and fungal communities in an alpine landscape. Plos One 6(5): e19950. doi:10.1371/journal.pone.0019950.

69. Thibault KM, White EP, Ernest KM (2004) Temporal dynamics in the structure and composition of a desert rodent community. Ecology 85:2649–2655.

70. Baez S, Collins ST, Lightfoot D, Koontz TL (2006) Ecology 87:2746–2754.

71. Collins SL, Smith MD (2006) Scale-dependent interaction of fire and grazing on community heterogeneity in tallgrass prairie. Ecology 87:2058–2067.

72. Feeley KJ, Davies SJ, Perez R, Hubbell SP, Foster RB (2011) Directional changes in the spceies composition of a tropical forest. Ecology 92:871–882.

73. Flohre A, Fischer C, Aavik T, Bengtsson J, Berendse F, et al. (2011) Agricultural intensification and biodiversity partitioning in European landscapes comparing plants, carabids and birds. Ecol Appl 21:1772–1781.

74. Chesson P, Huntly N (1997) The roles of harsh fluctuation conditions in the dynamics of ecological communities. Amer Natur 150:519–553.

75. Lennon JJ, Koleff P, Grenwood JJD, Gaston KJ (2001) The geographical structure of british birds distributions: diversity, spatial turnover and scale. J Animal Ecol 70:966–979.

15

Tillage, Mulch and N Fertilizer Affect Emissions of CO_2 under the Rain Fed Condition

Sikander Khan Tanveer, Xiaoxia Wen, Xing Li Lu, Junli Zhang, Yuncheng Liao*

College of Agronomy, Northwest A&F University Yangling, Shaanxi, P.R. China

Abstract

A two year (2010–2012) study was conducted to assess the effects of different agronomic management practices on the emissions of CO_2 from a field of non-irrigated wheat planted on China's Loess Plateau. Management practices included four tillage methods i.e. T_1: (chisel plow tillage), T_2: (zero-tillage), T_3: (rotary tillage) and T_4: (mold board plow tillage), 2 mulch levels i.e., M_0 (no corn residue mulch) and M_1 (application of corn residue mulch) and 5 levels of N fertilizer (0, 80, 160, 240, 320 kg N/ha). A factorial experiment having a strip split-split arrangement, with tillage methods in the main plots, mulch levels in the sub plots and N-fertilizer levels in the sub-sub plots with three replicates, was used for this study. The CO_2 data were recorded three times per week using a portable GXH-3010E1 gas analyzer. The highest CO_2 emissions were recorded following rotary tillage, compared to the lowest emissions from the zero tillage planting method. The lowest emissions were recorded at the 160 kg N/ha, fertilizer level. Higher CO_2 emissions were recorded during the cropping year 2010–11 relative to the year 2011–12. During cropping year 2010–11, applications of corn residue mulch significantly increased CO_2 emissions in comparison to the non-mulched treatments, and during the year 2011–12, equal emissions were recorded for both types of mulch treatments. Higher CO_2 emissions were recorded immediately after the tillage operations. Different environmental factors, i.e., rain, air temperatures, soil temperatures and soil moistures, had significant effects on the CO_2 emissions. We conclude that conservation tillage practices, i.e., zero tillage, the use of corn residue mulch and optimum N fertilizer use, can reduce CO_2 emissions, give better yields and provide environmentally friendly options.

Editor: Ben Bond-Lamberty, DOE Pacific Northwest National Laboratory, United States of America

Funding: The study was supported by the National Natural Science foundation of China (31071375 and 31171506). The funders had no role in study design, data collection and analysis, decision to publish, or preparation of the manuscript.

Competing Interests: The authors have declared that no competing interests exist.

* E-mail: yunchengliao@163.com or yunchengliao@nwsuaf.edu.cn

Introduction

Studies regarding soil CO_2 emissions have attracted significant attention because the concentration of CO_2 in the atmosphere is increasing very rapidly as a consequence of fossil fuel combustion and deforestation. The past two centuries of human activities have reportedly contributed as much as approximately half of the increase in CO_2 emissions [1], [2]. Global terrestrial ecosystems absorbed carbon at the rate of 1–4 Pg yr^{-1}, during 1980s and 1990s, which made up approximately 10–60% of the fossil fuel emissions [3], [4]. Currently, significant attention is given to CO_2 emissions from soils because this source significantly affects the global carbon cycle and the function of the terrestrial ecosystem [5]. Fluxes of greenhouse gases (CO_2, N_2O and CH_4) between the atmosphere and agricultural soils considerably influence the stock of anthropogenic greenhouse gases [6]. Agriculture is an important source of emissions for these different gases, and its contribution to climate change is approximately 20% on an annual basis [7]. It has been reported that soils have already contributed approximately 50 Pg of anthropogenic CO_2 to the atmosphere in the past, through cultivation processes [8].

Tillage is an integral part of agriculture which not only significantly affects crop production but is also considered one of the leading factors in soil degradation. This technique is a fundamental operation that has affected both the soil and the environment and is considered one of the most important sources of CO_2 emissions into the atmosphere [9] because humans have tilled the soil for crop production for thousands of years [10] and approximately 23–44% of total CO_2, is emitted into the atmosphere through soil preparation-related operations [11]. Approximately 30–50% of soil C has already been lost through the adaptations of intensive tillage practices [12], and major C losses from, soils in the form of CO_2 occur immediately after the tillage operations [13].

Agricultural management practices affect different soil processes (i.e., soil temperature, soil moisture and soil pH), and other ongoing soil decomposition processes, which ultimately result in the conversions of plant-derived C to soil organic matter and CO_2 [14]. Applications of inorganic as well as organic fertilizers [15] and different degrees of soil moisture and temperature strongly affect the fluxes of soil CO_2 [16], [17] & [18]. Similarly, the application of N fertilizer also affects soil CO_2 emissions [19]. Instead of burning crops residues, farmers, applications of inorganic fertilizers and use of green manures as well as organic manures can be of great use in maintaining soil fertility [20]. These practices can provide essential nutrients to crops and reductions in the burning of crops can reduce CO_2 emissions into the atmosphere [21].

Agricultural tillage practices can be helpful in the sequestering of atmospheric CO_2 [22], [23] and [24]. Conservation tillage has the potential to increase soil C and N [25] and other types of

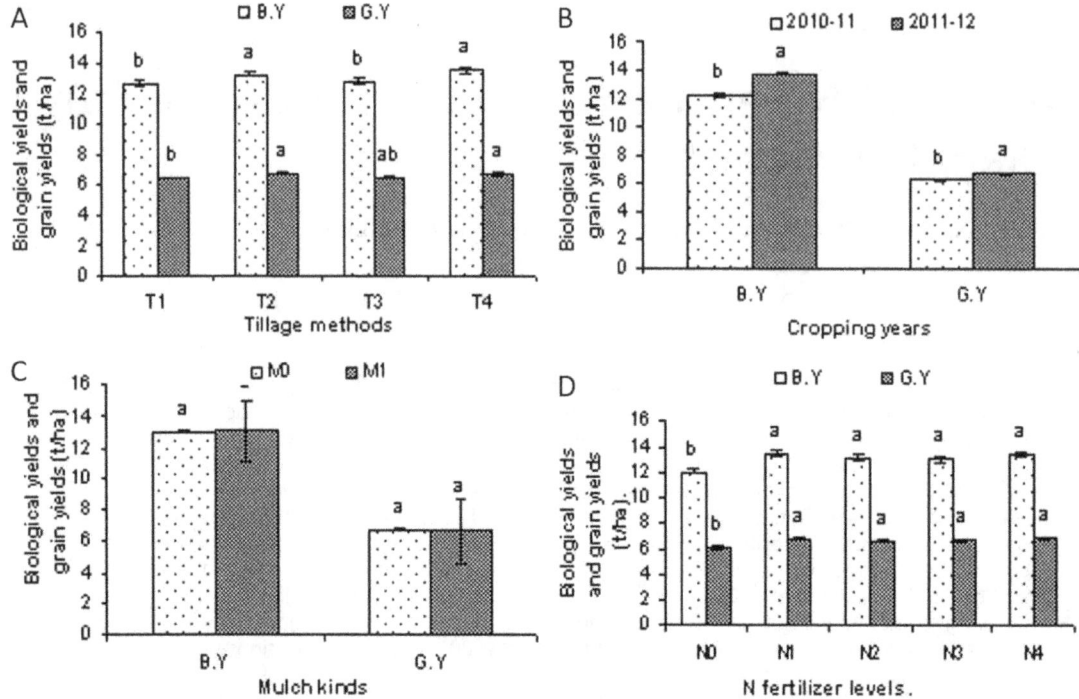

Figure 1. Wheat crop biological and grain yields as affected by different tillage methods, different mulch kinds and different N fertilizer levels during different cropping years (2010–12). (A), Wheat crop biological and grain yields as affected by different tillage methods i.e. T_1, chisel plow tillage., T_2, zero tillage., T_3, rotary tillage and T_4, mold board plow tillage., (B), Mean wheat crop biological and grain yields as affected by different tillage methods, mulch kinds and different N fertilizer levels during two cropping years (2010–12). (C), Wheat crop biological and grain yields as affected by different mulch kinds i.e. M_0, No mulch and M_1, corn residue mulch., (D), Wheat crop biological and grain yields as affected by different N fertilizer levels during two cropping years (2010–12) including, N_0, 0 kg N/ha., N_1, 80 kg N/ha, N_2, 160 kg N/ha., N_3, 240 kg N/ha and N_4, 320 kg N/ha.

conservation practices can be helpful in reducing the loss of soil organic carbon from the soils [26], [27]. Similarly, the retention of crop residue, nitrogen fertilization and no-tillage are generally supposed to enhance the soil organic carbon (SOC) stocks in the soil [28] because these farms, management practices not only increase crops biomass, but are also considered very important for the microbial decomposition of crop residues [29]. As far as N fertilization is concerned, some scientists have reported that increased N fertilization can depress CO_2 emissions [30], [31] however, others [32] have reported that N fertilization has no effect on SOC, while some other scientists [33] have reported that higher N fertilization improves the SOC of the soil.

The Loess Plateau has an area of approximately 63, 5000 km^2. It covers many provinces in China, and is home to millions of people. It is one of the most highly eroded areas of the world, and traditional agriculture, i.e., intensive tillage, is considered one of the leading man-made factors responsible for this erosion. However many crops residues are produced in this region. A small portion of these residues is used for forage or fuel consumption and the remaining residues are generally burned. Mold board plowing followed by harrowing is commonly used for the tillage operations in this region [34]. Few studies showing CO_2 emissions to the adaptation of different agronomic management practices have been previously reported from this region of China. In this area intensive tillage methods i.e. rotary tillage and mold board plow tillage methods are commonly used for land preparation. Commonly higher levels of N fertilizers are applied and crop residues are removed from the fields at the time of soil preparation.

The main aim of this two year study was to identify the effects of different tillage methods i.e. chisel plow and zero tillage in comparison with intensive tillage practices i.e. rotary tillage and mold board plow tillage methods, different N fertilizer levels and the application of corn residue mulch on CO_2 emissions. The results from this study can be of great help in improving the management of soils not only in this area of China but also in other regions of the world.

Results

Wheat crop yields

Significant variations in wheat crops biomass and grain yields were recorded during both study years (2010–12). When compared to the other tillage methods, there were better yields overall with the zero tillage planting method. No grain yields differences were recorded under different mulch treatments and similarly low yields were recorded for N_0 nitrogen fertilizer levels. Statistically equal grain yields were recorded for all of the other higher tested levels of N fertilizer (Fig. 1).

Soil CO_2 flux

All tested treatments had significant effects on the CO_2 emissions (Figs. 2, 3). The details are given below.

Tillage method effects on CO_2 emissions

Two years of combined data show that the rotary tillage and mold board plow tillage methods had their highest and statistically equal CO_2 emissions during the first week of planting the wheat

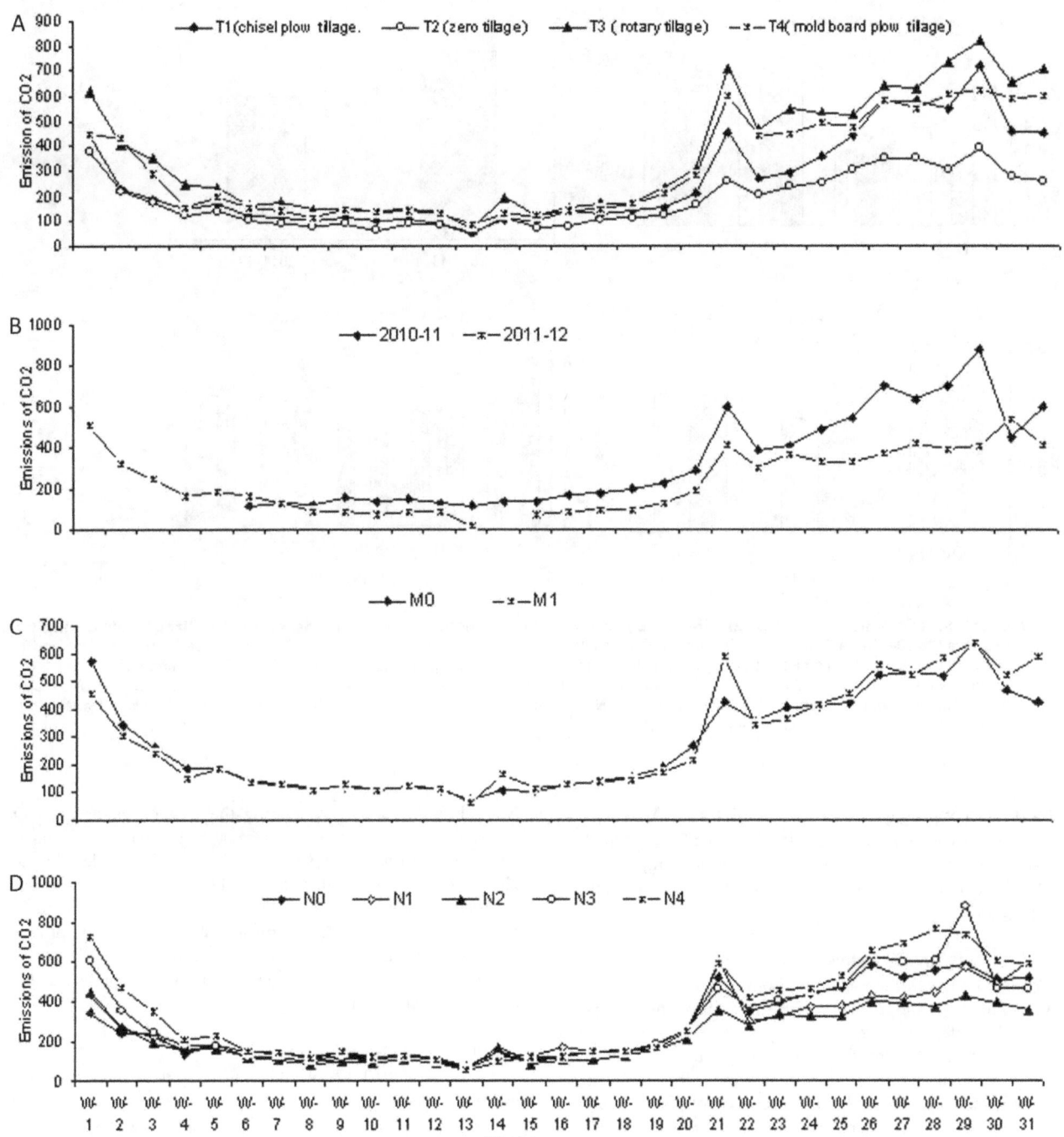

Figure 2. CO_2 emission trends as affected by different tillage methods, mulch kinds and N fertilizer levels during two cropping years (2010–12). (A). Emissions trends of CO_2 from different tillage methods i.e. T_1, chisel plow tillage, T_2, zero tillage, T_3, rotary tillage and T_4, mold board plow tillage., (B).CO_2 emission trends as affected by different kinds of tillage methods, different types of mulch and different N fertilizer levels, during two cropping years (2010–12). (C). CO_2 emission trends due to different mulch kinds i.e. M_0, no mulch and M_1, corn residue mulch., (D). CO_2 emissions trends due to different N fertilizer levels during two cropping years (2010–12) including, N_0, 0 kg N/ha., N_1, 80 kg N/ha, N_2, 160 kg N/ha, N_3, 240 kg N/ha and N_4, 320 kg N/ha.

crop, compared to the chisel plow tillage and zero tillage planting methods. When compared to all the other tillage methods, the lowest CO_2 emissions were recorded for the zero tillage planting method (Fig. 2; Fig. 3). Emission trends recorded during the whole wheat crop seasons show that although there were variations in the CO_2 emissions in response to different tillage methods, the overall highest emissions were recorded for the rotary tillage planting method, followed by mold board plow tillage. Although higher CO_2 emissions were recorded for the chisel plow tillage method, the lowest CO_2 emissions were generally recorded for zero tillage planting method (Fig. 2; Fig. 3).

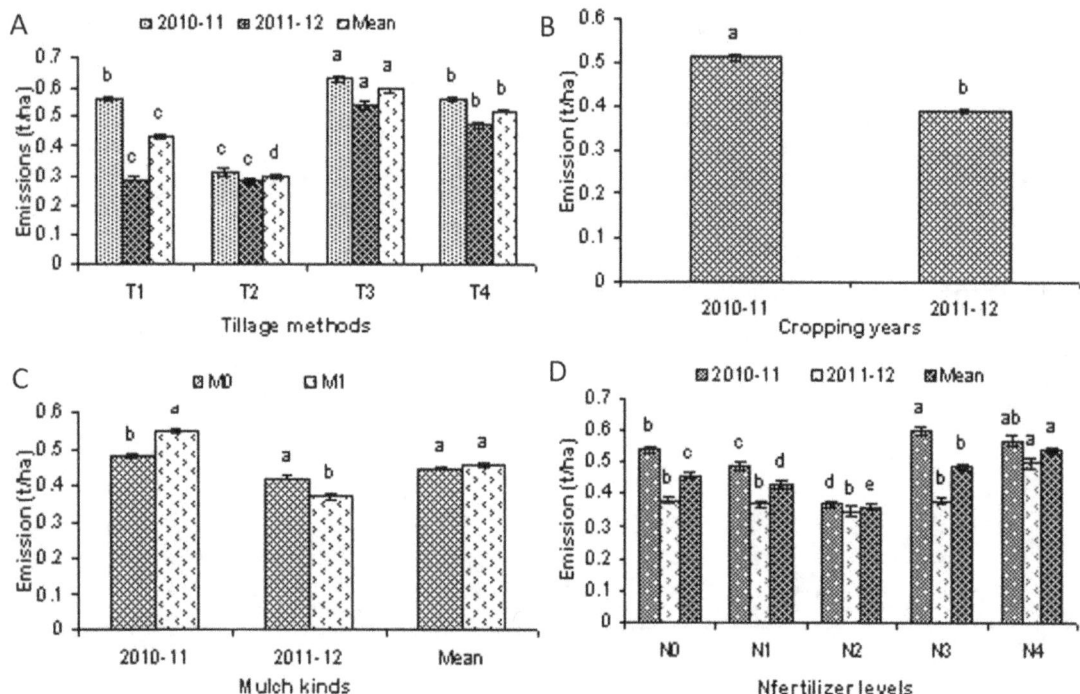

Figure 3. Total/mean emissions of CO$_2$ as affected by different tillage methods, mulch kinds, and N fertilizer levels during two cropping years (2010–12). (A).Emissions of CO$_2$ from different tillage methods i.e. T$_1$, chisel plow tillage., T$_2$, zero tillage.,T$_3$, rotary tillage and T$_4$., mold board plow tillage., (B). Total emissions of CO$_2$ as affected by different kinds of tillage methods, different kinds of mulch and different levels of N fertilizer, during two cropping years (2010–12)., (C), Emissions of CO$_2$ from the different mulch kinds i.e. M$_0$, no mulch and M$_1$, corn residue mulch., (D).Total/mean emissions of CO$_2$ as affected by different levels of N fertilizer, during two cropping years (2010–12) including, N$_0$, 0 kg N/ha., N$_1$, 80 kg N/ha., N$_2$,160 kg N/ha., N$_3$, 240 kg N/ha and N$_4$, 320 kg N/ha.

Effects of cropping years on CO$_2$ emissions

With the exception of one week (i.e., week 23), the highest weekly emissions of CO$_2$ were recorded during cropping year 2010–11 in comparison to the weekly emissions of CO$_2$ during cropping year 2011–12 (Figs. 2, 3).

Mulch effects on CO$_2$ emissions

Weekly CO$_2$ emissions during both wheat crops growing seasons (2010–11 and 2011–12) and the mean of the two years (2010–12) of data (Figs. 2, 3) show that there were more CO$_2$ emissions recorded from the corn residue-mulched treatments during cropping year 2010–11 than from the non-residue mulched treatments (Figs. 2, 3). However, fewer CO$_2$ emissions were recorded from the corn residue mulched treatments during cropping year 2011–12 and in comparison to the non residue-mulched treatments, over all less emissions of CO$_2$ were recorded from the corn residue mulched treatments (Figs. 2, 3). However the two year mean data show mixed types of emissions were recorded on a weekly basis following the applications of corn residue mulch or no mulch (Fig. 2).

N fertilizer level effects on CO$_2$ emissions

CO$_2$ emissions data recorded during both wheat cropping seasons, i.e., 2010–11 and 2011–12 and the two year (2010–12)) mean data show that there were significant differences in the weekly CO$_2$ emissions in response to different N fertilizer levels (Fig. 2; Fig. 3). This finding also indicates that the lowest CO$_2$ emissions were recorded for the N$_0$, N fertilizer level at the start of the wheat crop growing seasons compared to all the other higher

N fertilizer levels (Fig. 2). During the winter months the CO$_2$ emissions decreased under all treatments but when the temperatures rose again, higher CO$_2$ emissions were recorded (Fig. 2). However, when compared to all the other N fertilizer level treatments, the lowest overall CO$_2$ emissions were recorded for 160 kg N/ha (Fig. 2). CO$_2$ emission fluxes increased with the increase in crop growth and temperatures, so during the last weeks of wheat crop growth equal CO$_2$ emissions were recorded for all of the N fertilizer treatments (Fig. 2).

Cumulative CO$_2$ emissions

Two years (2010–12) of CO$_2$ emissions data show that on a mean cumulative basis, except in the case of corn residue mulch treatments, significant differences in the emissions of CO$_2$ were recorded for all of the different tillage methods and different N fertilizer level treatments (Fig. 3). Statistically significant variations in the total and mean CO$_2$ emissions were recorded for all of the tillage methods, and the emissions trend for the different tillage methods was T$_3$>T$_4$>T$_1$>T$_2$ (Fig. 3). Different N fertilizer levels had significant effects on the total and mean CO$_2$ emissions. The emissions trend for the different nitrogen fertilizer levels was N$_4$>N$_3$>N$_0$>N$_1$>N$_2$ (Fig. 3). On a cumulative basis, more CO$_2$ emissions were recorded during the cropping year 2010–11 than during cropping year 2011–12 (Fig. 3), and on the whole approximately 30% more CO$_2$ emissions were recorded during cropping year 2010–11 than during 2011–12 (Fig. 3).

CO$_2$ emissions varied for all of the tillage methods and N fertilizer levels. For the chisel plow tillage treatment, using 160 kg N/ha reduced the emissions of CO$_2$ (data not shown) and the CO$_2$

emissions trend in case of the chisel plow tillage method and different N fertilizer levels was $N_4>N_3>N_0>N_1>N_2$ (data not shown) (Table 1). For the Zero tillage planting method and N fertilizer level interactions, the CO_2 emission varied but the emissions trend was $N_0>N_3>N_2>N_1>N_4$ (data not shown) (Table 1). For the rotary tillage planting method and N fertilizer level interactions, the CO_2 emissions trend was $N_4>N_3>N_1>N_0>N_2$ (data not shown) (Table 1). Similarly for the mold board plow tillage method and N fertilizer level interactions, the CO_2 emissions trend was $N_0>N_4>N_1>N_3>N_2$ (data not shown) (Table 1).

With the exception of the mold board plow tillage method, corn residue mulch applications increased the CO_2 emissions in all of the other three tillage methods. Lower CO_2 emissions were recorded for all the tillage methods during cropping year 2011–12 in comparison to cropping year 2010–11 and throughout cropping year 2010–11, corn residue mulch applications increased the CO_2 emissions to 16.5% compared to the non residue mulched treatments. Similarly during cropping year, 2011–12, 12.6% fewer CO_2 emissions were recorded for the corn residue mulched treatments in comparison to the non-residue mulched treatments (data not shown) (Table 1).

For the N_0 (0 kg N/ha), N_1 (80 kg N/ha), N_2 (160 kg N/ha) and N_4 (320 kg N/ha), treatments, the use of corn residue mulch increased the CO_2 emissions by approximately 6.2%, 5.5%, 0.6% and 35.6%, respectively, and for N_4 (240 kg N/ha), the application of corn residue mulch reduced the CO_2 emissions by approximately 35.6% (data not shown) (Table 1).

Soil temperatures versus CO_2 emissions

Temperature changes for the top 5 cm of soil from the different treatments during the two years (2010–12) of study are given in Fig. 4. Although with the passage of time, there were variations in soil temperatures for the different tillage methods, however intermediary types of temperature changes were recorded following zero tillage planting in comparison to the other three tillage methods (Fig. 4). Similarly, although statistically comparable temperatures were recorded for both types of mulch treatments, slightly increased temperatures were recorded in the corn residue mulched treatments relative to the non-residue mulched treatments (Fig. 4). However, no soil temperature differences were recorded for the different N fertilizer level treatments (Fig. 4). Generally speaking, higher soil temperatures were recorded during the cropping year 2010–11 than during cropping year 2011–12 (Fig. 4). CO_2 emission trends showed that CO_2 emissions increased with an increase in soil temperatures and vice versa (Figs. 2, 4).

Soil moisture versus CO_2 emissions

When compared to all of the other tillage methods, the highest soil water contents were recorded in response to the rotary tillage planting method (Fig. 5). Similarly, higher water contents were recorded during the cropping year 2011–12 for the different crop growth stages relative to cropping year 2010–11 (Fig. 5). Corn residue mulch increased the soil moisture contents of the crop revival stage and on the booting stage compared to the non-residue mulched treatments (Fig. 5). Lower soil moisture contents were recorded for the 80 kg N/ha (N_1) and 160 kg N/ha (N_2), treatments compared to all of the other N fertilizer level treatments (Fig. 5). Two years of mean data show that the CO_2 emissions were lower in those tillage methods or N fertilizer levels treatments that had lower water contents (Fig. 2; Fig. 4).

Table 1. ANOVA (Mean Square Values) of biological yields, grain yields, soil organic carbon, and cumulative emissions of CO_2 during two cropping years (2010–12).

Source	D.F	B.Y	G.Y	SOC	CO$_2$
Tillage methods	3	6593271.9***	122260.74**	32.12145866***	9605737865***
Planting years	1	13496603.4***	29504910.64***	97.90534734***	8211670776***
Mulch kinds	1	132977.5NS	20786.14NS	3.74523078***	93546720 NS
N Fertilizer levels	4	1470235.5***	4549170.27***	3.78932599***	21321766841***
Tillage methods X Mulch kinds	3	6780289.1***	1690508.29***	26.55366456***	1455896530***
Tillage methods X N Fertilizer levels	12	1443071.8 NS	302485.89NS	5.42559025***	1418530119***
Tillage methods X planting years	3	3443481.8NS	1276931.92*	12.85179871***	1455478300***
Planting years X Mulch kinds	1	66692.7NS	418751.64NS	0.67995212 NS	2547564904***
Planting years X N Fertilizer levels	4	10758689.1***	3983648.62***	3.45638215***	720140735***
Mulch kinds X N Fertilizer levels	4	975872.1NS	461889.16NS	3.23681715***	3518560905***
Tillage methods X Planting years X Mulch kinds	3	1162751.4NS	190004.99NS	12.61344035***	406313591***
Tillage methods X Planting years X N fertilizer levels	12	12108232.1NS	369168.67NS	4.50848299***	959919429***
Tillage methods X Mulch kinds X N fertilizer levels	12	1785389.2NS	420989.67NS	6.20244131***	2881830882***
Tillage methods X Mulch kinds X Planting years X N fertilizer levels	16	2029198.1 NS	782603.91***	4.05394235***	1326809061***

*Significant at 0.05 probability levels.
**Significant at 0.01 probability levels.
***Significant at 0.001 probability levels.
B.Y, Biological yields., G.Y, Grain yields., SOC, Soil organic carbon.
X, indicates interactions between different factors i.e. Tillage's X Mulches indicates interactions between different tillage methods and mulch kinds.

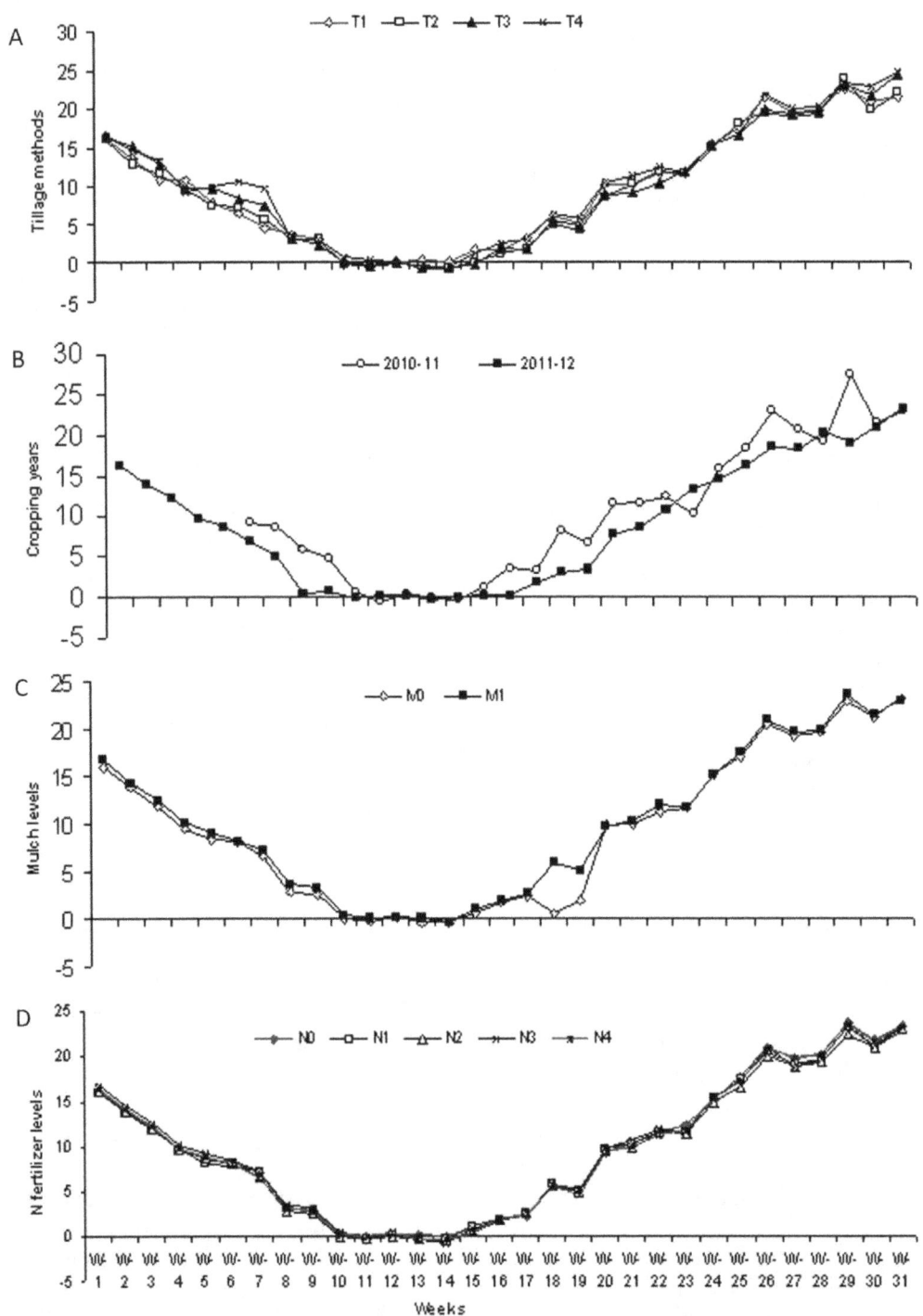

Figure 4. Changes in soil temperatures (0–5 cm depth) due to different tillage methods, corn residue mulch and N fertilizer levels during two cropping years (2010–12). (A). Changes in soil temperatures (0–5 cm depth) due to different tillage methods i.e. T_1, chisel plow tillage, T_2, zero tillage, T_3, rotary tillage and T_4, mold board plow tillage method., (B) Changes in soil temperatures (0–5 cm depth), as affected by different tillage methods, different mulch kinds and different N fertilizer levels during two cropping years (2010–12)., (C), Changes in soil temperatures (0–5 cm depth) during two cropping years (2010–12) due to different mulch kinds i.e. M_0, no mulch and M_1, corn residue mulch., (D). Changes in soil temperatures (0–5 cm depth) due to different N fertilizer levels during two cropping years (2010–12) including, N_0, 0 kg N/ha., N_1, 80 kg N/ha., N_2, 160 kg N/ha., N_3, 240 kg N/ha and N_4, 320 kg N/ha.

Soil organic carbon versus CO_2 emissions

Two years of mean data show that the tillage methods had significant effects on the SOC from 0–10 cm soil depths and that compared to all of the tillage methods, the highest SOC was recorded following chisel Plow tillage (Fig. 6). Higher SOC contents were recorded for all of treatments during cropping

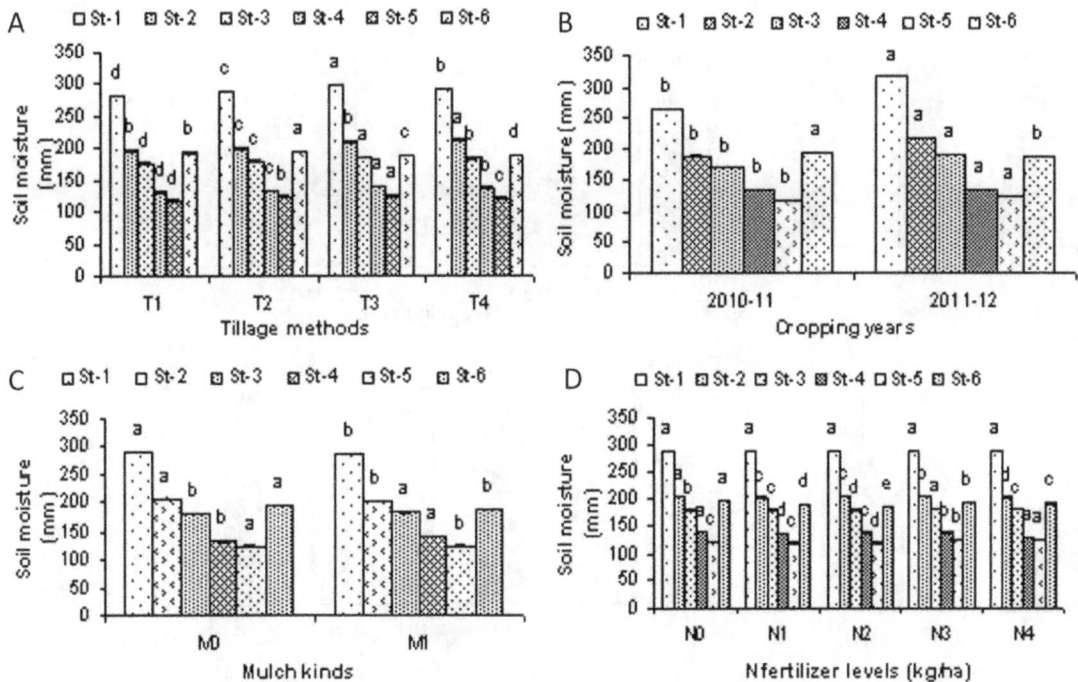

Figure 5. Total soil moisture contents (0–100 cm soil depth) as affected by different tillage methods, mulch kinds, and N fertilizer levels during two cropping years (2010–12). (A). Total soil moisture contents (0–100 cm soil depth) due to different tillage methods i.e. T_1, chisel plow tillage., T_2, zero tillage., T_3, rotary tillage and T_4, mold board plow tillage method., (B).Total soil moisture contents (0–100 cm soil depth) as affected by different tillage methods, different mulch kinds and different N fertilizer levels, during two cropping years (2010–12).(C), Total soil moisture contents (0–100 cm soil depth) as affected by different mulch kinds during the two cropping years (2010–12) i.e. M_0, no mulch and M_1, corn residue mulch., (D). Total soil moisture contents (0–100 cm soil depth) as affected by different N fertilizer levels, during two cropping years (2010–12) including, N_0, 0 kg N/ha., N_1, 80 kg N/ha, N_2, 160 kg N/ha., N_3, 240 kg N/ha and N_4, 320 kg N/ha., (*) Stage-1, (Crop sowing stage), Stage-2, (Crop revival stage), Stage-3, (Stem elongation stage), Stage-4, (Booting stage), Stage-5, (Grain formation stage) and Stage-6 (Crop harvesting stage).

year 2011–12 in comparison to cropping year 2010–11 (Fig. 6). The use of corn residue mulch increased the SOC compared to the non-mulched treatments (Fig. 6), and higher over all SOC contents were recorded in the cases of N_1 (80 kg N/ha) and N_2 (160 kg N/ha) in comparison to all of the other N fertilizer levels treatments (Fig. 6). The data show that CO_2 emissions were lower in those treatments that had higher SOC contents (Fig. 3; Fig. 6).

Effects of seasonal variations on CO_2 emissions

Seasonal temperature variations had significant effects on the CO_2 emissions. Because there were normal temperatures when the wheat crops were sown, higher CO_2 emissions were recorded, but reduced emissions were recorded with the decline in temperature during the winter seasons. With increased of crop growth and ascending of seasonal temperatures higher CO_2 emissions were recorded during both years (Figs. 2, 3, 4, 7, 8,).

Discussion

Soil CO_2 fluxes

Variations in seasonal temperatures had significant effects on soil temperatures, which ultimately affected the CO_2 emissions (Figs. 2, 3, 4, 7, 8). Higher CO_2 fluxes were recorded immediately after the tillage operations, which continued for a few days, and these emissions decreased with the passage of time (Fig. 2). Our results are in agreement with other findings [35]. Other investigators have reported changes in CO_2 emissions with seasonal variations. According to these investigators, seasonal variations in CO_2 emissions are controlled not only by the soil

temperatures and soil moistures but also by the tillage practices. Changes in CO_2 emissions from seasonal variations have been reported for almost all ecosystems. These emissions mainly depend both on the type of climate and the ecosystem [36].

Similar findings regarding variations in CO_2 fluxes with changes in soil temperatures and crop growth stages have been reported for rice crops [37]. In our study, the air and soil temperatures had significant effects on CO_2 emissions. Thus with the decrease in soil temperatures during the winter months, the CO_2 emissions decreased, and the CO_2 emissions again increased with the rise in soil temperatures during the summer months (Figs. 2, 3, 4, 7, 8).

Generally, the seasonal CO_2 emissions variations found in our experiment were similar to other findings [38]. Other investigators reported that these emission variations might be related to variations in autotrophic and in heterotrophic respiration because both are involved in soil CO_2 emissions. In addition a large amount of CO_2 is released from plant roots, during the continuation of the plant-energy system. Microbial and root respiration can also significantly contribute to CO_2 emissions.

Tillage method effects on CO_2 emissions

Higher CO_2 emissions were recorded immediately after the tillage operations, which continued for a few days (Fig. 2). Our results are in agreement with findings from other researchers, who reported that CO_2 emissions following tillage increased up to 2–15 times [39], [40], [41], [42] and [43]. According to these findings [39], instead of microbial activity, the basic reason for higher CO_2 emissions immediately after the tillage was actually the release of entrapped CO_2 from the soil pores as a result of physical

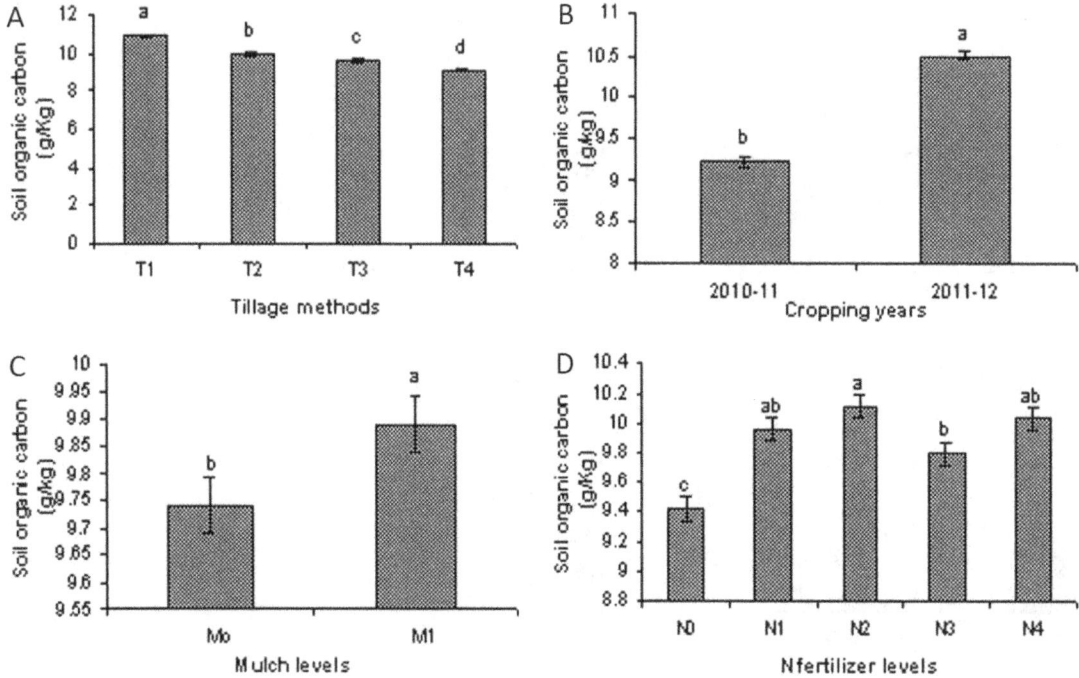

Figure 6. Soil organic carbon (SOC) contents (g/kg) as affected by different tillage methods, different mulch kinds and different N fertilizer levels during different cropping years (2010–12). (A), Soil organic carbon (SOC) contents as affected by different tillage methods i.e. T_1, chisel plow tillage., T_2, zero tillage., T_3, rotary tillage and T_4, mold board plow tillage method., (B), Mean soil organic carbon (SOC) contents as affected by different tillage methods, mulch kinds and different N fertilizer levels during two cropping years (2010–12)., (C), Soil organic carbon (SOC) contents as affected by different mulch kinds i.e. M_0, no corn residue mulch and M_1, corn residue mulch., (D), Soil organic carbon (SOC) contents as affected by different N fertilizer levels during two cropping years (2010–12) including, N_0, 0 kg N/ha., N_1, 80 kg N/ha., N_2, 160 kg N/ha., N_3, 240 kg N/ha and N_4, 320 kg N/ha.

operations. The other reasons for these higher emissions might be that (1) tillage operations break soil aggregates and expose their organic matter to microbial attack [44], [45]; (2) tillage operations encourage the mineralization of soil organic matter by incorporating crops residues into the soil [46]; and (3) tillage operations enhance soil aeration [43]. In our study, the tillage methods had significant effects on the CO_2 emissions and, overall, the rotary tillage and mold board plow tillage methods led to higher CO_2 emissions compared to the chisel plow and zero tillage planting methods (Fig. 2 and Fig. 3). Similar findings under different tillage systems have been previously reported [13]. These researchers reported significantly more CO_2 emission fluxes from the fields tilled by a mold board plow relative to the fields prepared by the chisel plow methods. According to these researchers, the basic reason for the greater emission fluxes from the mold board plow tillage method compared to the chisel plow tillage method was the depth and extent to which the soil was disturbed by using the different tillage implements. In our experiment, higher soil temperatures in the top 5 cm depth (Fig. 4) and generally higher moisture contents (Fig. 5) in the rotary tillage and mold board plow methods might be responsible for the higher emissions, in addition to the soil preparation depths. The tillage depths also resulted in a reduction of the SOC in the top 0–10 cm soil layers for the rotary tillage and mold board plow methods, compared to the chisel plow tillage and zero tillage planting methods (Fig. 6).

Corn residue mulch effects on CO_2 emissions

Weekly CO_2 emissions data (Figs. 2, 4) show that during cropping year 2010–11, the application of corn residue mulch caused an overall increase in CO_2 emissions compared to the non-

residue mulched treatments, but during the cropping year 2011–12, fewer CO_2 emissions were recorded in response to the application of corn residue mulch relative to the non-residue mulched treatments (Fig. 2 and Fig. 4). Although more CO_2 emissions fluxes were recorded in the corn residue mulched treatments in comparison to the non-residue mulched treatments, these might be due to the more microbial activities in the corn residue mulched treatments, which might have increased the SOC with the passage of time (Fig. 6). However, the two year mean data show that statistically non significant differences were recorded for CO_2 emissions following the use of the corn residue mulched or non mulched treatments. These findings might be explained by noting that both years CO_2 collecting chambers were fixed before the application of corn residue mulch. As a result there were fewer corn residues within the CO_2 collecting chambers from the corn residue mulched treatments. This smaller amount might be the reason why no CO_2 emission differences were recorded following the applications of different corn residue mulch treatments. The other reason might be that during the cropping year 2010–11, the application of corn residue mulch increased the CO_2 emissions but this mulch also increased the SOC contents of the corn residue mulched treatments, possibly resulting in the lower emission of CO_2 from the corn residue mulched treatments during the year 2011–12 (Fig. 3). A modeling study reported [47] that instead of applying higher rates of fertilizers, the use of crop residues or manure amendments would mitigate GHG emissions more efficiently. Similarly, it has been reported [48] that applications of straw increased the SOC sequestration in the soil which ultimately influenced the temporal patterns of CO_2 emissions from the soil.

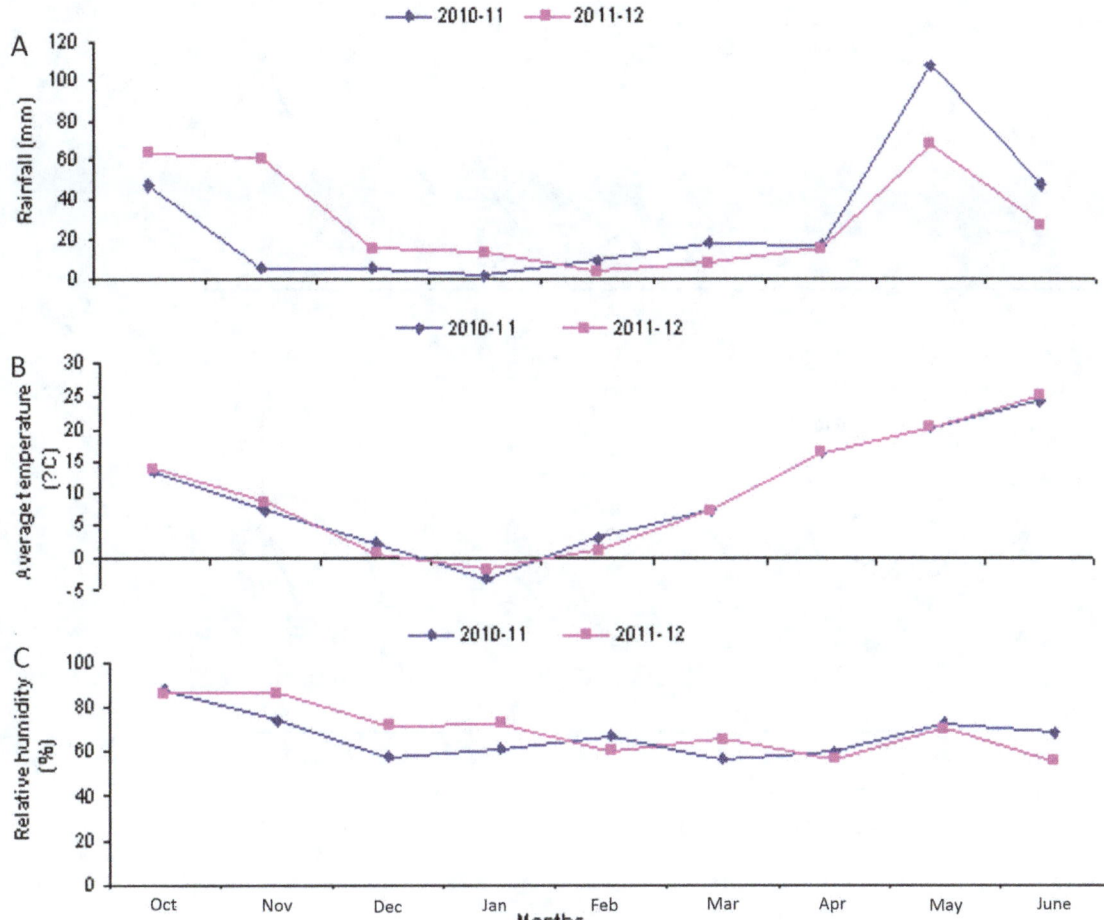

Figure 7. Monthly rainfalls, average temperatures and average relative humidity's of the study area during two wheat crop growing seasons (2010–12). (A).Rainfalls of the study area during two wheat crop growing seasons (2010–12), (B) Average temperatures of the study area during two wheat crop growing seasons (2010–12) (C) Average relative humidities of the study area during two wheat crop growing seasons (2010–12). (*). R.F-1, Rainfalls during cropping year 2010–11., R.F-2, rainfalls during cropping year 2011–12., Av.temp-1, Monthly average temperatures during the cropping year 2010–11., Av.temp-2, Monthly average temperatures during the cropping year 2011–12., R.H (%)-1, Average monthly relative humidities during the cropping year 2010–11., R.H (%)-2, Average monthly relative humidities during the cropping year 2011–12.

N fertilizer level effects on CO_2 emissions

Very few studies regarding CO_2 emissions in relation to the different tillage methods, corn residue mulch and N fertilizer levels have previously been reported in this region of China. It is expected that the application of inorganic N fertilizers along with organic materials will affect the mineralization of soil organic matter and crop productions, which will ultimately affect CO_2 emissions [8].

However variations in CO_2 emissions following fertilizer applications have been reported for different areas of China. Some scientists [15] have reported that the fertilizer applications suppresses CO_2 emissions and others [49] have reported that fertilizer application enhances CO_2 emissions. Moreover, some other scientists [19] have reported that fertilizer applications have no effects on CO_2 emissions.

Our study shows that the use of 80 kg and 160 kg N/ha, suppressed CO_2 emissions when compared with the 0 N fertilizer level, but that further increases in N fertilizer application rates enhanced CO_2 emissions. Higher emissions from the N_0, nitrogen fertilizer treatments might be explained by noting that plants under unfertilized N treatments are considered to respond to a relative shortage of N by increasing the plant's carbon allocation to

its structures and functions, which are responsible for N acquisition [50]. In our study, the use of different levels of N fertilizer relative to the no nitrogen fertilizer level significantly increased crop yields (Fig. 1).This result shows that the higher CO_2 emission fluxes in response to higher levels of N fertilizer might be explained by the increased use of C for microbial growth [51] and it might also be explained by the less efficient use of carbon by the microbial biomass, which resulted in a greater proportion of carbon loss in the form of CO_2 fluxes [52].

In our case, the use of corn residue mulch in combination with different levels of N fertilizer increased CO_2 emissions following the chisel plow, zero tillage and rotary tillage planting methods (Table 1). Our results regarding CO_2 emissions following the combination of N fertilizer and organic amendments are in agreement with the results of other scientists [53], [38] and [54]. These investigators reported a higher CO_2 emission flux from the treatments that utilized both fertilizers and organic manures. This finding shows that up to a certain extent the use of higher N fertilizer levels suppresses CO_2 emissions but in our case, higher levels of N fertilizer (i.e., 240 and 320 kg N/ha) enhanced CO_2 emissions. These results are contrary to the findings of other

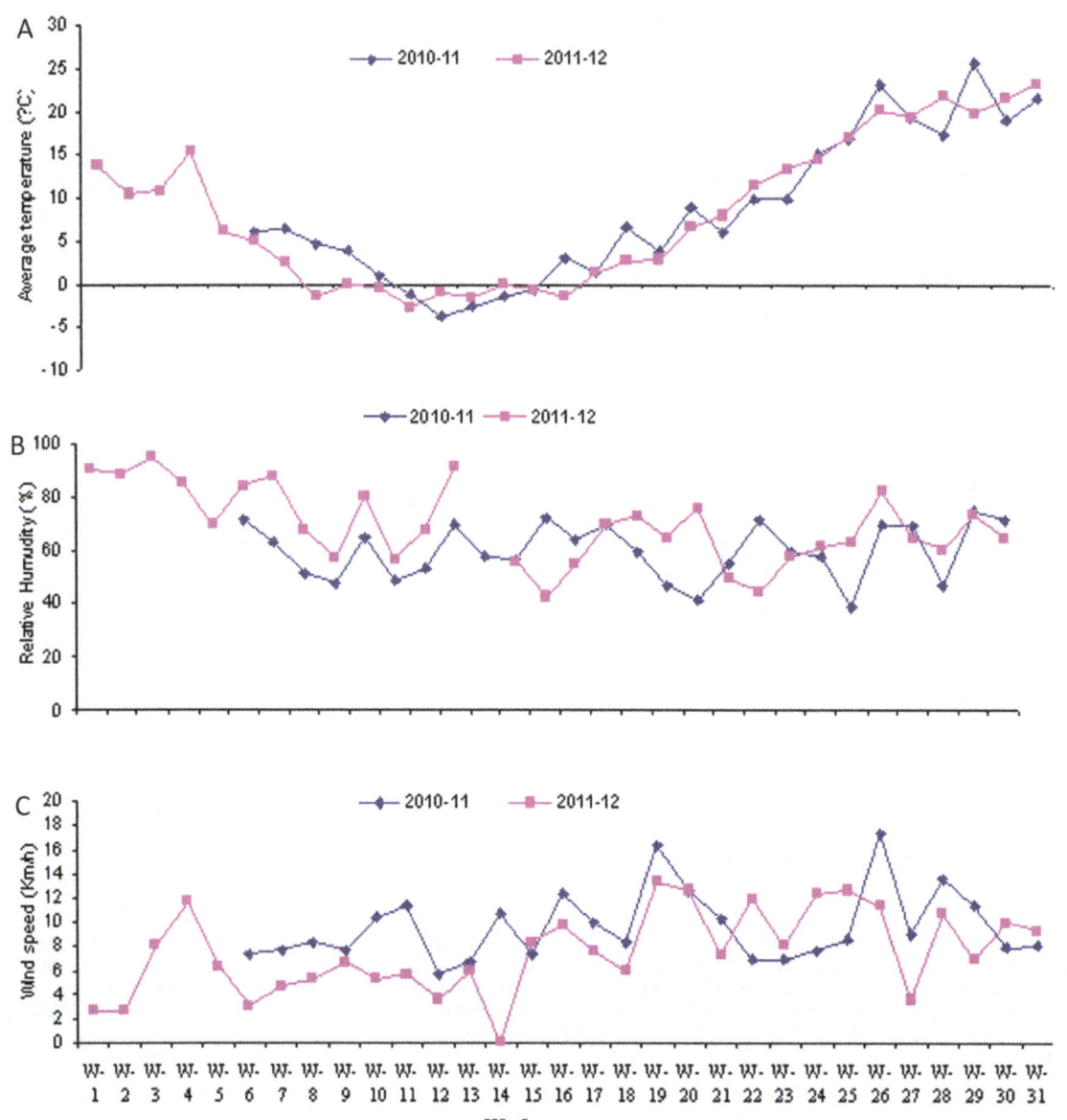

Figure 8. Average weekly temperatures, relative humidity's and wind speeds during the two wheat crop growing seasons (2010–12). (A) Average weekly temperatures of the study area during the two wheat crop growing seasons (2010–12), (B).Average weekly relative humidity's of the study area during the two wheat crop growing seasons (2010–12), (C) Average weekly wind speeds of the study area during the two wheat crop growing seasons (2010–12).

scientists [30] and [31]. These investigators reported no clear reasons for the CO_2 emission reductions.

Cropping year effects on CO_2 emissions

Two years of data on a weekly as well as on a cumulative basis show that there were more CO_2 emissions from all of the treatments during cropping year 2010–11 compared to the CO_2 emissions during cropping year 2011–12 (Figs. 2, 3). The main reasons for these lower emissions during cropping year 2011–12 were, on the whole, lower air temperatures, higher relative humidities and higher over all rainfall concentrations during the early periods of wheat crop growth relative to cropping year 2010–11 (Fig. 7). These factors all ultimately reduced the soil temperatures, which resulted in reduced CO_2 emissions during

the cropping year 2011–12 compared to cropping year 2010–11 (Figs. 4, 7, 8). Another reason for the lower emissions might be an increase in the SOC from the corn residue mulched treatments, which might have ultimately helped to reduce the CO_2 emissions during cropping year 2011–12 in comparison to cropping year 2010–11 (Figs. 2, 3, 6).

Effects of soil temperatures, soil moisture and soil organic carbon on CO_2 emissions

Respiration of ecosystem mainly depends on both the heterotrophic (microbe) and autotrophic (plant) activities and both of these factors are controlled by the prevailing environmental conditions (basically temperature and water availability), availability of carbohydrates and substrates and others [55], [56] and

Table 2. $Q10$ values of the different treatments during the different cropping years (2010–12).

Treatments		chisel plow		zero tillage		rotary tillage		mold board plow tillage	
		2010-11	2011-12	2010-11	2011-12	2010-11	2011-12	2010-11	2011-12
N_0	Y	$69.220e^{0.0516x}$	$47.81\,e^{0.0898x}$	$73.371e^{0.0736x}$	$74.836e^{0.807x}$	$140.49e^{0.0427x}$	$67.628\,e^{0.749x}$	$287.91e^{0.0707x}$	$78.941e^{0.1001x}$
	R^2	0.7076	0.8064	0.0823	0.7213	0.7472	0.7372	0.6076	0.7439
	$Q10$	1.68	2.43	2.09	1.90	2.54	2.55	2.72	2.72
N_1	Y	$79.76e^{0.07744x}$	$73.979e^{0.0816x}$	$57.341e^{0.0845x}$	$62.738e^{0.0996x}$	$142.39e^{0.0691x}$	$93.744e^{0.092x}$	$247.36e^{0.03385x}$	$129.37e^{0.0818x}$
	R^2	0.6672	0.7158	0.0869	0.752	0.8187	0.8074	0.5688	0.736
	$Q10$	2.10	2.26	2.33	2.71	1.63	2.11	2.03	2.72
N_2	Y	$80.442e^{0.0827x}$	$0.787e^{0.1012x}$	$114.12e^{0.0509x}$	$57.382e^{0.0891x}$	$90.813e^{0.0616x}$	$122.76e^{0.0669x}$	$108.97e^{0.0634x}$	$79.756e^{0.0873x}$
	R^2	0.5944	0.8201	0.528	0.7701	0.790	0.6534	0.8134	0.7563
	$Q10$	2.29	2.75	1.66	2.70	2.00	2.51	1.47	2.27
N_3	Y	$125.25e^{0.1085x}$	$70.222e^{0.0931x}$	$82.409e^{0.0478x}$	$55.391e^{0.1006x}$	$176.74e^{0.097x}$	$149.95e^{0.0859x}$	$103.92e^{0.0614x}$	$69.581e^{0.0843x}$
	R^2	0.7977	0.7054	0.7439	0.7937	0.6429	0.7876	0.852	0.6457
	$Q10$	2.96	2.47	2.33	2.02	1.85	1.95	1.89	2.32
N_4	Y	$55.139e^{0.0653x}$	$61.6e^{0.090x}$	$43.884e^{0.0903x}$	$60.312e^{0.8051x}$	$90.419e^{0.058x}$	$221.26e^{0.0624x}$	$116.85e^{0.0613x}$	$96.28e^{0.1086x}$
	R^2	0.6443	0.7961	0.7996	0.9015	0.6951	0.6654	0.7312	0.7969
	$Q10$	1.02	2.18	1.61	2.16	2.64	2.32	1.85	2.96

N_0, 0 kg N/ha., N_1, 80 kg N/ha., N_2, 160 kg N/ha., N_3, 240 kg N/ha and N_4, 320 kg N/ha.

Table 3. Q10 values of the different treatments during the different cropping years (2010–12).

Treatments	chisel plow		zero tillage		rotary tillage		mold board plow tillage	
	2010-11	2011-12	2010-11	2011-12	2010-11	2011-12	2010-11	2011-12
N_0+M	$Y=43.137e^{0.1316x}$	$Y=87.49e^{0.0778x}$	$Y=163.72e^{0.0363x}$	$Y=61.295e^{0.09x}$	$Y=62.946e^{0.1111x}$	$Y=99.78e^{0.0961x}$	$Y=183.66e^{0.0518x}$	$Y=55.625e^{0.0947x}$
	$R^2=0.8849$	$R^2=0.6632$	$R^2=0.4067$	$R^2=0.8372$	$R^2=0.7113$	$R^2=0.8096$	$R^2=0.753$	$R^2=0.8216$
	Q10=3.72	Q10=2.69	Q10=2.47	Q10=2.10	Q10=1.79	Q10=2.96	Q10=1.85	Q10=2.58
N_1+M	$Y=254.21e^{0.0484x}$	$Y=0.0988e^{0.0516x}$	$Y=84.882e^{0.0413x}$	$Y=49.272e^{0.0965x}$	$Y=175.25e^{0.0557x}$	$Y=76.273e^{0.0858x}$	$Y=112.84e^{0.0703x}$	$Y=51.454e^{0.0957x}$
	$R^2=0.7689$	$R^2=0.8367$	$R^2=0.7109$	$R^2=0.7881$	$R^2=0.5183$	$R^2=0.8149$	$R^2=0.695$	$R^2=0.7376$
	Q10=1.62	Q10=2.52	Q10=1.44	Q10=2.22	Q10=3.04	Q10=2.58	Q10=1.68	Q10=2.60
N_2+M	$Y=72.653^{0.0809x}$	$Y=45.736e^{0.0926x}$	$Y=80.339e^{0.0722x}$	$Y=61.262e^{0.0907x}$	$Y=163.02e^{0.0519x}$	$Y=104.06e^{0.083x}$	$Y=93.668e^{0.0666x}$	$Y=46.012e^{0.1166x}$
	$R^2=0.8237$	$R^2=0.7309$	$R^2=0.8477$	$R^2=0.7992$	$R^2=0.4896$	$R^2=0.7173$	$R^2=0.6503$	$R^2=0.816$
	Q10=2.25	Q10=2.87	Q10=1.52	Q10=2.18	Q10=1.95	Q10=2.36	Q10=2.02	Q10=3.17
N_3+M	$Y=1.015e^{0.0898x}$	$Y=46.994e^{0.1054x}$	$Y=164.85e^{0.0502x}$	$Y=30.192e^{0.1197x}$	$Y=163.9e^{0.405x}$	$Y=40.179e^{0.0983x}$	$Y=75.02e^{0.0742x}$	$Y=72.535e^{0.1075x}$
	$R^2=0.8604$	$R^2=0.8214$	$R^2=0.4050$	$R^2=0.8784$	$R^2=0.6464$	$R^2=0.84$	$R^2=0.7363$	$R^2=0.8221$
	Q10=2.45	Q10=1.11	Q10=2.06	Q10=2.18	Q10=1.68	Q10=2.29	Q10=1.95	Q10=2.93
N_4+M	$Y=8.69e^{0.10179x}$	$Y=69.932e^{0.0859x}$	$Y=51.822e^{0.0641x}$	$Y=36.672e^{0.1073x}$	$Y=287.97e^{0.0937x}$	$Y=121e^{0.1050x}$	$Y=127.88e^{0.0702x}$	$Y=73.455e^{0.1177x}$
	$R^2=0.8342$	$R^2=0.7264$	$R^2=0.7247$	$R^2=0.8364$	$R^2=0.6465$	$R^2=0.854$	$R^2=0.7656$	$R^2=0.8821$
	Q10=2.76	Q10=2.36	Q10=1.65	Q10=2.74	Q10=1.50	Q10=2.67	Q10=2.10	Q10=3.24

N_0+M, 0 kg N/ha + corn residue mulch., N_1+M, 80 kg N/ha + corn residue mulch., N_2+M, 160 kg N/ha + corn residue mulch., N_3+M, 240 kg N/ha + corn residue mulch., N_4+M, 320 kg N/ha + corn residue mulch.

[57]. Many studies have shown that seasonal variations in CO_2 emissions were mainly caused by the soil temperature, soil moisture or the combination of both these factors [58], [36].

Our study also indicates that soil temperature was an important driving force for the increased CO_2 emissions, which is also supported by the Q10 values given in Table 2 and Table 3. It has also been reported that the CO_2 evolution rate significantly increases with an increase in temperature and moisture [37]. Our results are also in agreement with many other field studies, which have shown strong relationships between soil temperatures and CO_2 flux rates [59], [60], [17] and [18]. Additionally, our results agree with the findings of [61], who reported a stronger polynomial for temperature and moisture interaction ($r^2 = 0.89$) than for temperature alone ($r^2 = 0.47$). Many previous studies have reported that changes in crop management practices, i.e., the appropriate use of tillage operations, proper fertilization, crop residue applications and crop rotations can be helpful for managing soil organic matter, e.g.,. [62].

Our study also shows that the results at the end of cropping year 2010–11 revealed more SOC in the top 0–10 cm of soil than that at the end of cropping year 2011–12. This result shows that the adaptation of different management practices, i.e., the application of crop residues, increased the SOC contents, especially following chisel plow tillage and zero tillage planting compared with rotary tillage and mold board plow tillage. SOC sequestration is a long term processes and various results have been previously reported, i.e. [63] and [64], have reported that conservation tillage is a recommended management practice for agricultural ecosystems that can enhance the pool of soil organic carbon (SOC), in the soil. Following an analysis of global data, NT reportedly sequestered carbon at an average rate of 0.57 Mg C ha^{-1} compared with the mold board plow [65]. It has also been reported that increases in the SOC pools can be credited to either reductions in the CO_2 efflux from the soil or to increases in the C inputs [65]. When comparing the soil surface across different tillage systems, conservation tillage systems retain more crop residues, which ultimately result in the formation of more SOC [66] and [67]. In addition to this finding, the decomposition process of surface applied plants residues as a part of conservation tillage is slow compared to conventional tillage systems because of lower contact with the soil.

In our study a negative but highly significant correlation coefficient (r) value i.e. $-0.19403**$ was recorded between the SOC and cumulative CO_2 emission. This finding might help in the reduction of CO_2 emissions from the chisel plow tillage and zero tillage planting methods compared to the rotary tillage and mold board plow methods (Figs. 2 and 3). In our experiment, increases in the SOC contents from different tillage methods, especially chisel plow tillage and zero tillage, might be related to the lower disturbance of the soil, retention of more crop residues on the soil surface and reductions in the efflux of CO_2 from the soil. Similar results have been reported by [65], [66] and [67].

Conclusions

Intensive tillage and higher N fertilization are not only detrimental to the soil but are also destructive for the entire environment. Adaptations of appropriate tillage methods, crop residue applications and proper fertilization are beneficial for the soil as well as for the environment. These practices are also economically beneficial for resource-poor farmers. The findings from our study clearly indicate that the tillage methods significantly affected CO_2 emissions and the zero tillage planting method emitted the lowest CO_2 compared with the other three tillage methods. No significant differences in CO_2 emissions were recorded for the applications of corn residue mulch, but the applications of corn mulch significantly improved the soil organic carbon (SOC) contents of the soil for all of the tillage systems. In addition, corn residue mulch application reduced the weed infestation by up to 40% (data not shown). Therefore, the application of corn residue mulch can be helpful for reducing the use of herbicides, which will also be helpful in establishing a healthy environment. Applications of different N fertilizer levels also significantly affected CO_2 emissions and, overall, the lowest emissions were recorded for the 160 kg N/ha treatment. Higher CO_2 emissions were recorded immediately after the tillage operations. This study also indicated that both soil temperatures and moistures strongly affected CO_2 emissions and that compared with the other tillage methods, zero tillage planting gave better grain yields. The lowest N fertilizer use gave equal yields, for the two year mean, as did the application of higher N fertilizer levels. These results clearly indicate that proper changes in farm management practices i.e., the adoption of zero tillage, crops residue application and optimum use of N fertilizers can reduce CO_2 emissions from soils. Therefore this type of long term study can be further helpful in reducing the emissions of CO_2 from soils, which will be helpful in reducing the use of inorganic fertilizers. These practices will be helpful in reducing production costs and will be beneficial for the entire environment.

Materials and Methods

Experimental site

A two-year (2010–12)), field study was conducted at the experimental area of Northwest A&F University, Shaanxi Province, northwestern China (latitude of 34°20′ N, longitude of 108° 04′E and elevation of 466.7 m above sea level) on the Eum-Orthrosols (Chinese soil Taxonomy) soil, with a mean bulk density of approximately 1.29 g cm^3. The soil in the top 40 cm had an SOC of approximately 14.26 g/kg, total nitrogen 0.74 g/kg and the pH was approximately 7.85. This area is under the corn-wheat rotation system. During both years, the wheat crop was planted after harvesting the corn crop. Both fertilizers, i.e., phosphorous in the form of calcium phosphate ($Ca_2 (PO_4)_3$ with 16% P and nitrogen in the form of urea with N≥46%, were applied to the corn crop at the rate of 750 kg/ha and 375 kg/ha, respectively. The total rainfall during the wheat crop growing season (October-June) was 231.6 mm and 242.7 mm during, cropping years 2010–11 and 2011–12 respectively (Fig. 7).

Experimental design and treatments

A factorial experiment having a strip split-split arrangement, with tillage methods in the main plots, mulch levels in the sub plots and N-fertilizer levels in the sub-sub plots with three replicates, was used for this study. Different tillage methods, i.e., chisel plow tillage (T_1), zero tillage (T_2), rotary tillage (T_3) and mold board plow tillage (T_4) methods, were kept in the main plots, different mulch kinds, i.e., M_0 (no residue mulch) and M_1 (corn residue mulch), in the sub plots, while different nitrogen fertilizer rates, i.e., 0, 80, 160, 240 and 320 kg N/ha, were kept in the sub-sub plots. The area was uniform in terms of fertility. The total experimental plot area (3300 m^2) was equally divided into four main tillage treatments. The area of each tillage treatment (33 m×25 m) was further sub-divided into sub–plots for mulch treatments, and finally the sub plots were further divided into sub-sub plots, and each sub-sub individual plot had an area of 3 m×25 m. Treatments were randomized within each sub-plot.

Chisel plow tillage (T$_1$), was performed using a chisel plow. Following fertilizer applications, a chisel plow with a shank spacing of approximately 40 cm apart and 30–35 cm deep was used once. Later on, fertilizers were mixed by using the rotavator for up to a 5 cm depth. For zero tillage (T$_2$) fertilizers were mixed by using the rotavator up to a 5 cm depth due to the lack of availability of a proper zero tillage drill. For rotary tillage (T$_3$), the seed bed was prepared by using the rotavator up to a depth of 15–20 cm, and in the case of the mold board. plow tillage method (T$_4$), the soil was plowed up to a 20–30 cm depth by using the mold board plow, followed by the rotavator for the final seed bed preparation.

Urea fertilizer with N≥46%, was used as the source of the nitrogen, and phosphorous (P) fertilizer in the form of calcium phosphate (Ca$_2$ (PO$_4$)$_3$ with 16% P was equally applied to all of the treatments at a rate of 750 kg/ha at the time of soil preparation. The treatments arrangements were kept the same during both years (2010–12) of study. Previously harvested air-dried corn crops residues were used as the source of corn residue mulch during both years. When the wheat crop was at the 3–4 leaf stage, mulch was applied at a rate of 750 g/m^2. The field was flat with a uniform topography. This area is rain fed, and a wheat-corn rotation is the main cropping system. No irrigation was applied to either crop. No changes were made to the areas with different tillage treatments, and a corn crop was planted after the wheat crop harvest by using the same tillage methods. The wheat crop was harvested using a combine harvester, and after harvesting the wheat crop, the corn crop was planted using a corn planter.

Winter wheat (C.V Shaan mai −139) was planted on October 17, 2010 and October 18, 2011 by using wheat drills. The line to line distance was maintained at approximately 16 cm apart. The seed had a 13% moisture contents and a 85% germination rate during both years. During cropping season 2010–11, a seed rate of about @ 190–200 kg/ha was used, while during cropping season 2011–12, a seed rate of approximately 205–210 kg/ha was employed. Experimental treatments were separated from each other by making boundaries between the treatments. Both years an herbicide application i.e. carfentrazone-ethyl (C$_{15}$H$_{14}$C$_{12}$F$_3$N$_3$O$_3$), was used to control the weeds. At physiological maturity, which occurred on June 8, 2011 and on June 10, 2012, three samples were randomly selected from each treatment and manually harvested by using the 1 m^2 quadrants to calculate the grain yields. Finally the wheat crop was harvested using a combine harvester.

Measurements

Meteorological factors. Meteorological data for the study area are given in Fig. 7 and Fig. 8, which indicate that during cropping year 2011–12, more rains were concentrated during the early period of crop growth relative to 2010–11, during which more rains were concentrated in the later crop growth stages. Higher average seasonal temperatures were recorded during the year 2010–11 than in 2011–12. However, more relative humidity was recorded during 2011–12 than in 2010–11. Weekly meteorological data showed higher, average temperatures during the year 2010–11 than during 2011–12. On the other hand, higher weekly relative humidities were recorded during the cropping year 2011–12 than during 2010–11. Similarly, higher weekly basis more wind speeds were recorded during the cropping year 2010–11 than during 2011–12.

Monitoring CO$_2$ emissions. Because of the high number of treatments, both years of CO$_2$ emissions data were recorded three times per week and one day of CO$_2$ emissions data was used as one replicate for statistical analysis. CO$_2$ emissions data were recorded using a GXH-3010E1 portable gas analyzer. This gas

analyzer is made by the Beijing Huayun Yiqi Company, and the CO$_2$ emission was recorded by using the method described by Gao et al [68]. During both years (2010–12) CO$_2$ emissions data from all of treatments was recorded 6840 times, which included 3120 times during the cropping year 2010–11 and 3720 times during cropping year 2011–12.

One round PVC chamber (21 cm in diameter and 13.5 cm in height), having total area of approximately 0.0047 m^3 was permanently fixed in the center of each treatment plot. The chamber was completely fixed in the soil up to a 4.5 cm depth. The plants growing with in the chambers were removed. As a consequence of some technical problems with the Gas analyzer, during wheat growing season 2010–11, the CO$_2$ emissions data were recorded after the wheat had been planted for one month, and during cropping season, 2011–12, CO$_2$ emissions were recorded starting during the first week of wheat planting. However, due to severe snow fall, data were not recorded on the 14th week during this year. Taking data from wheat planting until harvest for 2010–11, the CO$_2$ emissions data were recorded for 26 weeks, and during cropping year 2011–12, these data were recorded for 31 weeks. Every week, the data were collected 3 times depending upon the environmental condition, i.e., if the field was too wet from rainfall, then the data recording was stopped. Each time the data collection started at 900 a.m., and each sequence of CO$_2$ flux measurements took at least 4 hours. Due to the large number of treatments the data were randomly collected from different treatments. The main purpose of this randomization was to minimize the effects of different days as well as changing soil temperatures on the emissions of CO$_2$. The GXH-3010E1 gas analyzer was attached to the data collector chambers with intake and an outtake tubes. These tubes were made up of soft plastic pipes and each one was approximately one meter in length. At the time of data recording, first the CO$_2$ data, i.e., X$_1$ were recorded without closing/covering the chamber, and then the chamber top was tightly closed with a cover that had a fixed small fan. The gas within the chamber was mixed for three minutes with this fan. After this CO$_2$ emission, data (X$_2$) were recorded using the gas analyzer.

Chambers tops were closed for only three minutes at the time of data recording. To avoid any chemical change in the soil, these chambers were kept open for the whole remaining time. Along with CO$_2$ data, soil temperatures data were also recorded from each treatment from depths of 5 cm. For this purpose, thermometers were permanently fixed in each plot each year for the whole crop growth period. The CO$_2$ emission rate was calculated by using equation (1) as described by Gao et al. [68].

$$F = k(X_2 - X_1)H/\Delta t \qquad (1)$$

Where F is the CO$_2$ emission in mg/(m^2.h); K is a constant with a value of 1.80 (25°C) and X$_1$ and X$_2$ are the CO$_2$ emissions rates from the chambers before and after covering of the chambers. H is the height of the chambers in meters and Δt is the time in hours (h). The cumulative emission of CO$_2$ was calculated using the following relationship, as described by Wilson H.M and Alkaisi W.W (2008) [69],

$$CO_2 - C\left(kgha^{-1}\right)$$
$$= \sum_{i=first}^{n=last} X_i + X_{i+1} * N + X_{i+2} * N + \ldots + X_{i+n} * N \qquad (2)$$

where (i) is the first week of the growing season when the first CO2 emission rate was taken, (n) is the last week of the growing season when the last CO2 emission rate was taken, X is the CO2 rate (Kg ha−1 day−1), and N is the number of days between two consecutive CO2 emission rates measurements. Finally, these CO2 emission rates for the whole wheat crop growing period (taken between 9.00 A.M until approximately 1.00 P.M) were converted into tons/ha.

Soil moisture measurements. Both years soil moisture contents were measured during different crop growth stages, i.e., before planting the wheat crops, at the 5 leaf stage (Zadoks stage 1.0–1.9), stem elongation stage (Zadoks stage 3.0–3.9), booting stage (Zadoks stage 4.0–4.9), grain formation stage (Zadoks stage 7.0–7.9) and harvesting stage. For this purpose, soil samples were collected from each treatment from 0–100 cm soil depths with three replicates, with increments of 0–10 cm, 10–20 cm, 20–30 cm. 30–40 cm, 40–60 cm, 60–80 cm and 80–100 cm. Soil samples were collected in aluminum boxes using a hand auger, and fresh soil samples were immediately transported to the laboratory. After recording the fresh weights, these samples were dried in an oven at 105°C for at least 48 hours, and the soil water contents were then measured using equation (3) given below.

$$\text{Soil water content (\%)} = \frac{\text{fresh weight of soil} - \text{dry weight of soil}}{\text{dry weight of soil}} \times 100 \quad (3)$$

Soil water contents (mm) were calculated as gravimetric moisture contents using the equation (soil water ×B.D× thickness of soil layer). Soil bulk densities from the different tillage treatments and from the different soil depths were measured using the core method. Total soil water contents in the 0–100 cm soil depths of the different treatments were measured on the basis of the bulk densities of these different soil layers.

Soil organic carbon (SOC) measurements. Both year soil SOC samples were collected from the top 0–10 cm soil depth after the wheat harvest. Soil samples were collected from two randomly chosen points from each plot using a hand augur. The samples from each treatment were then mixed together to make a composite sample of each treatment. These samples were then air-dried at room temperature, crushed gently and passed through a 2 mm sieve for further chemical analysis. Soil organic carbon was determined by the oxidation method with $K_2 Cr_2 O_7$ -H_2SO_4. For chemical analysis, 0.5 grams of soil was digested with 5 mL of 1 M $K_2 Cr_2 O_7$ and 5 mL of concentrated H_2SO_4 and was heated at 175°C for 5 minutes followed by titration of the digests with $FeSO_4$ [70].

Statistics

Annual data collected for the CO_2 emission rates and for other related parameters over the whole 2-year period were subjected to an analysis of variance (ANOVA) by using the factorial experiment with the strip-split-split arrangement having the tillage methods in the main plots, mulch in the sub plots and N-fertilizer levels in the sub-sub plots. The SAS analytical software package GLM (8.01) was used for the analyses. Mean values and standard errors (SE) were calculated for each treatment, and an ANOVA was used to assess the treatment effects on the measured variables. Means were declared statistically significant at a 0.05 probability level, or P≤(0.05), using the DUNCAN test (DNMRT).

Author Contributions

Conceived and designed the experiments: SKT YL. Performed the experiments: SKT XLL. Analyzed the data: SKT JLZ. Wrote the paper: SKT XXW YL.

References

1. Post WM, Peng TH, Emanuel WR, King AW, Dale VH, et al (1990) The global carbon cycle. Am Sci 78: 310–326.
2. Houghton RA, Skole DL (1990) Carbon, In: Turner BL, Clark WC, Kates RW, Richards JF, Mathews JT, Meyer WB, editors. The earth as Transformed by Human Action. Cambridge University Press. 393–408.
3. Houghton RA (2007) Balancing the global carbon budget. Ann Rev Earth Planet Sci 35: 313–47.
4. Denman KL, Brasseur G, Chidthaisong A, Ciais P, Cox PM, et al. (2007) In Climate Change 2007: The Physical Science Basis. Contribution of Working Group 1 to the Fourth Assessment Report of the International Panel on Climate Change. Solomon, S., et al, editors. Cambridge Univ. Press. 499–587.
5. Valentini R, Matteucci G, Dolman AJ, Schulze ED, Rebmann C, et al. (2000) Respiration as the main determinant of carbon balance in European forests. Nature 404: 862–864.
6. IPCC (2007) Climate Change 2007: The Physical Science Basis. Cambridge University Press. 996.
7. Cole CV, Duxbury J, Freney J, Heinemeyer O, Minami K, et al. (1997) Global estimates of potential mitigation of greenhouse gas emissions by agriculture. Nut. Cycl. Agroecosyst. 49: 221–228.
8. Paustian K, Collins HP, Paul EA (1997) Management controls on soil carbon. In: Paul EA, Paustian K, Elliot ET, Cole CV, editors. Soil Organic Matter in Temperate Agro ecosystems (Long-term Experiments in North America). CRC Press, Boca Raton, FL. 15 49.
9. Lal R (1997) Long–term tillage and maize monoculture effects on a tropical alfisol in western Nigeria.1. Crop yield and soil physical properties. Soil and Tillage Res 42: 145–160.
10. Ahmadi H, Mollazade K (2009) Effect of plowing depth and soil moisture content on reduced secondary tillage. Agricultural Engineering International: The CIGR E Journal 11: 1–9.
11. Koga N, Tsuruta H, Tsuji H, Nakanoa H (2003) Fuel consumption – derived CO2 emission under conventional and reduced tillage cropping systems in northern Japan, Agriculture Ecosystems and Environment 99: 213–219.
12. Schlesinger WH (1986) Changes in soil carbon storage and associated properties with disturbance and recovery. In: Trabalka JR, Reichle DE, editors. The changing Carbon Cycle- A global analysis. Springer Verlag, New York. 194–220.
13. Reicosky DC, Lindstrom MJ (1993) Fall tillage method: Effect on short – term carbon dioxide flux from soil. Agron J 85(6): 1237–1243.
14. Franzluebbers AJ, Hons FM, Zubberer DA (1995) Tillage induced seasonal changes in soil physical properties affecting soil CO2 evolution under intensive cropping. Soil Tillage Res 34: 41–60.
15. Ding W, Cai Y, Cai Z, Zheng X (2006) Diel pattern of soil respiration in N-amended soil under maize cultivation. Atmos Environ 40: 3294–3305.
16. Ren X, Wang Q, Tong C, Wu J, Wang K, et al. (2007) Estimation of soil respiration in a paddy ecosystem in the subtropical region of China. Chin Sci Bull 52 (19): 2722–2730.
17. Iqbal J, Ronggui H, Lijun D, Lan L, Shan L, et al. (2008) Differences in soil CO2 flux between different land use types in mid-subtropical China. Soil Bio Biochem 40: 2324–2333.
18. Liu H, Zhao P, Lu P, Wang YS, Lin YB, et al. (2008) Greenhouse gas fluxes from soils of different land-use types in a hilly area of south China. Agric Ecosyst Environ 124(1–2): 125–135.
19. Lee DK, Doolittle JJ, Owens VN (2007) Soil carbon dioxide fluxes in established switch grass land managed for biomass production. Soil Biol Biochem 39: 178–186.
20. Ladd JN, Amato M, Zhou LK, Schultz JE (1994) Differential effects of rotation, plant residue and nitrogen fertilizer on microbial biomass and organic matter in an Australian alfisol. Soil Biol Biochem 26: 821–831.
21. Edmeades DC (2003) The long-term effects of manures and fertilizers on soil productivity: a review. Nutr Cycl Agroecosys 66: 165–180.
22. Kern JS, Johnson MG (1993) Conservation tillage impacts on natural soil and atmospheric carbon levels. Soil Sci SOC Am J 57: 200–210.
23. Reeves DW (1997) The role of soil organic matter in maintaining soil quality in continuous cropping system. Soil Tillage Res 43: 131–167.
24. Smith P, Powlson DS, Glendining MJ, Smith JU (1998) Preliminary estimates of potential for carbon mitigation in European soils through no-till farming. Global Chang Biol 4: 679–685.
25. Schlesinger WH (1999) Carbon sequestration in soils. Science 248: 2095.
26. Johnson MG, Levine ER, Kern JS (1995) Soil organic matter, distribution, genesis, and management to reduce greenhouse gas emissions. Water Air Soil Pollut 82: 593–615.
27. Lal R, Follett RF, Kimble J, Cole CV (1999) Managing US crop land to sequester carbon in soil. J. Soil Water Conserv 54: 374–381.

28. Paustian K, Andren O, Janzen H, Lal R, Smith P, et al. (1997) Agricultural soils as a C sink to offset CO_2 emissions. Soil Use Manage 13: 230–244.

29. Green CJ, Blackmer AM, Horton R (1995) Nitrogen effects on conservation of carbon during corn residue decomposition in soil. Soil Sci Soc Am J 59: 453–459.

30. Kowallenko CG, Ivarson KG, Cameron DR (1978) Effect of moisture content, temperature and nitrogen fertilization on carbon dioxide evolution from field soils. Soil Biol Biochem J 10: 417–423.

31. Fogg K (1988) The effect of added nitrogen on the rate of organic matter decomposition. Biol Rev 63: 433–472.

32. Halvorson AD, Wienhold BJ, Black AL (2002) Tillage, nitrogen, and cropping system effects on soil carbon sequestration. Soil Sci Soc Am J 63: 906–912.

33. Liang BC, Mackenzie AF (1992) Changes in the soil organic carbon and nitrogen after six years of corn production. Soil Sci 153: 307–313.

34. Wang XB, Cai DX, Hoogmoed WB, Oenema O, Perdok UD (2007) Developments in conservation tillage in rainfed regions of North China. Soil & Tillage Res 93: 239–250.

35. Zhang H, Wang X, Feng Z, Pang J, Lu F, et al. (2011) Soil temperature and moisture sensitivities of soil CO_2 efflux before and after tillage in a wheat field of Loess Plateau, China. J Environ Sci 23: 79–86.

36. Lou YZ, Zhou X (2006) Soil respiration and the environment. Academic, San Diego. 110–111.

37. Iqbal J, Hu R, Lin S, Hatano R, Feng M, et al. (2009) CO_2 emission in a subtropical red paddy soil (Ultisol) as affected by straw and N- fertilizer applications: A case study in Southern China. Agriculture Ecosyst and Environ 131: 292–302.

38. Zou J, Huang Y, Zheng X, Wang Y, Chen Y (2004) Static opaque chamber – based technique for determination of net exchange of CO_2 between terrestrial ecosystem and atmosphere. Chin Sci Bull 49 (4): 381–388.

39. Reiccosky DC, Dugas WA, Torbert HA (1997) Tillage- induced soil carbon dioxide loss from different cropping system. Soil Tillage Res 41: 105–118.

40. Calderon FJ, Jackson LE, Scow KM, Rolston DE (2001) Short term dynamics of nitrogen, microbial activity, and phospholipid fatty acids after tillage. Soil Sci Soc Am J 65: 118–126.

41. Reiccosky DC (2002) Long term effect of moldboard plowing on tillage-induced CO_2 loss. In: Kimble JM. Lal R, Follett RF editors. Agricultural practices and policies for carbon sequestration in soil. CRC Press, Boca, 87–96.

42. La Scala N Jr, Lopes A, Panosso AR, Camara FT, Periera GT (2005) Soil CO_2 efflux following rotary tillage of a tropical soil. Soil Tillage Res 84: 222–225.

43. Alvaro-Fuentes J, Cantero-Martinez C, Lopez MV, Arrue JL (2007) Soil carbon dioxide fluxes following tillage in semiarid Mediterranean agro-ecosystems. Soil Tillage Res 96: 331–341.

44. Beare MH, Cabrera ML, Hendrix PF, Coleman DC (1994) Aggregate–protected and unprotected organic matter pools in conventional and no-tillage soils. Soil Sci Soc Am J 58: 787–795.

45. Kladivko EJ (2001) Tillage systems and soil ecology. Soil Tillage Res 61: 61–76.

46. Zhang HX, Zhou X, Lu F, Pang J, Feng Z, et al (2011) Seasonal dynamics of soil CO_2 efflux in a conventional tilled wheat field of the Loess Plateau, China. Ecol Res. 26: 735–743. DOI 10. 1007/s11284-011-0832-5.

47. Li H, Qiu J, Wang L, Tang H, Li C, et al. (2010) Modelling impacts of alternative farming management practices on greenhouse emissions from a winter wheat – maize rotation system in China. Agriculture, Ecosyst and Environ 135: 24–33.

48. Jacinthe PA, Lal R, Kimble JM (2002) Carbon budget and seasonal carbon dioxide emission from a Central Ohio Luvisol as influenced by wheat residue amendment. Soil & Tillage Res 67: 147–157.

49. Xiao Y, Xie G, Lu G, Ding X, Lu Y (2005) The value of gas exchange as a service by rice paddies in suburban Shanghai, PR China. Agric Ecosyst Environ 109: 273–283.

50. Chapin FS III (1991) Effects of multiple environmental stress upon nutrient availability In: Mooney HA, Winner WA, Pell EJ, editors. Integrated Responses of Plants to Environmental Stress. Academic New York. 67–86.

51. Fisk MC, Fahey TJ (2001) Microbial biomass and nitrogen cycling responses to fertilization and litter removal in young northern hardwood forests. Biogeochemistry 53: 201–223.

52. Anderson TH (1994) Physical analysis of microbial communities in soil: applications and limitations. In: Ritz K, Dighton J, Giller KE, editors. Beyond the Biomass: Compositional and Functional Analysis of Soil Microbial Communities. John Wiley and Sons. New York. 67–76.

53. Galantini J, Rosell R (2006) Long-term fertilization effects on soil organic matter quality and dynamics under different production systems in semiarid Pampean soils. Soil Tillage Res 34: 41–60.

54. Ding W, Meng L, Yin Y, Cai Z, Zheng X (2007) CO_2 emission in an intensively cultivated loam as affected by long-term application of organic manure and nitrogen fertilizers. Soil Biol Biochem 39: 669–679.

55. Raich JW, Schlesinger WH (1992) Modeling soil organic matter in organic amended and N-fertilized, long-term plots. Soil Sci Soc Am J 56: 476–488.

56. Davidson EA, Belk E, Boone RD (1998) Soil water content and temperature as independent or confounded factors controlling soil respiration in a temperate mixed hardwood forest. Glob Chang Biol 4: 217–227.

57. Reichstein M, Tenhunen JD, Roupsard O, Ourcival JM, Rambal S, et al. (2002) Ecosystem respiration in two Mediterranean evergreen Holm oak forests: drought effects and decomposition dynamics. Funct Ecol 16: 27–39.

58. Buchmann N (2000) Biotic and abiotic factors controlling soil respiration rates in *Picea abies* stands. Soil Biol Biochem 32: 1625–1635.

59. Saiz G, Black K, Reidy B, Lopez S, Farrell EP (2007) Assessment of soil CO_2 efflux and its components using a process-based model in a young temperate forest site. Geoderma 139: 79–89.

60. Bauer J, Herbst M, Huisman JA, Weihermuller L, Vereecken H (2008) Sensitivity of simulated soil heterotrophic respiration to temperature and moisture reduction functions. Geoderma 145: 17–27.

61. Bowden RD, Newkirk KM, Rullo GM (1998) Carbon dioxide and methane fluxes by a forest soil under laboratory-controlled moisture and temperature conditions. Soil Biol. Biochem 30: 1591–1597.

62. Haynes RJ (2000) Labile organic matter as an indicator of organic matter quality in arable and pastoral soils in New Zealand. Soil Biol Biochem 32: 211–219.

63. Lal R, Kimble JM (1997) Conservation tillage for carbon sequestration. Nutr. Cycling Agroecosyst 49: 243–253.

64. Lal R, Kimble JM, Follet RF, Cole CV (1998) The potential for US crop land to Sequester Carbon and Mitigate the Greenhouse Effect. Chelsea: Ann Arbor Science.

65. West TO, Post WM (2002) Soil organic carbon sequestration rates by tillage and crop rotation: a global data analysis. Soil Sci. Soc. Am J 66: 1930–1946.

66. Drury CF, Tan CS, Welacky TW, Oloya TO, Hamill AS, et al. (1999) Red clover and tillage influence on soil temperature, water content and corn emergence. Agron J 91: 101–108.

67. Hutchinson JI, Campbell CA, Desjardins RL (2007) Some perspectives on carbon sequestration in agriculture. Agric Meteorol 142: 288–302.

68. Gao C, Sun X, Cao J, Luan Y, Hao H, et al. (2008) A method and apparatus of measurement of carbon dioxide flux from soil surface from situ [J]. Journal of Beijing Forestry University 30 (2): 102–105. (In Chinese with English abstract).

69. Wilson HM, Al-kaisi MM (2008) Crop rotation and nitrogen fertilization effect on soil CO_2 emissions in Central Lova. Applied Soil Ecology 39: 264–270.

70. Bao SD (2008) Soil Agricultural Chemistry Analysis [M]. Beijing: China Agricultural press (In Chinese).

16

Profiling Nematode Communities in Unmanaged Flowerbed and Agricultural Field Soils in Japan by DNA Barcode Sequencing

Hisashi Morise, Erika Miyazaki, Shoko Yoshimitsu, Toshihiko Eki*

Molecular Genetics Laboratory, Division of Bioscience and Biotechnology, Department of Environmental and Life Sciences, Toyohashi University of Technology, Tempaku-cho, Toyohashi, Aichi, Japan

Abstract

Soil nematodes play crucial roles in the soil food web and are a suitable indicator for assessing soil environments and ecosystems. Previous nematode community analyses based on nematode morphology classification have been shown to be useful for assessing various soil environments. Here we have conducted DNA barcode analysis for soil nematode community analyses in Japanese soils. We isolated nematodes from two different environmental soils of an unmanaged flowerbed and an agricultural field using the improved flotation-sieving method. Small subunit (SSU) rDNA fragments were directly amplified from each of 68 (flowerbed samples) and 48 (field samples) isolated nematodes to determine the nucleotide sequence. Sixteen and thirteen operational taxonomic units (OTUs) were obtained by multiple sequence alignment from the flowerbed and agricultural field nematodes, respectively. All 29 SSU rDNA-derived OTUs (rOTUs) were further mapped onto a phylogenetic tree with 107 known nematode species. Interestingly, the two nematode communities examined were clearly distinct from each other in terms of trophic groups: Animal predators and plant feeders were markedly abundant in the flowerbed soils, in contrast, bacterial feeders were dominantly observed in the agricultural field soils. The data from the flowerbed nematodes suggests a possible food web among two different trophic nematode groups and plants (weeds) in the closed soil environment. Finally, DNA sequences derived from the mitochondrial cytochrome oxidase c subunit 1 (COI) gene were determined as a DNA barcode from 43 agricultural field soil nematodes. These nematodes were assigned to 13 rDNA-derived OTUs, but in the COI gene analysis were assigned to 23 COI gene-derived OTUs (cOTUs), indicating that COI gene-based barcoding may provide higher taxonomic resolution than conventional SSU rDNA-barcoding in soil nematode community analysis.

Editor: Jose Luis Balcazar, Catalan Institute for Water Research (ICRA), Spain

Funding: This work was supported in part by grants from the research program of Toyohashi University of Technology (to TE). The funders had no role in study design, data collection and analysis, decision to publish, or preparation of the manuscript. No additional external funding was received for this study.

Competing Interests: The authors have declared that no competing interests exist.

* E-mail: eki@ens.tut.ac.jp

Introduction

Nematodes are one of the most abundant metazoans on the Earth and are universally found in freshwater, terrestrial and marine environments [1] and even in the deep sea [2]. They exhibit various feeding types and survival strategies, for example, free-living bacterial and fungal feeders, predators, or animal and plant parasites. Nematodes are involved in the recycling of organic materials in soil and affect plant growth [3,4]. Therefore, nematode feeding activity contributes to maintaining the integrity of soil food web. It is also well known that nematode plant parasites inhibit plant growth and crop production in farmland soils and can cause serious damage in agriculture [5]. Nematodes are suitable indicators for monitoring soil environments and the dynamics of nematode populations reflects nutrient conditions or toxicity in the soils (e.g., reviewed in [6–8]). For instance, Urzelai et al. examined the nematode communities in contaminated soils and suggested a possible good indicator of plant parasite index for monitoring recovery processes after perturbation [9]. Heininger and the colleagues found that nematode communities in the sediments were affected by the levels of pollution and showed that the genera composition of nematodes is useful as an indicator for assessing sediment pollution [10]. Nematode communities in various soils have been extensively studied in agricultural and environmental sciences, mainly by the morphology and/or feeding habit-based classification. These analyses require excellent taxonomic skills as well as a great deal of knowledge and experience, and exhibit serious limitations in terms of taxonomic identification and low sample throughput.

To overcome these bottlenecks in traditional morphology-based analyses, several PCR- or sequence-dependent molecular biological methods (e.g., denaturing gradient gel electrophoresis (DGGE) [11–13] and DNA barcoding) and spectroscopic techniques [14] have been developed for nematode taxonomic studies (also reviewed in [15]). Especially, DNA barcoding is based on interspecific differences of nucleotide sequences from a particular DNA region (i.e., DNA barcode sequences) and has been used as a powerful tool for taxonomic identification of eukaryotes and/or for their community analysis in various ecosystems [16–18]. In these studies, DNA sequences from the small subunit (SSU) or large subunit (LSU) rDNAs, mitochondrial cytochrome c oxidase 1 (COI) gene, and the internal transcribed sequence (ITS) of rDNA have been preferentially used as DNA barcodes. DNA barcoding has been an effective tool for taxonomic and community studies of soil animals including nematodes [19,20]. The operational taxonomic units (OTUs) are generated by multiple-alignment of nematode DNA barcode sequences and nematodes aligned to the same OTU are assumed to be in the same taxonomic group. In addition, numbers of OTUs and abundance of nematodes in each OTU provide us with important qualitative and quantitative information on the nematode community: The former represents the number of taxonomic groups (i.e., variation of the species) and the latter shows the proportion of nematodes in each OTU within the entire nematode population present. In addition to taxonomic studies, a DNA barcode technique has been applied to analyzing the community structures of terrestrial and marine nematodes in various environments [21–28].

In addition to the assessment of soil environments [9,10], soil nematode community analyses have previously showed their utility in biological assessments of environmental soils such as agricultural lands [7,29–36]. However, these previous studies have been dependent on traditional morphology-based classification of nematodes by microscopic observations. So far, only a few studies using DNA barcoding have been reported for analyzing the soil nematode community [22,24,25,28].

Here, we have performed DNA barcoding community analysis of nematodes living in two different soil environments in Japan. By using the improved flotation-sieving method, nematodes were isolated efficiently from the soils. DNA fragments of SSU rDNA were amplified by PCR for direct sequencing, resulting in SSU rDNA-derived OTUs (designated as rOTUs). Taxonomic classification of nematodes in each rOTU was determined by assigning 29 rOTUs to the phylogenetic tree containing 107 reference nematode SSU rDNA sequences. The taxonomic analysis of the rOTUs revealed a strikingly contrasting nematode population in the two soil environments: plant feeding and animal predatory nematodes were most abundant in the unmanaged flowerbed soil, whereas bacterial feeding nematodes were most abundant in the soybean-cultivated field soil. Finally, community analysis of the agricultural field soil nematodes was performed using two DNA barcode genes (SSU rDNA and the COI gene). The COI barcode analysis provided a more detailed nematode community structure rather than the conventional SSU rDNA-based analysis.

Materials and Methods

Study sites and soil sampling

Soil samples were collected between November 2010-January 2011 (under comparable climatic conditions) from a flowerbed framed in by concrete blocks and an agricultural field cultivated with soybean at the campus of Toyohashi University of Technology (Toyohashi Tech.), Toyohashi, Japan (137°24′E, 34°42′N [longitude: 137.4086, latitude: 34.7017]). The area of the flowerbed and the agricultural field were 1.19 m^2 (0.7×1.7 m) and 2.38 m^2 (0.7×3.4 m), respectively. The distance between two sampling sites is approximately 400 m. The flowerbed was unmanaged and weed-strewn for several years with no addition of fertilizer or tillage regime. The experimental field was cultivated with soybean after tillage, soil neutralization, and application of chemically synthesized fertilizer. Soil on the site was sampled to a depth of 15–20 cm using a 2.5 cm-diameter soil sampling auger (Fujiwara Scientific Co., Tokyo, Japan) and more than 3 independent soil samples were taken and mixed from each site. Average pH and water content of soils were 6.9±0.2 and 26.8±1.9% (flowerbed soil) and 8.1±0.1 and 22.8±0.6% (agricultural field soil), respectively. pH was measured using soil

suspension with distilled water (soil:water = 1:2.5) and water content was determined by measuring the weight of dried soil (after drying at 65°C overnight). After removal of over-sized contaminants (e.g., stones) by passing soil samples through a 0.7 mm sieve, ca. 10 g of fresh weight soil was directly used for nematode isolation.

Nematode isolation and DNA preparation

In the early stages of this study, nematodes were isolated from soil samples by a standard Baermann funnel method [37], however, the nematode isolation procedure was revised to increase the nematode isolation efficiency. An improved flotation sieving method using colloidal silica [38] was used in this study with some modifications. In brief, 10 g soil was made up to 40 ml with distilled water in a conical tube and gently mixed. After centrifugation at 1800× g for 5 min, the supernatant was poured off and 20 ml of 50% colloidal silica (LUDOX TM-50, Sigma-Aldrich, St. Louis, MO, USA) was added to the remaining soil pellet, followed by centrifugation at 1800× g for 15 min. The resulting supernatant was passed through a 160 μm nylon sieve (47 mm diameter, Millipore, Billerica, MA, USA) equipped with a filter holder (type KP-47S, ADVANTEC, Tokyo, Japan) and the flow-through fraction was then passed through a second, 20 μm nylon sieve (Millipore). Nematodes trapped on both sieves were eluted into water in a watch glass. Individual soil nematodes were randomly picked up and transferred into DNA LoBind tubes (Eppendorf, Hamburg, Germany) under a SZX16 stereomicroscope (Olympus, Tokyo, Japan). Single nematodes in individual tubes were subjected to subsequent DNA preparation [24]. Individual nematode in a tube was heated at 99°C for 3 min in 20 μl of 0.25 M NaOH and then neutralized by adding 4 μl of 1 M HCl, 10 μl of 0.5 M Tris-HCl (pH 8.0), and 5 μl of 2% Triton X-100. DNA samples were stored at −20°C until use. Each nematode isolated from flowerbed and agricultural field soils was designated as SnTUT_k01 and SnTUT_h01 with a two-digit serial number (e.g., SnTUT_k01_01), respectively.

PCR amplifications

The 18S small subunit ribosomal RNA (SSU) gene fragments (approximately 900 bp) were amplified in 20 μl-reactions containing 10 μl of 2× PCR buffer for KOD FX Neo, 4 μl of 2 mM dNTPs, 0.4 unit of KOD FX Neo DNA polymerase (Toyobo, Tokyo, Japan), 2 μl of DNA sample, using 0.3 μM each of primers SSU18A-4F (5′-GCTTGTCTCAAAGATTAAGCCATG-CATG-3′) and SSU26Rplus4 (5′- AAGACATTCTTGG-CAAATGCTTTCG-3′). Amplification was initiated with a 2-min denaturation at 94°C followed by 35 cycles each involving denaturation at 94°C for 10 s, annealing at 55°C for 30 s, and extension at 68°C for 1 min. Prior to the experiments, several sets of primers were tested to determine if PCR products were reliably amplified from the nematode DNAs. A pair of PCR primers (SSU18A-4F and SSU26Rplus4) for the 900 bp-region of SSU rDNA (corresponding to the positions 963–1865 in the sequence of C. elegans rDNA cluster [accession no. X03680]) were selected and generated from the primers reported by Blaxter et al. [21] with modifications. The DNA fragments (approximately 400 bp) of mitochondrial COI gene were generated using primers COI_jb3-F (5′- GATTTTTTGGTCAYCCGGARGT-3′) and COI_jb5-R (5′- GYAACTACATAATAAGTRTCRTG-3′). Amplification for COI was initiated with a 2 min denaturation at 94°C, then a touch-down for 5 cycles with each cycle consisting of denaturation at 98°C for 10 s, annealing at 50, 49, 48, 47, and 46°C for 30 s, and extension at 68°C for 30 s, followed by 35 cycles of 98°C for 10 s, annealing at 45°C for 30 s, and extension at 68°C for 30 s.

We initially tested the standard PCR primer set of LCO1490 and HCO2198 [17], however, PCR amplifications of the COI gene fragment from several soil nematode DNAs failed using these primers. Therefore, we generated new primers from several nematode COI gene sequences in the database and found that our newly designed primer set (as described above), derived from the I3-M11 partition in marine nematodes [23], can amplify COI fragments more efficiently than other primers. The forward and reverse primer sequences correspond to the positions 8572–8593 and 8967–8989 of the *C. elegans* mitochondrion genome nucleotide sequence (accession no. NC_001328). PCR amplifications were visually assessed by 1% agarose gel electrophoresis using 5 μl of each reaction.

DNA sequencing

PCR products were purified from the reaction mixture using Illustra GFX PCR DNA and gel band purification kits (GE Healthcare UK Ltd., Buckinghamshire, England) following the manufacturer's protocol. When nonspecific amplification products were detected in PCR reactions, the specific products were purified from agarose gels with the kit. DNA sequences of both strands of purified PCR fragment were determined by direct sequencing. Sequencing reactions in both directions were performed using the BigDye Terminator v3.1 cycle sequencing kit with the same primers as used for PCR. Sequencing products were subsequently purified using BigDye XTerminator purification kit and analyzed in an ABI3130 genetic analyzer (Applied Biosystems, Foster City, CA, USA) following the manufacturer's instructions. Two inner primers for SSU rDNA (SSUF22: 5′-TCCAAGGAAGGCAGCAGGC-3′, SSUR09: 5′-AGCTG-GAATTACCGCGGCTG-3′) were used for re-sequencing PCR products where the chromatogram contained ambiguous sequence. Chromatograms were trimmed and assembled using the ATGC software (version 4 for Macintosh, Genetyx Co., Tokyo, Japan) to obtain consensus sequences. For assembling sequences, unidirectional sequences were only used when high quality chromatograms with no double peak patterns and high fluorescence signals were obtained. Double peaks in both strands derived from multiple SSU rDNA alleles were confirmed by re-sequencing of the PCR product. DNA sequences (approximately 700–760 bp of SSU rDNA and 370 bp of COI) were deposited at DDBJ under the following accession numbers, AB728324-AB728482.

Identification of operational taxonomic units (OTUs)

DNA sequence homologies among the nematode sequences were systematically examined using the GENETYX-MAC (version 16) and ATSQ software (Genetyx Co., Tokyo, Japan) to identify the operational taxonomic units (OTUs) containing almost identical sequences of the SSU rDNA and COI genes. Sequences with more than 99.5% identify were defined to be in the same OTU. The OTUs derived from SSU rDNA and COI gene sequences were designated as rOTU and cOTU, respectively. Each rOTU obtained from the flowerbed and agricultural field soil nematodes was named as K01rOTU and H01rOTU plus a serial two-digit number, respectively. The cOTUs were further refined by aligning the COI gene sequences of 43 nematodes isolated from agricultural field soil whose rOTUs were clarified.

Phylogenetic analysis

FASTA format files containing SSU rDNA sequences of 16 K01rOTUs, 13 H01rOTUs, 107 publicly available reference nematodes and 4 outgroup species (Table S1) were prepared and these sequences were aligned using the ClustalW2 algorithm [39] in the software SeaView (version 4.2.12 for Macintosh) [40]. After

sequence trimming, alignment of these sequences was subsequently performed using ClustalW2. A phylogenetic tree was constructed from the alignment file of SSU rDNA sequences using the neighbor-joining algorithm with Kimura two-parameter distance (bootstrap: 1000 replicates) in the SeaView program. The resultant tree file was used for drawing a cladogram using the GENETYX-Tree software (Genetyx Co., Tokyo, Japan). 107 reference nematodes were selected from previous taxonomic studies of terrestrial and marine nematodes to cover the orders in phylum Nematoda [21,24,26]. The outgroup consisted of *Dilta littoralis* (Arthropoda), *Gordius aquaticus* (Nematomorpha), *Priapulus caudatus* (Priapulida) and *Thulinia stephaniae* (Tardigrada). Phylogenetic analyses of K01rOTUs and H01rOTUs were also performed to draw unrooted trees as described above.

Results

Isolation of nematodes from the flowerbed and agricultural field soils

The soils for nematode isolation were sampled from two different sites on the campus: an isolated and unmanaged flowerbed surrounded by concrete blocks (Fig. 1A), and an agricultural field cultivated with soybean (Fig. 1B) (see Materials and Methods for details about two sampling sites). Nematodes used for these experiments were isolated from the soils at the depth of 15–20 cm by the improved method based on centrifugal flotation and sieving in place of a standard Baermann funnel method [37]. Prior to the study, we examined numbers and morphological variation of nematodes isolated by these two methods, and found that the former improved method allowed us to isolate an increased number of various kinds of nematodes compared with the Baermann funnel method. For example, although 10±8.2 nematodes were isolated from 10 g of flowerbed soil by the conventional funnel method, higher numbers of nematodes (40.7±4.2) were consistently recovered from the same soil by the improved flotation and sieving method. Nematode density in the soybean-cultivated field soil was 3 to 4 fold higher than that in the flowerbed soil (data not shown).

SSU rDNA barcode sequencing of soil nematodes and phylogenetic analysis

DNA samples were prepared by lysing nematodes with alkaline and heat treatments. 80 and 56 DNA samples were prepared from the nematodes isolated from the flowerbed and agricultural field soils, respectively. 68 and 48 nematode DNA sequences of 900 bp-SSU rDNA fragments from the flowerbed and agricultural field soils were successfully determined. Finally, the operational taxonomic units (OTUs) were generated by multiple sequence

Figure 1. Photograph of the flowerbed and agricultural field sites for soil sampling. The unmanaged flowerbed framed by concrete walls (A) and the soybean-cultivated field (B) in the campus of Toyohashi University of Technology from autumn till winter in 2010.

alignments with these SSU rDNA sequences, and 16 and 13 rOTUs (named as OTUs derived from SSU rDNA sequences) were obtained from flowerbed and agricultural field nematodes, respectively (Table 1). Each rOTU generated from the flowerbed and agricultural field nematodes was named as K01rOTU and H01rOTU with a two-digit number such as K01rOTU01 or H01rOTU01, respectively. The sequence identity between K01rOTU02a and K01rOTU02b was very close to the border value of 99.5% (i.e., 99.47%, 4 bp difference in 763 bp), and both rOTUs were distinguished by the names of K01rOTU with the same two-digit number (02) with a distinct letter. Several allelic sites were detected in the sequences from the flowerbed and agricultural field nematodes (Tables S2 and S3). These sites were reproducibly found as double peaks in the PCR product sequence chromatograms and were determined to be derived from different alleles.

The 29 rOTUs obtained from the nematodes in two different soils were further assigned into their taxonomic positions in phylum Nematoda by a phylogenetic analysis. The cladogram of the resulting SSU rDNA-based phylogenetic tree was shown in Figure 2. Twelve out of sixteen K01rOTUs from the flowerbed nematodes were mapped to the orders of Dorylamida (6 rOTUs), Mononchida (2 rOTUs), Enoplida (2 rOTUs), and Tylenchida (2 rOTUs). In contrast, 10 out of 13 H01rOTUs from the agricultural soil nematodes were assigned to the orders of Rhabditida (5 rOTUs), Araeolaimida (3 rOTUs), and Dorylamida (2 rOTUs). These clearly indicate distinct nematode communities in the two soils.

Two soil nematode communities with different trophic types

To further investigate the taxonomic differences between the two soil nematode populations, the 29 rOTUs from both samples were classified by their sequence homologies and the resulting unrooted phylogenetic tree of the rOTUs is shown together with predicted trophic types in Figure 3. One pair of rOTUs (i.e., K01rOTU05 and H01rOTU04) was common between the two nematode communities, whereas all other rOTUs identified were not common between the two soils. According to a previous study on nematode feeding behaviors [41], we determined feeding habits (trophic types) for each rOTU as those of the nematode species with the highest homologies to the rOTU sequence by the BLAST search (Tables S2 and S3). Many nematodes from the flowerbed soil were assigned to animal predation and plant feeding trophic types. On the contrary, nematodes from the agricultural field soil (closed squares) were mostly related to bacterial feeding nematodes (bacteriovores) (Fig. 3). For example, the SSU rDNA barcode sequence of K01rOTU01, containing a large number (18) of members, completely matches those of the representative plant-parasitic nematode *Xiphinema* species [42] by homology search (100% sequence identity in the entire 766 bp-region). Two related rOTUs of K01rOTU2a and K01rOTU2b containing a total of 24 isolates shared the highest sequence homologies with the SSU rDNA sequences of the predatory nematode species of the genus Mylonchulus (Table S2) [43]. About 60% of isolates (42 out of 68) from the flowerbed soil belong to these three K01rOTUs (01, 02a and 02b). Based on the SSU rDNA sequence homology search, 6 and 4 nematodes in the next largest K01rOTUs (K01rOTU03 and K01rOTU04) were also assigned to the genera Tripylella and Coslenchus, with predacious and plant feeding behaviors [41], respectively.

In the soybean-cultivated field soil, the six H01rOTUs containing more than 5 isolates were assigned to genera in the families Cephalobidae (H01rOTU01, H01rOTU03, and H01rOTU04), Rhabditidae (H01rOTU02 and H01rOTU05), and Plectidae (H01rOTU06). According to a previous study [41,44], these six H01rOTUs, representing approximately 60% of total sequenced nematodes from the agricultural field soil, were strongly suggested to have bacterial feeding habits (Table S3).

SSU- and COI-barcode analyses of the nematode community in the agricultural field soil

In this study, we used a SSU rDNA sequence as a DNA barcode for analyzing the nematode community because SSU rDNA sequences have been widely used in previous studies and a large number of SSU rDNA sequences of nematode species are available in the public database. The mitochondrial COI gene sequence has also been a popular choice as a DNA barcode for phylogenetic analysis and community analysis in animals. Although there are limited numbers of nematode COI gene sequences deposited in the database, some studies have suggested that mitochondrial COI gene barcoding was useful for community analysis of nematodes [22,23,26]. Therefore, we performed a phylogenetic analysis of soil nematodes from the agricultural field soil using the COI gene barcode. By using the newly-designed primers (see Materials and Methods for details), PCR products of approximately 400 bp were obtained in the reactions in a relatively reproducible fashion. Then, we have determined the partial COI gene sequences from 48 nematodes (in Table S3) from the soybean-cultivated field in addition to their SSU rDNA barcode sequences. 43 out of 48 COI gene sequences (approximately 370 bp) were successfully recovered and aligned for further analysis. Twenty-three COI gene-derived OTUs (designated as H01cOTUs) were generated from the agricultural field nematodes (Table S4). By comparing with the 13 H01rOTUs generated in this study, almost twice as many H01cOTUs were generated from the same set of nematode DNAs, suggesting that the COI gene-based barcoding provides the nematode community analysis with higher resolution than the conventional SSU rDNA-based barcoding. Many of the H01cOTUs (16 of 23 H01cOTUs) only contained a single nematode as a member. The H01rOTUs and H01cOTUs obtained from 43 individual nematodes are summarized in Table 2. Eight out of 13 H01rOTUs include multiple H01cOTUs. Most of the H01cOTUs correspond to distinct H01rOTUs, however, 5 nematodes assigned to H01cOTU02 were separately classified to 3 different H01rOTUs: one isolate each for H01rOTU03 and H01rOTU04, and 3 isolates to H01rOTU05, respectively. This complicated pattern in SSU rDNA- and COI gene-based OTUs has been also observed in the previous taxonomic study of tardigrade specimens by Blaxter et al. [22], and may correspond to particularly variable single, or diverging taxonomic groups. This pattern may reflect divergence of matrilineal lines due to asexual reproduction in nematodes as described previously [22].

Discussion

DNA barcode analysis is a powerful tool for clarifying the composition and taxonomic identity of organisms in the environment. Previous studies of DNA barcoding have been mainly focused on taxonomic identification and classification of organisms. In this study, we have performed DNA barcode-based analyses of the nematode community living in two ordinary Japanese soil environments: an unmanaged and isolated flowerbed and a typical agricultural field cultivated with soybean.

Since efficient isolation of the nematodes living in soils is very important for further community analysis, prior to the study, we referred to previous studies on improved isolation of soil

Table 1. Summary of SSU rDNA barcode analysis of nematodes isolated from flowerbed and agricultural field soils.

rOTU[a]	No. of isolates	Nematodes belonging to each rOTU[b]	Feeding types[c]	Accession no.
K01rOTU01	18	SnTUT_k01_01, 05, 06, 08, 11, 12, 16, 18, 19, 24, 26, 29, 44, 52, 54, 65, 69, 75	plant feeding (1d)	AB728372–AB728389
K01rOTU02a	18	SnTUT_k01_03, 07, 14, 21, 32, 41, 42, 45, 51, 56, 63, 64, 66, 67, 68, 74, 76, 80	animal predation (5a)	AB728390–AB728407
K01rOTU02b	6	SnTUT_k01_22, 43, 55, 58, 59, 62	animal predation (5a)	AB728408–AB728413
K01rOTU03	6	SnTUT_k01_23, 33, 46, 49, 60, 61	animal predation (5)	AB728414–AB728419
K01rOTU04	4	SnTUT_k01_28, 40, 53, 73	plant feeding (1e)	AB728420–AB728423
K01rOTU05	3	SnTUT_k01_02, 20, 48	bacterial feeding (3)	AB728424–AB728426
K01rOTU06	3	SnTUT_k01_25, 36, 71	animal predation (5)	AB728427–AB728429
K01rOTU07	2	SnTUT_k01_70, 77	plant feeding (1d)	AB728430, AB728431
K01rOTU08	1	SnTUT_k01_34	plant feeding (1d), animal predation (5), omnivorous (8)	AB728432
K01rOTU09	1	SnTUT_k01_35	bacterial feeding (3)	AB728433
K01rOTU10	1	SnTUT_k01_37	plant feeding (1), hyphal feeding (2)?	AB728434
K01rOTU11	1	SnTUT_k01_39	animal predation (5), omnivorous (8)	AB728435
K01rOTU12	1	SnTUT_k01_47	-	AB728436
K01rOTU13	1	SnTUT_k01_50	-	AB728437
K01rOTU14	1	SnTUT_k01_57	-	AB728438
K01rOTU15	1	SnTUT_k01_72	-	AB728439
H01rOTU01	8	SnTUT_h01_02, 15, 21, 29, 38, 41, 43, 46	bacterial feeding (3)	AB728324–AB728331
H01rOTU02	6	SnTUT_h01_05, 19, 24, 26, 27, 40	bacterial feeding[d]	AB728332–AB728337
H01rOTU03	6	SnTUT_h01_22, 25, 30, 48, 54, 55	bacterial feeding (3)	AB728338–AB728343
H01rOTU04	5	SnTUT_h01_08, 23, 47, 50, 52	bacterial feeding (3)	AB728344–AB728348
H01rOTU05	5	SnTUT_h01_03, 14, 16, 33, 37	bacterial feeding (3)	AB728349–AB728353
H01rOTU06	5	SnTUT_h01_10, 13, 42, 51, 53	bacterial feeding (3)	AB728354–AB728358
H01rOTU07	3	SnTUT_h01_20, 44, 45	bacterial feeding (3)	AB728359–AB728361
H01rOTU08	2	SnTUT_h01_09, 35	bacterial feeding (3)	AB728362, AB728363
H01rOTU09	2	SnTUT_h01_04, 18	animal predation (5a)	AB728364, AB728365
H01rOTU10	2	SnTUT_h01_31, 32	-	AB728366, AB728367
H01rOTU11	2	SnTUT_h01_36, 39	hyphal feeding (2) or plant feeding (1e)	AB728368, AB728369
H01rOTU12	1	SnTUT_h01_01	-	AB728370
H01rOTU13	1	SnTUT_h01_17	animal predation (5)	AB728371

[a]Codes "K01", "H01" and "rOTU" represent "flowerbed sample with experimental code" , "agricultural field sample with experimental code" and "SSU rDNA-derived OTU", respectively. Two-digit serial numbers were assigned in order of the number of nematodes in the rOTU.
[b]Nematodes isolated from flowerbed and agricultural field soils were designated as SnTUT_k01_a two-digit serial number and SnTUT_h01_a two-digit serial number, respectively.
[c]Feeding types were derived from those of nematode species with the highest sequence homologies by Blast search according to the references in Yeates et al. (1993) [41]. Numbers and letters in parentheses indicate feeding types in the reference.
[d]Feeding type for H01rOTU02 was derived from the other reference [44].

nematodes with colloidal silica extraction and/or sieving [38,45–49], and further modified a flotation and sieving method in place of the commonly used Baermann funnel method [37]. By using this method, higher numbers (ca. 3–4 fold) of total nematodes were obtained when compared with the Baermann funnel method. In addition, nematodes closely related to the family Hemicycliophoridae with relatively poor locomotion were successfully isolated only by the improved method (e.g., two nematodes in K01rOTU07) (Table S2). Due to improved recovery of nematodes, the nematode population isolated in this study is considered to be suitable for the community analysis.

Nematode community analyses are useful for assessing soil environments and several extensive biological assessments have been conducted for various soil environments in Europe, the United States, Asia, Australia and New Zealand [7,29–36]. These analyses were performed based on the traditional morphological classification of nematodes under the microscope. In spite of recent progress in DNA barcoding, a limited number of studies on terrestrial nematode community analyses have been reported. To date, SSU rDNA barcoding to assess the community structure of soil nematodes have been undertaken from a barley field soil [25], a hill farm grassland soil [24], and moss ecosystems in the UK

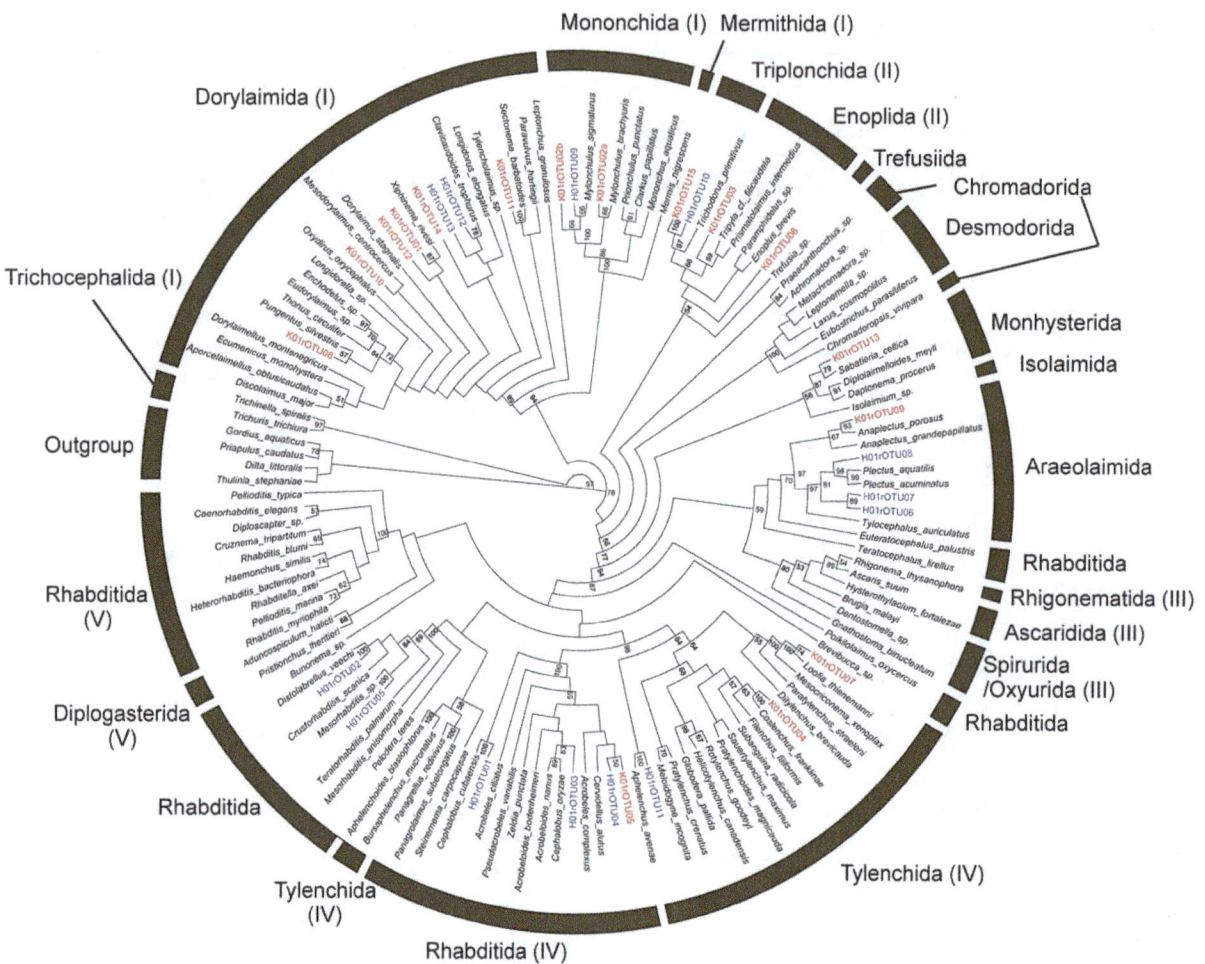

Figure 2. Neighbor-joining tree of SSU rDNA barcode sequences of soil nematodes and reference nematode species. SSU rDNA barcode sequences of 29 rOTUs from flowerbed and agricultural field soil nematodes (designated with K01rOTU and H01rOTU, and shown in red and blue, respectively) were analyzed with the corresponding SSU rDNA sequences of 107 reference nematodes (Table S1) and the resultant tree is displayed as a cladogram. Orders corresponding to the reference nematode species are indicated on the outside of the cladogram. The clade numbers (I–V) in the previous phylogenetic tree [21] were also indicated in parenthesis. Numbers on nodes are bootstrap values (>50%). *Dilta littoralis* (Arthropoda), *Gordius aquaticus* (Nematomorpha), *Priapulus caudatus* (Priapulida), *Thulinia stephaniae* (Tardigrada) were used as outgroup species.

[22]. In addition, Powers and colleagues applied the SSU-based DNA barcoding to analyze nematode communities in a tropical rainforest in Costa Rica and found potentially novel taxonomic groups [28]. Although DNA barcode-mediated taxonomic identifications of soil nematode species have been reported in Japan [50–52], nematode community analyses in Japanese soils with DNA barcoding have not been undertaken. In this study, we first analyzed the nematode communities in two different types of soils in Japan by DNA barcoding, and found contrasting community structures in these soils. The flowerbed soil was dominated by animal predators and plant feeders that occupied more than 80% of total nematodes (Fig. 3 and Table 1), strongly suggesting a possible food web including two nematode populations and plants (weeds) in the unmanaged and isolated flowerbed soil environment. Plant feeding nematodes propagate by living with weeds in the flowerbed and predators likely prey on these plant-parasitic nematodes in a possible food chain.

In the soybean-cultivated field soil, bacterivores were the dominant nematode type isolated (Fig. 3 and Table 1). Previous studies have examined nematode communities in agricultural soils.

Okada and Harada analyzed the taxonomy of nematodes in a Japanese soybean field with different treatments by morphological classification in order to examine the effects of tillage and fertilizer [32]. In this study, two abundant nematode genera Acrobeloides (bacterial feeder) and Pratylenchus (root feeder) were observed as well as three other predominant bacterial feeders of the families Rhabditidae, Plectidae, and Prismatolaimidae. Baird and Bernard also examined the population and dynamics of nematodes in soybean-wheat cropping regimes and reported that approximately 80% of total nematodes were almost evenly classified to three groups, plant-parasites, Dorylaimida (the order including plant-parasites, omnivores and fungivores) and bacterivores [53]. Neher et al. characterized the soil nematode communities in three different ecosystems (agricultural land, forest and wetland) [54]. They also showed abundant bacterial feeding nematodes in the families Cephalobidae and Rhabditidae in agricultural soils, similar to those observed in the soybean-cultivated field soil in our study. From these previous studies and our own current study, we have shown that abundant bacterial feeding nematodes are characteristically observed in agricultural soils.

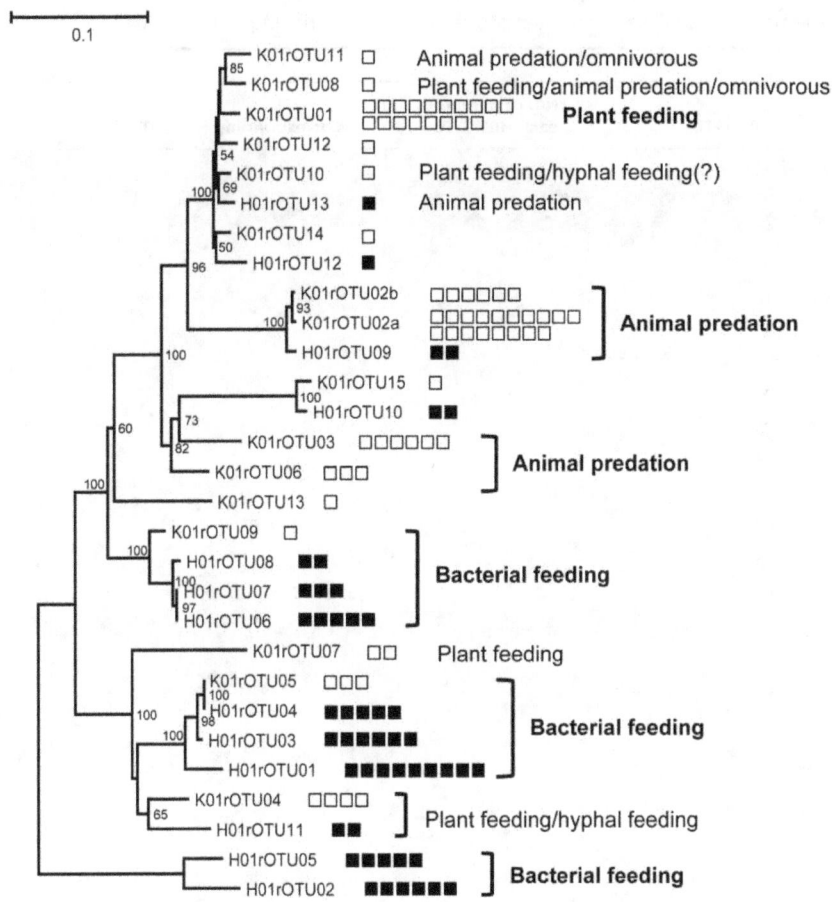

Figure 3. An unrooted phylogram of rOTUs from soil nematodes isolated from the flowerbed and agricultural field with predicted trophic types. Twenty-nine SSU rDNA barcode sequences of K01rOTUs (flowerbed samples) and H01rOTUs (agricultural field samples) were aligned for preparing a phylogenetic tree. The numbers of soil nematodes belonging to each rOTU correspond to the numbers of open (flowerbed samples) and closed (agricultural field samples) squares at the right. Numbers on nodes are bootstrap values (>50%). Trophic types indicated in the rOTUs were derived from those of the nematode species with the highest homology in SSU rDNA barcode sequences (Table 1), and the feeding types for the rOTUs containing a large number of nematode members are shown in boldface. Bar: 0.1 substitutions per site.

Interestingly, Neher et al. found nematodes from the families Criconematidae, Hoplolaimidae, and Oxydiridae (plant-parasites) and Nygolaimidae (predators) as major populations in undisturbed wetland soil [54]. Unlike in agricultural soils, plant feeder- and predator-dominant community structure observed in the wetland was commonly found in the unmanaged flowerbed in our study (Fig. 3). The distinct composition of nematodes in these ecosystems may reflect the difference in microbial biomass. Previous studies showed that a high abundance of bacterivores is associated with an increase in microbial organisms [33] and soil organic matter [55]. Yeates and King also examined the nematode populations in native and improved (fertilized) grasslands in Australia and found a significant increase in the proportion of bacterial feeding nematodes in the improved soils [34]. Although we did not measure microbial biomass and nutrients in our soils, a fertilized and soybean-cultivated field soil likely contains more abundant bacteria that are enough to support a large number of bacterivores rather than a weed-grown flowerbed soil without any supply of fertilizer. Unlike agricultural field soils with abundant biomass, bacterial feeding nematodes are unlikely to flourish in soils with limited biomass such as wetland and isolated flowerbed soils, however, plant-parasites that can live with some weeds and their predators could survive and grow in these soils.

In this study, we have compared the classification of a set of soil nematodes from the agricultural field by SSU rDNA- and COI gene-barcoding. As shown in Tables 2 and S4, COI gene sequences from 43 soil nematodes were classified to 23 cOTUs, a significant increase compared to the 13 rOTUs from SSU rDNA barcode sequencing. Nassonova et al. [56] recently reported that the COI-based barcoding of amoebae provided a higher taxonomic resolution than the SSU rDNA-based analysis, and it may be in agreement with our results. On the other hands, Blaxter et al. [22] compared the SSU rDNA- and COI-OTUs obtained from 82 tardigrades from moss ecosystems and showed that the larger number of SSU rDNA-OTUs (23) was detected than 17 of COI-OTUs. The disagreement among the studies may be derived from the difference of examined organisms or the different COI gene regions used. We also found some issues with using the COI gene-barcoding approach. For example, PCR with the conventional primer set (LCO1490 and HCO2198) from the M1-M6 region of COI gene [17] worked poorly with soil nematode DNAs under our experimental conditions and we had to develop novel primers (COI_jb3-F and COI_jb5-R) from another region of the COI gene (I3-M11 region) [23]. Derycke et al. also reported inefficient PCR with the conventional COI gene primers for barcode analysis of marine nematodes and developed new primers

Table 2. Summary of SSU rDNA- and COI gene-barcode analyses using nematodes from the agricultural field soil.

H01rOTU[a]	Name of nematode	H01cOTU	No. of nematodes in each H01rOTU	Corresponding H01cOTUs[b]
01	SnTUT_h01_02	01	8	01, 19, 21
	SnTUT_h01_15	01		
	SnTUT_h01_21	01		
	SnTUT_h01_29	01		
	SnTUT_h01_38	19		
	SnTUT_h01_41	21		
	SnTUT_h01_43	01		
	SnTUT_h01_46	01		
02	SnTUT_h01_05	03	5	03, 15
	SnTUT_h01_19	15		
	SnTUT_h01_24	03		
	SnTUT_h01_26	03		
	SnTUT_h01_27	03		
03	SnTUT_h01_22	05	5	*02*[c], 05
	SnTUT_h01_25	05		
	SnTUT_h01_30	05		
	SnTUT_h01_48	02		
	SnTUT_h01_54	05		
04	SnTUT_h01_08	10	5	*02*, 06, 10, 16, 23
	SnTUT_h01_23	16		
	SnTUT_h01_47	02		
	SnTUT_h01_50	23		
	SnTUT_h01_52	06		
05	SnTUT_h01_03	02	4	*02*, 12
	SnTUT_h01_16	12		
	SnTUT_h01_33	02		
	SnTUT_h01_37	02		
06	SnTUT_h01_10	11	5	04, 11
	SnTUT_h01_13	04		
	SnTUT_h01_42	04		
	SnTUT_h01_51	04		
	SnTUT_h01_53	04		
07	SnTUT_h01_20	08	3	08, 22
	SnTUT_h01_44	22		
	SnTUT_h01_45	08		
08	SnTUT_h01_09	07	2	07
	SnTUT_h01_35	07		
09	SnTUT_h01_18	14	1	14
10	SnTUT_h01_32	17	1	17
11	SnTUT_h01_36	18	2	18, 20
	SnTUT_h01_39	20		
12	SnTUT_h01_01	09	1	09
13	SnTUT_h01_17	13	1	13

[a]Two-digit number of each H01rOTU is indicated. The description of two-digit number was omitted in the following rows containing nematodes belonging to the same H01rOTU.
[b]Nematodes in H01rOTU in the left-most column belong to the H01cOTUs with two-digit numbers shown.
[c]Nematodes in H01cOTU02 were separately assigned to H01rOTU03-H01rOTU05 and are indicated in italics.

from the I3-M11 region [23]. Although we improved the primers and reaction conditions and succeeded in increasing the efficiency of PCR amplification in this study, this primer set is not still perfect for COI gene-barcoding studies. Amplification of the COI gene requires complicated PCR cycles and amplification is less efficient when compared with amplification of SSU rDNA fragments. Further improvements to the method may be required to optimize PCR for preparing COI gene fragments from various nematode species.

Recently, high-throughput next generation DNA sequencing has been used to accelerate DNA barcode-based community analyses for various eukaryotes to assess ecosystem biodiversity [57,58]. This analysis primarily depends on cluster analysis with large numbers of sequences of PCR-amplified DNA fragments from organisms in the community and enables researchers to analyze a large number of organisms at once unlike a single isolate-based DNA barcode sequencing. Metagenomic analyses for nematode diversity were extensively performed using the next generation sequencer (the 454 GS FLX) by Porazinska et al. [59–61]. In these analyses, huge amounts of nucleotide sequence data were generated from PCR-amplified SSU rDNA fragments (ca. 400 bp in size). Unlike conventional DNA barcode analyses, thousands of OTUs have been obtained by clustering the sequences (more than 200 bp in each length). Although the nematode communities from multiple soil samples can be achieved in a high-throughput fashion, considerable issues still remain in these metagenomic analyses such as contamination by interspecific and intraspecific chimeric sequences and unclassified sequences and possible biased amplification of PCR products [60]. In place of traditional morphology-based methods, a conventional isolate-based DNA barcoding is sufficient to determine limited but accurate community structures of soil nematodes in ecosystems and agricultural soils for probing and assessing soil environments.

Supporting Information

Table S1 Nematode species used in the phylogenetic analysis and accession numbers.

Table S2   Summary of H01rOTUs of nematodes isolated from a flowerbed soil.

Table S3   Summary of H01rOTUs of nematodes isolated from an agricultural field soil.

Table S4   Summary of COI gene barcode analysis of nematodes isolated from an agricultural field soil.

Acknowledgments

The authors thank Prof. Masahiko Saigusa (Research Center for Agrotechnology and Biotechnology, Toyohashi University of Technology) for technical comments and support, and other members of the Eki laboratory for encouragement.

Author Contributions

Conceived and designed the experiments: HM TE. Performed the experiments: HM EM SY. Analyzed the data: HM TE. Contributed reagents/materials/analysis tools: HM EM. Wrote the paper: TE.

References

1. Boag B, Yeates GW (1998) Soil nematode biodiversity in terrestrial ecosystems. Biodivers Conserv 7: 617–630.
2. Vanreusel A, De Groote A, Gollner S, Bright M (2010) Ecology and biogeography of free-living nematodes associated with chemosynthetic environments in the deep sea: a review. PLoS One 5: e12449.
3. Ferris H, Venette RC, Scow KM (2004) Soil management to enhance bacterivore and fungivore nematode populations and their nitrogen mineralisation function. Appl Soil Ecol 25: 19–35.
4. Procter DLC (1990) Global overview of the functional roles of soil-living nematodes in terrestrial communities and ecosystems. J Nematol 22: 1–7.
5. Neher DA (2010) Ecology of plant and free-living nematodes in natural and agricultural soil. Annu Rev Phytopathol 48: 371–394.
6. Eki T (2011) Ecotoxicology of free-living soil nematode *Caenorhabditis elegans*. In: Visser JE, editor. Ecotoxicology around the Grobe. Hauppauge, NY: Nove Science Publishers. pp. 1–51.
7. Ritz K, Trudgill DL (1999) Utility of nematode community analysis as an integrated measure of the functional state of soils: perspectives and challenges— discussion paper. Plant Soil 212: 1–11.
8. Sochová I, Hofman J, Holoubek I (2006) Using nematodes in soil ecotoxicology. Environ Int 32: 374–383.
9. Urzelai A, Hernández AJ, Pastor J (2000) Biotic indices based on soil nematode communities for assessing soil quality in terrestrial ecosystems. Sci Total Environ 247: 253–261.
10. Heininger P, Hoss S, Claus E, Pelzer J, Traunspurger W (2007) Nematode communities in contaminated river sediments. Environ Pollut 146: 64–76.
11. Foucher ALJL., Bongers T, Noble LR, Wilson MJ (2004) Assessment of nematode biodiversity using DGGE of 18S rDNA following extraction of nematodes from soil. Soil Biol Biochem 36: 2027–2032.
12. Okada H, Oba H (2008) Comparison of nematode community similarities assessed by polymerase chain reaction-denaturing gradient gel electrophoresis (DGGE) and by morphological identification. Nematology 10: 689–700.
13. Waite IS, O'Donnell AG, Harrison A, Davies JT, Colvan SR, et al. (2003) Design and evaluation of nematode 18S rDNA primers for PCR and denaturing gradient gel electrophoresis (DGGE) of soil community DNA. Soil Biol Biochem 35: 1165–1173.
14. Barthès BG, Brunet D, Rabary B, Ba O, Villenave C (2011) Near infrared reflectance spectroscopy (NIRS) could be used for characterization of soil nematode community. Soil Biol Biochem 43: 1649–1659.
15. Abebe E, Mekete T, Thomas WK (2011) A critique of current methods in nematode taxonomy. Afr J Biotechnol 10: 312–323.
16. Blaxter ML (2004) The promise of a DNA taxonomy. Philos Trans R Soc Lond B Biol Sci 359: 669–679.
17. Hebert PD, Cywinska A, Ball SL, deWaard JR (2003) Biological identifications through DNA barcodes. Proc Biol Sci 270: 313–321.
18. Valentini A, Pompanon F, Taberlet P (2009) DNA barcoding for ecologists. Trends Ecol Evol 24: 110–117.
19. Hamilton H, Strickland MS, Wickings K, Bradford MA, Fierer N (2009) Surveying soil faunal communities using a direct molecular approach. Soil Biol Biochem 41: 1311–1314.
20. Wu T, Ayres E, KLi G, Bardgett RD, Wall DH, et al. (2009) Molecular profiling of soil animal diversity in natural ecosystems: Incongruence of molecular and morphological results. Soil Biol Biochem 41: 849–857.
21. Blaxter ML, De Ley P, Garey JR, Liu LX, Scheldeman P, et al. (1998) A molecular evolutionary framework for the phylum Nematoda. Nature 392: 71–75.
22. Blaxter M, Mann J, Chapman T, Thomas F, Whitton C, et al. (2005) Defining operational taxonomic units using DNA barcode data. Philos Trans R Soc Lond B Biol Sci 360: 1935–1943.
23. Derycke S, Vanaverbeke J, Rigaux A, Backeljau T, Moens T (2010) Exploring the use of cytochrome oxidase c subunit 1 (COI) for DNA barcoding of free-living marine nematodes. PLoS One 5: e13716.
24. Floyd R, Abebe E, Papert A, Blaxter M (2002) Molecular barcodes for soil nematode identification. Mol Ecol 11: 839–850.
25. Griffiths BS, Donn S, Neilson R, Daniell TJ (2006) Molecular sequencing and morphological analysis of a nematode community. Appl Soil Ecol 32: 325–337.
26. Holterman M, van der Wurff A, van den Elsen S, van Megen H, Bongers T, et al. (2006) Phylum-wide analysis of SSU rDNA reveals deep phylogenetic relationships among nematodes and accelerated evolution toward crown Clades. Mol Biol Evol 23: 1792–1800.
27. Holterman M, Holovachov O, van den Elsen S, van Megen H, Bongers T, et al. (2008) Small subunit ribosomal DNA-based phylogeny of basal Chromadoria (Nematoda) suggests that transitions from marine to terrestrial habitats (and *vice versa*) require relatively simple adaptations. Mol Phylogenet Evol 48: 758–763.
28. Powers TO, Neher DA, Mullin P, Esquivel A, Giblin-Davis RM, et al. (2009) Tropical nematode diversity: vertical stratification of nematode communities in a Costa Rican humid lowland rainforest. Mol Ecol 18: 985–996.
29. Bongers T, Ferris H (1999) Nematode community structure as a bioindicator in environmental monitoring. Trends Ecol Evol 14: 224–228.

30. Liang W, Li Q, Jiang Y, Neher DA (2005) Nematode faunal analysis in an aquic brown soil fertilised with slow-release urea, Northeast China. Appl Soil Ecol 29: 185–192.

31. Neher DA (2001) Role of nematodes in soil health and their use as indicators. J Nematol 33: 161–168.

32. Okada H, Harada H (2007) Effects of tillage and fertilizer on nematode communities in a Japanese soybean field. Appl Soil Ecol 35: 582–598.

33. Wardle DA, Yeates GW, Watson RN, Nicholson KS (1995) Development of decomposer food-web, trophic relationships, and ecosystem properties during a three-year primary succession in sawdust. Oikos 73: 155–166.

34. Yeates GW, King KL (1997) Soil nematodes as indicators of the effects of management on grasslands in the New England tablelands (NSW): comparison of native and improved grasslands. Pedobiologia 41: 526–536.

35. Yeates GW, Bongers T (1999) Nematode diversity in agroecosystems. Agric Ecosyst Environ 74: 113–135.

36. Yeates G (2003) Nematodes as soil indicators: functional and biodiversity aspects. Biol Fertil Soils 37: 199–210.

37. Baermann G (1917) Eine einfache Methode zur Auffindung von Ancylostomum (Nematoden) Larven in Erdproben. Geneeskd Tijdschr Ned Indie 57: 131–137.

38. Griffiths BS, Boag B, Neilson R, Palmer L (1990) The use of colloidal silica to extract nematodes from small samples of soil or sediment. Nematologica 36: 465–473.

39. Larkin MA, Blackshields G, Brown NP, Chenna R, McGettigan PA, et al. (2007) Clustal W and Clustal X version 2.0. Bioinformatics 23: 2947–2948.

40. Gouy M, Guindon S, Gascuel O (2010) SeaView version 4: A multiplatform graphical user interface for sequence alignment and phylogenetic tree building. Mol Biol Evol 27: 221–224.

41. Yeates GW, Bongers T, De Goede RG, Freckman DW, Georgieva SS (1993) Feeding habits in soil nematode families and genera-an outline for soil ecologists. J Nematol 25: 315–331.

42. Oliveira CM, Hubschen J, Brown DJ, Ferraz LC, Wright F, et al. (2004) Phylogenetic relationships among *Xiphinema* and *Xiphidorus* nematode species from Brazil inferred from 18S rDNA sequences. J Nematol 36: 153–159.

43. Jairajupuri MS, Azumi MI (1978) Some studies on the predatory behavior of *Mylonchulus dentatus*. Nematol Mediterr 6: 205–212.

44. Carta LK (2000) Bacterial-feeding nematode growth and preference for biocontrol isolates of the bacterium *Burkholderia cepacia*. J Nematol 32: 362–369.

45. Brown DJF, Boag B (1988) An examination of methods used to extract virus-vector nematodes (Nematoda: Longidoridae and Trichodoridae) from soil samples. Nematol Medit 16: 93–99.

46. Coolen WA, D'Herde CJ (1977) Extraction de *Longidrous* et *Xiphinema* spp. du sol par centrifugation en utilsant du silice colloidal. Nematologica Mediterranea 5: 195–206.

47. McSorley R, Parrado JL (1987) Nematode losses during centrifugal extraction from two soil types. Nematropica 17: 147–161.

48. McSorley R, Frederick JJ (2004) Effect of extraction method on perceived composition of the soil nematode community. Appl Soil Ecol 27: 55–63.

49. Rodríguez-Kábana R, Pope MH (1981) A simple incubation method for the extraction of nematodes from soil. Nematropica 11: 175–186.

50. Olia M, Ahmad W, Araki M, Minaka N, Oba H, Okada H. (2008) *Actus salvadoricus* Baqri and Jairajpuri (Mononchida: Mylonchulidae) from Japan with comment on the phylogenetic position of the genus *Actus* based on 18S rDNA sequences. Jpn J Nematol 38: 57–69.

51. Olia M, Ahmad W, Araki M, Minaka N (2009) Molecular characterization of some species of *Mylonchulus* (Nematode: Mononchida) from Japan and comments on the status of *Paramylonchulus* and *Paknylonchulus*. Nematology 11: 337–342.

52. Sakai H, Takeda A, Mizukubo T (2011) First report of *Xiphinema brevicolle* Lordello et Costa, 1961 (Nematoda, Longidoridae) in Japan. Zookeys: 21–40.

53. Baird SM, Bernard EC (1984) Nematode population and community dynamics in soybean-wheat cropping and tillage regimes. J Nematol 16: 379–386.

54. Neher DA, Wu J, Barbercheck ME, Anas O (2005) Ecosystem type affects interpretation of soil nematode community measures. Appl Soil Ecol 30: 47–64.

55. Griffiths BS, Ritz K, Wheatley RE (1994) Nematodes as indicators of enhanced microbiological activity in a Scottish organic farming. Soil Use Manage 10: 20–24.

56. Nassonova E, Smirnov A, Fahrni J, Pawlowski J (2010) Barcoding amoebae: comparison of SSU, ITS and COI genes as tools for molecular identification of naked lobose amoebae. Protist 161: 102–115.

57. Taylor HR, Harris WE (2012) An emergent science on the brink of irrelevance: a review of the past 8 years of DNA barcoding. Mol Ecol Resour 12: 377–388.

58. Timmermans MJ, Dodsworth S, Culverwell CL, Bocak L, Ahrens D, et al. (2010) Why barcode? High-throughput multiplex sequencing of mitochondrial genomes for molecular systematics. Nucleic Acids Res 38: e197.

59. Porazinska DL, Giblin-Davis RM, Faller L, Farmerie W, Kanzaki N, et al. (2009) Evaluating high-throughput sequencing as a method for metagenomic analysis of nematode diversity. Mol Ecol Resour 9: 1439–1450.

60. Porazinska DL, Giblin-Davis RM, Esquivel A, Powers TO, Sung W, et al. (2010) Ecometagenetics confirm high tropical rainforest nematode diversity. Mol Ecol 19: 5521–5530.

61. Porazinska DL, Sung W, Giblin-Davis RM, Thomas WK (2010) Reproducibility of read numbers in high-throughput sequencing analysis of nematode community composition and structure. Mol Ecol Resour 10: 666–676.

17

Isolation and Characterization of Carbendazim-degrading *Rhodococcus erythropolis* djl-11

Xinjian Zhang[1⊙], **Yujie Huang**[1⊙], **Paul R. Harvey**[1,2], **Hongmei Li**[1], **Yan Ren**[1], **Jishun Li**[1], **Jianing Wang**[1], **Hetong Yang**[1]*

1 Shandong Provincial Key Laboratory of Applied Microbiology, Biotechnology Center of Shandong Academy of Sciences, Jinan, Shandong Province, People's Republic of China, **2** CSIRO Sustainable Agriculture National Research Flagship and CSIRO Ecosystem Sciences, Glen Osmond, South Australia, Australia

Abstract

Carbendazim (methyl *1H*-benzimidazol-2-yl carbamate) is one of the most widely used fungicides in agriculture worldwide, but has been reported to have adverse effects on animal health and ecosystem function. A highly efficient carbendazim-degrading bacterium (strain djl-11) was isolated from carbendazim-contaminated soil samples via enrichment culture. Strain djl-11 was identified as *Rhodococcus erythropolis* based on morphological, physiological and biochemical characters, including sequence analysis of the 16S rRNA gene. *In vitro* degradation of carbendazim (1000 mg·L^{-1}) by djl-11 in minimal salts medium (MSM) was highly efficient, and with an average degradation rate of 333.33 mg·L^{-1}·d^{-1} at 28°C. The optimal temperature range for carbendazim degradation by djl-11 in MSM was 25–30°C. Whilst strain djl-11 was capable of metabolizing cabendazim as the sole source of carbon and nitrogen, degradation was significantly ($P<0.05$) increased by addition of 12.5 mM NH_4NO_3. Changes in MSM pH (4–9), substitution of NH_4NO_3 with organic substrates as N and C sources or replacing Mg^{2+} with Mn^{2+}, Zn^{2+} or Fe^{2+} did not significantly affect carbendazim degradation by djl-11. During the degradation process, liquid chromatography-mass spectrometry (LC-MS) detected the metabolites 2-aminobenzimidazole and 2-hydroxybenzimidazole. A putative carbendazim-hydrolyzing esterase gene was cloned from chromosomal DNA of djl-11 and showed 99% sequence homology to the *mhel* carbendazim-hydrolyzing esterase gene from *Nocardioides* sp. SG-4G.

Editor: Stephen J. Johnson, University of Kansas, United States of America

Funding: This work was supported by the grant 2010DFA32330 from the Ministry of Science and Technology of China (http://www.most.gov.cn), the grant 2012GNC11004 and Taishan Research Fellowship from Department of Science and Technology of Shandong province (http://www.sdstc.gov.cn/). The funders had no role in study design, data collection and analysis, decision to publish, or preparation of the manuscript.

Competing Interests: The authors have declared that no competing interests exist.

* E-mail: yanght@sdas.org

⊙ These authors contributed equally to this work.

Introduction

Carbendazim (methyl *1H*-benzimidazol-2-yl carbamate, MBC) is a systemic benzimidazole fungicide widely used in many countries to control a broad range of fungal diseases of agricultural crops [1]. MBC is the hydrolytic product and active component of some other widely used benzimidzaole fungicides such as benomyl and thiophanate methyl [2,3]. MBC is relatively stable in soil and water and is reported to have an environmental half-life of up to 12 months [4]. The soil persistence and the plant systemic nature of MBC can in turn, lead to the contamination of water and plant products [5]. This causes serious concerns because MBC is a suspected mutagen, teratogen and carcinogen and is reported to be toxic to mammalian liver, endocrine and reproductive tissues [6,7]. Residual MBC in soil has also been reported to alter the taxonomic structure of soil bacterial communities and may therefore adversely affect microbial-mediated ecosystem functions [8].

There is an increasing demand to remediate soils contaminated with MBC because of the prolonged use of the fungicide in agriculture, its environmental persistence and adverse impacts on animal health. Degradation rates of MBC by physical and abiotic chemical processes are reported to be slow, with microbial metabolism thought to be the principal degradative process in natural soils [4,9,10]. Only a limited number of MBC-degrading bacterial strains have been previously reported [4,5,11] and highly efficacious, ecologically competitive microbes are required to remediate a range of MBC contaminated environments. A gene-enzyme system for MBC degradation has been previously reported in *Nocardioides* sp. [4], but mechanisms utilized by other MBC-degrading microbes are yet to be elucidated. In this study, we describe the isolation of a highly efficacious MBC-degrading *Rhodococcus erythropolis* strain djl-11, conditions affecting MBC bio-degradation by this strain and sequence characterization of the djl-11 MBC-hydrolyzing esterase gene.

Materials and Methods

Chemicals and growth media

Analytical-grade carbendazim (MBC), 2-aminobenzimidazole (2-AB) and 2-hydroxybenzimidazole (2-HB) were purchased from Sigma-Aldrich Inc. All other chemicals and solvents were of highest analytical-reagent grade.

Liquid minimal salts medium (MSM) consisted of 1.0 g NH_4NO_3, 1.0 g NaCl, 1.5 g K_2HPO_4, 0.5 g KH_2PO_4, 0.2 g $MgSO_4·7H_2O$ per liter. Unless otherwise stated, MSM was adjusted to pH 7.0 and MBC was added at a final concentration of

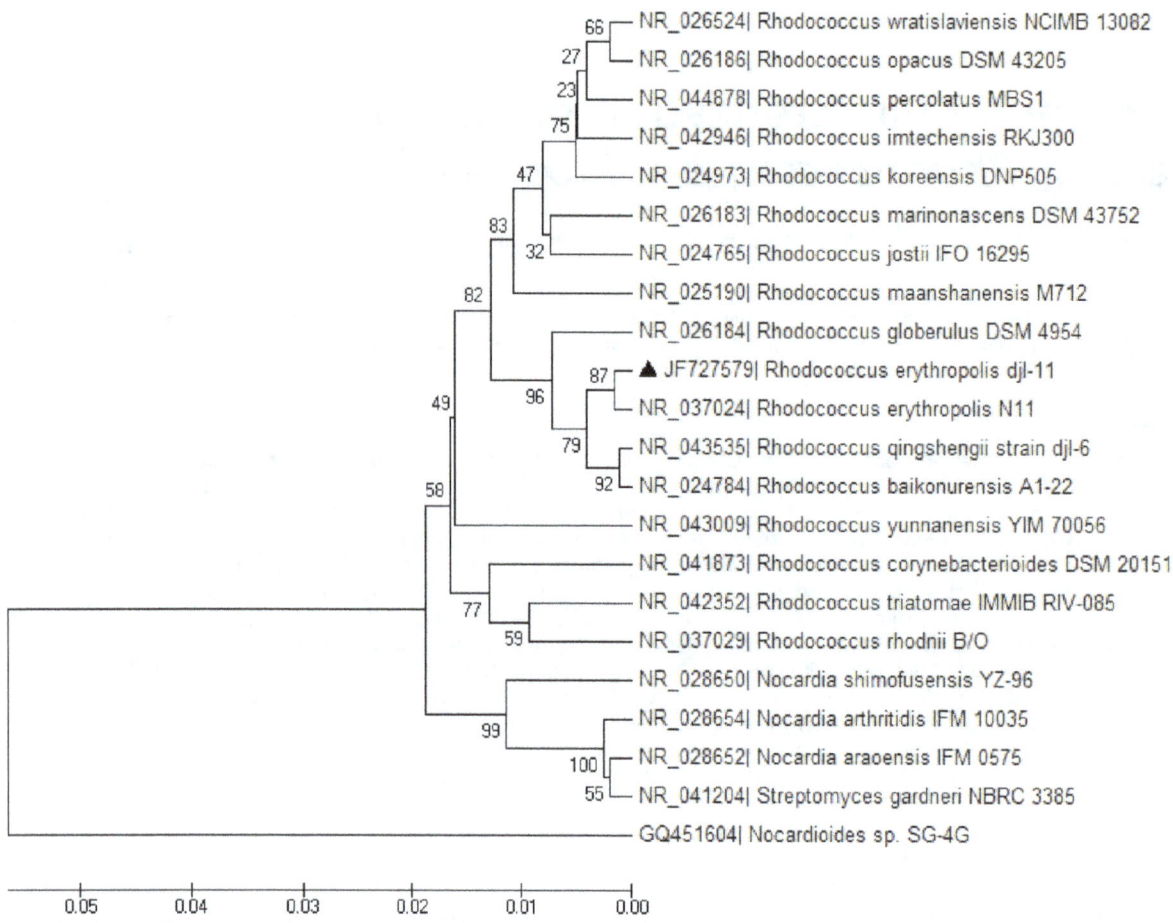

Figure 1. Phylogenetic tree based on the 16S rDNA sequence of strain djl-11 and related species. Strain djl-11 is marked with "▲". The percentages of replicate trees in which the associated species clustered together in the bootstrap test (1000 replicates) are shown next to the branches. Genbank accession numbers are shown.

1000 mg·L^{-1} in powdered form. Solid carbendazim- amended MSM contained 15 g agar L^{-1}, the carbendazim solution added to the cooled medium after autoclaving. Luria-Bertani (LB) medium was used for general bacterial growth.

Isolation of MBC degrading microorganisms

Soil samples were taken from vineyards in Rizhao (Shandong Province, China) with a 10-year history of repeated MBC applications. To select for MBC-degrading microbes, 10 g of soil was placed in a 500 mL Erlenmeyer flask containing 100 mL of MSM supplemented with 1000 mg·L^{-1} MBC (*i.e.* MSM-C$_{1000}$) as the sole carbon source and incubated at 28°C on a rotary shaker (150 rpm). After 7 days, 5 mL of culture was inoculated to 100 mL fresh MSM-C$_{1000}$ and incubated under the same conditions for another 7 days. After 5 sequential rounds of enrichment (*i.e.* 35 days exposure to MSM-C$_{1000}$), 100 μL of culture was plated onto MSM-C$_{1000}$ agar and incubated at 28°C for 5 days. Colonies showing transparent halos indicative of MBC-degradation were streaked onto fresh MSM-C$_{1000}$ agar plates to confirm MBC degradation [4] and single cell colonies were purified for further analyses.

Identification of MBC-degrading bacteria

Identification of MBC-degrading bacteria was based on morphological, physiological and biochemical characterization

according to Bergey's Manual of Systematic Bacteriology [12]. Molecular taxonomy was based on PCR amplification and DNA sequencing of the 16S rRNA gene with the universal primers 8F (5′-AGAGTTTGATCCTGGCTCAG-3′) and 1541R (5′-AAG-GAGGTGATCCAGCCGCA-3′) according to established protocols [13]. PCR primer synthesis and DNA sequencing (Applied Biosystems) were conducted at Sangon Biotech Co. Ltd., (Shanghai, China). The resulting nucleotide sequences were compared to those in GenBank using a BLAST search.

Microbial MBC degradation

MBC biodegradation was quantified by monitoring decreasing concentrations of the fungicide in liquid culture over time. Strain djl-11 was grown in LB broth at 28°C on a rotary shaker (150 rpm) for 24 h. Cells were collected by centrifugation (6000 g for 5 min.), washed twice and re-suspended to an OD$_{600}$ = 0.8 (Lambda Bio Spectrophotometer, Perkin Elmer, USA) in sterile water. The cell suspension (approx. 1×10^8 cells·mL^{-1}) was used to inoculate (1% v/v) 100 mL flasks of MSM-C$_{1000}$ and incubated at 28°C on a rotary shaker (150 rpm). Uninoculated MSM-C$_{1000}$ served as the negative control and each treatment was replicated 3 times. Culture samples were collected at 6 h intervals over a 72 h growth period, and 4 volumes of acetone were added to each sample. Samples were mixed well and stored at 4°C until analyzed by

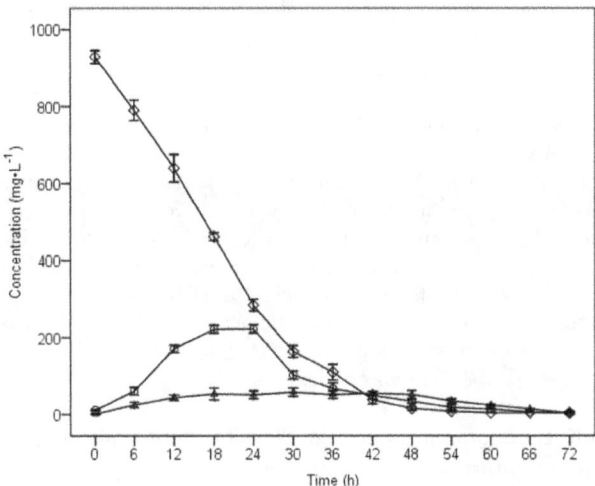

Figure 2. Chemical kinetics of MBC, 2-AB and 2-HB in liquid culture of strain djl-11. ◇, MBC; ○, 2-AB; △, 2-HB. Values are the means of three replicates with standard deviation.

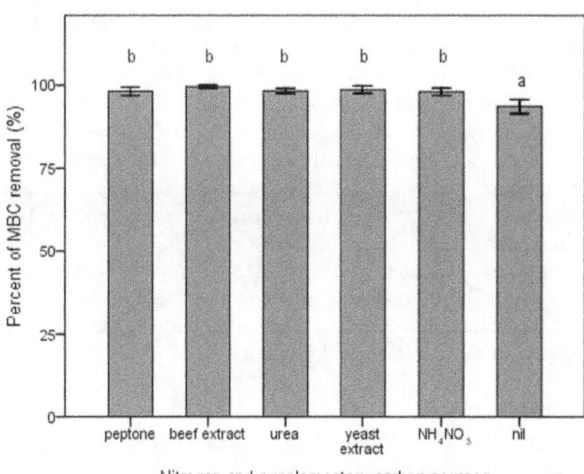

Figure 3. Effects of omitting and substituting NH₄NO₃ with organic substrates (N and C sources) on MBC biodegradation. The means of three independent experiments were plotted with error bars indicating standard deviations. Different letters above each column indicate significant differences among treatments (P<0.05).

liquid chromatography (LC) and LC-mass spectrometry (LC-MS).

To study the effects of MBC concentration on degradation by djl-11, cell suspensions were prepared as described above, inoculated (1% v/v) to 100 mL of MSM containing 200, 400, 600, 800 and 1000 mg·L^{-1} MBC and incubated at 28°C on a rotary shaker (150 rpm) for 48 h. Uninoculated flasks of MSM comprising the 5 MBC concentrations served as the negative controls. Each treatment was replicated 3 times. All samples were collected and prepared as described above for subsequent LC and LC-MS analyses.

Identification and quantification of MBC and its metabolites

Quantitative analysis of MBC, 2-AB, and 2-HB was conducted with an Agilent series LC system (Agilent Technologies, USA). Chromatographic separation was achieved on an Eclipse XDB-C18 column (150 mm×4.6 mm, 5-μm particle size) at 25°C. MBC and its metabolites were monitored at 270 nm using an acetonitrile-water mixture (16:84 [v/v] containing 0.1% [v/v] formic acid) as a mobile phase, at a flow rate of 1 mL·min^{-1}. Metabolites were qualitatively analyzed by a LC-MS mass spectrometer (Agilent Technologies, USA). The separated substrate and metabolites were ionized with positive polarity and scanned within a mass range of 29 to 500 m/z.

Effects of organic substrates, bivalent cations, pH and temperature on MBC degradation

Four experiments were established to determine the individual effects of organic substrates (N and C sources), cations, pH and temperature on MBC degradation by djl-11. Unless otherwise indicated, bacterial inoculum was prepared, cultured in MSM-C$_{1000}$ (48 h) and MBC degradation analyzed using the methods described above. Each treatment of the 4 experiments was replicated 3 times.

To examine the effects of organic substrates as alternative sources of nitrogen and supplementary sources of carbon on MBC biodegradation, 0.1% peptone, 0.1% beef extract, 0.1% urea, and 0.1% yeast extract (w/v) were respectively added to MSM-C$_{1000}$ in the absence of NH₄NO₃. MSM with no nitrogen source was

included as the negative control. Similarly, substitutions of the 810.8 μM MgSO₄ in MSM-C$_{1000}$ with equimolar amounts of ZnSO₄, MnSO₄, CuSO₄, CaSO₄ or FeSO₄ were used to determine the effects of alternative bivalent cations on MBC biodegradation.

The effect of initial MSM-C$_{1000}$ pH on MBC biodegradation was observed on a scale of pH 4 to pH 9 at increments of 1 pH unit. Optimal culture incubation temperatures for biodegradation of MBC were examined using 5°C increments on a scale of 20°C to 40°C.

Cloning of the MBC-hydrolyzing esterase gene

A putative MBC-hydrolyzing esterase gene was amplified by PCR from chromosomal DNA of djl-11. The primers MheI-F (5′-gcatggccaacttcgtcctcg-3′) and MheI-R (5′-gcgcccagcgccgccagc-3′) were designed according to the sequence of a MBC-hydrolyzing esterase encoding gene *mheI* (GenBank accession GQ454794) [4]. For PCR amplification, approximately 270 ng of djl-11 gDNA,

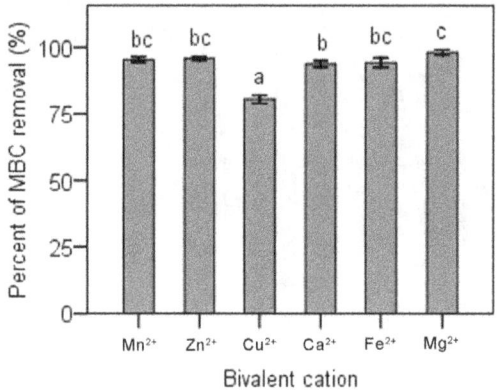

Figure 4. Effects of metal ions on MBC biodegradation. The means of three independent experiments were plotted with error bars indicating standard deviations. Different letters above each column indicate significant differences among treatments (P<0.05).

5 μL of 10× PCR buffer (Mg^{2+} free), 3 μL of 25 mM MgCl$_2$, 4 μL of 2.5 mM dNTPs, 20 pmol of each primer, and 1.25 U of TaKaRa Taq polymerase (TaKaRa Bio Inc., Dalian) were added to a final volume of 50 μL. PCR amplification was carried out as follows: 4 min at 95°C, 30 cycles of 40 sec at 95°C, 30 sec at 58°C and 40 sec at 72°C, plus a final extension step of 8 min at 72°C. The amplified product was purified using an Agarose gel DNA purification kit (TaKaRa Bio Inc., Dalian), inserted into the T-A cloning vector pMD18-T (TaKaRa Bio Inc., Dalian) and sequenced on an Applied Biosystems DNA analyzer. PCR primer synthesis and DNA sequencing were conducted at Sangon Biotech Co. Ltd. (Shanghai, China). The nucleotide sequence was compared to those in GenBank using a BLAST search.

Statistical analysis

Biodegradation of MBC by strain dj1-11 was assessed by comparing differences in MBC concentration between treatments, each consisting of 3 replicates. All data were analyzed by analyses of variance (ANOVA), using SPSS 16.0 statistical software (SPSS Inc., USA). Pairwise comparisons of means were used to compute Fisher's least significant difference values (LSD, P = 0.05).

Ethics statement

We confirm that the owner of vineyards gave permissions to take soil samples from the fields. We confirm that no endangered or protected species were involved in field studies.

Results

Isolation and identification of MBC-degrading strain dj1-11

Enrichment cultures established from MBC-contaminated soils were plated onto MSM-C$_{1000}$ agar to select for putative MBC degrading microbes. Bacteria representing different colony morphologies were purified and confirmed to have MBC degradative function via plate-clearing assays. Strain dj1-11, qualitatively assessed as the most effective MBC degrader, was selected for further study.

Strain dj1-11 was a gram-positive, non-motile, rod-shaped bacterium that formed orange colonies on LB agar after 72 h at 28°C. In physiological and biochemical tests, dj1-11 tested positive for catalase, urease and acetoin produciton, but negative for oxidase, starch hydrolysis and nitrate reductase. Strain dj1-11 was

able to utilise citrate, mannose, sodium benzoate and maltose as sole carbon sources and acetylamine and asparagine as sole sources of carbon and nitrogen.

A 1.5 kb 16S rRNA fragment was amplified from strain dj1-11, sequenced and showed 99% homology to *Rhodococcus erythropolis* 16S rRNA. Phylogenetic analysis (Figure 1) based on 16S rDNA sequences revealed that strain dj1-11 clustered with *Rhodococcus* species and was most closely related to *R. erythropolis* N11. Molecular taxonomy, cellular and colony morphologies and physiological and biochemical characteristics identified dj1-11 as *R. erythropolis*. Strain dj1-11 and its 16s rRNA sequence were deposited in the China General Microbiological Culture Collection Center (Accession No. CGMCC4554) and GenBank (Accession No. JF727579), respectively.

MBC biodegradation and metabolite identification in liquid culture

Seed cultures of dj1-11 cells (1% v/v) provided with MBC (1000 mg·L^{-1}) as a sole carbon-source degraded approximately 95% of the fungicide in 48 h, with the remaining MBC completely degraded by 72 h (Figure 2). The average degradation rate of MBC by dj1-11 was 333.33 mg·L^{-1}·d^{-1} in MSM-C$_{1000}$ at 28°C. Varying the concentration of MBC (200–1000 mg·L^{-1}) at time of dj1-11 inoculation had no significant effect on overall degradation, with 95% of the fungicide removed after 48 h growth in all treatments (data not shown). No significant MBC degradation was observed in any of the non-inoculated controls, regardless of MBC concentration at time of inoculation.

Two major metabolite peaks were detected during the growth of dj1-11 on MBC, these degradation intermediates being identified as 2-aminobenzimidazole (2-AB) and 2-hydroxybenzimidazole (2-HB) by LC and LC-MS using authentic standards.

Effects of organic substrates, bivalent cations, pH and temperature on MBC degradation

Strain dj1-11 degraded approximately 90% of MBC from MSM-C$_{1000}$ in the absence of NH$_4$NO$_3$, indicating its capability to utilize MBC as a sole source of carbon and nitrogen (Figure 3). Omission of NH$_4$NO$_3$ however, resulted in significantly less (P<0.05) MBC degradation compared to that degraded in the presence of this nitrogen-source (Figure 3). Substitution of NH$_4$NO$_3$ in MSM-C$_{1000}$ with equivalent amounts of organic nitrogen and supplementary carbon sources (*i.e.* peptone, urea, beef extract or yeast extract) had no significant effect on MBC degradation by dj1-11.

Substitution of Mg^{2+} in MSM-C$_{1000}$ with equimolar amounts of Mn^{2+}, Zn^{2+} or Fe^{2+} had no significant effect on MBC degradation by dj1-11 (Figure 4). In contrast, MBC degradation was significantly decreased (P<0.05) when Mg^{2+} in MSM-C$_{1000}$ was substituted with equimolar Ca^{2+} or Cu^{2+} (Figure 4). Substitution with Cu^{2+} resulted in the lowest overall MBC bio-degradation, significantly less (P<0.05) than that observed in the MSM-C$_{1000}$ containing Mg^{2+} (Figure 4).

Temperature had a significant (P<0.05) effect on MBC degradation by dj1-11 (Figure 5), with optimal degradation detected in the 25°C to 30°C range. MBC degradation was significantly (P<0.05) lower at 20°C and inhibited further at elevated temperatures of 35°C to 40°C (Figure 5). In contrast, MSM-C$_{1000}$ pH (range pH 4–9) at time of inoculation has no significant effect on biodegradation of MBC by dj1-11 (data not shown).

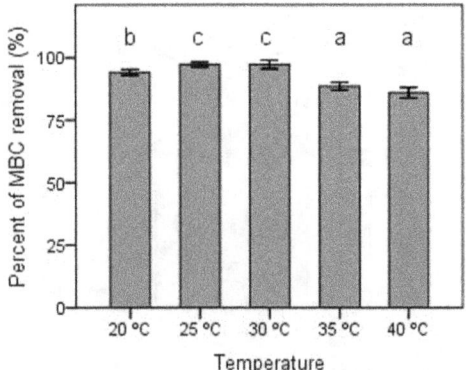

Figure 5. Effect of temperature on MBC biodegradation. The means of three independent experiments were plotted with error bars indicating standard deviations. Different letters above each column indicate significant differences among treatments (P<0.05).

MBC-hydrolyzing esterase gene of strain dj1-11

Primers designed from the previously reported MBC-hydrolyzing esterase gene *mheI* amplified a dj1-11 DNA sequence (*Mhe*) consisting of a 729 bp open reading frame starting with the ATG codon, ending with the stop codon TGA and encoding 242 amino acids residues. The predicted amino acid sequence corresponded to a 26.285 kDa protein with an isoelectric point of 6.27. *R. erythropolis* dj1-11 Mhe exhibited 99% amino acid sequence identity with MBC-hydrolyzing esterase encoded by *mheI* from *Nocardioides* sp. strain SG-4G (GenBank accession number GQ454794). The djl-11 *Mhe* DNA sequence of was deposited in GenBank (accession number HQ874282).

Discussion

At present, only a limited number of bacterial strains capable of degrading MBC have been reported [4,5,11]. Strains from the bacterial genus *Rhodococcus* were most often reported, such as *R. erythropolis* [14], *Rhodococcus qingshengii* [15,16], *Rhodococcus jialingiae* [11]. In this study, strain *R. erythropolis* djl-11 capable of catabolizing and utilizing MBC as the sole carbon and nitrogen sources was isolated. Strain djl-11 showed high MBC-degrading efficacy, with 99% of 1000 mg·L^{-1} MBC being degraded within 72 h. In comparison, *R. qingshengii* djl-6 utilized 100 mg·L^{-1} MBC as the sole carbon source, with an average MBC degradation rate of only 55 mg·L^{-1}·d^{-1} [15].

Varying the concentration of MBC (200–1000 mg·L^{-1}) at time of inoculation had no significant effect on MBC degradation by strain djl-11, with 95% of the fungicide removed after 48h. In contrast, previous researches on MBC degradation by *Bacillus pumilus* NY97-1 [17] and *Pseudomonas* sp. CBW [5] reported enhanced MBC degradation as concentrations of the fungicide increased. However, MBC concentrations (1–300 mg·L^{-1}) in these studies were much lower than those exposed to djl-11, and degradation by *Pseudomonas* sp. CBW was significantly inhibited above MBC concentration of 100 mg·L^{-1} [5].

MBC degradations by *Bacillus pumilus* NY97-1 and *Pseudomonas* sp. CBW were significantly influenced by factors such as pH, temperature and nutrient composition of the culture media [5,17]. In contrast, MBC degradations by strain djl-11 were not significantly affected by varying the initial pH ranging from 4–9, replacing Mg^{2+} with Mn^{2+}, Zn^{2+} or Fe^{2+} or substituting NH$_4$NO$_3$ with organic substrates (peptone, urea, beef and yeast extracts), the latter providing alternative and additional sources of nitrogen and carbon, respectively. MBC degradation by djl-11 was however,

reduced by 5–15% (P<0.05) in the absence of NH$_4$NO$_3$ when using the fungicide as a sole nitrogen and carbon source or at culture temperature ±5–10°C of the optimum for growth (30°C). Whilst significant, these reductions in djl-11 MBC degrading efficacy are relatively small in comparison with other bacterial strains [5,17], indicating the robustness of the process by djl-11 and the potential for MBC bioremediation in different environments.

The metabolites 2-AB and 2-HB were identified during the growth of strain djl-11 on MBC, supporting previous studies of MBC catabolism by *Nocardioides* sp. SG-4G [4] and *Pseudomonas* sp. CBW [5]. Whilst MBC degradation by *R. qingshengii djl-6* [15] and *R. jialingiae* djl-6-2 [11] also produced 2-AB, the intermediate benzimidazole (BI) was also detected either in the presence [11] or absence [15] of 2-HB. Notably, BI was not detected during growth of *R. erythropolis* djl-11, *Nocardioides* SG-4G [4] or *Pseudomonas* CBW [5] on MBC. As 2-AB and 2-HB exhibit relatively benign toxicity [18], no attempt was made to define the downstream metabolites. It was proposed that MBC was first converted to 2-AB, which was then transformed to 2-HB, 1,2-diaminobenzene, catechol, and finally to carbon dioxide by *Pseudomonas* sp. CBW [5].

Till now, only one gene, *mheI* from *Nocardioides* sp. SG-4G, which encodes the first enzyme of the pathway that detoxifies MBC by hydrolyzing it to 2-AB was cloned and reported [4]. In this study, MBC-hydrolyzing esterase (Mhe) gene from *R. erythropolis* djl-11 was cloned, and Mhe exhibited 99% amino acid identity to that of *Nocardioides* sp. SG-4G MheI esterase, suggesting that both strains utilize enzymatic hydrolysis as the first step in catabolism and detoxification of MBC to 2-AB. BLAST searches for *Mhe* in all available *Rhodococcus* genomes, including 3 other *R. erythropolis* strains, did not detect any homologous loci with high similarities. The evolutionary origin of Mhe gene and its frequency among other MBC-degrading bacterial strains need to be further elucidated.

Acknowledgments

The authors would like to thank Dr. Jinpeng Yuan for his excellent assistance in the LC-MS analysis and elucidation.

Author Contributions

Conceived and designed the experiments: XZ YH HY. Performed the experiments: XZ YH HL YR. Analyzed the data: XZ PRH JW JL. Contributed reagents/materials/analysis tools: YH HL YR. Wrote the paper: XZ PRH HY.

References

1. Chen Y, Zhou MG (2009) Characterization of Fusarium graminearum isolates resistant to both carbendazim and a new fungicide JS399–19. Phytopathology 99: 441–446.
2. Sandahl M, Mathiasson L, Jonsson JA (2000) Determination of thiophanate-methyl and its metabolites at thrace level in spiked natural water using the supported liquid membrane extraction and the microporous membrane liquid-liquid extraction techniques combined on-line with highperformance liquid chromatography. J Chromatogr A 893: 123–131.
3. Boudina A, Emmelin C, Baaliouamer A, Grenier-Loustalot MF, Chovelon JM (2003) Photochemical behaviour of carbendazim in aqueous solution. Chemosphere 50: 649–655.
4. Pandey G, Dorrian SJ, Russell RJ, Brearley C, Kotsonis S, et al. (2010) Cloning and biochemical characterization of a novel carbendazim (methyl-1H-benzimidazol-2-ylcarbamate)-hydrolyzing esterase from the newly isolated Nocardioides sp. strain SG-4G and its potential for use in enzymatic bioremediation. Appl Environ Microbiol 76: 2940–2945.
5. Fang H, Wang Y, Gao C, Yan H, Dong B, et al. (2010) Isolation and characterization of Pseudomonas sp. CBW capable of degrading carbendazim. Biodegradation 21: 939–946.
6. Selmanoglu G, Barlas N, Songur S, Kockaya EA (2001) Carbendazim-induced haematological, biochemical and histopathological changes to the liver and kidney of male rats. Hum Exp Toxicol 20: 625–630.
7. Farag A, Ebrahim H, ElMazoudy R, Kadous E (2011) Developmental toxicity of fungicide carbendazim in female mice. Birth Defects Res B Dev Reprod Toxicol 92: 122–130.
8. Wang YS, Huang YJ, Chen WC, Yen JH (2009) Effect of carbendazim and pencycuron on soil bacterial community. J Hazard Mater 172: 84–91.
9. Kiss A, Virag D (2009) Photostability and photodegradation pathways of distinctive pesticides. J Environ Qual 38: 157–163.
10. Yarden O, Salomon R, Katan J, Aharonson N (1990) Involvement of fungi and bacteria in enhanced and nonenhanced biodegradation of carbendazim and other benzimidazole compounds in soil. Can J Microbiol 36: 15–23.
11. Wang Z, Wang Y, Gong F, Zhang J, Hong Q, et al. (2010) Biodegradation of carbendazim by a novel actinobacterium Rhodococcus jialingiae djl-6-2. Chemosphere 81: 639–644.
12. Goodfellow M (1989) Genus Rhodococcus. In: Williams ST, Sharpe ME, Holt JG, editors. Bergey's Manual of Systematic Bacteriology. Baltimore, MD.: Williams & Wilkins. 2362–2371.
13. Baker GC, Smith JJ, Cowan DA (2003) Review and re-analysis of domain-specific 16S primers. J Microbiol Methods 55: 541–555.
14. Holtman MA, Kobayashi DY (1997) Identification of *Rhodococcus erythropolis* isolates capable of degrading the fungicide carbendazim. Appl Microbiol Biotechnol 47: 578–582.

15. Xu J, Gu X, Shen B, Wang Z, Wang K, et al. (2006) Isolation and characterization of a carbendazim-degrading Rhodococcus sp. djl-6. Curr Microbiol 53: 72–76.

16. Xu JL, He J, Wang ZC, Wang K, Li WJ, et al. (2007) Rhodococcus qingshengii sp. nov., a carbendazim-degrading bacterium. Int J Syst Evol Microbiol 57: 2754–2757.

17. Zhang LZ, Qiao XW, Ma LP (2009) Influence of environmental factors on degradation of carbendazim by *Bacillus pumilus* strain NY97-1. Int J Environ Pollut 38: 309–317.

18. Stringer A, Wright MA (1976) The toxicity of benomyl and some related 2-substituted benzimidazoles to the earthworm *Lumbricus terrestris*. Pestic Sci 7: 459–464.

18

Survival of *Escherichia coli* O157:H7 in Soils from Jiangsu Province, China

Taoxiang Zhang, Haizhen Wang*, Laosheng Wu, Jun Lou, Jianjun Wu, Philip C. Brookes, Jianming Xu*

College of Environmental and Natural Resource Sciences, Zhejiang Provincial Key Laboratory of Subtropical Soil and Plant Nutrition, Zhejiang University, Hangzhou, China

Abstract

Escherichia coli O157:H7 (*E. coli* O157:H7) is recognized as a hazardous microorganism in the environment and for public health. The *E. coli* O157:H7 survival dynamics were investigated in 12 representative soils from Jiangsu Province, where the largest *E. coli* O157:H7 infection in China occurred. It was observed that *E. coli* O157:H7 declined rapidly in acidic soils (pH, 4.57 – 5.14) but slowly in neutral soils (pH, 6.51 – 7.39). The survival dynamics were well described by the Weibull model, with the calculated t_d value (survival time of the culturable *E. coli* O157:H7 needed to reach the detection limit of 100 CFU g^{-1}) from 4.57 days in an acidic soil (pH, 4.57) to 34.34 days in a neutral soil (pH, 6.77). Stepwise multiple regression analysis indicated that soil pH and soil organic carbon favored *E. coli* O157:H7 survival, while a high initial ratio of Gram-negative bacteria phospholipid fatty acids (PLFAs) to Gram-positive bacteria PLFAs, and high content of exchangeable potassium inhibited *E. coli* O157:H7 survival. Principal component analysis clearly showed that the survival profiles in soils with high pH were different from those with low pH.

Editor: Dongsheng Zhou, Beijing Institute of Microbiology and Epidemiology, China

Funding: This work was financially supported by the Major Program of National Natural Science Foundation of China (41130532) and the National Natural Science Foundation of China (40971255). The funders had no role in study design, data collection and analysis, decision to publish, or preparation of the manuscript.

Competing Interests: The authors have declared that no competing interests exist.

* E-mail: mywhz@163.com (HW); jmxu@zju.edu.cn (JX)

Introduction

Applications of animal manure as fertilizers or soil amendments to agricultural soils are routine, world-wide. In the UK, the annual amount of animal manure applied to land was recently estimated at 4.3×10^5 tons dry weight [1]. Though animal manure can provide nutrients, a variety of pathogenic bacteria may survive in the manure, which in turn may serve as a primary hazardous material for environmental contamination and as a public health threat [2]. For example, *Escherichia coli* O157:H7 (*E. coli* O157:H7), which can cause severe hemorrhagic colitis and haemolytic uraemia in humans, can persist in soil for days to more than 1 year following manure application to land [3]. It was reported that 20 people were infected with *E. coli* O157:H7 through manure-contaminated soil after camping on a field in Scotland that was previously grazed by sheep [4]. Increasing evidence shows that soil and animal manures are the main transport agents of *E. coli* O157:H7 to contaminate fresh vegetables, fruits and drinking water [1], [5], [6]. Therefore, it is important to understand the nature of *E. coli* O157:H7 survival and its infective risk in soil or soil-related (manure) environments.

In America, about 63,000 human cases of *E. coli* O157:H7 infections have been reported every year [7]. Many studies have focused on the survival of *E. coli* O157:H7 in soil, manure, water, and vegetables [1], [2], [4], [5], [8], [9]. However, very little attention has been paid to *E. coli* O157:H7 survival and its potential environmental contamination risk in the soils in the areas where outbreaks of *E. coli* O157:H7 infection have occurred. Previous studies found that the survival time of *E. coli* O157:H7 in soils depends on soil type, physicochemical properties, indigenous

microorganisms, etc. [2], [9–15]. Franz et al. (2008) pointed out that higher amounts of dissolved organic carbon and dissolved organic nitrogen were the best predictors for long *E. coli* O157:H7 survival time in organically managed soils [2]. In several experiments, *E. coli* O157:H7 persisted longer in silty clay soils than in sandy soils [2], [10], [11]. *E. coli* O157:H7 survived for up to 77, 226, and 231 days at 5, 15, and 25°C in manure-amended autoclaved soil, respectively [13]. *E. coli* O157:H7 survived significantly longer under anaerobic than under aerobic conditions in manure and slurry [14]. Yao et al. (2012) and van Elsas et al. (2012) proved *E. coli* O157:H7 survival was affected by indigenous microorganisms in soil [9], [15]. Different *E. coli* O157:H7 survival rates indicate the different potential risks of the pathogen contamination under various soil environments. Consequently, a better understanding of *E. coli* O157:H7 survival in soils will help in reducing the potential risk of pathogen contamination and avoiding human infection from the pathogen.

In the present study, experiments were carried out to investigate *E. coli* O157:H7 survival in 12 soils taken from Jiangsu Province, China. In 1987, researchers firstly detected *E. coli* O157:H7 from the fecal samples of patients in Jiangsu Province, where the largest *E. coli* O157:H7 outbreak in China occurred [16], [17]. Later, Xu et al.(1990) found that the biochemical reactions of five strains of *E. coli* O157:H7, which were isolated from 486 stool specimens of patients with diarrhea in Xuzhou City, Jiangsu Province, were almost identical with those of the well-known *E. colt* O157:H7 (strain EDL933) [18]. Recently, numerous researchers reported that *E. coli* O157:H7 (strain EDL933) had been detected in excrement, sewage, foods, and soils from many provinces of

China, including Jiangsu Province [16], [19–21]. Thus, *E. coli* O157:H7 (strain EDL933) was selected as a representative strain in this study.

Most *E. coli* O157:H7 outbreaks occur in summer [17–19]. The summer mean temperature in Jiangsu Province is about 26°C [22]. Furthermore, the water content under –33 kPa, indicating the water holding capacity of the soil and representing the highest available water contents in soil, was generally used to simulate field conditions [23]. Hence, our simulation experiments also used incubation conditions of 25±1°C and water content under –33 kPa. The aims of this study were to (1) investigate *E. coli* O157:H7 survival dynamics in soils, (2) identify the relationships between *E. coli* O157:H7 survival time and soil physicochemical properties and microbial community structure, and (3) understand the possible risks of pathogen contamination to prevent further disease outbreaks from *E. coli* O157:H7.

Materials and Methods

Ethics Statement

The samples were not collected from national parks, protected areas or private land. Hence, no specific permission was required to obtain these samples. The sampling did not cause any disturbance to the environment or to protected species at the sampling sites.

Soils

The 12 soils (S1–S12) used in this study were taken from Jiangsu Province, China (32.05°N – 34.70°N). Three replicates were sampled at each soil site from the surface horizon (0–20 cm) and each sample was a composite of several individual soil cores taken at 5-m interval. After sampling, the 36 individually bulked fresh soil samples were immediately taken to the laboratory in coolers containing ice packs. The samples were then hand-picked to remove discrete plant residues, sieved <2 mm, homogenized thoroughly, and then stored at 4 °C. According to the protocols of the Agricultural Chemistry Committee of China [24], a subsample of the sieved soil from each sample was air-dried for physical and chemical analyses, including pH, total organic carbon (OC), exchangeable potassium (K), humic acid, fulvic acid, sand, clay and silt. Total nitrogen (TN) was determined following digestion using a Kieldahl apparatus (BÜCHI Labortechnik AG, Flawil, Switzerland). Total dissolved organic carbon (DOC) and total dissolved nitrogen (TDN) were measured by a Multi N/C TOC analyzer (Analytic Jena AG, Jena, Germany). Exchangeable potassium (K) was extracted with 1 M ammonium acetate and measured by Flame Atomic Absorption Spectrophotometry (Analytik Jena AG, Jena, Germany). The soil water content at –33 kPa was determined using a pressure membrane apparatus (Soil Moisture Equipment Corp, CA,USA) as described by Richards [23].

Another portion of each sieved fresh sample was frozen at –80°C and then freeze-dried (FreeZone Freeze Dry Systems LABCONCO Corp, MO, USA) for phospholipid fatty acids (PLFAs) analysis [25]. The PLFAs are the specific components of cell membranes that are only found in intact (viable) cells [26]. Many studies have widely used PLFAs to express the biomass and composition of microbial communities in soils [9], [27], [28]. Thus, PLFAs were used to determine the effects of soil indigenous microorganisms on *E. coli* O157:H7 survival in this study. According to Ding et al. (2009) [29] and Ying et al. (2013) [30], the PLFA biomarkers i15:0, a15:0, 15:0, i16:0, 16:1ω7c, i17:0, a17:0, cy17:0, 17:0, 18:1ω7c, cy19:0ω8c were used to represent bacterial biomass; i15:0, a15:0, i16:0, i17:0, and a17:0 were used

to indicate Gram-positive (G$^+$) bacteria biomass; 16:1ω7c, 18:1ω5c, cy17:0, and cy19:0 indicated Gram-negative (G$^-$) bacterial biomass. Polyenoic, unsaturated PLFA 18:2ω6,9c was used to represent fungal biomass. The fatty acids 10Me 16:0, 10Me 17:0, 10Me 18:0 were used to indicate actinomycetic biomass. Based on the studies of the Institute of Soil Science, Academia Sinica [31], the 12 soils used were classified into acidic soils (pH < 6.5) and neutral soils (pH > 6.5). Selected soil properties are shown in Table 1.

Incubation experiments

The preparation of *E. coli* O157:H7 (strain EDL933) and inoculums were described previously by Wang et al. (2013) [32]. *E. coli* O157:H7 cells in sterile deionized water were inoculated into soils to achieve a cell density of about 10^7 colony-forming units per gram soil oven-dry weight (CFU g^{-1}). The incubation experiments were conducted on each of the three soil replicates taken from each of the twelve sampling sites (a total of 36 soil samples). All the inoculated and uninoculated control soil samples were incubated in the dark at 25±1°C and kept at soil moisture of –33 kPa.

The inoculated soils were sampled at 0, 0.25, 1, 2, 3, 5, 7, 10, 15, 20, 25, 30 and 35 days after inoculation (DAI), and *E. coli* O157:H7 was extracted by 0.1% peptone buffer (Lab M, Lancashire, UK). The resulting soil suspension was then subjected to 10-fold serial dilutions and *E. coli* O157:H7 enumerated at the last three of the serial dilutions. The detection limit of the plating technique is about 100 CFU g^{-1}. The sampling was stopped after plate counts of zero appeared twice in succession during the incubation.

Statistical analysis

The survival data was converted to log$_{10}$ (CFU g^{-1}) and then analyzed by the Weibull survival model (Equation 1), as described by Wang et al. (2013) [32] and Ma et al. (2013) [33].

$$\log 10(Nt) = \log 10(N0) - (\frac{t}{\delta})^p \qquad (1)$$

N_t represents the number of surviving cells remaining at time t, N_0 is the initial cells of the inoculum population; p is the shape parameter; δ is the scale parameter that represents the time needed for the first decimal reduction. The time when N_t reaches the detection limit (100 CFU g^{-1}) of the culturable *E. coli* O157:H7, t_d, can also be calculated from Equation (1).

In addition, principal component analysis (PCA) of the parameters (p and δ) and t_d values were performed using the R software vegan package v 2.0–5 [34] to visualize survival patterns of *E. coli* O157:H7 in the test soils. Stepwise multiple-linear regression analysis was carried out by using SPSS 13.0 for Windows (SPSS Inc., IL, USA) to better understand how soil properties affected the survival of *E. coli* O157: H7 in soils. Analysis of variance (ANOVA) was also carried out to test the differences at the 5% significant level in the survival parameters (p and δ) and t_d values among soils by SPSS statistic software.

Results

Survival of *E. coli* O157:H7 in soils

The *E. coli* O157:H7 population declined by 1.14 log$_{10}$ (CFU g^{-1}) within the first day after inoculation in all the test soils. After 1 day, two different survival dynamics of *E. coli* O157:H7 were observed (Fig. 1). *E. coli* O157:H7 declined rapidly in the acidic

Table 1. Physical, chemical, and biological properties of the soils used in this study.

Soil Code	Location	pH	OC#	TN	C/N	K	DOC	TDN	FA	HA	HA/FA	PLFA-T	PLFA-F	PLFA-A	PLFA-B	PLFA-G⁻	PLFA-G⁺	G⁻/G⁺	Clay	Silt	Sand
			g kg⁻¹			mg.kg⁻¹			%			n mol g⁻¹							%		
S1	Lianyungang	4.57	20.3	1.51	13.44	89.10	51.97	7.46	6.40	43.25	6.75	35.83	0.58	4.24	16.63	7.29	7.67	0.95	17.13	16.72	66.15
S2	Xuzhou	7.27	25.4	2.44	10.41	194.86	28.55	14.97	6.26	22.28	3.56	67.19	1.68	8.52	22.95	10.47	10.77	0.97	31.73	35.19	33.08
S3	Xuyi	6.77	34.4	3.14	10.96	207.22	88.06	6.70	4.68	17.41	3.72	125.46	1.61	14.61	42.25	13.12	27.47	0.47	40.52	32.74	26.74
S4	Yangzhou	4.90	11.6	1.24	9.35	53.72	29.41	2.94	7.67	51.72	6.74	38.45	1.09	4.31	15.71	7.70	7.75	0.99	23.51	48.91	27.58
S5	Nanjing	6.61	8.1	1.01	8.02	123.16	19.83	5.74	23.09	16.91	0.73	21.39	0.21	2.75	9.63	2.50	5.69	0.43	41.12	39.69	19.19
S6	Nanjing	7.39	15.2	1.63	9.32	106.51	30.64	10.05	21.05	3.36	0.16	85.11	1.85	10.31	37.54	19.96	16.71	1.19	35.69	34.66	29.65
S7	Yangzhou	5.14	16.1	1.71	9.42	144.06	61.28	5.66	8.88	38.57	4.34	43.93	1.17	5.43	19.68	8.83	10.62	0.83	27.48	44.04	28.48
S8	Jurong	5.03	15.4	1.42	10.85	98.37	42.58	3.41	11.88	37.14	3.12	46.21	1.31	5.37	18.75	9.50	8.94	1.06	33.77	40.24	25.99
S9	Nanjing	4.66	51.5	3.40	15.15	165.56	134.27	29.42	18.46	16.42	0.89	106.54	1.64	10.91	44.16	22.62	20.81	1.09	18.01	39.13	42.86
S10	Nanjing	6.51	13.7	1.30	10.54	67.67	26.85	3.70	5.62	14.38	2.56	27.91	0.42	3.45	11.87	4.71	6.14	0.76	20.57	47.80	31.63
S11	Jiangyin	4.68	19.4	1.43	13.56	108.82	33.67	5.49	7.16	43.40	6.05	26.6	0.91	3.47	11.67	5.50	6.01	0.92	33.06	39.91	27.03
S12	Jurong	6.78	61.4	5.00	12.28	325.53	115.06	14.27	4.30	20.99	4.89	121.5	3.40	23.21	44.41	16.48	25.9	0.63	24.61	40.51	34.88

Total soil organic carbon (OC); total soil nitrogen (TN); the ratio of OC to TN (C/N); exchangeable potassium (K); dissolved organic carbon (DOC); total dissolved nitrogen (TDN); the ratio of OC to TN (C/N); total phospholipid fatty acids (PLFA-T); fungus phospholipid fatty acids (PLFA-F); actinomyces phospholipid fatty acids (PLFA-A); bacteria phospholipid fatty acids (PLFA-B); Gram-negative bacteria phospholipid fatty acids (PLFA-G⁻); Gram-positive bacteria phospholipid fatty acids (PLFA-G⁺); the ratio of PLFA-G⁻ to PLFA-G⁺ (G⁻/G⁺);

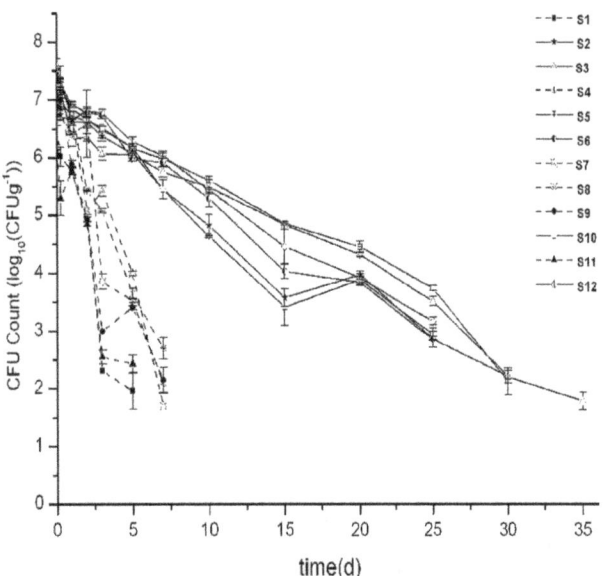

Figure 1. Survival of *E. coli* O157:H7 in the tested soils. Solid line: neutral soils; dashed line: acidic soils.

Table 2. Statistical measures and fitted parameter values of the Weibull model describing the survival of *E. coli* O157:H7 in soils.

Soil Code.	t_d #	δ	p	R^2
S1	4.57±0.23e*	1.01±0.04ef	0.97±0.06ab	0.89±0.01
S2	32.07±0.43c	3.25±0.58cd	0.72±0.05def	0.97±0.01
S3	34.34±0.86a	5.76±0.66b	0.90±0.06ebc	0.97±0.01
S4	8.09±0.35d	0.97±0.09ef	0.80±0.04cd	0.99±0.01
S5	34.03±1.60ab	6.78±1.00a	1.01±0.11ab	0.97±0.01
S6	31.55±0.89c	3.50±0.38c	0.77±0.04cde	0.99±0.01
S7	7.01±0.23d	1.36±0.37ef	1.00±0.17ab	0.91±0.01
S8	7.15±0.22d	1.57±0.14e	1.08±0.06a	0.96±0.01
S9	7.00±0.22d	0.55±0.05f	0.64±0.01f	0.91±0.01
S10	31.96±1.02c	2.56±0.08d	0.67±0.02ef	0.96±0.02
S11	5.16±0.12e	1.03±0.07ef	0.88±0.04bc	0.84±0.01
S12	32.77±1.16bc	5.22±0.43b	0.90±0.04bc	0.98±0.01

\# Survival time to reach detection limit (t_d); time needed for first decimal reduction in *E. coli* O157:H7 population (δ); shape parameter (p).
*Significant differences (p < 0.05) indicated by different letters.

soils (S1, 4, 7, 8, 9, 11) and was not detectable after 7 days. In the neutral soils (S2, 3, 5, 6, 10, 12), *E. coli* O157:H7 declined much more slowly during 1 – 3 days, and then entered a relatively rapid decline period, reaching the detection limit after 30 days.

Modeling of survival data

The statistical measures and parameter values of the fitted Weibull model describing the survival of *E. coli* O157:H7 in the test soils are presented in Table 2. The *E. coli* O157:H7 survival dynamics in all soils were accurately described by the model, with R^2 ranging from 0.84 to 0.99 (P < 0.001), while the first decimal reduction time (δ) and the shape parameter (p) varied in the test soils (Table 2). Survival times (t_d) of *E. coli* O157:H7 in the 12 soils were calculated from the Weibull model (Equation 1). The survival times (t_d) varied considerably among the soils, with t_d of 6.50±1.34 days in acidic soils with pH of 4.57 to 5.14 and significantly longer (32.79±1.16 days, P < 0.05) in neutral soils (pH, 6.57–7.39). A long survival time of 32.07 days was detected in soil from Xuzhou (S2), where the biggest infection ever reported in China.

The relationship between soil properties and the survival of *E. coli* O157:H7

For all soils, stepwise regression analysis showed that soil pH, the ratio of G⁻ bacteria PLFAs to G⁺ bacteria PLFAs (G⁻/G⁺); exchangeable K and OC were the key factors affecting the survival of *E. coli* O157:H7. Soil pH (*P* < 0.001) and OC (*P* < 0.05) had positive effects on *E. coli* O157:H7 survival time (t_d). In contrast, a high G⁻/G⁺ ratio, and high exchangeable potassium concentration decreased the survival time. The results also suggested that soil pH and G⁻/G⁺ were the most important factors determining the survival of *E. coli* O157:H7 in the test soils (Table 3).

Discussion

The results indicated that *E. coli* O157:H7 survival times (t_d) mainly depended on the initial soil pH. Principal component analysis (PCA) clearly showed a substantial difference in *E. coli* O157:H7 survival times between acidic and neutral soils (Fig. 2).

In the acidic soils, *E. coli* O157:H7 could not be detected by 7 days after inoculation (Fig.1). However, in the neutral soils, *E. coli* O157:H7 declined much slowly than in the acidic soils and survived for up to 38 days. There have been previous reports which linked *E. coli* O157:H7 survival times (t_d) with soil pH [9], [12], [32], [35], However, our data showed an exceptionally highly statistically significant correlation between these two parameters. It therefore appears that the different physical, chemical and biological properties of acidic and neutral soils resulted in the remarkable differences in *E. coli* O157:H7 survival. Stepwise multiple regression analysis indicated that *E. coli* O157:H7 survival time was significantly and positively correlated with soil pH and OC (Table 3). In addition, the initial G⁻/G⁺ and exchangeable K were negatively correlated with *E. coli* O157:H7 survival time (t_d).

There may be several reasons why soil pH had the most significant positive correlation with *E. coli* O157:H7 survival time (t_d). Firstly, most soil microbial species are generally adapted to neutral or slightly alkaline environments and the quantity of soil bacteria increases from pH 4 to 7, and they can readily adapt to changing pH within this range [36], [37]. Secondly, pH can affect the adsorption of bacteria to soil minerals. The adsorption decreased gradually with increasing pH [38], [39]. For example, Zhao et al. (2013) found that more *E. coli* was absorbed by soil colloids when solution pH decreased from 9 to 4 [38]. Also, Cai et al. (2013) showed that large decreases in the viability of *E. coli* O157:H7 can be caused by the sorption of *E. coli* O157:H7 to soil minerals [40]. Therefore, a large amount of *E. coli* O157:H7 adsorbed to soil minerals would result in a high loss of viability of *E. coli* O157:H7 in low pH soils. Also, the low biological availability of phosphorus and organic nitrogen and the high toxicity of Al and Mn in the soil with low pH [1], [41–43], might indirectly affect *E. coli* O157:H7 survival and activity in the acidic soils.

Because of competition for nutrients and niche space, as well as predation, the survival of introduced pathogens can be affected by the diverse coexisting populations of the indigenous microorganisms [9], [15]. The stepwise multiple regression analysis showed

Table 3. Stepwise multiple-linear regression analysis of soil properties and the survival time (t_d) of E. coli O157:H7 in the test soils.

Regression equations #	R^2	F value	T value of the partial regression coefficient	
				T value
			Constant	−13.208
t_d = −34.188+12.360pH −17.856(G⁻/G⁺) −0.076K+0.301OC	0.998	451.782***	pH	32.277***
			G−/G+	−10.970**
			K	−6.450***
			OC	6.494***

Survival time to reach detection limit (t_d), the ratio of Gram-negative bacteria phospholipid fatty acids (PLFAs) to Gram-positive bacteria PLFAs (G⁻/G⁺); exchangeable potassium (K); total soil organic carbon (OC); correlation is significant at the 0.001 probability level(***).

that the initial G^-/G^+ ratio was negatively correlated with E. coli O157:H7 survival time (t_d) (Table 3). van Elsas et al. (2012) and Ma et al. (2013) also indicated that survival of the invading E. coli O157:H7 was negatively correlated with soil microbial diversity, due to the competition between native bacterial communities and the introduced species for nutrients and niche space [15], [44]. The G^- bacteria are known to out-compete G^+ bacteria for nutrients in soil [45]. Furthermore, the production of bacteriocins to kill closely related species drives the negative interspecies interactions in bacterial systems, which plays an important role in determining the fate of invading bacteria [46]. E. coli O157:H7 belongs to G^- bacteria, therefore we surmised that G^- bacterial species in soil would have a stronger antagonism to E. coli O157:H7 than G^+ bacteria. Another study confirmed that *Bacteroidetes, Gammaproteobacteria* and *Firmicutes*, mainly G^- bacteria, could inhibit E. coli O157:H7 survival due to antagonism [47]. Therefore, the greater direct or indirect antagonism of G^- bacteria than G^+ bacteria to E. coli O157:H7 might account for the statistically significant negative correlation between E. coli O157:H7 survival time (t_d) and the initial G^-/G^+ ratio (Table 3).

E. coli O157:H7 survival times in the test soils were decreased by increasing exchangeable K. A recent study revealed that t_d values decreased significantly ($P< 0.05$) with increasing electrical conductivity and concentrations of individual soil cations, e.g. K^+, Na^+, Ca^{2+}, and Mg^{2+}. This could interfere with ion transport, enzyme activity, and crucial protein synthesis in E. coli O157:H7, and finally result in reduced E. coli O157:H7 survival in soils [33]. In addition, our results showed that soil OC was significantly correlated with E. coli O157:H7 survival time (t_d). Soil organic matter is a major energy source for microorganisms, and can provide carbon sources for the growth and survival of E. coli in soil and water [48]. Furthermore, abundant organic carbon can decrease the competitive pressure between organisms through providing easily available energy sources in soil, and thus possibly enhance the persistence of E. coli O157:H7 [15]. Soil organic carbon also helps to improve soil structure by forming multi-pored aggregates which serves as microbial habitats [49]. Therefore, the abundance of organic carbon in soil can provide more nutrients, water, air and biological niches for E. coli O157:H7 and decrease the competition with indigenous microorganisms, slowing the decline of E. coli O157:H7 [3], [15].

Conclusions

This research has enhanced our understanding of the survival of E. coli O157:H7 in soils. E. coli O157:H7 could survive for 32.79 ± 1.16 days in neutral soils, and only 6.5 ± 1.34 days in acidic soils. Special attention should be paid to the different survival times of E. coli O157:H7 in acidic and neutral soils when evaluating the environmental risk associated with it. The survival of E. coli O157:H7 in soils might relate to the interactions between numerous physical, chemical and biological factors. Our findings suggested that high soil pH and organic carbon could prolong E. coli O157:H7 survival times (t_d), while the initial G^-/G^+ ratio and exchangeable K were negatively correlated with survival times. Soil pH governs the fate of E. coli O157:H7 directly and indirectly through the adsorption/desorption of soil minerals, nutrition availability, and metal toxicity. However, the experiment was carried out under laboratory conditions. In order to better understand the survival dynamics of E. coli O157:H7 and provide precise information to assess the possible risk of contamination by this pathogen, further research should be done under natural conditions.

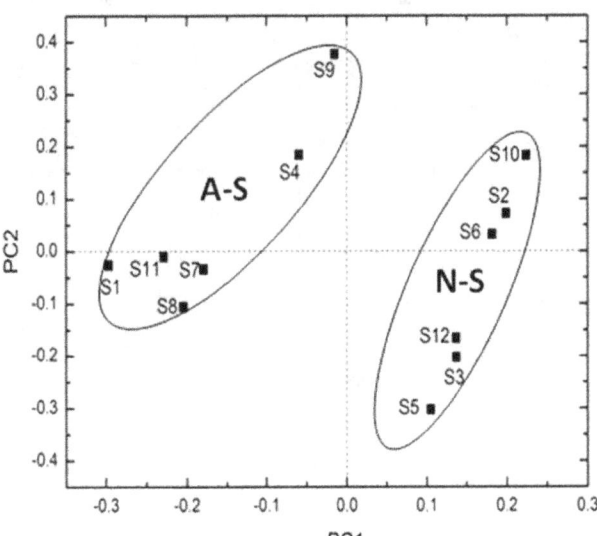

**Figure 2. Principal component analysis (PCA) of the survival parameters (δ, p and t_d). δ, p and t_d are the same as shown in Table 2. A-S: acidic soils; N-S: neutral and slight alkaline soils.

Author Contributions

Conceived and designed the experiments: JX HW. Performed the experiments: TZ. Analyzed the data: TZ HW JL. Contributed reagents/

materials/analysis tools: JX HW LW JW. Wrote the paper: TZ HW. Designed/drew figures: TZ HW JL. Edited the manuscript: JX TZ HW LW JW PCB.

References

1. Ogden ID, Fenlon DR, Vinten AJA, Lewis D (2001) The fate of *Escherichia coli* O157 in soil and its potential to contaminate drinking water. Int J Food Microbiol 66: 111–117.

2. Franz E, Semenov AV, Termorshuizen AJ, de Vos OJ, Bokhorst JG, et al. (2008) Manure-amended soil characteristics affecting the survival of *E. coli* O157:H7 in 36 Dutch soils. Environ Microbiol 10: 313–327.

3. Williams AP, Avery LM, Killham K, Jones DL (2007) Survival of *Escherichia coli* O157:H7 in the rhizosphere of maize grown in waste amended soil. J Appl Microbiol 102: 319–326.

4. Mukherjee A, Cho S, Scheftel J, Jawahir S, Smith K, et al. (2006) Soil survival of *Escherichia coli* O157:H7 acquired by a child from garden soil recently fertilized with cattle manure. J Appl Microbiol 101: 429–436.

5. Ongeng D, Muyanja C, Ryckeboer J, Geeraerd AH, Springael D (2011) Rhizosphere effect on survival of *Escherichia coli* O157:H7 and *Salmonella enteric serovar typhimurium* in manure-amended soil during cabbage (Brassica oleracea) cultivation under tropical field conditions in sub-Saharan Africa. Int J Food Microbiol 149: 133–142.

6. Doyle MP, Erickson MC (2008) Summer meeting 2007–the problems with fresh produce: an overview. J Appl Microbiol 105: 317–330.

7. Scallan E, Hoekstra RM, Angulo FJ, Tauxe RV, Widdowson MA, et al. (2011) Foodborne illness acquired in the United States—major pathogens. Emerg Infect Dis. 17: 7.

8. Wang G, Doyle MP (1998) Survival of enterohemorrhagic Escherichia coli O157: H7 in water. J Food Protect 6: 662–667.

9. Yao ZY, Wei G, Wang HZ, Wu LS, W JJ, et al. (2013) Survival of *Escherichia coli* O157:H7 in soils from vegetable fields with different cultivation patterns. Appl Environ Microb 79: 1755–1756.

10. Vidovic S, Block HC, Korber DR (2007) Effect of soil composition, temperature, indigenous microflora, and environmental conditions on the survival of *Escherichia coli* O157:H7. Can J Microbiol 53: 822–829.

11. Mubiru DN, Coyne MS, Grove JH (2000) Mortality of *Escherichia coli* O157:H7 in two soils with different physical and chemical properties. J Environ Qual 29: 1821–1825.

12. Ongeng D, Muyanja C, Geeraerd AH, Springael D, Ryckeboer J (2011) Survival of *Escherichia coli* O157:H7 and *Salmonella enteric serovar Typhimurium* in manure and manure-amended soil under tropical climatic conditions in Sub-Saharan Africa. J Appl Microbiol 110: 1007–1022.

13. Jiang XP, Morgan J, Doyle MP (2002) Fate of *Escherichia coli* O157:H7 in manure- amended soil. Appl Environ Microb 68: 2605–2609.

14. Semenov AV, van Overbeek L, Termorshuizen AJ, van Bruggen AHC (2011) Influence of aerobic and anaerobic conditions on survival of *Escherichia coli* O157:H7 and *Salmonella enteric serovar Typhimurium* in Luriae-Bertani broth, farm-yard manure and slurry. J Environ Manage 92: 780–787.

15. van Elsas JD, Chiurazzia M, Mallona CA, Elhottova D, Krištufekb V, et al. (2012) Microbial diversity determines the invasion of soil by a bacterial pathogen. Proc Natl Acad Sci USA 109: 1159–1164.

16. Li HW, Jing HQ, Pang B, Zhao GF, Yang JC, et al. (2002) Study on diarrhea disease caused by enterohemorrhagic *Escherichia coli O157:H7* in Xuzhou city, Jiangsu province in 2000. Chin J Epidemiol 23: 119–122. (in Chinese)

17. Liu JB, Yang JC, Jing HQ, Xu JG (2007) Epidemiological investigation of enterohemorrhagic *Escherichia coli* O157:H7 infection status in Xuzhou city of Jiangsu province from 1999 to 2006. Dis Surveill 22: 516–518. (in Chinese)

18. Xu JG, Quan TS, Xiao DL, Fan TR, Li LM, et al. (1990) Isolation and characterization of *Escherichia coli* O157:H7 strains in China. Curr Microbiol 20: 299–303.

19. Liu JB, Jing HQ, Xu JG (2007) Research progress in enterohemorrhagic *Escherichia coli O157:H7* infection. Chin J School Docto 21: 720–722. (in Chinese)

20. Zhu WG, Xie SQ, Hong JX, Cui JF, Guo XF (2006) Detection of *Escherichia coli* O157:H7 by multiplex polymerase chain reaction. Chin J Zoono 22: 428–432. (in Chinese)

21. Wu EP, Liu LZ, Zhang Y (2006) Investigation into the infectiou of *Escherichia coli* O157:H7 in Zhengzhou in 2005. Mod Prev Med 13: 2455–2556. (in Chinese)

22. Miao QL, Pan WZ, Xu XZ (2008) Characteristic analysis of summer temperature in Nanjing during 56 years. J Tro Meteorol 24: 737–742. (in Chinese)

23. Richards LA (1949) Methods of measuring soil moisture tension. Soil Sci 68: 95–112.

24. Agricultural Chemistry Committee of China (1983) Conventional Methods of Soil and Agricultural Chemistry Analysis: Science Press, Beijing, China. (in Chinese)

25. Wu YP, Ma B, Zhou L, Wang HZ, Xu JM, et al. (2009) Changes in the soil microbial community structure with latitude in eastern China, based on phospholipid fatty acid analysis. Appl Soil Ecol 43: 234–240.

26. Zelles L. (1999) Fatty acid patterns of phospholipids and lipopolysaccharides in the characterisation of microbial communities in soil: a review. Biol Fertil Soils 29: 111–129.

27. Swallow M, Quideau SA (2013) Moisture effects on microbial communities in boreal forest floors are stand-dependent. Appl Soil Ecol 63: 120–126.

28. Dangi SR, Stahl PD, Wick AF, Ingram LJ, Buyer JS (2012) Soil microbial community recovery in reclaimed soils on a surface coal mine site. Soil Sci Soc Am J 76: 915–924.

29. Ding N, Guo HC, Hayat T, Wu YP, Xu JM. (2009) Microbial community structure changes during Aroclor 1242 degradation in the rhizosphere of ryegrass (Lolium multiflorum L.). FEMS Microbiol Ecol 70:305–314.

30. Ying JY, Zhang LM, Wei WX, He JZ. (2013) Effects of land utilization patterns on soil microbial communities in an acid red soil based on DNA and PLFA analyses. J Soil Sediment 13:1223–1231.

31. Institute of Soil Science, Academia Sinica (1990) Soils of China. Science Press, Beijing, China. 514–532. (in Chinese)

32. Wang HZ, Zhang TX, Wei G, Wu LS, Wu JJ, Xu JM. (2013) Survival of *Escherichia coli* O157: H7 in soils under different land use types. Environ Sci Pollut Res. DOI 10.1007/s11356-013-1938-9.

33. Ma JC, Ibekwe AM, Crowley DE, Yang CH (2012) Persistence of *Escherichia coli* O157:H7 in major leafy green producing soils. Environ Sci Technol 46: 12154–12161.

34. Oksanen J, Blanchet FG, Kindt R, Legendre P, Minchin PR, et al. (2012) vegan: Community ecology package. R package version 2.0-5. http://CRAN.R-project.org/package = vegan.

35. Benjamin MM, Datta AR (1995) Acid tolerance of enterohemorrhagic *Escherichia coli*. Appl Environ Microb 61: 1669–1672.

36. Fierer N, Jackson RB (2006) The diversity and biogeography of soil bacterial communities. Proc Natl Acad Sci USA 103: 626–631.

37. Lauber CL, Hamady M, Knight R, Fierer N (2009) Pyrosequencing-based assessment of soil pH as a predictor of soil bacterial community structure at the continental scale. Appl Environ Microb 75: 5111–5120.

38. Zhao WQ, Liu X, Cai P, Huang QY (2013) Mechanisms of bacterial pathogens adsorption on red soil colloids. Acta Pedol Sin 50: 1–9. (in Chinese)

39. Yee N, Fein JB, Daughne CJ (2000) Experimental study of the pH, ionic strength, and reversibility bacteria–mineral adsorption. Geochim Cosmochim Ac 64: 609– 617.

40. Cai P, Huang QY, Walker SL (2013) Deposition and survival of *Escherichia coli* O157:H7 on Clay Minerals in a Parallel Plate Flow System. Environ Sci Technol 47: 1896–1903.

41. Devau N, Cadre EL, Hinsinger P, Jaillard B, Gérard F (2009) Soil pH controls the environmental availability of phosphorus: experimental and mechanistic modelling approaches. Appl Geochem 24: 2163–2174.

42. Curtin D, Campbell CA, Jalil A (1998) Effects of acidity on mineralization: pH-dependence of organic matter mineralization in weakly acidic soils. Soil Biol Biochem 30: 57–64.

43. Aciego Pietri JC, Brookes PC (2008) Relationships between soil pH and microbial properties in a UK arable soil. Soil Biol Biochem 40: 1856–1861.

44. Ma JC, Ibekwe AM, Yang C, Crowley DE (2013) Influence of bacterial communities based on 454-pyrosequencing on the survival of *Escherichia coli* O157: H7 in soils. FEMS Microbial Ecol. 84: 542–554.

45. Birk JJ, Steiner C, Teixiera WC, Zech W, Glaser B. (2009) Microbial response to charcoal amendments and fertilization of a highly weathered tropical soil [M]// Amazonian Dark Earths: Wim Sombroek's Vision. Springer Netherlands,: 309–324.

46. Majeed H, Gillor O, Kerr B, Riley MA (2011) Competitive interactions in *Escherichia coli* populations: the role of bacteriocins. ISME J 5: 71–81.

47. Westphal A, Williams ML, Baysal-Gurel F, LeJeune JT, Gardener BBM (2011) General suppression of *Escherichia coli* O157:H7 in sand-based dairy livestock bedding. Appl Environ Microb 77: 2113–2121.

48. van Elsas JD, Semenov AV, Costa R, Trevors JT (2011) Survival of *Escherichia coli* in the environment: fundamental and public health aspects. ISME J 5: 173–183.

49. Simonetti G, Francioso O, Nardi S, Berti A, Brugnoli E, et al. (2011) Characterization of humic carbon in soil aggregates in a long-term experiment with manure and mineral fertilization. Soil Sci 76: 880–890.

19

Injury Profile SIMulator, a Qualitative Aggregative Modelling Framework to Predict Injury Profile as a Function of Cropping Practices, and Abiotic and Biotic Environment. II. Proof of Concept: Design of IPSIM-Wheat-Eyespot

Marie-Hélène Robin[1,2], **Nathalie Colbach**[3], **Philippe Lucas**[4], **Françoise Montfort**[4], **Célia Cholez**[1,2], **Philippe Debaeke**[1,5], **Jean-Noël Aubertot**[1,5]*

1 Institut National de la Recherche Agronomique, Unité Mixte de Recherche 1248 Agrosystèmes et agricultures, Gestion des ressources, Innovations et Ruralités, Castanet-Tolosan, France, 2 Université de Toulouse, Institut National Polytechnique de Toulouse, Ecole d'Ingénieurs de Purpan, Toulouse, France, 3 Institut National de la Recherche Agronomique, Unité Mixte de Recherche 1347 Agroécologie, Dijon, France, 4 Institut National de la Recherche Agronomique, Unité Mixte de Recherche 1099 Biologie des Organismes et des Populations appliquée à la Protection des Plantes. Le Rheu, France, 5 Université Toulouse, Institut National Polytechnique de Toulouse, Unité Mixte de Recherche 1248 Agrosystèmes et agricultures, Gestion des Ressources, Innovations et Ruralités, Castanet-Tolosan, France

Abstract

IPSIM (Injury Profile SIMulator) is a generic modelling framework presented in a companion paper. It aims at predicting a crop injury profile as a function of cropping practices and abiotic and biotic environment. IPSIM's modelling approach consists of designing a model with an aggregative hierarchical tree of attributes. In order to provide a proof of concept, a model, named IPSIM-Wheat-Eyespot, has been developed with the software DEXi according to the conceptual framework of IPSIM to represent final incidence of eyespot on wheat. This paper briefly presents the pathosystem, the method used to develop IPSIM-Wheat-Eyespot using IPSIM's modelling framework, simulation examples, an evaluation of the predictive quality of the model with a large dataset (526 observed site-years) and a discussion on the benefits and limitations of the approach. IPSIM-Wheat-Eyespot proved to successfully represent the annual variability of the disease, as well as the effects of cropping practices (Efficiency = 0.51, Root Mean Square Error of Prediction = 24%; bias = 5.0%). IPSIM-Wheat-Eyespot does not aim to precisely predict the incidence of eyespot on wheat. It rather aims to rank cropping systems with regard to the risk of eyespot on wheat in a given production situation through *ex ante* evaluations. IPSIM-Wheat-Eyespot can also help perform diagnoses of commercial fields. Its structure is simple and permits to combine available knowledge in the scientific literature (data, models) and expertise. IPSIM-Wheat-Eyespot is now available to help design cropping systems with a low risk of eyespot on wheat in a wide range of production situations, and can help perform diagnoses of commercial fields. In addition, it provides a proof of concept with regard to the modelling approach of IPSIM. IPSIM-Wheat-Eyespot will be a sub-model of IPSIM-Wheat, a model that will predict injury profile on wheat as a function of cropping practices and the production situation.

Editor: Matteo Convertino, University of Florida, United States of America

Funding: This study was carried out within a PhD project co-funded by INRA and INPT EI Purpan, by the project MICMAC design (ANR-09-STRA-06) supported by the French National Agency for Research, and by the Programme "Assessing and reducing environmental risks from plant protection products (pesticides)", funded by the French Ministry in charge of Ecology and Sustainable Development (project "ASPIB"). The funders had no role in study design, data collection and analysis, decision to publish, or preparation of the manuscript.

Competing Interests: The authors have declared that no competing interests exist.

* E-mail: Jean-Noel.Aubertot@toulouse.inra.fr

Introduction

Stem base diseases on cereals and grasses are widespread in many eco-regions of the world and cause important production and economic losses. The most detrimental foot and root pathogens on cereals in temperate areas are *Pseudocercosporella herpotrichoides*; *Fusarium* spp, *Rhizoctonia cerealis* and *Gaeumannomyces graminis* [1]. Eyespot caused by the necrotrophic and soil-borne fungi *Oculimacula yallundae* and *O. acuformis*, anamorph *Pseudocercosporella herpotrichoides* [2–4] is considered to be the most important

stem base disease of cereals in temperate countries [5]. Under cool and wet conditions in autumn and spring, both species sporulate and infect the stem bases of their hosts. Without any host crops (cereals, ryegrass), the pathogen survives on previously infected stubble, on which splash-dispersed conidia and air-dispersed ascospores are produced [6]. Injuries interfere with the circulation of nutrients and water through the base of the stem [7] leading to a weakening and possibly to a breakage of the stem base, causing lodging before harvest [5,8]. Relative yield losses of up to 50%

have been reported for the most severe attacks on winter wheat with lodging [2,7,9–11].

In the past, the control of eyespot has relied largely on chemical protection [12]. However, due to the development of resistance to the main available fungicides in *O. yallundae* and *O. acuformis* populations, adaptation of the entire cropping system to control eyespot on wheat is a sound alternative [13,14]. Furthermore, growing concerns about the impact of pesticides on the environment and human health has led to attempts to limit pesticide use [15,16]. Most governments of developed countries have launched national action plans to reduce pesticide use. For instance, the French government has set as a goal to reduce pesticide use by 50% by 2018 if possible [17]. The European Union has proposed to encourage the use of low-pesticide farming as one of its priorities by the Sustainable Use Directive (SUD) (http://eurlex.europa.eu/LexUriServ/LexUriServ.do?uri = OJ:L:2009:309:0071:0086:FR:PDF, accessed November 2012).

In addition, the USA decided to support and develop Integrated Pest Management (IPM) nationwide in order to reduce pesticide use [18]. It appears necessary therefore to combine various methods (cultural, genetic and chemical) in IPM strategies [19] to control eyespot on wheat. The main cultural practices that can partly control eyespot through a specific adaptation are: a low host frequency in the crop sequence, infected stubble management through adapted tillage, a late sowing date and low sowing rate [10,20,29] The genetic control of eyespot consists of using resistant cultivars. There are several known sources of resistance to eyespot, but only three resistance genes have been described so far [21–23].

IPM strategies, based on these control methods, have to be developed, adapted and applied to a wide range of physical, chemical, biological and socio-economic contexts. However, it is extremely difficult to describe the entirety of the cropping practices*environment*crop*pest system because of the tremendous number of interactions [24]. Modelling is certainly the best way to handle such a level of complexity and to help design sustainable innovative cropping systems less reliant on pesticides.

However, crop models do not deal with injuries caused by pests [25] and few pest models integrate the effects of cultural practices because of the difficulty of describing their numerous consequences on the agroecosystem [26] Thus, different models have been developed to represent eyespot injuries on wheat [27–29] or the associated damage [30] Among these, only one model takes into account the effect of the cropping system (crop succession, tillage, sowing date, sowing rate, total nitrogen fertiliser and its form) on injuries caused by eyespot [29]. However, this model does not take into account soil and climate, along with some cultural practices that can greatly influence the disease development (e.g. cultivar choice). There is therefore a need for a model that predicts as exhaustively as possible the effect of cropping practices on eyespot on wheat in a given production situation.

In this article, we will define the production situation as the physical, chemical and biological components, except for the crop, of a given field (or agroecosystem), its environment, as well as socio-economic drivers that affect farmers' decisions (adapted from [31,32], [33]). In this definition, "environment" refers to climate and the fraction of the territory that can influence pest dynamics through dispersal of harmful or beneficial organisms. In a given production situation, a farmer can design several cropping systems according to his goals, his perception of the socio-economic context and his environment, farm features, his knowledge and cognition. However, it is assumed that a given cropping system in a given production situation, such as defined above, should lead to a unique injury profile. In IPSIM, production situations are partly described by three components: soil, climate, and the biological environment of the field [33]. In the approach used here, the farmer's decision-making process and socio-economic drivers are not taken into account.

The conceptual bases of IPSIM have been described in detail by Aubertot and Robin [33]. The generic hierarchical aggregative modelling framework of IPSIM aims at predicting an injury profile as a function of cropping practices, soil, and climate and the biological field environment for any mono-specific crop production (arable crop, perennial or protected crops). In order to test whether this modelling approach could be successfully applied to represent injuries caused by a single pest, a model, named IPSIM-Wheat-Eyespot, has been developed according to the conceptual framework of IPSIM. It aims at predicting the final incidence of eyespot on wheat as a function of the production situation and cropping practices. IPSIM-Wheat-Eyespot gathers available knowledge in the scientific literature (models, experimental results) and expertise and will help design cropping systems with low risk of eyespot on wheat and perform diagnoses of commercial wheat fields. IPSIM-Wheat-Eyespot will be used as a sub-model for IPSIM-Wheat, a model that will predict the injury profile on winter wheat (i.e. the distribution of injuries caused by the most important detrimental pests on wheat [34]). This paper presents the method used to develop IPSIM-Wheat-Eyespot using the conceptual modelling framework of IPSIM [33], an evaluation of its predictive quality and a discussion on the limitations and benefits of the model.

Materials and Methods

Design of IPSIM-Wheat-Eyespot

1. General Approach. IPSIM-Wheat-Eyespot is based on the DEX method, and is implemented with the software DEXi [35]. DEX is a method for qualitative hierarchical multi-attribute decision modelling and support, based on a breakdown of a complex decision problem into smaller and less complex sub-problems, characterised by indicators (or attributes) that are organised hierarchically into a decision tree. These attributes are characterised by their name, a description and a scale. DEXi is generally used to evaluate and analyse decision problems, e.g. [36]. However, the DEX method has been used here in an original way to model complex agroecosystems. IPSIM-Wheat-Eyespot is therefore a hierarchical and qualitative multi-criteria model, allowing the prediction of eyespot injury according to various factors with sometimes opposite effects. IPSIM-Wheat-Eyespot has the following features (derived from [37]):

i)   Processes are hierarchically organised into a tree of attributes that constitutes the structure of the model;

ii) Terminal attributes of the tree (i.e. leaves or *basic attributes*) are input variables of the model and must be specified by users; the "trunk" of the tree (i.e. the final aggregated attribute) is the main model output variable (final eyespot incidence on wheat); internal nodes are called *aggregated attributes*;

iii) All model attributes are qualitative variables (nominal or ordinal) rather than quantitative variables. They take only discrete symbolic values, usually represented by words rather than numbers: e.g. "ploughing, stubble disking, rotary harrowing" for nominal variables, "low, medium, high" for ordinal variables;

iv) The aggregation of values up the tree is defined by aggregating tables for each aggregated attribute based on "if-then" decision rules. These aggregating tables can be seen as equivalents of parameters for quantitative numerical models, whereas the tree of attributes can be viewed as the equivalent of their mathematical structure.

IPSIM-Wheat-Eyespot was designed in 3 steps [37]: (1) identification and organisation of the attributes, (2) definition of attribute scales, and (3) definition of aggregating tables.

2. Identification and Organisation of Attributes. IPSIM-Wheat-Eyespot aims at predicting the incidence of eyespot on wheat in a given field according to a set of input variables. The spatial scale addressed is the field and the temporal scale is the wheat growing season, although some input variables encompass the crop sequence (up to the pre-preceding crop). IPSIM-Wheat-Eyespot is a static deterministic model.

The hierarchical structure presented in Figure 1 represents the breakdown of factors affecting eyespot final incidence into specific explanatory variables, represented by lower-level attributes. This figure represents the adaptation to eyespot of the model structure presented in Figure 2 by Aubertot and Robin [33].

In all, IPSIM-Wheat-Eyespot has 21 attributes, of which 14 are basic (i.e. input variables) and 7 aggregated. The 14 basic attributes are presented as the terminal leaves of the tree and their levels are aggregated into higher levels according to aggregating tables. They represent input variables of the model. Some of them (e.g. those representing the interactions at the territory level) could be omitted since they do not influence the final output. However, they were kept because these basic attributes will be necessary for the modelling of the whole injury profile on wheat. The aggregated attributes are internal nodes. They represent state variables or the output variable of IPSIM-Wheat-Eyespot. They are determined by lower-level basic attributes [38]. The output of IPSIM-Wheat-Eyespot is represented by the attribute "Final eyespot incidence" (eyespot incidence at the "milky grain", stage 7: development of fruit on BBCH scale [39]) which is determined by three main factors: cropping practices, soil and climate and the biological environment of the considered field. This is reflected by the hierarchical structure of the model, which consists of three sub-trees of attributes (Figure 1) split into one main part and two smaller ones. The main sub-tree, "Effect of cropping practices", illustrates the complexity of the effects of cropping practices and the need to consider a

combination of practices in order to evaluate the final eyespot incidence. It uses indicators based on tactical (with a short time-frame) or strategic decisions (with a longer time-frame [40]). These decisions can affect the agroecosystem at several stages.

i) Eyespot is considered as a highly endocyclic disease (as defined in [33]). Upstream, some cropping practices affect the quantity of the endo-inoculum (initial pathogen population present in the field). Crop sequence and tillage determine the vertical distribution of infected stubble and have proven to be of major importance for eyespot control [41–45]. Nevertheless, the effects of tillage on the disease are controversial in the literature. According to several authors [1,41,46–50], minimum tillage is highly favourable to eyespot development in the presence of preceding host-crop residues in the top layer, whereas ploughing significantly reduces its incidence by burying host-crop residues. These results conflict with those that show that eyespot was more severe after soil inversion than after non-inversion under moist, cool conditions [44–47,51–55]. The possible explanation of this apparent contradiction is that non-inversion is more favourable to antagonistic micro-organisms than ploughing (the microbiological activity is higher at the soil surface than in the top 20 cm soil layer and the weather in some experiments, such as those in Italy, was probably too dry for antagonistic biota to flourish on crop debris and thus to control eyespot [1].

ii) Action by escape consists of shifting periods of highest crop susceptibility away from the main periods of pathogen contamination. This is achieved by altering the wheat sowing date. In the case of eyespot, "escape strategies" cannot really be considered. However, early sowing increases the probability of autumn contamination through primary infection, due to the longer time available for eyespot to develop and to affect stems [42].

iii) During the crop cycle, some cropping practices can mitigate infection through crop status by increasing crop competitiveness and/or by creating less favourable conditions for pest development. Low plant density can limit pathogen development through several mechanisms, such

Attribute
```
Final incidence of eyespot
 ├Effects of cropping practices
 │ ├Primary inoculum management: interaction between crop sequence and tillage
 │ │ ├Preceding crop
 │ │ ├Pre-preceding crop
 │ │ ├Tillage after harvest of the previous crop
 │ │ └Tillage after harvest of the pre-previous crop
 │ ├Escape: effects of the sowing date
 │ ├Mitigation through crop status
 │ │ ├Cultivar choice
 │ │ ├Level of N fertilisation
 │ │ └Sowing rate
 │ └Chemical control: use of fungicide
 ├Effects of soil and climate
 │ ├Soil
 │ └Climate
 │   ├Autumn/winter
 │   └Spring
 └Interactions with the territory
   ├Beneficial sources
   └Primary inoculum sources
```

Figure 1. Hierarchical structure of IPSIM-Wheat-Eyespot (screenshot of the DEXi software). Bolded and non-bold terms represent aggregated and basic attributes, respectively.

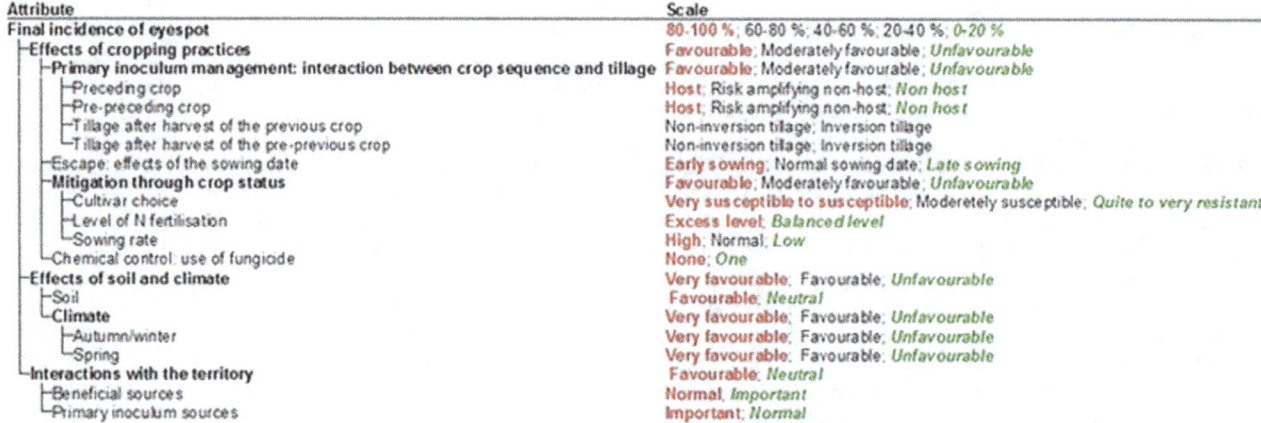

Figure 2. Attribute scales of IPSIM-Wheat-Eyespot (screenshot of the DEXi software). All the scales are ordered from values detrimental to the crop (i.e. favourable to eyespot) on the left-hand side to values beneficial to the crop on the right-hand side (i.e. unfavourable to eyespot). In the DEXi software, this difference is clearly visible because, by convention, values beneficial to the user are coloured in green, detrimental in red, and neutral in black.

as restricting the contact between plant organs and infectious propagules and lowering the humidity within the canopy. This results in a control of soil-borne diseases like eyespot by low plant density and/or a high shoot number per plant [20]. In addition, low densities increase distances between plants, which limits secondary pathogen cycles, and leads to a drier microclimate. Excessive use of nitrogen fertilisers produces lush crops and favours eyespot through direct and indirect effects [56,57]. However, in the case of eyespot, nitrogen availability in the soil seems to be a minor factor for the development of the disease [10,20,29].

Use of disease-resistant cultivars provides an economic, environmentally friendly and effective strategy to control disease. However, not all resistant cultivars have been assessed in integrated cropping systems [58] and cultivars do not share the same susceptibilities to different diseases [59]. Eyespot resistance is generally not complete and its expression depends widely on environmental factors [22].

iv)   Lastly, a fall-back solution (use of fungicide) can be used when alternative practices are not sufficient. However, several studies have provided evidence for reduced susceptibility to fungicides in populations of *O. yallundae* and *O. acuformis* [60]. For the sake of simplicity, resistance to fungicide in pathogen populations was not taken into account in IPSIM-Wheat-Eyespot.

The two other sub-trees describe the biological environment of the considered field, as well as soil and climate. These sub-trees are not affected by cropping practices. Among these factors, climate is the main factor affecting eyespot development [9,43].

3. Definition of the Attribute Scales. The second step in the design of a DEXi model is the choice of ordinal or nominal scales for basic and aggregated indicators. Sets of discrete values were defined for all attributes of the model and described by symbolic value scales defined by words. These values were defined according to the knowledge available in the international literature and some expertise when needed. IPSIM-Wheat-Eyespot uses at most a three-grade value scale (i.e. "Unfavourable", "Favourable", "Very favourable") for the aggregated and basic attributes.

This scale refers to the disease. The value "Favourable" means that the attribute is favourable to the development of the disease and therefore potentially detrimental to the crop.

Some values for basic indicators can be specified using quantitative values that are then translated into qualitative values. For instance, the translation into qualitative values of the sowing date, sowing density or N rate is performed using experimental references or expertise. This translation takes into account the regional context. For example, a sowing date classified as "Early" in the south of France might be classified as "Normal" in northern France. This classification actually depends on the sowing date distribution in the considered region.

Other attributes are directly qualitatively estimated. For instance, the indicators "inversion tillage or non-inversion tillage" or "preceding and pre-preceding crop" are nominal variables and directly monitored as such in experiments [61,62]. The level of cultivar resistance has been described using the official list provided by the French National Seed Station (Groupe d'Etude et de contrôle des Variétés et des Semences; http://cat.geves.info/Page/ListeNationale; accessed November 2012) and published by Arvalis-Institut du végétal (http://www.arvalisinfos.fr/_plugins/WMS_BO_Gallery/page/getElementStream.html?id=13504&prop=file; accessed November 2012). In this list, cultivars are rated for their susceptibility to eyespot on a 0–9 scale, from very susceptible to resistant.

For the climate attribute, a three-value scale ("Unfavourable"; "Favourable"; "Very favourable") was defined using climatic models [27,43] and data from the INRA Climatik database.

All the scales in Figure 2 are ordered from values detrimental to the crop (i.e. favourable to the disease) on the left-hand side to values beneficial to the crop on the right-hand side (i.e. unfavourable to the disease). In the DEXi software, this difference is clearly visible because, by convention, values beneficial to the user are coloured in green, detrimental in red, and neutral in black. The scales for the "tillage after preceding crop" and "tillage after pre-preceding crop" attributes appear in black since their effects on the disease cannot be defined independently from the crop sequence.

Initial input attribute values (either quantitative or qualitative) are translated into qualitative appreciation, according to two to three scales defined on the basis of available information in the literature, models or expertise. Sometimes, a two-value scale is enough to represent the value of an indicator (e.g. chemical control was applied or not; or the soil has either been ploughed or not after the preceding harvest). However, other attributes usually need a three-value scale to describe the diversity of cropping practices or environment (e.g. the sowing rate attribute requires three grades to describe farmers' practices: the sowing rate can be low, normal or high).

4. Definition of Aggregating Tables.   The third step in the design of a DEXi model is the choice of aggregating tables determining the aggregation of attributes in the tree and their interactions. For each aggregated attribute in the model, a set of "*if-then*" rules define the value of the considered attribute as a function of the values of its immediate descendants in the model. The rules that correspond to a single aggregated attribute are gathered together and conveniently represented in tabular form. In this way, each table defines a mapping of all value combinations of lower-level attributes into the values of the aggregate attribute. Figure 3 shows decision rules that correspond to the "mitigation through crop status" aggregated attribute and define the value of this attribute for the 18 possible combinations of the three cultivar choices, the 2 levels of fertilisation and the 3 sowing densities. For example, if the cultivar is quite resistant, the level of N fertilisation balanced and the sowing rate low, then the "mitigation through crop status" attribute will be unfavourable to eyespot (the final incidence will decrease). However, even if the sowing rate and the N application rate are both high, the "mitigation through crop status" attribute during wheat growth will control eyespot significantly because the "cultivar choice" attribute is much more influential than the two other attributes (Figure 3).

The aggregating tables of IPSIM-Wheat-Eyespot have been established using knowledge available in the international literature and summarised in Table 1, and expert knowledge when needed. All aggregating tables of the model are presented in figures S1, S2, S3, S4, S5, S6.

5. Attribute Weights.   The influence of each basic and aggregated attribute on the value of the output variable can be characterised with weights. The higher the weight, the more important the attribute. Table 2 summarises the weights of each of the 19 attributes of the model, providing an overview of the model's structure. IPSIM-Wheat Eyespot has 3 levels of aggregation (Figure 1), the third one being the leaves (i.e. the model input basic attributes). The "local" and "global" weights are normalised in two different ways. "Local" weights are given to each aggregated attribute separately so that the sum of weights of its immediate descendants in the hierarchy equals 100%. The "global" weights are calculated at a given level of aggregation and express the influence of each attribute at that aggregation level. They are obtained by multiplying the local weight of a given attribute at a given level of aggregation, by local weighting of its ascendants. For instance, the value of the "soil and climate" attribute is completely defined by the "Climate" attribute (100%, local weight), but this attribute only contributes 53% to the definition of the value of "Eyespot incidence" (global weight at the second level of aggregation). Local and global weights are identical at the first level of aggregation, since in this case there is only one level of aggregation. Global weights of basic attributes are shown in bold in Table 2 in order to ease their identification, since they are distributed among the second and third levels of aggregation of IPSIM-Wheat-Eyespot. The sum of global weights at the third level is only 76%. This is because some basic attributes are directly embedded in the model at the second level of aggregation. The sum of global basic attribute weights is logically equal to 100%. Table 2 can be seen as an equivalent of a sensitivity analysis that would aim at identifying the most influential input (and state) variables of a quantitative model.

6. Simulations with DEXi.   The qualitative final attribute value (final incidence of eyespot) is calculated by DEXi. The

	Cultivar choice	Level of N fertilisation	Sowing rate	Mitigation through crop status
1	Very susceptible to susceptible	Excess level	High	**Favourable**
2	Very susceptible to susceptible	Excess level	Normal	**Favourable**
3	Very susceptible to susceptible	Excess level	Low	**Favourable**
4	Very susceptible to susceptible	Balanced level	High	**Favourable**
5	Very susceptible to susceptible	Balanced level	Normal	**Favourable**
6	Very susceptible to susceptible	Balanced level	Low	**Favourable**
7	Moderetely susceptible	Excess level	High	**Moderately favourable**
8	Moderetely susceptible	Excess level	Normal	**Moderately favourable**
9	Moderetely susceptible	Excess level	Low	**Moderately favourable**
10	Moderetely susceptible	Balanced level	High	**Moderately favourable**
11	Moderetely susceptible	Balanced level	Normal	**Moderately favourable**
12	Moderetely susceptible	Balanced level	Low	**Moderately favourable**
13	Quite to very resistant	Excess level	High	**Unfavourable**
14	Quite to very resistant	Excess level	Normal	**Unfavourable**
15	Quite to very resistant	Excess level	Low	**Unfavourable**
16	Quite to very resistant	Balanced level	High	**Unfavourable**
17	Quite to very resistant	Balanced level	Normal	**Unfavourable**
18	Quite to very resistant	Balanced level	Low	**Unfavourable**

Figure 3. Aggregating table for the "Mitigation through crop status" aggregated attribute (screenshot of the DEXi software). Aggregation rules for the 18 possible combinations of the 3 cultivar choices, the 2 levels of fertilisation and the 3 sowing rates.

Table 1. Available knowledge in the scientific literature describing the effects of cropping practices and the production situation on the incidence of eyespot on wheat.

Factor	Direction of the effect	Intensity of the effect	Impact on eyespot development	References
Tillage	+/−	++	Contradictory results. For some authors, reduced soil tillage decreased eyespot infection. For others, eyespot was often more severe after ploughing than after non-inversion tillage.	[1,11,41,42,44–55]
Preceding and pre-preceding crop	+	++	Preceding and pre-preceding host crops are known to favour eyespot. However, the interaction between tillage and the crop sequence has to be taken into account.	[9,10,29,41–42,55,61,62]
Sowing date	+	++	Eyespot has always been reported to be more severe in early sown crops.	[10,20,27,29,42,55]
N fertilisation rate	+	+	High nitrogen availability generally favoured the disease. However these results were questioned.	[9,10,20,29,56,57]
Sowing rate	+	+	Prevalence was increased by high plant density and/or low shoot number per plant.	[20,29]
Cultivar choice	+	+++	The use of varieties with resistance could obviate the need for fungicide.	[10,21,22,42,58,59]
Cultivar mixture	0	0	No significant difference was found between the disease level in mixtures and the mean of disease level of the mixture components in pure stands.	[70–72]
Climate	+	++	Eyespot strongly depends on climate. Infections require periods of at least 15 h with T° between 4°C and 13°C and HR>80% (from October to April).	[9,27,28,29,43]

Cropping practices and climate can be favourable (+), unfavourable (−) or neutral (0) to the development of eyespot. The intensity of the considered factor is summarised with 4 classes: 0, no effect; +, slight; ++, significant; +++, crucial.

calculation consists in computing all aggregated attribute values according to: (i) the structure of the tree; (ii) a set of input variables (basic attribute values) defining a simulation unit; and (iii) the aggregating tables for the aggregation of attributes. An example of output results obtained for two simulation units is provided in Figure 4 (input basic attributes and calculated aggregated attribute values for the simulation of two systems: an organic and a high-input one).

Table 2. Respective weights of the attributes of IPSIM-Wheat-eyespot.

Attributes defining the final incidence of eyespot	Local level 1	Local level 2	Local level 3	Global level 1	Global level 2	Global level 3
1 Effects of cropping practices	47			47		
1.1 Primary inoculum management		21			10	
1.1.1 Preceding crop			40			4
1.1.2 Pre-preceding crop			12			1
1.1.3 Tillage after the preceding crop			40			4
1.1.4 Tillage after pre-preceding crop			8			1
1.2 Escape: effects of sowing date		9			4	
1.3 Mitigation through crop status		26			12	
1.3.1 Cultivar choice			100			12
1.3.2 Level of N fertilisation			0			0
1.3.3 Sowing rate			0			0
1.4 Chemical control		44			21	
2 Effects of soil and climate	53			53		
2.1 Soil		0			0	
2.2 Climate		100			53	
2.2.1 Autumn/winter			29			15
2.2.2 Spring			71			38
3 Interactions with the rest of the territory	0			0		

The "local" and "global" weights are calculated for each aggregated attribute separately and are distributed in 3 levels of aggregation. Bold and non-bold terms represent basic attributes and aggregated terms, respectively.

Option	Organic system	High input system
. Final incidence of eyespot	20-40 %	60-80 %
. . Effects of cropping practices	Unfavourable	Moderately favourable
. . . Primary inoculum management: interaction between crop sequence and tillage	Unfavourable	Favourable
. . . . Preceding crop	Non host	Host
. . . . Pre-preceding crop	Non host	Host
. . . . Tillage after harvest of the previous crop	Inversion tillage	Non-inversion tillage
. . . . Tillage after harvest of the pre-previous crop	Inversion tillage	Non-inversion tillage
. . . Escape: effects of the sowing date	Late sowing	Early sowing
. . . Mitigation through crop status	Unfavourable	Favourable
. . . . Cultivar choice	Quite to very resistant	Very susceptible to susceptible
. . . . Level of N fertilisation	Balanced level	Balanced level
. . . . Sowing rate	High	Normal
. . . Chemical control: use of fungicide	None	One
. . Effects of soil and climate	Very favourable	Very favourable
. . . Soil	Favourable	Favourable
. . . Climate	Very favourable	Very favourable
. . . . Autumn/winter	Very favourable	Very favourable
. . . . Spring	Very favourable	Very favourable
. . Interactions with the territory	Neutral	Neutral
. . . Beneficial sources	Normal	Normal
. . . Primary inoculum sources	Normal	Normal

Figure 4. Example of 2 simulations carried out with IPSIM-Wheat-Eyespot (screenshot of the DEXi software).

Evaluation of the Predictive Quality of IPSIM-Wheat-Eyespot

1. Description of the Dataset Used. Data representative of a wide range of climate patterns, soils and cropping practices are needed to assess the predictive quality of the model. A large dataset was therefore developed to assess the predictive quality of IPSIM-Wheat-Eyespot. A national survey was conducted to identify relevant data from various research and development institutes. The required datasets had to provide information for input attributes of IPSIM-Wheat-Eyespot (description of cropping practices, soil and climate) and its output (eyespot incidence at the "milky grain", stage 7: development of fruit on BBCH scale [39]). The dataset obtained is summarised in Table 3. It comprises results from multifactorial trials from 1980 to 1994 in 7 contrasting regions in France, which were set up to analyse the effects of various cropping practices on foot and root winter wheat diseases on different soils and with differing climate patterns. Various cultivars were combined with different crop sequences, conventional and reduced tillage, low or high plant densities, early or late sowing dates, low or high N fertilisation, in various areas of production where eyespot epidemics are observed. Most of these trials were specific studies on foot diseases [20,41,61,62], so the experimental conditions were suited to ensure the presence of eyespot (i.e. infected wheat present in the crop sequence and only susceptible cultivars). Other data originated from a regional agronomic diagnosis [63] performed in cereal fields from 1987 to 1994 in 19 French regions to analyse the effects of cultural practices on the incidence and severity of foot and root disease complexes [64]. In this survey, data were collected on 894 cereal fields in a wide range of production situations.

For some situations, the pre-preceding crop (3 possible types of crop in the model: "host", "non-host" and "risk amplifying") and the associated tillage after the harvest of this crop (2 possible values

in the model) were not observed. Instead of ignoring these precious data, simulations were performed for the 3*2 possibilities and only cases for which the 6 simulations led to similar output values were kept for evaluating the model. In all, 526 site-years were used for the evaluation of the model and they represented a large number of combinations of cropping practices and production situations (19 French regions over 9 years).

The data presented in Table 3 were transformed into qualitative values and used as input basic attributes to feed IPSIM-Wheat-Eyespot.

2. Evaluation of the Predictive Quality of IPSIM-Wheat-Eyespot. The evaluation consisted in comparing simulated and observed values. Since the model predicts classes of incidence, observed incidences at wheat stage 7 were transformed into observed incidence classes using the same discretisation as the model (i.e. 0–20%, 20–40%, 40–60%, 60–80%, 80–100%). However, one might want to predict incidences rather than classes of incidence. In order to test the predictive quality of IPSIM-Wheat-Eyespot for incidences, its output main variable was transformed into a numerical value by replacing the predicted incidence class by the centre of the class. The model was therefore evaluated in two ways: first, on its ability to predict incidence classes, and second on its ability to predict eyespot incidences.

For incidence classes, the deviation of the model was characterised by calculating the number of classes of difference between observed and simulated classes. The distribution of simulated classes was displayed according to observed incidence classes. This information was summarised by a multinomial distribution in 9 difference classes (from −4 to +4) since the model has 5 incidence classes. The proportion of situations for which the model correctly predicted the observed incidence class was taken as an indicator of the quality of prediction of the model. In addition, a non-parametric Wilcoxon test was performed to test

Table 3. Main features of the datasets used for the evaluation of IPSIM-Wheat-Eyespot's predictive quality.

Cropping practice	Design	Year	Location	Number of site-years	References
Crop sequence	Multifactorial field trials	1981–1982	Toulouse (Midi-Pyrénées)	11	[61]
Crop sequence including various durations of continuous cereal cropping	Multifactorial field trials	1980–1994	Grignon (Ile-de-France)	29	[62]
Tillage (soil structure)	Multifactorial field trials	1992–1993	Péronne (Picardie)	8	[73]
Tillage (crop residue vertical distribution)	Multifactorial field trials	1992–1993	Chartres (Centre), Grignon (Ile-de-France)	12	[41]
Sowing date, sowing rate, N fertilisation	Multifactorial field trials	1992–1994	Chartres, La Verrière (Ile-de-France), Le Rheu (Bretagne), Nancy (Lorraine), Dijon (Bourgogne)	95	[20]
Tillage, previous crop, fertilisation, sowing rate, sowing date, cultivar choice and use of fungicide	Diagnoses in cereal fields	1987–1994	19 French regions	370	[64]
Crop sequence	Multifactorial field trials	1981–1982	Toulouse (Midi-Pyrénées)	11	[61]
Crop sequence including various durations of continuous cereal cropping	Multifactorial field trials	1980–1994	Grignon (Ile-de-France)	29	[62]
Tillage (soil structure)	Multifactorial field trials	1992–1993	Péronne (Picardie)	8	[73]
Tillage (crop residue vertical distribution)	Multifactorial field trials	1992–1993	Chartres (Centre), Grignon (Ile-de-France)	12	[41]
Sowing date, sowing rate, N fertilisation	Multifactorial field trials	1992–1994	Chartres, La Verrière (Ile-de-France), Le Rheu (Bretagne), Nancy (Lorraine), Dijon (Bourgogne)	95	[20]
Tillage, previous crop, fertilisation, sowing rate, sowing date, cultivar Diagnoses in cereal fields choice and use of fungicide		1987–1994	19 French regions	370	[64]

whether the distribution of errors was zero-centred (in that case, the model can be considered unbiased).

For incidences, the predictive quality of IPSIM-Wheat-Eyespot was characterised using three common statistical criteria [65]: bias (Equation 1), Root Mean Square Error of Prediction (RMSEP, Equation 2), and efficiency (Equation 3).

$$Bias = \frac{1}{n} \sum_{i=1}^{i=n} \left(Y_i^{obs} - Y_i^{sim} \right) \qquad (1)$$

where n is the total number of considered situations, Y_i^{obs} the observed value for situation i, and Y_i^{sim} is the corresponding value simulated by the model. The bias measures the average difference between observed and simulated values. If the model underestimates the considered variable, the bias is positive. Conversely, if the model overestimates the variable, the bias is negative.

$$RMSEP = \sqrt{\frac{1}{n} \sum_{i=1}^{i=n} \left(Y_i^{obs} - Y_i^{sim} \right)^2} \qquad (2)$$

RMSEP quantifies the prediction error when the model parameters have not been estimated using the observations Y_i^{obs} used in the calculation of this criterion.

$$EF = 1 - \frac{\sum_{i=1}^{i=n} \left(Y_i^{obs} - Y_i^{sim} \right)^2}{\sum_{i=1}^{i=n} \left(Y_i^{obs} - \bar{Y} \right)^2} \qquad (3)$$

Where $\bar{Y}$ is the mean of observed data. Nash and Sutcliffe [66] defined the efficiency as a normalised statistic that determines the relative magnitude of the residual variance ("noise") compared with the measured data variance ("information"). The efficiency defines the ability of a model to predict the value of a variable. The efficiency can range from $-\infty$ to 1. If the model perfectly predicts the observations, the efficiency is maximum and is equal to 1. Efficiency values lower than 0 indicate that the mean observed value is a better predictor than the simulated values, which indicate a poor predictive quality of the model. Values between 0 and 1 are generally viewed as acceptable levels of performance. The closer the model efficiency is to 1, the better is the fit between observed and simulated data [65].

Results

Evaluation of the Quality of Prediction for Final Incidence Classes

The high number of observed site-years in the dataset (526) permitted a reliable evaluation of the predictive quality of IPSIM-Wheat-Eyespot. Residuals were distributed around 0 (Figure 5), indicating that the predicted values were close to observations. Nearly half (47.1%) of the simulated classes encompassed the observed values and 80.4% had at most a difference of one class only. In addition, there are nearly as many negative as positive differences of exactly one class. The Wilcoxon test performed over the 9 class differences (from -4 to $+4$) proved that the model was significantly biased (simulated final incidence classes lower than observations, $p < 1.0 \ 10^{-10}$). Figure 6 illustrates the distribution of class differences between observed and predicted final eyespot incidences. The overall predictive quality of IPSIM-Wheat-Eyespot was judged fair, even if slightly biased. The predictive

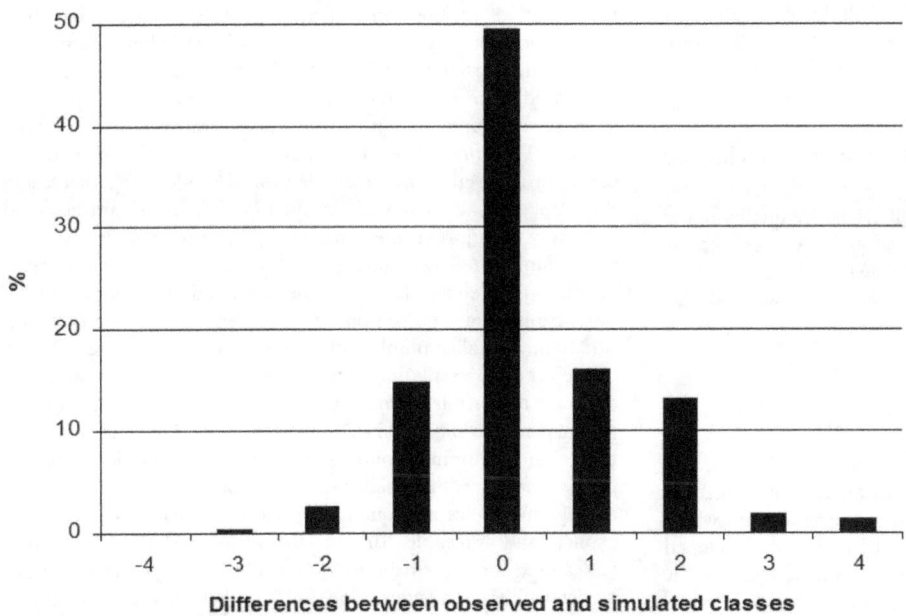

Figure 5. Evaluation of the predictive quality of IPSIM-Wheat-Eyespot. Residuals distribution: number of classes of difference between observed and simulated final eyespot classes (0–20%, 20–40%, 40–60%, 60–80%, 80–100%; 526 fields, over 9 years and 19 French regions).

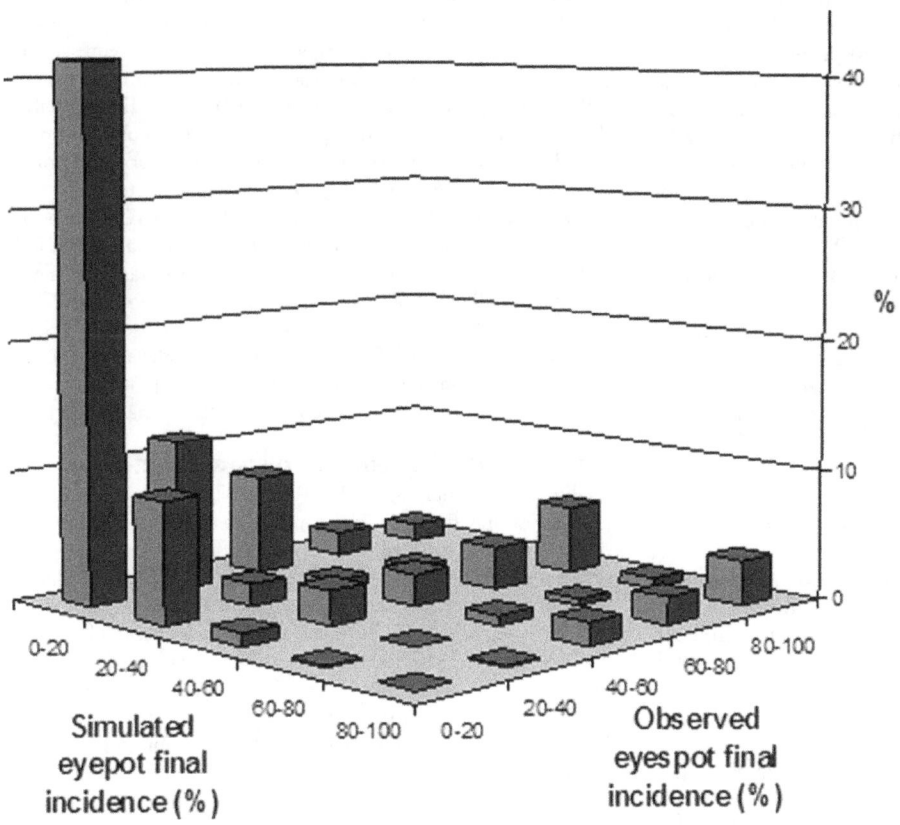

Figure 6. Evaluation of the predictive quality of IPSIM-Wheat-Eyespot Distribution of class differences between observed and predicted final eyespot incidences. (526 fields, over 9 years and 19 French regions).

quality was good for the lowest class (52% of all the observations in the dataset): 80% of the observed values between 0 and 20% were correctly simulated. The model underestimated final incidences for observations higher than 20%.

Evaluation of the Quality of Prediction for Final Incidence Values

For these 526 output values, the overall predictive quality of the model was correct. The model's predictive quality was good as its efficiency value was correct: 0.51. The Root Mean Square Error of Prediction error was quite high, 24%. The bias was positive (5.0%), so the model slightly underestimated final eyespot incidences.

Discussion

Interests and Limitations of IPSIM-Wheat-Eyespot

Several studies have been conducted to analyse the effects of cropping practices on the development of eyespot on wheat [10,29]. However, only one statistical model had been developed in order to predict the incidence of eyespot as a function of few cropping practices [29]. IPSIM-Wheat-Eyespot offers new possibilities for the design of innovative cropping systems since it is the first functional model to encompass simultaneously the effects of soil, climate and the cropping system and to represent the effects of interactions among these many factors.

The development of IPSIM-Wheat-Eyespot was made possible using (1) a schematic representation of the relationships between cropping practices, the production situation and injuries, (2) the translation of this conceptual scheme into a simulation model, and (3) a combination of data from a wide range of production situations (many regions and years) to test its predictive quality.

1. Conceptual Bases of IPSIM-Wheat-Eyespot. The conceptual scheme of IPSIM-Wheat-Eyespot is innovative because i) it encompasses a temporal scale longer than the cropping season (effect of the crop sequence in interaction with tillage over two years); ii) the main cropping practices that can affect the disease are represented; iii) interactions between practices as well as interactions between practices and climate are taken into account. As compared to the conceptual scheme of IPSIM [33], the spatial scale considered was limited to the field because of the lack of interactions at larger scales. In addition, the conceptual scheme of IPSIM-Wheat-Eyespot does not take into account socio-economic drivers, farmer's goals and cognition since it does not aim at simulating decisions. However, this original conceptual model can help design innovative cropping systems less susceptible to eyespot. The information provided by IPSIM-Wheat-Eyespot should be combined with other sources of information (references, other models, or expertise) in order to design new cropping systems, especially since damage (i.e. crop loss) caused by the disease are not represented.

2. Hierarchical Tree of Attributes and Aggregating Tables. The qualitative nature of the DEX method is well suited to the modelling of complex systems for which no high level of precision is required. The DEXi software tool [36] offered a suitable environment for the organisation of available knowledge and a rapid development of IPSIM-Wheat-Eyespot. The main breakthrough of the IPSIM platform is to allow the handling of complexity in a simple way [33]. The work presented in this paper provides a proof of concept for this innovative modelling approach in the field of crop protection for a single disease. A major innovation of this modelling approach is to be able to aggregate attributes of different natures (e.g. cultivar choice, a nominal variable and fertilisation rate, a quantitative variable) to describe

the impact of various components of cropping systems and their interactions on eyespot incidence. IPSIM-Wheat-Eyespot is actually the first model which can overcome the lack of data on the relationships between cropping practices and a single pest in a given production situation to help design strategies to control the disease. The qualitative DEXi approach may lead to a loss of precision and sensitivity in the developed model [67]. Increasing the number of attribute scales at the top of the decision tree could be a way to improve the sensitivity [68]. However, it results in more complicated aggregating tables which are consequently more difficult to define. Due to the tremendous complexity of interactions between cropping practices and the production situation, a smaller number of indicator states have been chosen to keep the representation of the complex underlying mechanisms as simple as possible. A correct definition of aggregating tables is of primary importance in DEXi models [68]. The choice of the nature and the number of qualitative scales is also crucial and will partly determine the quality of prediction. The choices of both aggregating tables and qualitative scales of attributes have to be explicit and traceable. Indeed, the scales and the aggregating tables used for the attributes of IPSIM-Wheat-Eyespot could not be determined in a generic way but have been specifically defined according to experimental results, models available in the literature and expert knowledge if need be. Unfortunately, literature to analyse some attributes may not exist, lack certain features, or controversial. For instance, the impact of soil type on eyespot incidence is very poorly described in the international literature and the relationships between tillage and eyespot are subject to much controversy [43]. For these cases, expert knowledge had to be used to complete some aggregating tables. In addition, the model runs using simple "if–then" rules, which are "shallow" in the sense that they only define direct relationships between conditions and consequences, but do not represent any "deeper" (or mechanistic) biological, physical, chemical processes [69]. Since the early stages of development of IPSIM-Wheat-Eyespot, it has been clear that precision was not an objective of the model. It appears more important to focus on accuracy rather than precision when modelling such a complex system.

Table 2 reveals the overall behaviour of IPSIM-Wheat-Eyespot. This is also an additional value of the IPSIM approach: the model is transparent and can be easily discussed. For instance, it is clear that the overall effect of fungicide on the disease is low (21%). This is because fungicide does not always control the disease efficiently [59]. The main factor influencing the disease is the spring weather (38%). This is consistent with Matusinsky et al. [43] who showed that the disease was very dependent on the climatic conditions during spring.

3. Predictive Quality of IPSIM-Wheat-Eyespot. The quality of the analysis of the IPSIM-Wheat-Eyespot predictive quality not only depends on the model itself (hierarchical structure of attributes and aggregating tables) but also on the diversity of the data used, which must reflect a wide range of production situations. These data should represent a variety of soil, climate and cropping practices, but also of final incidences. The dataset used in this study satisfied the three former conditions, but did not fully satisfy the latter. The observed final eyespot incidences were generally quite low, so the predictive quality of IPSIM-Wheat-Eyespot could not be extensively evaluated for high levels of incidence.

The main difference with other models is that IPSIM-Wheat-Eyespot is based on qualitative variables and not quantitative ones. The use of qualitative data requires greater attention to the description of the adopted hypotheses, because qualitative data are more difficult to interpret objectively [68]. This is particularly the case for the transformation of quantitative variables that have to be translated into qualitative input of IPSIM-Wheat-Eyespot (e.g. sowing density expressed in kg/ha or number of seeds m^{-2} and translated into "low", "normal" or "high"). Thus, IPSIM-Wheat-Eyespot can be used in 2 ways. On the one hand, some users can provide directly qualitative basic attributes (i.e. input variables of the model) if they want to test the performance of some technical options in given production situation. On the other hand, other users might want to run the model for real or putative situations where both the production situation and cropping practices are characterised with quantitative, qualitative or nominative data. In this case, an algorithm should be developed in order to rigorously translate these data into appropriate basic attributes based on national or international official references (e.g. a given cultivar will be classified as "very susceptible to susceptible"; "moderately susceptible"; or "resistant" according to official national or international seed classification); regional references (e.g. a given sowing date will be classified as "early"; "normal"; "late" as a function of regional references established by extension services); knowledge available in the literature (e.g. a given crop will be classified as "host", "non-host", or "risk-amplifying crop" according to published scientific articles); references produced by models (e.g. a given weather scenario can be classified as "very favourable"; "favourable"; or "unfavourable" according to a published model).

IPSIM-Wheat-Eyespot proved to fairly represent the variability of the 526 "site-years" used to test its predictive quality. This indicates that the model is already operational and can represent the effects of a wide range of production situations*cropping practices combinations for eyespot epidemics to help design cropping systems less susceptible to the disease. This is remarkable since, unlike most models, no fitting procedure was used.

Prospects

1. Improvements to the Model. Further refinements could be added in the future. They should keep the balance between: i) modelling of the effects of cropping practices and the production situation on eyespot epidemics as accurately as possible, and ii) keeping the model as simple as possible. In addition to the design of a model, the approach presented in this article allowed us to structure the available knowledge in the literature about the effects of cropping practices and the production situation on eyespot epidemics (Table 1). Aggregating tables derived from Table 1 could be easily adapted according to future advances in the knowledge of underlying mechanisms responsible for the disease. In the same way, the model structure could easily be modified to integrate new knowledge. For instance, the model does not yet take into account the effects of cultivar mixtures, whereas some authors have described a reduction of eyespot by cultivar mixtures [70–72]. However, this cropping practices is not currently widespread, data are sparse and there is no consensus in the literature on this matter.

IPSIM-Wheat-Eyespot requires the provision of qualitative basic attributes. This is a benefit for the *ex-ante* design of innovative cropping systems. However, this requires translating nominative or quantitative variables used to describe cropping practices and the production situation into *ad hoc* qualitative variables. In order to avoid subjectivity when translating these variables, some reference values have to be used. Such values were gathered for several French regions (data not shown) in order to design an algorithm that translates nominative or quantitative variables describing cropping practices and the production situation into relevant basic attributes of the model. This algorithm can be easily adapted to any location where wheat is grown and eyespot is present,

provided that relevant reference values are available. At last, aggregating tables could be adjusted to improve IPSIM-Wheat-Eyespot's predictive quality using statistical procedures, as done for parameter estimation for quantitative models.

2. Future Use of the Model. The main breakthrough of the IPSIM framework, with a simple hierarchical aggregative structure, is to allow the handling of complexity in a simple way. The input variables of models developed with IPSIM, such as IPSIM-Wheat-Eyespot, are easily obtained [33]. IPSIM-Wheat-Eyespot will help design cropping systems with a lower risk of eyespot on wheat. In order to do so, simulation plans will be defined to assess the performance of cropping practices in a given production situation with regard to the control of the disease. It is obvious that this simulation work will have to be combined with other sources of information such as other models, expert knowledge, diagnoses in commercial fields or experiments to propose innovative sustainable cropping systems.

The model, along with the interface that translates nominative and quantitative variables into relevant qualitative input variables for IPSIM-Wheat-Eyespot (Microsoft® Office Excel 2003), is now available upon request. This model can now be used as a communication, organisation, training and teaching tool for researchers, extension engineers, advisers, teachers or even farmers. Appropriation and adaptation of the model by technicians, advisers or farmers could be useful to exchange knowledge and experience (building up from their technical know-how).

The model presented in this paper only takes into account one pest among the biocenosis of a wheat field. Nevertheless, it is necessary to consider the entirety of the major pests when designing cropping systems because farmers have to manage combinations of pest populations, leading to injury profiles, which can in turn lead to quantitative or qualitative damage and ultimately economic losses. In addition to being a model specific to a given disease, IPSIM-Wheat-Eyespot can also be seen as the first sub-model of IPSIM-Wheat, a model that will predict injury profiles on wheat as a function of cropping practices and the production situation.

Supporting Information

Figure S1 Aggregating table used for the calculation of the value of the aggregative attribute "Final incidence of Eyespot" (screenshot of the DEXi software).

Figure S2 Aggregating table used for the calculation of the value of the aggregative attribute "Effects of cropping practices" (screenshot of the DEXi software).

Figure S3 Aggregating table used for the calculation of the value of the aggregative attribute "Effects of soil and climate" (screenshot of the DEXi software).

Figure S4 Aggregating table used for the calculation of the value of the aggregative attribute "Primary inoculum management: interaction between crop sequence and tillage" (screenshot of the DEXi software).

Figure S5 Aggregating table used for the calculation of the value of the aggregative attribute "Mitigation through crop status" (screenshot of the DEXi software).

Figure S6 Aggregating table used for the calculation of the value of the aggregative attribute "Climate" (screenshot of the DEXi software).

Acknowledgments

We thank Alain Cavelier (INRA, Rennes), Xavier Coquil (INRA, Nancy), Claire Thierry (INRA, Nancy), Michel Bertrand (INRA, Versailles-Grignon) for providing data. We also acknowledge Marc Délos (French Ministry of Agriculture), David Gouache (ARVALIS-Institut du Végétal), Claude Maumené (ARVALIS-Institut du Végétal), Sabrina Gaba (INRA, Dijon), Marie Gosme (INRA, Versailles-Grignon), Robert Faivre (INRA, Toulouse) and Bruno Coulomb (INRA, Toulouse) for their useful advices.

Author Contributions

Conceived and designed the experiments: NC PL FM. Performed the experiments: NC PL FM. Analyzed the data: JNA MHR CC. Contributed reagents/materials/analysis tools: JNA MHR. Wrote the paper: JNA MHR PD NC.

References

1. Montanari M, Innocenti G, Toderi G (2006) Effects of cultural management on the foot and root disease complex of durum wheat. J Plant Pathol 88: 149–156.
2. Lucas JA, Dyer PS, Murray TD (2000) Pathogenicity, host-specificity, and population biology of Tapesia spp., causal agents of eyespot disease of cereals. Advances in Botanical Research Incorporating. Adv Bot Pathol 33: 225–258.
3. Crous PW, Groenewald JZ, Gams W (2003) Eyespot of cereals revisited: ITS phylogeny reveals new species relationships. Eur J Plant Pathol 109: 841–850.
4. Ray RV, Jenkinson P, Edwards SG (2004) Effects of fungicides on eyespot, caused predominantly by Oculimacula acuformis, and yield of early-drilled winter wheat. Crop Prot 23: 1199–1207.
5. Ray RV, Crook MJ, Jenkinson P, Edwards SG (2006) Effect of eyespot caused by Oculimacula yallundae and O. acuformis, assessed visually and by competitive PCR, on stem strength associated with lodging resistance and yield of winter wheat. J Exp Bot 57: 2249–2257.
6. Dyer PS, Nicholson P, Lucas JA, Peberdy JF (1996) Tapesia acuformis as a causal agent of eyespot disease of cereals and evidence for a heterothallic mating system using molecular markers. Mycol Res 100: 1219–1226.
7. Scott PR, Hollins TW (1974) Effects of eyespot on the yield of winter wheat. Ann App Biol 78: 269–279.
8. Clarkson JDS (1981) Relationship between eyespot severity and yield loss in winter-wheat. Plant Pathol 30: 125–131.
9. Fitt BDL, Goulds A, Polley RW (1988) Eyespot (Pseudocercosporella herpotrichoides) epidemiology in relation to prediction of disease severity and yield loss in winter-wheat - a review. Plant Pathol 37: 311–328.
10. Fitt BDL, Goulds A, Hollins TW, Jones DR (1990) Strategies for control of eyespot (Pseudocercosporella herpotrichoides) in UK winter wheat and winter barley. Ann App Biol 117: 473–486.
11. Clarkson JP, Lucas JA (1993) Screening for potential antagonists of Pseudocercosporella herpotrichoides, the causal agent of eyespot disease of cereals 1. Bacteria. Plant Pathol 42: 543–551.
12. Russell PE (2005) A century of fungicide evolution. J Agr Sci 143: 11–25.
13. Leroux P, Gredt M (1997) Evolution of fungicide resistance in the cereal eyespot fungi Tapesia yallundae and Tapesia acuformis in France. Pestic Sci 51: 321–327.
14. Leroux P, Gredt M, Albertini C, Walker AS (2006) Characteristics and distribution of strains resistant to fungicides in the wheat eyespot fungi in France. 8ème Conference Internationale sur les Maladies des Plantes, Tours, France, 5 et 6 Décembre. pp. 574–583.
15. Stoate C, Boatman ND, Borralho RJ, Carvalho CR, Snoo GRd, et al. (2001) Ecological impacts of arable intensification in Europe. J Environ Manage 63: 337–365.
16. Bell EM, Sandler DP, Alavanja MC (2006) High pesticide exposure events among farmers and spouses enrolled in the agricultural health study. J Agri Saf Health 12: 101–116.
17. Paillotin G (2008). Ecophyto2018. Chantier 15 «agriculture écologique et productive» Rapport final du Président du Comité opérationnel. Ministère de l'agriculture et de la pêche. Paris, France. 138 pp. http://agriculture.gouv.fr/IMG/pdf/Rapport_Paillotin_.pdf. Accessed November 2012.
18. Epstein L, Bassein S (2003) Patterns of pesticide use in California and the implications for strategies for reduction of pesticides. Annu Rev Phytopathol 41: 351–375.
19. Birch ANE, Begg GS, Squire GR (2011) How agro-ecological research helps to address food security issues under new IPM and pesticide reduction policies for global crop production systems. J Exp Bot 62: 3251–3261.
20. Colbach N, Saur L (1998) Influence of crop management on eyespot development and infection cycles of winter wheat. Eur J Plant Pathol 104: 37–48.
21. Doussinault G, Delibes A, Sanchez-Monge R, Garcia-Olmedo F (1983) Transfer of a dominant gene for resistance to eyespot disease from a wild grass to hexaploid wheat. Nature 303: 698–700. United Kingdom.
22. Wei L, Muranty H, Zhang H (2011) Advances and prospects in wheat eyespot research: contributions from genetics and molecular tools. J Phytopathol 159: 457–470.
23. Murray TD, Delapena RC, Yildirim A, Jones SS (1994) A new source of resistance to Pseudocercosporella-herpotrichoides, cause of eyespot disease of wheat, located on chromosome-4v of Dasypyrum-villosum. Plant Breeding 113: 281–286.
24. Savary S, Mille B, Rolland B, Lucas P (2006) Patterns and management of crop multiple pathosystems. Eur J Plant Pathol 115: 123–138.
25. Bergez JE, Colbach N, Crespo O, Garcia F, Jeuffroy MH, et al. (2010) Designing crop management systems by simulation. Eur J Agron 32: 3–9.
26. Aubertot JN, Doré T, Ennaifar S, Ferré F, Fourbet JF et al. (2005) Integrated Crop Management requires to better take into account cropping systems in epidemiological models. Proceedings of the 9th International Workshop on Plant Disease Epidemiology. 11–15 April. Landerneau, France.
27. Payen D, Rapilly F, Galliot M (1979) Effect of the sowing date on the climatic potentialities of winter infections of soft winter wheat by eyespot in France. Comptes Rendus des Séances de l'Académie d'Agriculture de France 7 : 473–481.
28. Délos M (1995) Top: a model forecasting eyespot development. Phytoma 474: 26–28.
29. Colbach N, Meynard JM, Duby C, Huet P (1999) A dynamic model of the influence of rotation and crop management on the disease development of eyespot. Proposal of cropping systems with low disease risk. Crop Prot 18: 451–461.
30. Willocquet L, Aubertot JN, Lebard S, Robert C, Lannou C, et al. (2008) Simulating multiple pest damage in varying winter wheat production situations. Field Crop Res 107: 12–28.
31. Savary S, Willocquet L, Elazegui FA, Teng PS, Pham Van D, et al. (2000) Rice pest constraints in tropical Asia: characterization of injury profiles in relation to production situations. Plant Dis 84: 341–337.
32. Breman H, de Wit CT (1983) Rangeland Productivity and Exploitation in the Sahel. Science 221: 1341–1347.
33. Aubertot J-N, Robin M-H (2013) Injury Profile Simulator, a Qualitative Aggregative Modelling Framework to Predict Crop Injury Profile as a Function of Cropping Practices, and the Abiotic and Biotic Environment. I. Conceptual Bases. PLoS ONE 8(9): e73202. doi:10.1371/journal.pone.0073202.
34. Wiese MV (1987) Compendium of wheat diseases. Second edition. American Phytopathological Society. St Paul, Minessota, USA. 112 pp.
35. Bohanec M (2009) DEXi: program for multi-attribute decision making, Version 3.02. Jozef Stefan Institute, Ljubljana. Available: http://www-ai.ijs.si/MarkoBohanec/dexi.html. Accessed November 2012
36. Griffiths BS, Ball BC, Daniell TJ, Hallett PD, Neilson R, et al. (2010) Integrating soil quality changes to arable agricultural systems following organic matter addition, or adoption of a ley-arable rotation. App Soil Ecol 46: 43–53.
37. Bohanec M (2003) Decision support. In: Mladenić D, Lavrač N, Bohanec M, Moyle S (Eds.). Data Mining and Decision Support: Integration and Collaboration, Kluwer Academic Publishers (2003). pp. 23–35.
38. Bohanec M, Cortet J, Griffiths B, Znidarsic M, Debeljak M, et al. (2007) A qualitative multi-attribute model for assessing the impact of cropping systems on soil quality. Pedobiologia 51: 239–250.
39. Lancashire PD, Bleiholder H, Boom Tvd, Langeluddeke P, Stauss R, et al. (1991) A uniform decimal code for growth stages of crops and weeds. Ann Appl Biol 119: 561–601.
40. Kropff MJ, Teng PS, Rabbinge R (1995) The challenge of linking pest and crop models. Agricultural Systems 49: 413–434.
41. Colbach N, Meynard JM (1995) Soil tillage and eyespot - influence of crop residue distribution on disease development and infection cycles. Eur J Plant Pathol 101: 601–611.
42. Meynard JM, Dore T, Lucas P (2003) Agronomic approach: cropping systems and plant diseases. C R Biol 326: 37–46.
43. Matusinsky P, Mikolasova R, Klem K, Spitzer T (2009) Eyespot infection risks on wheat with respect to climatic conditions and soil management. J Plant Pathol 91: 93–101.
44. Vanova M, Matusinsky P, Javurek M, Vach M (2011) Effect of soil tillage practices on severity of selected diseases in winter wheat. Plant Soil Environ 57: 245–250.
45. Jenkyn JF, Gutteridge RJ, Bateman GL, Jalaluddin M (2010) Effects of crop debris and cultivations on the development of eyespot of wheat caused by Oculimacula spp. Ann Appl Biol 156: 387–399.
46. Cox J, Cock LJ (1962) Survival of Cercosporella herpotrichoides on naturally infected straws of wheat and barley. Plant Pathol 11: 65–66.
47. Herrman T, Wiese MV (1985) Influence of cultural practices on incidence of foot rot in winter wheat. Plant Dis 69: 948–950.

48. Smiley RW, Collins HP, Rasmussen PE (1996) Diseases of wheat in long-term agronomic experiments at Pendleton, Oregon. Plant Dis 80: 813–820.

49. Innocenti G, Montanari M, Marenghi A, Toderi G (2000) Influence of cropping systems on eyespot in winter cereals. Ramulispora herpotrichoides in cereali vernini in diverse situazioni colturali. Atti, Giornate fitopatologiche, Perugia, 16–20 aprile, Volume 2: 241–246.

50. Bailey KL, Gossen BD, Lafond GP, Watson PR, Derksen DA (2001) Effect of tillage and crop rotation on root and foliar diseases of wheat and pea in Saskatchewan from 1991 to 1998: univariate and multivariate analyses. Can J Plant Sci 81: 789–803.

51. Jenkyn JF, Christian DG, Bacon ETG, Gutteridge RJ, Todd AD (2001) Effects of incorporating different amounts of straw on growth, diseases and yield of consecutive crops of winter wheat grown on contrasting soil types. J Agr Sci 136: 1–14.

52. Jalaluddin M, Jenkyn JF (1996) Effects of wheat crop debris on the sporulation and survival of Pseudocercosporella herpotrichoides. Plant Pathol 45: 1052–1064.

53. Anken T, Weisskopf P, Zihlmann U, Forrer H, Jansa J, et al. (2004) Long-term tillage system effects under moist cool conditions in Switzerland. Soil Till Res 78: 171–183.

54. Prew RD, Ashby JE, Bacon ETG, Christian DG, Gutteridge RJ, et al. (1995) Effects of incorporating or burning straw, and of different cultivation systems, on winter wheat grown on two soil types, 1985–91. J Agr Sci 124: 173–194.

55. Burnett FJ, Hughes G (2004) The development of a risk assessment method to identify wheat crops at risk from eyespot. HGCA Project Report: 87 pp.

56. Agrios GN (2005) Plant pathology. Fifth edition. Elsevier Academic Press. San Diego, USA. 948 pp.

57. Datnoff LE, Elmer WH, Huber DM (2007) Mineral nutrition and plant disease. American Phytopathological Society Press. St Paul, Minessota, USA. 278 pp.

58. Loyce C, Meynard JM, Bouchard C, Rolland B, Lonnet P, et al. (2008) Interaction between cultivar and crop management effects on winter wheat diseases, lodging, and yield. Crop Prot 27: 1131–1142.

59. Zhang XY, Loyce C, Meynard JM, Savary S (2006) Characterization of multiple disease systems and cultivar susceptibilities for the analysis of yield losses in winter wheat. Crop Prot 25: 1013–1023.

60. Parnell S, Gilligan CA, Lucas JA, Bock CH, van den Bosch F (2008) Changes in fungicide sensitivity and relative species abundance in Oculimacula yallundae

and O. acuformis populations (eyespot disease of cereals) in Western Europe. Plant Pathol 57: 509–517.

61. Colbach N, Lucas P, Cavelier N (1994) Influence of crop succession on foot and root diseases of wheat. Agronomie 14: 525–540.

62. Colbach N, Huet P (1995) Modelling the frequency and severity of root and foot diseases in winter wheat monocultures. Eur J Agron 4: 217–227.

63. Dore T, Clermont-Dauphin C, Crozat Y, David C, Jeuffroy MH, et al. (2008) Methodological progress in on-farm regional agronomic diagnosis. A review. Agron Sustain Dev 28: 151–161.

64. Cavelier A, Cavelier N, Colas AY, Montfort F, Lucas P (1998) ITITECH: a survey to improve the evaluation of relationships between cultural practices and cereal disease incidence. Brighton Crop Protection Conference: Pests & Diseases. Volume 3: Proceedings of an International Conference, Brighton, UK, 16–19 November. 1023–1028.

65. Wallach D, Makowski D, Jones J (2006). Working Dynamic Crop Models: Evaluation, Analysis, Parameterization and Application. Elsevier. 447p.

66. Nash JE, Sutcliffe JV (1970) River flow forecasting through conceptual models part I — A discussion of principles. J Hydrol 10: 282–290.

67. Sadok W, Angevin F, Bergez JE, Bockstaller C, Colomb B, et al. (2009) MASC, a qualitative multi-attribute decision model for ex ante assessment of the sustainability of cropping systems. Agron Sustain Dev 29: 447–461.

68. Pelzer E, Fortino G, Bockstaller C, Angevin F, Lamine C, et al. (2012) Assessing innovative cropping systems with DEXiPM, a qualitative multi-criteria assessment tool derived from DEXi. Ecol Indic 18: 171–182.

69. Bohanec M, Messéan A, Scatasta S, Angevin F, Griffiths B, et al. (2008) A qualitative multi-attribute model for economic and ecological assessment of genetically modified crops. Ecol Model 215: 247–261.

70. Vilichmeller V (1992) Mixed cropping of cereals to suppress plant-diseases and omit pesticide applications. Biol Agric Hortic 8: 299–308.

71. Mundt CC, Brophy LS, Schmitt MS (1995) Choosing crop cultivars and cultivar mixtures under low versus high disease pressure - a case-study with wheat. Crop Prot 14: 509–515.

72. Saur L, Mille B (1997) Disease progress of Pseudocercosporella herpotrichoides in mixed stands of winter wheat cultivars. Agronomie 17: 113–118.

73. Colbach N (1995) Modélisation de l'influence des systèmes de culture sur les maladies du pied et des racines du blé tendre d'hiver. Doctorat de l'INA P-G, Paris. 258 p.

20

Evaluating Status Change of Soil Potassium from Path Model

Wenming He[1,2], Fang Chen[1,3]*

1 Key Laboratory of Aquatic Botany and Watershed Ecology, Wuhan Botanical Garden, Chinese Academy of Sciences, Moshan, Wuchang, Wuhan, Hubei Province, China, 2 Graduate University of Chinese Academy of Sciences, Beijing, China, 3 International Plant Nutrition Institute, Wuhan, China

Abstract

The purpose of this study is to determine critical environmental parameters of soil K availability and to quantify those contributors by using a proposed path model. In this study, plot experiments were designed into different treatments, and soil samples were collected and further analyzed in laboratory to investigate soil properties influence on soil potassium forms (water soluble K, exchangeable K, non-exchangeable K). Furthermore, path analysis based on proposed path model was carried out to evaluate the relationship between potassium forms and soil properties. Research findings were achieved as followings. Firstly, key direct factors were soil S, ratio of sodium-potassium (Na/K), the chemical index of alteration (CIA), Soil Organic Matter in soil solution (SOM), Na and total nitrogen in soil solution (TN), and key indirect factors were Carbonate (CO_3), Mg, pH, Na, S, and SOM. Secondly, path model can effectively determine direction and quantities of potassium status changes between Exchangeable potassium (eK), Non-exchangeable potassium (neK) and water-soluble potassium (wsK) under influences of specific environmental parameters. In reversible equilibrium state of $wsK \underset{\beta}{\overset{\alpha}{\rightleftharpoons}} neK \underset{\chi}{\overset{\gamma}{\rightleftharpoons}} eK$, K balance state was inclined to be moved into β and χ directions in treatments of potassium shortage. However in reversible equilibrium of $wsK \underset{\theta}{\overset{\varepsilon}{\rightleftharpoons}} eK \underset{\omega}{\overset{\lambda}{\rightleftharpoons}} neK$, K balance state was inclined to be moved into θ and λ directions in treatments of water shortage. Results showed that the proposed path model was able to quantitatively disclose moving direction of K status and quantify its equilibrium threshold. It provided a theoretical and practical basis for scientific and effective fertilization in agricultural plants growth.

Editor: Raffaella Balestrini, Institute for Plant Protection (IPP), CNR, Italy

Funding: The study was supported by the National Natural Science Foundation of China (41171243), the National Key Technology R&D Program (2012BAD15B01) and the Cooperated Program with the International Plant Nutrition Institute (IPNI-HB-33). The funders had no role in study design, data collection and analysis, decision to publish, or preparation of the manuscript.

Competing Interests: The authors have declared that no competing interests exist.

* E-mail: fchen@ipni.ac.cn

Introduction

The status and transformation of potassium in the soil is of great significance to crop growth [1,2]. Investigations on potassium status gained a numerous achievements[3,4]. The effectiveness of potassium in soil is controlled by four forms, e.g., mineral potassium, non-exchangeable potassium (neK), exchangeable potassium (eK) and water-soluble potassium (wsK) which can be transformed into each other [5–8]. Absorption capacity of crops and application of farming fertilizer have influence on the potassium forms, thus impact on the release and fixation of potassium in the soil [9]. Within these different forms of soil potassium, there exists a complex and dynamic chemical balance, and it greatly depends on the situation of each form and its environment conditions [10–13]. However, there are few experimental studies on ternary systems involving K^+ in spite of the plant nutritional importance of K^+ in soils [3,5,14]. Research indicated that the amount of K ions retained on the solid phase of soil was much larger than that dissolved in the soil solution. It suggested that a root segment become easily permeable to Ca^{2+} followed by Na^+ and K^+ with aging [15,16]. Thus, it is indispensable to take into account the cation exchange processes

in modeling the response of the soil solution to fertilizer application and transport of cationic solutes in soils.

The transform of soil non-exchangeable potassium to soil exchangeable potassium and soil soluble potassium is a slowly process. It is hard to be determined by using the routine method of ion exchange, although it shows that sodium tetraphenylborate method is more accurate than conventional methods to reflect changes in soil potassium through long-term field fixed experiments of potash fertilization [17,18]. Path analysis is a statistical technique that distinguishes coefficient and causation by partitioning coefficients into direct and indirect effects. Researchers used path analysis to analysis phosphorus retention capacity in allophanic and non-allophanic andisols and soil organic matter effects on phosphorus sorption is successful [19,20]. The method of path coefficients, which was proposed by Wright (1921), was effective in disclosing relationships between variables by diagrams [21–25]. However, path coefficients approach has not yet applied to the study on soil potassium. The potassium absorb efficiency of plant root is affected by multiple factors [26], so it is not suitable to make estimation by use a single parameter [27–29]. The above mentioned studies focused on qualitatively descriptive analysis of single factor or multifactor, and failed to quantitatively disclose the dynamic process of potassium status [30].

In this study, we conducted soil K ions exchange experiments for soil samples and further calculated their selectivity coefficients, and explored potassium status on basis of path model to determine dynamic changes of potassium status and specific environmental parameters. The model assumptions were related to the adsorption mechanism at molecular level. The first step, we attempted to examine different indicators of potassium in the soils, and then to analyze direct and indirect correlations between water-soluble potassium (wsK), non-exchangeable potassium (neK) and exchangeable potassium (eK), and further to calculate path coefficients of different potassium status, finally to construct path model for describing the changes of potassium status. The model was calibrated and validated by using 48 cotton rhizosphere/nonrhizosphere soils samples to test its accuracy. We quantitatively investigated on interaction process of plant-soil-microorganisms and relationship between environmental parameters and potassium release through path coefficient model, and the model was used to predict the multi-component ion exchange equilibrium in soil.

Materials and Methods

Hereby, I, along with coauthor, confirm that no specific permissions were required for our experiment locations/activities since this experiment field belongs to our institute and for scientific research only. And the field studies did not involve endangered or protected species.

In this study, meadow soil was selected as experimental agents in order to investigate potassium nutrient supply and physiological mechanism (Table 1). Rhizosphere soil samples were collected from 90 days cotton plant. When collecting soil samples, we firstly loosed root zone soil to collect rhizosphere soil, and uprooted whole plant cotton, and then gently shook off root zone soil, and finally get root surface adhesion soil. The non-rhizosphere soil samples were collected 10–15 cm depth from surface.

For purpose of this study, we proposed a detail research scheme to evaluate status changes of soil potassium from path model (Fig. 1). We conducted soil K ions exchange experiments for soil samples, and further calculated their selectivity coefficients, and explored potassium status on basis of path model to determine dynamic changes of potassium status and specific environmental parameters. Soil samples were **N**on-**R**hizosphere soil in **OPT**imum of **K** of **H**igh efficiency genotype cotton (NROPTH); **N**on-**R**hizosphere soil in **OPT**imum of **K** of **L**ow efficiency genotype

cotton (NROPTL); **N**on-**R**hizosphere soil in **S**hortage of **K** of **H**igh efficiency genotype cotton (NRSKH); **N**on-**R**hizosphere soil in **S**hortage of **K** of **L**ow efficiency genotype cotton (NRSKL); **N**on-**R**hizosphere soil in **S**hortage of **W**ater of **H**igh efficiency genotype cotton (NRSWH); **N**on-**R**hizosphere soil in **S**hortage of **W**ater of **L**ow efficiency genotype cotton (NRSWL); **N**on-**R**hizosphere soil in **S**hortage of **W**ater and **K** of **H**igh efficiency genotype cotton (NRSWKH); **N**on-**R**hizosphere soil in **S**hortage of **W**ater and **K** of **L**ow efficiency genotype cotton (NRSWKL); **R**hizosphere soil in **OPT**imum of **K** of **H**igh efficiency genotype cotton (ROPTH); **R**hizosphere soil in **OPT**imum of **K** of **L**ow efficiency genotype cotton (ROPTL); **R**hizosphere soil in **S**hortage of **K** of **H**igh efficiency genotype cotton (RSKH); **R**hizosphere soil in **S**hortage of **K** of **L**ow efficiency genotype cotton (RSKL); **R**hizosphere soil in **S**hortage of **W**ater of **H**igh efficiency genotype cotton (RSWH); **R**hizosphere soil in **S**hortage of **W**ater of **L**ow efficiency genotype cotton (RSWL); **R**hizosphere soil in **S**hortage of **W**ater and **K** of **H**igh efficiency genotype cotton (RSWKH); **R**hizosphere soil in **S**hortage of **W**ater and **K** of **L**ow efficiency genotype cotton (RSWKL). Firstly, we divided experiment soil into 8 treatments, and then implemented 256 soil samples test (Fig. 2). We measured 33 parameters of the different attributes of 256 soil samples. Those parameters were selected based on standards of Hashimoto and Kang research methods of phosphorus[19,31]. Humic acid (NHA) in soil which can be extracted by NaOH solution (0.1 $mol \cdot L^{-1}$, pH = 3), humic acid (PHA) in soil which can be extracted by $Na_2P_2O_4$ solution (0.05 $mol \cdot L^{-1}$, pH = 9.2), humus linked to iron (HMi) and humic linked to clay (HMc). Key experimental sampling test datasets were listed in Table 2–Table 6. Secondly, statistical analysis was further implemented to investigate relationship between environmental parameters and potassium status. Datasets were normalized, followed by correlation analysis (Table 7), 8 out of 16 parameters were selected thus to determine the most important parameters. Path coefficients of absolute value of represented the size effect on potassium morphology change. The size of the "+" meant the same as the arrow direction with arrow, "−" represents in contrast to the arrow direction with arrow. Finally, path model of potassium status was constructed on basis of direct and indirect correlated parameters. In this model, "e1" represents unknown variable and its impact factor, the straight line with arrows stands for direct impact factor, and double arc arrow is direction of interaction between parameters.

Table 1. Properties of mead soil for experiments.

Soil properties		Average value
Mean compact density ($g \cdot cm^{-3}$)		1.54
Particle composition	Clay (<0.002 mm, %)	16.8
	Sand (2-0.05 mm, %)	61.7
	Silt (0.05-0.002 mm, %)	21.4
Mineral composition	Smectite (%)	4.0
	Vermienlite (%)	25.0
	Intergrade mineral hydromica (1.4 nm, %)	49.0
	Kaolinite (%)	19.0
Cations composition	exchangeable Ca ($mmol \cdot kg^{-1}$)	35.4
	exchangeable Mg ($mmol \cdot kg^{-1}$)	18.1
	exchangeable Na ($mmol \cdot kg^{-1}$)	2.0
	FeO (noncrystalline iron extracted with Tamm's solution, $g \cdot kg^{-1}$)	49.0

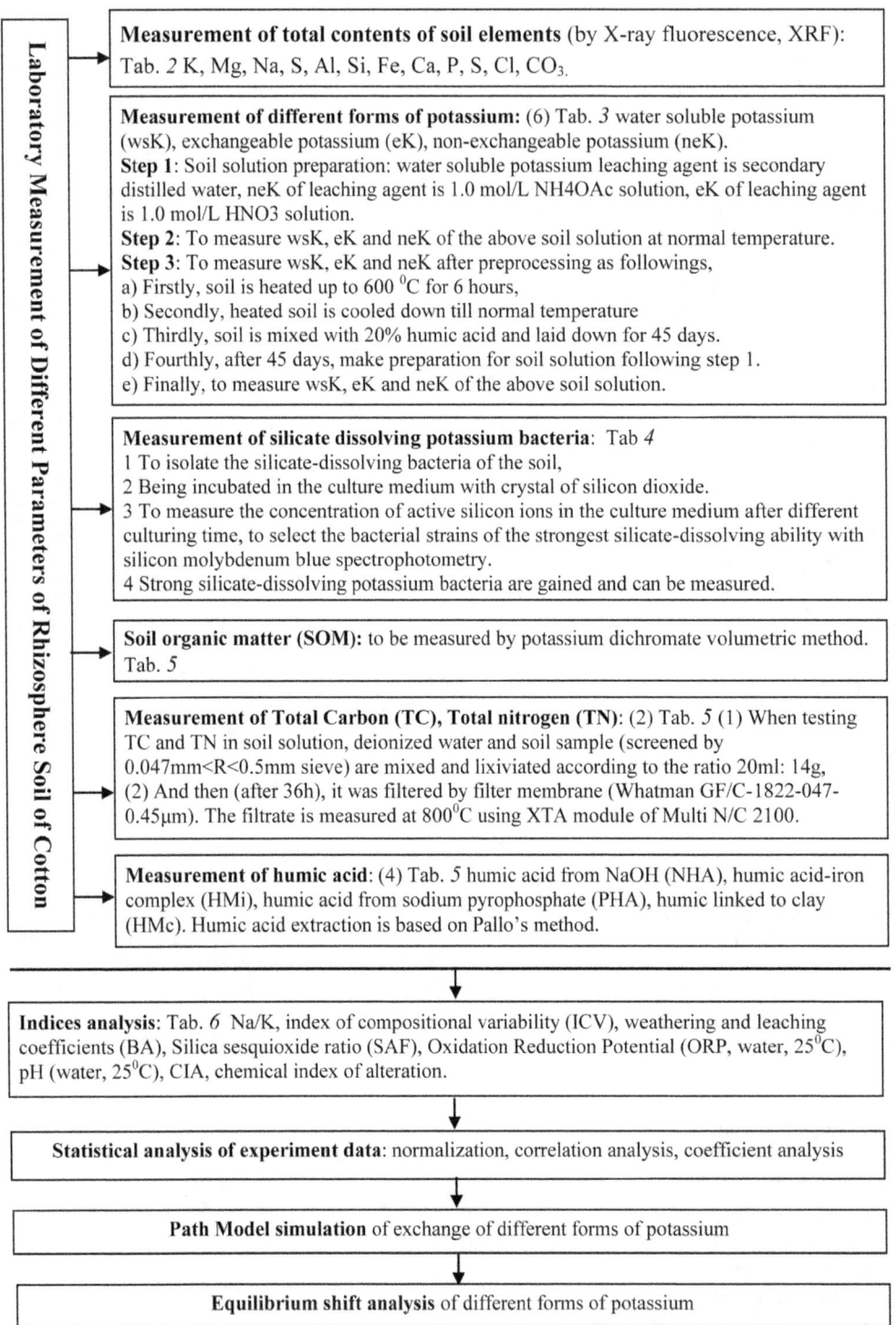

Figure 1. Workflow of evaluating status changes of soil potassium from Path Model.

Results

Path analysis of water-soluble potassium (wsK)

For purpose of path analysis of water-soluble potassium, direct and indirect impact factors of the path analysis were derived from multiple linear regressions coefficients of wsK, and correlation coefficients between soil properties. Direct coefficient variables of wsK include neK, eK, PHA, SOM, S, Na and Na/K; while

indirect coefficient variables of wsK include CIA, Na, Mg, CO_3 and pH. Impact factor of path size of water-soluble potassium was determined by pH, CIA, Oxidation Reduction Potential (ORP), S, Na (Fig. 3 and Table 8). In the path model of wsK, direct effects of soil properties on the normalized value of wsK (ZwsK) were represented by single-headed arrows, while coefficients between soil properties were represented by double-headed arrows. Direct and indirect effects were indicated by value and marked with "±".

	OPT	SW	SK	SWK
HE&HP	Soil moisture (35%) KCl (6.83g)	Soil moisture (25%) KCl (6.83 g)	Soil moisture (35%) KCl (0g)	Soil moisture (25%) KCl (0g)
	Soil moisture (35%) KCl (6.83g)	Soil moisture (25%) KCl (6.83 g)	Soil moisture (35%) KCl (0g)	Soil moisture (25%) KCl (0g)
	Soil moisture (35%) KCl (6.83g)	Soil moisture (25%) KCl (6.83 g)	Soil moisture (35%) KCl (0g)	Soil moisture (25%) KCl (0g)

	OPT	SW	SK	SWK
LE&LP	Soil moisture (35%) KCl (6.83g)	Soil moisture (25%) KCl (6.83 g)	Soil moisture (35%) KCl (0g)	Soil moisture (25%) KCl (0g)
	Soil moisture (35%) KCl (6.83g)	Soil moisture (25%) KCl (6.83 g)	Soil moisture (35%) KCl (0g)	Soil moisture (25%) KCl (0g)
	Soil moisture (35%) KCl (6.83g)	Soil moisture (25%) KCl (6.83 g)	Soil moisture (35%) KCl (0g)	Soil moisture (25%) KCl (0g)

Figure 2. Diagram of experimental design to evaluate status changes of soil potassium.

Table 2. Total contents of soil elements of Laboratory measurement for different parameters of 256 soil samples (R, rhizosphere soil; NR, non-rhizosphere).

Sample	Na%	Mg%	Al%	Si%	Fe%	Mg%	Ca%	Pppm	CO₃%	Sppm	Clppm
NR1-16	0.44	0.39	6.49	36.86	3.35	1.49	0.37	491.01	49.74	219.34	119.8
R17-32	0.45	0.37	6.53	37.16	3.34	1.49	0.39	507.57	49.38	243.65	228.2
NR33-48	0.44	0.39	6.5	36.92	3.33	1.49	0.38	494.36	49.72	218.56	122.07
R49-64	0.44	0.39	6.51	36.94	3.34	1.49	0.39	483.04	49.64	251.93	217.78
N65-80	0.44	0.39	6.46	36.93	3.32	1.47	0.36	488.97	49.77	220.97	68.15
R81-96	0.47	0.42	6.3	36.38	3.29	1.46	0.4	515.89	50.45	248.77	80.32
NR97-112	0.44	0.38	6.49	37.04	3.35	1.48	0.37	506.34	49.62	218.38	82.48
R113-128	0.44	0.38	6.53	37.09	3.35	1.48	0.4	518.35	49.48	243.94	89.5
NR129-144	0.44	0.39	6.38	36.84	3.29	1.48	0.36	481.3	49.95	224.13	152.4
R145-160	0.45	0.39	6.39	36.94	3.27	1.47	0.38	512.73	49.84	237.7	201.2
NR161-176	0.43	0.38	6.49	37.09	3.32	1.49	0.37	484.59	49.58	225.39	142.88
R177-192	0.45	0.38	6.51	37.01	3.33	1.48	0.39	496.2	49.57	236.79	189.7
NR193-208	0.44	0.39	6.43	36.92	3.31	1.47	0.36	502.75	49.83	223.08	71.6
R209-224	0.45	0.38	6.43	37.08	3.32	1.46	0.39	491.22	49.64	245.74	81.48
NR225-240	0.44	0.38	6.46	37	3.31	1.47	0.38	509.11	49.7	228.9	90.5
R241-256	0.44	0.37	6.46	37.28	3.34	1.47	0.39	498.72	49.4	239.9	87.4

Table 3. Different forms of potassium of Laboratory measurement for different parameters of 256 soil samples (R, rhizosphere soil; NR, non-rhizosphere).

Sample	K(g/Kg)	wsK(ug/g)	neK(ug/g)	eK(ug/g)	wsK(ug/g, (weathered)	neK(ug/g) (weathered)	eK(ug/g) weathered
NR1-16	14.93	27.1	134.9	58.64	67.57	174.48	62.9
R17-32	14.9	12.03	54.21	14.32	59.36	155.15	66.59
NR33-48	14.9	20.01	134.13	50.84	76.97	177.27	55.87
R49-64	14.87	9.22	58.27	12.21	123.72	170.67	71.24
N65-80	14.7	24.8	121.55	40.27	53.02	190.22	59.93
R81-96	14.57	5.65	39.17	6.28	61.31	153.27	57.35
NR97-112	14.75	41.28	101.36	27.46	56.24	167.15	69.44
R113-128	14.8	6.05	42.83	5.23	58.41	149.91	52.24
NR129-144	14.77	38.55	158.29	56.41	68.09	190.55	65.03
R145-160	14.73	15.92	78.62	23.18	60.89	167.13	52.79
NR161-176	14.87	88.78	162.24	61.11	70.46	181.02	55.25
R177-192	14.83	18.61	70.07	19.27	64.82	185.07	59.89
NR193-208	14.73	13.26	99.74	30.21	63.65	171.35	62.95
R209-224	14.63	6.41	38.74	7.63	61.96	162.21	52.09
NR225-240	14.73	25.27	105.8	29.54	64.59	182.24	59.77
R241-256	14.73	6.56	44.8	5.31	72.96	154.08	60.7

The direct effects of soil properties on the ZwsK were termed path coefficients and were standardized partial regression coefficients for each of the soil properties in the multiple linear regression against the ZwsK. Indirect effects of soil properties on the ZwsK were determined from the product of the simple coefficient between soil properties and the path coefficient (i.e., one two-headed arrow and one single-headed arrow). The coefficient between the ZwsK and soil property was the sum of the entire path connecting two variables, as described by

$$
\begin{aligned}
r_{neK,wsK} = {} & P_{wsK,neK} + r_{neK,PHA} \cdot P_{wsK,PHA} + \\
& r_{neK,S} \cdot P_{wsK,S} + r_{neK,Na/K} \cdot P_{wsK,Na/K} + \\
& r_{neK,CIA} \cdot P_{wsK,CIA} + r_{neK,Na} \cdot P_{wsK,Na} + \\
& r_{neK,SOM} \cdot P_{wsK,SOM} + r_{neK,eK} \cdot P_{wsK,eK}
\end{aligned}
\tag{1}
$$

Table 4. Silicate dissolving potassium bacteria of Laboratory measurement for different parameters of 256 soil samples (R, rhizosphere soil; NR, non-rhizosphere).

Sample	Bacteria of seedling stage ($\times 10^4$ CFUg^{-1})	Bacteria of budding stage ($\times 10^4$ CFUg^{-1})	Bacteria of wadding stage ($\times 10^4$ CFUg^{-1})
NR1-16	0.8	2.1	1.6
R17-32	1.5	5.2	1.4
NR33-48	0.7	4.3	1.7
R49-64	1.1	3.4	1.9
N65-80	0.9	17	14
R81-96	1.0	87	30
NR97-112	1.2	26	19
R113-128	1.9	39	24
NR129-144	0.1	1.2	0.5
R145-160	0.4	1.3	0.7
NR161-176	0.9	4.3	2.2
R177-192	0.3	1.7	0.6
NR193-208	0.8	1.9	1.6
R209-224	0.7	7.1	4.0
NR225-240	0.1	2.3	0.9
R241-256	1.2	6.3	2.8

Table 5. Soil organic matter, humic acid, Total Carbon (TC), Total nitrogen (TN) of Laboratory measurement for different parameters of 256 soil samples (R, rhizosphere soil; NR, non-rhizosphere).

Sample	TC%	TN%	SOM(%)	PHA%	NHA%	Hmi%	HMc%
NR1-16	4.51E-03	1.71E-03	7.88	4.53	2.59	1.47	0.92
R17-32	4.41E-03	5.74E-03	5.67	4.84	3.52	0.33	0.33
NR33-48	5.34E-03	2.02E-03	7.32	3.48	3.67	0.42	0.37
R49-64	3.74E-03	1.04E-02	5.54	4.11	2.55	0.45	0.32
N65-80	4.50E-03	1.64E-03	8.01	5.33	3.89	1.35	1.24
R81-96	3.64E-03	4.24E-03	5.79	5.26	4.58	0.42	0.23
NR97-112	3.01E-03	3.93E-03	8.53	3.62	3.28	0.66	0.41
R113-128	2.80E-03	1.31E-02	5.70	4.41	3.68	0.46	0.27
NR129-144	3.52E-03	3.46E-03	8.30	4.46	3.70	0.35	0.28
R145-160	3.22E-03	2.02E-02	5.38	5.00	3.82	0.30	0.30
NR161-176	3.18E-03	3.97E-03	7.84	4.90	1.75	0.88	0.45
R177-192	3.12E-03	2.31E-02	6.79	6.89	3.95	0.34	0.26
NR193-208	3.57E-03	3.24E-03	8.45	4.57	3.82	0.25	0.19
R209-224	3.79E-03	2.67E-02	5.79	5.43	4.22	0.44	0.36
NR225-240	4.30E-03	3.83E-03	8.01	4.94	4.61	0.52	0.32
R241-256	4.37E-03	2.75E-02	6.23	4.70	4.08	0.36	0.18

Table 6. Weathering indices of soil minerals of Laboratory measurement for different parameters of 256 soil samples (R, rhizosphere soil; NR, non-rhizosphere).

Sample	ORP(eV)	pH	ba	Na/K	ICA	saf	CIA
NR1-16	17.30	7.27	0.45	0.29	1.03	8.73	76.08
R17-32	18.65	6.84	0.44	0.30	1.03	8.77	75.82
NR33-48	13.80	7.20	0.45	0.29	1.03	8.75	76.02
R49-64	30.24	6.5	0.45	0.30	1.04	8.74	75.95
N65-80	22.66	6.79	0.45	0.30	1.03	8.8	76.18
R81-96	22.68	6.73	0.48	0.32	1.08	8.86	75.08
NR97-112	10.22	6.94	0.44	0.30	1.03	8.78	76.21
R113-128	65.80	6.09	0.45	0.30	1.03	8.74	76
NR129-144	18.50	6.93	0.45	0.30	1.04	8.88	75.88
R145-160	80.84	5.79	0.46	0.31	1.04	8.90	75.67
NR161-176	31.46	6.48	0.44	0.29	1.02	8.80	76.16
R177-192	62.82	5.85	0.45	0.30	1.03	8.75	75.92
NR193-208	19.33	6.87	0.45	0.30	1.03	8.83	76.06
R209-224	52.48	6.19	0.45	0.31	1.04	8.86	75.76
NR225-240	13.72	6.98	0.45	0.30	1.03	8.83	75.99
R241-256	20.54	6.86	0.45	0.30	1.04	8.87	75.84

where, r_{ij} is the simple correlation between the ZwsK and a parameter of soil property, P_{ij} is the path coefficient between the ZwsK and a parameter of soil property, and $r_{ij}P_{ij}$ is the indirect effect of a parameter of on the ZwsK. An uncorrelated residue (e_{wsk}) that represented the unexplained part of an observed variable in the path model was

$$e_{wsk} = (1 - R_{wsk}^2)^{0.5} = 0.2 \qquad (2)$$

where, R_{wsk}^2 is the coefficient of determination of the multiple regression equation between the ZwsK and the eight soil properties. Those parameters of soil properties were neK, PHA, S, Na/K, CIA, Na, SOM, eK. Backward regression analysis was performed with the ZwsK as the dependent variable and the eight soil properties were regarded as independent variables. Backward regression was a multiple regression procedure in which all the independent variables were entered into the regression equation at the beginning, and variables that did not contribute significantly to the fit of regression model were successively eliminated until only statistically significant variables remain. Fig. 4 and Fig. 5 were similar expressions.

In the path of water-soluble potassium, path contribution coefficients were quite different. Coefficient values were 2.10, −1.52 and 0.20 for wsK to neK, neK to wsK, and other variables to wsK respectively. It disclosed that most part of wsK was changed into neK (with values of 2.10). In the path of neK to wsK, coefficient was negative (with value of −1.52); it indicated that there was an inverse change from wsK to neK, a lot of wsK was changed into neK. Those dynamic changes indicated that wsK and eK reached equilibrium in soil solution, and this equilibrium represented its quick reaction process.

These changes impacted greatly on nutrients dynamic equilibrium of rhizosphere soil. This changed nutrients recycling process of the soil environment since continuous-release of humic acid had direct effects on absorption and uptakes of those variables including neK, eK and PHA, SOM, S, Na and Na/K. It was clear that rhizosphere effects were likely to result in changes in root exudates, pH and ORP (eV). Consequently, it led to changes in pH values and population, quality and activity of microorganisms while pH change was caused by root exuded organic acids. Under conditions of different pH values, humic acid significantly correlated to adsorption and desorption of potassium. Humic acid of potassium adsorption and desorption presented an upward trend when it increased in initial concentration (pH was 4.0–8.0), however, desorption rate declined. The migration distance also showed very significant linear relationship within ranges of migrations between water-soluble potassium and exchangeable potassium. Organic matter nutrients and chemical efficiency of humic acid altered different variables such as physical, chemical, and biotic in root-soil interface.

Path analysis of non-exchangeable potassium (neK)

For path analysis of non-exchangeable potassium, we further investigated its direct and indirect impact factors. Results showed that direct coefficient variables of neK include eK, wsK, SOM, S, Na, Na/K, TN, CIA; and indirect coefficient variables of neK include Mg, CO₃. Impact factor of path size of neK was determined by eK, CO3, pH (Fig. 4 Table 9). Table 9 showed there was significant correlation between eK and neK (with coefficient of 0.81). In the Path of non-exchangeable potassium, contribution coefficient was 0.01 for wsK to neK, although wsK was determined by general effects of N, Na, S and CIA. The SOM was also an important contributor to eK and wsK of rhizosphere soil. Coefficients were 0.69 and 0.55 between SOM and eK, SOM and wsK, respectively.

The correlation between the ZneK and a soil property was the sum of the entire path connecting two variables, as described by

Table 7. Correlation analysis of environmental parameters of soil potassium.

	Z(SOM)	Z(CIA)	Z(Na/K)	Z(TC)	Z(TN)	Z(wsK)	Z(neK)	Z(eK)	Z(bacteria)	Z(ORP)	Z(pH)	Z(PHA)	Z(HMi)
Z(CIA)	0.61a												
Z(Na/K)	−0.52a	0.829b											
Z(TN)	−0.61a												
Z(wsK)	0.58a		−0.52a										
Z(neK)	0.80b	0.56a	−0.66b		−0.64b	0.77b							
Z(ek)	0.74b	0.51a	−0.67b		−0.62a	0.73b	0.98b						
Z(bacteria)	0.55a						0.53a						
Z(ORP)	−0.59a			−0.56a	0.66b								
Z(pH)	0.58a			0.66b	−0.68b					−0.95b			
Z(NHA%)											−0.57a		
Z(PHA%)		−0.51a	0.64			−0.65b		−0.51a					
Z(HMi%)								0.54a					
Z(HMc%)													0.93b
Z(CO₃)		−0.62b	0.56					−0.50a					
Z(K)		0.53a	−0.79b									−0.77b	

(Note: statistically, probabilities of a and b: a, $p<0.01$; b, $p<0.05$)

$$r_{neK,eK} = P_{neK,eK} + r_{neK,SOM} \cdot P_{neK,SOM} +$$
$$r_{neK,S} \cdot P_{neK,S} + r_{neK,Na/K} \cdot P_{neK,Na/K} + \quad (3)$$
$$r_{neK,TN} \cdot P_{neK,TN} + r_{neK,Na} \cdot P_{neK,Na} +$$
$$r_{neK,SOM} \cdot P_{neK,SOM} + r_{neK,wsK} \cdot P_{neK,wsK}$$

where, r_{ij} is the simple correlation coefficient between the ZneK and a soil property, P_{ij} is the path coefficient between the ZneK and a soil property, and $r_{ij}P_{ij}$ is the indirect effect of a soil property on the ZneK. An uncorrelated residue (e) that represents the unexplained part of an observed variable in the path model was

$$e_{nek} = (1 - R_{nek}^2)^{0.5} = 0.01 \quad (4)$$

where R_{nek}^2 is the coefficient of determination for the multiple regression equation between the ZneK and those soil properties. Those soil properties were wsK, SOM, S, Na/K, TN, Na, SOM, eK.

It disclosed that plant nutrition was indirectly influenced by nutrients release of microbial decomposition and organic matter. It was particularly obvious in nutrition release of low molecular weight organics in rhizosphere soil. High contents of associated ions (Na⁺) were limit factors that control crop yields and soil nutrients use. Ammonium nitrogen was likely to reduce fixation of potassium ions of soil, and increase risk of potassium leaching. NH₄⁺ and K⁺ were competitive ions for absorption of ORP (eV) of eK and neK in rhizosphere soil, and different status of nitrogen effects on efficient use of potassium. Therefore, suitable proportion of N and K is important for plant. Soil nitrogen availability determines potassium absorption of root surface and cell membrane. Root membrane could directly affected by NH₄⁺, NO₃⁻, K⁺ when it absorbs potassium ions, and indirect affected by potential balance from assimilation of NH₄⁺. Strong affinity of K⁺, high efficiency and fast speed of absorption of potassium are likely to improve efficient absorption of potassium.

Path analysis of exchangeable potassium (eK)

As to exchangeable potassium, it was quite different situation in comparison with those of water soluble and non-exchangeable potassium. It showed that direct coefficient variables of eK include neK, wsK, SOM, S, Na, Na/K, TN, CIA, HMi, PHA; and indirect coefficient variables of eK include Mg, CO₃. Impact factor of path size of eK was determined by neK, CO₃, pH, ORP, Na (Table 10).

From Fig. 5, it showed that neK was closely related to wsK (with correlation coefficients of 0.72), and there was an inverse proportion between ZneK and ZwsK (the normalized value of neK and wsK) in condition of changes of micro-bioorganic and geochemical environmental factors in rhizosphere soil. Path contribution coefficients were 1.10 and 0.04 for neK to eK, and wsK to eK. It suggests that neK is more likely to be changed into eK while the possibility of wsK to eK is relatively low.

The correlation between the normalized value of eK (ZeK) and a soil property is the sum of the entire path connecting two variables, as described by

$$r_{neK,eK} = P_{eK,neK} + r_{neK,SOM} \cdot P_{eK,SOM} +$$
$$r_{neK,Na/K} \cdot P_{eK,Na/K} + r_{neK,Na} \cdot P_{eK,Na} +$$
$$r_{neK,HMi} \cdot P_{eK,HMi} + r_{neK,TN} \cdot P_{eK,TN} + \quad (5)$$
$$r_{neK,wsK} \cdot P_{eK,wsK} + r_{neK,S} \cdot P_{eK,S}$$

where r_{ij} is the simple correlation coefficient between the ZeK and a soil property, P_{ij} is the path coefficient between the ZeK and a soil property, and $r_{ij}P_{ij}$ is the indirect effect of a soil property on the ZeK. An uncorrelated residue (e) that represents the unexplained part of an observed variable in the path model was

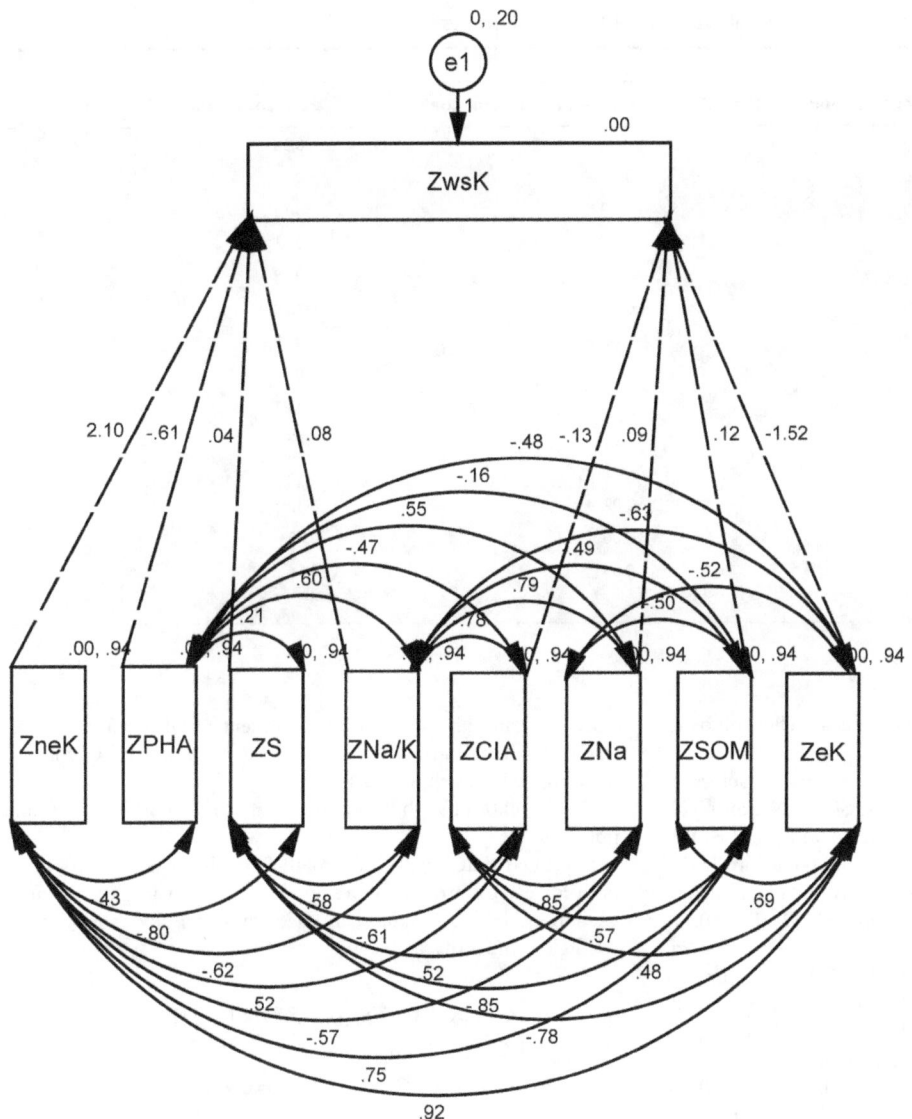

Figure 3. Path model of water-soluble potassium (wsK, solid line for double arrow, dashed-line for single arrow, and single arrow flows into ZwsK).

$$e_{ek} = \left(1 - R_{ek}^{2}\right)^{0.5} = 0.01 \qquad (6)$$

Where, R_{ek}^{2} is the coefficient of determination of the multiple regression equation between the ZwsK and the eight soil properties. Those soil properties are neK, SOM, Na/K, Na, HMi, TN, wsK, S. Backward regression analysis was performed with the ZeK as the dependent variable and the eight soil properties as independent variables. Backward regression is a multiple regression procedure in which all the independent variables are entered into the regression equation at the beginning, and variables that do not contribute significantly to the fit of the regression model are successively eliminated until only statistically significant variables remain.

Cation exchange capacity (CEC) or the number of exchangeable cation depends on the type and number of clay and organic matter in the soil. To a certain extent, TN, SOM, HMi and the PHA and the CIA are able to well reflect type and number of clay

and organic matte in rhizosphere soil. Proper pH value of solution, hot water pressure, concentration increases of magnesium carbonate and ion are useful to exchangeable potassium (neK) when it is absorbed by crops. A plenty of CO_2 could be produced by root respiration and soil microbe respiration, and is fixed in form of CO_3^{2-}. For stable pH values, it is beneficial for organic acid to reach in equilibrium of absorption in soil solution. The exchangeable potassium and fixed potassium which comes from layers of secondary clay mineral are released when there occurs hydrolysis and cation substitution resulted by chemical element H^+. Those variables, such soil organic matter (ZSOM), iron-binding Humic (ZHMi), and water-soluble nitrogen (ZTN), not only provide nutrition for cotton root, but also changed soil chemical compositions and absorption characteristics of potassium in rhizosphere soil. Additionally, concomitant phenomenon (ZS) appears between S (the lost chemical element in soil) and potassium. Antagonism occurred between Na and K, and it

Table 8. Direct and indirect correlation coefficients of water-soluble potassium (*wsK*).

Variables	Direct correlation coefficients	Indirect correlation coefficients	Path contrition coefficients
Z(PHA)	−0.65	−0.32	−0.38
Z(S)	−0.56	0.17	−0.8
Z(Na)	−0.55	−1.15	−0.78
Z(Na/K)	−0.52	−0.01	−0.01
Z(TN)	−0.41	0.23	−0.38
Z(ORP)	−0.22	0.02	−1.28
Z(Mg)	−0.14	−0.79	−0.37
Z(CO₃)	−0.07	−0.53	−0.24
Z(pH)	0.14	−0.35	−1.87
Z(HMi)	0.39	0.2	0.29
Z(K)	0.41	0.02	−0.01
Z(CIA)	0.47	2.02	−1.61
Z(SOM)	0.58	0.06	0.25
Z(eK)	0.73	0.05	0.14
Z(neK)	0.77		

implied that sodium salt was able to have significant effect on plant (ZNa/K, ZNa).

From above analysis, we conclude that key direct factors of wsK, neK, and eK were S, Na/K, the CIA, SOM, Na, and TN, and key indirect factors were CO_3, Mg, pH, Na, S, and SOM. Whereas, direct impact factors of water soluble potassium (wsK) were PHA, S, Na, Na/K, K, the CIA, SOM, eK and neK. Each of its impact value was −0.65, −0.56, −0.55, −0.52, 0.41, 0.47, 0.58, 0.73, and 0.77 respectively. Indirect impact factors were Na, Mg, CO_3, pH, PHA, S, HMi, TN and CIA. Each of its impact value was −1.15, −0.79, −0.53, −0.35, −0.32, 0.17, 0.2, 0.23, and 2.02, respectively. Non exchangeable potassium direct impact factors were: S, Na/K, Na, TN, the CIA, wsK, SOM and eK. Each of its impact value was −0.86, −0.66, −0.64, −0.61, 0.56, 0.77, 0.8, and 0.98 respectively. Indirect impact factors were CO_3, eK, S, SOM, K, pH, Na and Mg. Each of its impact value was 1.18, 0.76, 0.19, −0.11, −0.14, −0.16, −0.34, and −0.67 respectively. Exchangeable potassium direct impact factors were S, Na/K, Na, TN, HMi, wsK, SOM and neK. Each of its impact value was −0.83, −0.67, −0.62, −0.55, 0.54, 0.73, 0.74, and 0.98 respectively. Indirect impact factors were CO_3, S, SOM, K, pH, Na, Mg and neK. Each of its impact value was −1.7, −0.28, 0.18, 0.21, 0.23, 0.46, 0.95, and 1.49 respectively. Absolute values of path coefficients represent the change ability of the potassium morphology, and the bigger value means the stronger ability to change potassium morphology. And "+"with and the arrow direction represent change direction of potassium morphology. In this way, we can quantitatively describe the K elements morphological changes.

Path model of soil potassium status changes of cotton

It is of great importance to further investigate on the equilibrium movement in soil. The following part will discuss the shift of dynamic balance among wsK, eK and neK, and their flow path. To test the reliability of the models, we collect another 48 cotton soil samples in the same experiment, using stepwise regression method, observation of potassium mobility and morphological change.

We carried out a plot experiment of potash application with high K-efficiency genotype (HEG) and low K-efficiency genotype

(LEG) cotton. The soil in each treatment weights 8.5 kg. Detail fertilizer schemes are listed in Table 11. LEG and HEG cottons' planting effects is shown in Fig. 6.

(1) Equilibrium shift in the path for non-exchangeable potassium

Supposed that $V_{wsK \to nek}$ is path for wsK to neK, $V_{ek \to nek}$ is path for eK to neK, therefore, equilibrium shift in path for non-exchangeable potassium can be expressed by contribution coefficient functions as below,

$$\begin{cases} y_{wsK} = 0.60 \cdot x_{neK} - 0.37 \cdot x_{PHA} + 5.87 \times 10^{-7} \left(R^2 = 0.87\right) \\ y_{neK} = 1.67 \cdot x_{wsK} + 0.62 \cdot x_{PHA} - 9.87 \times 10^{-7} \\ y_{eK} = 0.98 \cdot x_{neK} + 1.23 \times 10^{-6} \left(R^2 = 0.92\right) \\ y_{neK} = 1.02 \cdot x_{eK} - 1.25 \times 10^{-6} \end{cases} \quad (7)$$

Equilibrium occurs when X_{wsK}, X_{PHA}, X_{eK} satisfy with Function 8,

$$\begin{cases} 1.67 \cdot x_{wsK} + 0.62 \cdot x_{PHA} - 9.87 \times 10^{-7} = 1.02 \cdot x_{eK} - 1.25 \times 10^{-6} \\ x_{wsK} + 0.37 \cdot x_{PHA} = 0.61 \cdot x_{eK} - 1.57 \times 10^{-7} \end{cases} \quad (8)$$

Here, y_{nek} is the equilibrium threshold. And the value of x_{wsK}, x_{PHA}, x_{eK} is determined by soil characteristics. Calculated results from Equation 7 were listed in Table 12. In the table, column 1 and 2 were calculated from Equations 7. We used the wsK, eK, neK to construct dynamic equilibrium equations ($wsK \underset{\beta}{\overset{\alpha}{\rightleftharpoons}} neK \underset{\gamma}{\overset{\chi}{\rightleftharpoons}} eK$). The column 3 was normalized values of neK and the column 4 was the direction of balance movement of neK. These three column datasets were presented in Fig. 6.

In the reversible equilibrium of $wsK \underset{\beta}{\overset{\alpha}{\rightleftharpoons}} neK \underset{\chi}{\overset{\gamma}{\rightleftharpoons}} eK$, soil samples RSWKL, RSWH, RSWL, NRSWKL, NRSWL, K balance cycle were moved in α, χ direction movement; soil

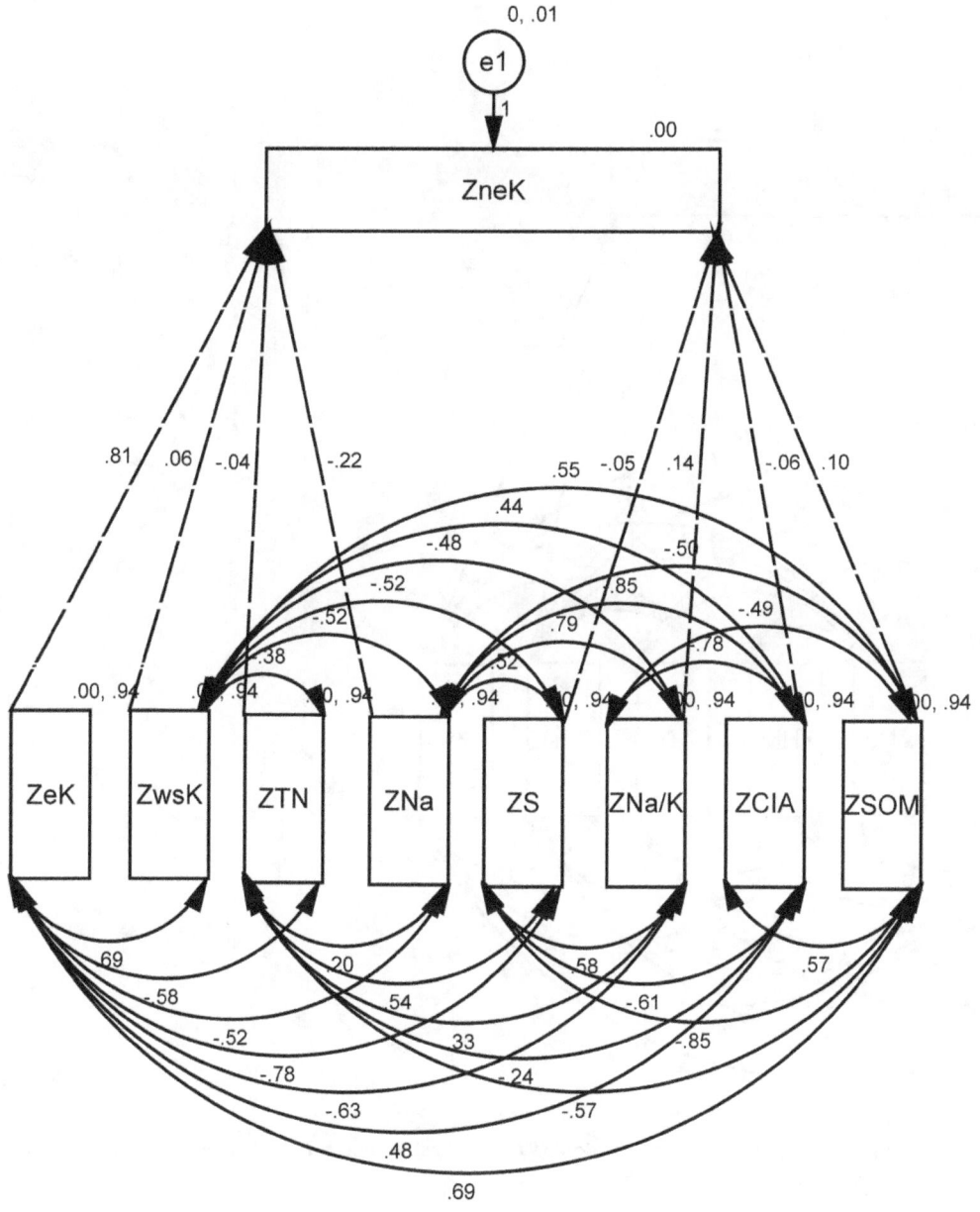

Figure 4. Path model of non-exchangeable potassium (neK, solid line for double arrow, dashed-line for single arrow, and single arrow flows into ZwsK).

samples RSWKH, RSKH, RSKL, ROPTH, ROPTL, NRSWKH, NRSWH, NROPTH, NROPTL, K balance cycle were moved in β and γ direction movement; soil samples NRSKH, NRSKL, K balance cycle were moved in β, χ direction movement.

In Fig. 6, neK reaches its equilibrium in following soil types which includes LEG in water stress treatment (SW), HEG after potassium stress treatment (SK), LEG after optimum fertilization treatment (OPT), and LEG after water stress treatment (SW). However, all the value in those soils with HEG in SW, LEG in SK, HEG in SW and SK, the values were all greater than the equilibrium, wsK reached its saturation. For other treatment soils, when wsK was lower than the equilibrium and PHA and eK was greater than equilibrium, neK changed into wsK. It proves that

eK equales to total K in matured crops. When eK is below a certain level in soil, plants can no longer get eK. When eK was at low levels, it was absorbed by other ions with stronger absorption. This reduces chance of potassium to enter into the soil solution. In this condition, neK begins to release through low concentrations of wsK and eK. However, the release capacity of neK depends on situations of crop types

(2) Equilibrium shift in the path for exchangeable potassium

Supposed that $V_{wsK \to eK}$ is path for wsK to eK, $V_{neK \to eK}$ is path for neK to eK, therefore, equilibrium shift in path for exchangeable potassium can be expressed by contribution coefficient functions as below,

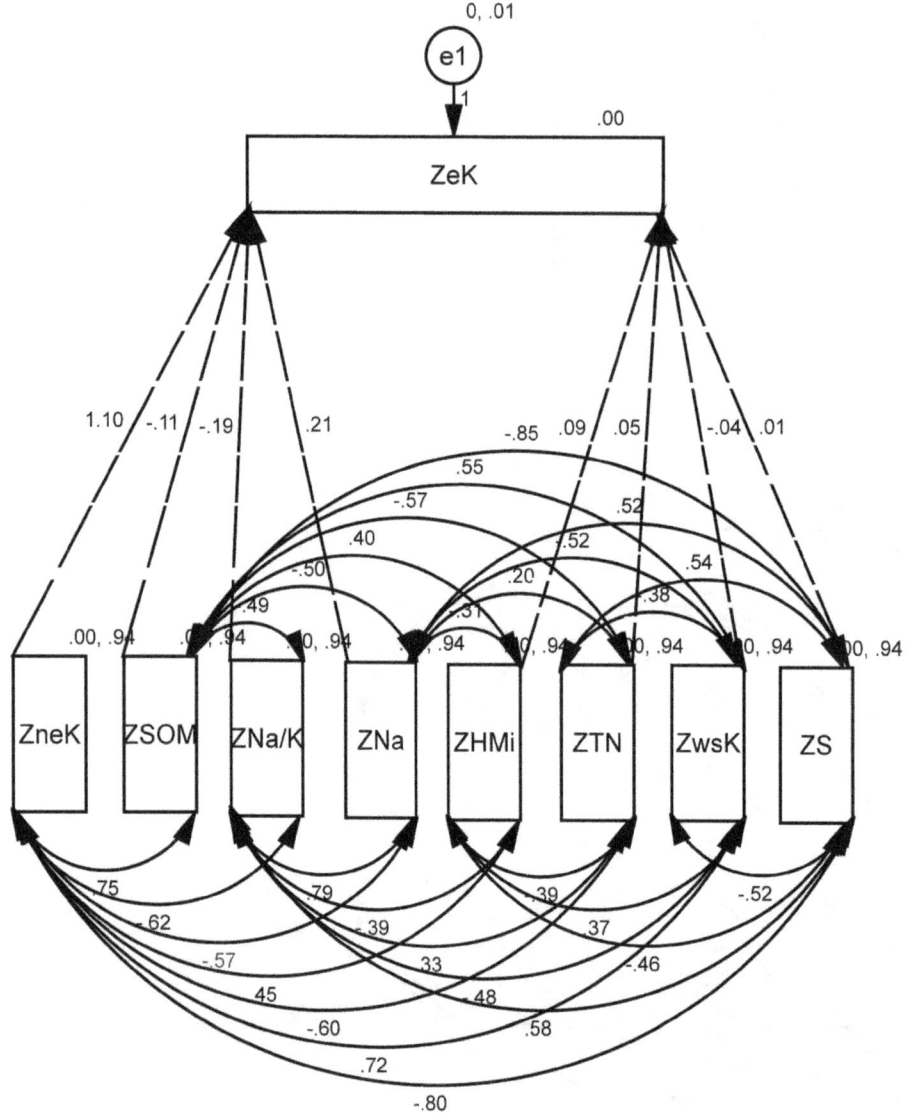

Figure 5. Path model of exchangeable potassium (eK, solid line for double arrow, dashed-line for single arrow, and single arrow flows into ZwsK).

$$\begin{cases} y_{wsk} = 0.98 \cdot x_{eK} + 3.328 \times 10^{-16} \quad (R^2 = 0.96) \\ y_{eK} = 1.02 \cdot x_{wsk} - 3.39 \times 10^{-16} \\ y_{neK} = 0.60 \cdot x_{eK} - 0.37 \cdot x_{PHA} + 7.89 \times 10^{-17} \quad (R^2 = 0.70) \\ y_{eK} = 1.67 \cdot x_{neK} + 0.62 \cdot x_{PHA} - 1.32 \times 10^{-16} \end{cases} \quad (9)$$

Equilibrium occurs when x_{wsK}, x_{PHA}, x_{neK} satisfy with Functions 11,

$$x_{wsk} = 1.64 \cdot x_{neK} + 0.61 \cdot x_{PHA} + 2.07 \times 10^{-16} \quad (10)$$

Here, y_{ek} is the equilibrium threshold. And the value of x_{wsK}, x_{PHA}, x_{neK} is determined by soil characteristics. Calculated results from Equations 9 and 10 are listed in Table 12. In this table, column 5 and 6 are calculated from Equations 9 and 10. We use

the wsK, eK, neK constructs the dynamic equilibrium equations ($wsK \underset{\theta}{\overset{\varepsilon}{\rightleftarrows}} eK \underset{\omega}{\overset{\lambda}{\rightleftarrows}} neK$).The column 7 is normalized values of neK and the column 8 is the direction of movement of balance of eK. These three column datasets are described in Fig. 7.

In the reversible equilibrium of $wsK \underset{\theta}{\overset{\varepsilon}{\rightleftarrows}} eK \underset{\omega}{\overset{\lambda}{\rightleftarrows}} neK$, soil samples SWH, NRSWL, K balance cycle were moved in ε, λ direction movement; soil samples RSWKL, RSWH, RSWL, K balance cycle were moved in θ, λ direction movement; soil samples RSWKH RSKH, RSKL ROPTH, ROPTL, NRSWKH, NRSKH, NRSKL, NROPTH, NROPTL K balance in soil samples, K balance cycle were moved in θ, ω direction movement respectively.

As is shown in Fig. 7, neK reaches its equilibrium in following soil plots, including LEG (SW), HEG (SK), LEG (OPT), and LEG (SW). And the exchangeable K reaches its equilibrium in soil types of LEG (SW). For other treatment soils, such as LEG (SW and

Table 9. Direct and indirect correlation coefficients of non-exchangeable potassium (neK).

Variables	Direct correlation coefficients	Indirect correlation coefficients	Path contrition coefficient
Z(S)	−0.86	0.19	−0.25
Z(Na/K)	−0.66	0.02	0.03
Z(TN)	−0.64	0.03	−0.03
Z(Na)	−0.61	−0.34	−0.37
Z(PHA)	−0.46	0.03	0.16
Z(ORP)	−0.41	0.10	−0.42
Z(Mg)	0.04	−0.67	−0.34
Z(CO3)	0.10	1.18	0.56
Z(K)	0.44	−0.14	0.2
Z(pH)	0.46	−0.16	−0.50
Z(HMi)	0.48	0.07	0.06
Z(CIA)	0.56	0.05	−0.07
Z(wsK)	0.77	0.01	0.01
Z(SOM)	0.8	−0.11	−0.14
Z(eK)	0.98	0.76	0.69

SK), LEG (SW), LEG (SK), HEG (SW and SK), HEG (OPT), neK changes to eK more or less. In soil of HEG (SW and SK), eK changes to wsK when wsK was lower than equilibrium. In soil of LEG (SW and SK), neK changes to eK when wsK of soil was greater than that of crops. Soils of both genotypes cotton (HEG and LEG) can efficiently absorb exchangeable potassium.

During a lot of vermiculite in the soil, those negative charges caused by vermiculite isomorphous substitution are near p-site, the electrostatic attraction of potassium ion became bigger, the adsorption capacity much more than other type (2:1) minerals, therefore, wsK and neK are fixed and the soil changed into deficiency. In general, drying accelerated the fixation of wsK

adsorption, but neK still moved in the direction of eK in the reversible equilibrium in HEG rhizosphere soil because rhizosphere soil exchangeable potassium content was low to meet the needs of the cotton growth in the SW and SWK plant. When moisture was adequate, redox reaction of soil became strengthened and oxidation reduction potential was reduced. For example, Fe^{2+} and Mn^{2+} ions were increased rapidly, Fe^{2+} and Mn^{2+} replaced p-site potassium of soil colloids and part of i-site potassium, and this improved the content of rapidly-available potassium and biological effective parts. In addition, the dissolved iron and manganese mineral alteration and also increased the release of mineral K.

Table 10. Direct and indirect correlation coefficients in exchangable potassium (eK).

Variables	Direct correlation coefficients	Indirect correlation coefficients	Path contrition coefficient
Z(S)	−0.83	−0.28	0.36
Z(Na/K)	−0.67	−0.03	−0.04
Z(TN)	−0.62	−0.04	0.04
Z(Na)	−0.55	0.46	0.54
Z(PHA)	−0.51	−0.06	−0.24
Z(ORP)	−0.38	−0.16	0.6
Z(Mg)	0.07	0.95	0.49
Z(CO3)	0.11	−1.7	−0.81
Z(pH)	0.47	0.23	0.71
Z(K)	0.48	0.21	−0.3
Z(CIA)	0.51	−0.07	0.1
Z(HMi)	0.54	−0.09	−0.09
Z(wsK)	0.73	−0.01	−0.02
Z(SOM)	0.74	0.18	0.2
Z(neK)	0.98	1.49	1.44

Table 11. Fertilizing rates of different treatments (unit: g).

Treatments	OPT	SW	SK	SWK
Urea	6.65	6.65	6.65	6.65
Na$_2$PO$_4$.2H$_2$O	11.72	11.72	11.72	11.72
KCl	6.83	6.83	0	0
CaCO$_3$	2.13	2.13	2.13	2.13
H$_3$BO$_3$	1.00	1.00	1.00	1.00
ZnSO$_4$.7H2O	1.87	1.87	1.87	1.87
MgSO$_4$.7H2O	2.61	2.61	2.61	2.61

OPT, optimum fertilization treatment; *SW*, water limited; *SK*, potassium limited; *SWK*, water & potassium limited.

Discussion and Conclusions

In this study, changes of potassium status in soil were investigated through our proposed path model on basis of laboratory experimental datasets. Through this investigation, following findings were achieved.

Firstly, changes of potassium status in the rhizosphere soil were controlled by different environmental variables. wsK, neK and eK were controlled by some common factors and have their speculiar characteristics respectively. Those parameters, such as soil organic matter (SOM), Na and Na/K and S, were significant direct coefficient factors which control dynamic process of wsK, neK and eK. CO$_3$ and Mg are significant indirect coefficient factors which control dynamic of wsK, neK and eK. The pH is the primary factor which controls path of the wsK, neK and eK. Organic acids of HEG root system secretion increased soil acidity; the amount of fixed potassium was reduced. PHA surface is porous structure, PHA increased the adsorption of wsK in the experiment, and HMi could promote the moving to wsK, eK in the K form exchange. Both of them were negative correlation significantly negative correlation. First, under acid condition, hydroxyl aluminum and aluminum ions can occupy the potassium fixation point stronger. At the same time, the larger diameter of hydroxy aluminium ions

entered into the interlayer of mineral, formed as the "island" and played a supporting role, provided the potassium ion diffusion of mobile channel in the interlayer. These reduced the soil potassium fixation. Second, water protonated in acidic condition, the radius of H$_3$O$^+$ 12–13.3 nm were similar to K$^+$, H$_3$O$^+$ can produced competitive adsorption in soil where fixed on K$^+$ parts, so potassium reduced fixation. Those dynamic processes, such as eK to wsK, low concentration wsK and eK to neK, are likely to occur, in condition that wsK is lower than equilibrium value and PHA and eK are greater than equilibrium values. eK is changed neK when wsK is greater than equilibrium value. In soil of HEG (SW and SK), eK is changed into wsK when plant wsK is lower than soil wsK. In soil of LEG (SW and SK), neK is changed into eK when plant wsK is higher than soil wsK. Both genotype cottons can effectively absorb eK in treatment of stress of potassium (SK).

Secondly, it discloses that SOM, S, TN, HMi and PHA in rhizosphere soil determined dynamic balance of potassium status which includes adsorption and desorption, precipitation and dissolution, complexion and chelating. Special characteristics of humus, e.g., acidic and hydrophilic, cationic exchange, complexing capacity and high absorption capacity, could improves the sustained-release effect of potassium fertilizer. Many variables, pH, ORP, etc., are closely related to transformation and utilization of potassium fertilizer. The pH change in soil is mainly caused by coupling effects of nutrient uptake by plant roots and the secretion of organic acid and absorption imbalance of positive ion and negative ion leads to rhizosphere pH change, however, different factors affect imbalance of absorption of positive ion and negative ion. Dynamic equilibrium of potassium is affected by equilibrium constants, temperature, and products of this equilibrium system (wsK). Therefore, it is useful for cotton planting to improve to appropriate temperature, to alter SOM, TN and pH in soil, and to reduce the equilibrium constant to positive reaction. Meanwhile, it is good to choose HEG cotton for better use soil wsK. It is important to understand ways to slow down the fast response and accelerate speed of slow response in potassium balance system.

Finally, path model can effectively determine direction and quantities of potassium status changes between eK, neK and wsK under influences of specific environmental parameters. The proposed path model was able to quantitatively disclose moving

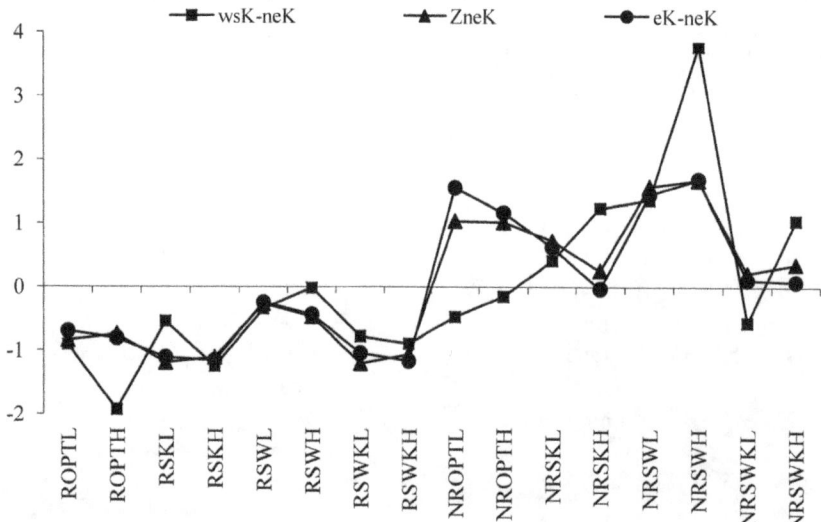

Figure 6. Dynamic equilibrium of non-exchangeable potassium in rhizosphere/non-rhizosphere soils.

Table 12. Dynamic equilibrium of exchangeable/non-exchangeable potassium (*neK/eK*) in rhizosphere/non-rhizosphere soils.

Variables	$wsK \underset{\beta}{\overset{\alpha}{\rightleftharpoons}} neK \underset{\chi}{\overset{\gamma}{\rightleftharpoons}} eK$			Movement direction at balance state	$wsK \underset{\theta}{\overset{\varepsilon}{\rightleftharpoons}} eK \underset{\omega}{\overset{\lambda}{\rightleftharpoons}} neK$			Movement direction at balance state
	$V_{wsk \to neK}$	$V_{ek \to neK}$	ZneK		$V_{wsk \to eK}$	$V_{nek \to eK}$	ZeK	
RSWKH	−0.89	−1.16	−1.06	β,χ	−0.78	−1.39	−1.14	θ, ω
RSWKL	−0.78	−1.04	−1.2	α, χ	−0.78	−1.51	−1.02	θ, λ
RSWH	−0.02	−0.44	−0.47	α, χ	−0.19	−0.5	−0.44	θ, λ
RSWL	−0.35	−0.25	−0.27	α, χ	−0.32	−0.28	−0.24	θ, λ
RSKH	−1.25	−1.16	−1.11	β, χ	−0.8	−1.79	−1.14	θ, ω
RSKL	−0.55	−1.11	−1.19	β, γ	−0.82	−1.19	−1.09	θ, λ
ROPTH	−1.93	−0.81	−0.75	β, χ	−0.65	−2.11	−0.79	θ, ω
ROPTL	−0.91	−0.70	−0.84	β, γ	−0.51	−1.48	−0.69	θ, λ
NRSWKH	1.05	0.08	0.36	β, χ	0.14	1.43	0.08	ε, λ
NRSWKL	−0.56	0.11	0.22	α, χ	−0.45	0.54	0.11	θ, ω
NRSWH	3.78	1.69	1.68	β, χ	3.24	1.28	1.66	ε, λ
NRSWL	1.37	1.45	1.58	α, χ	0.79	2.73	1.42	θ, λ
NRSKH	1.24	−0.03	0.26	β, χ	0.92	0.17	−0.03	ε, λ
NRSKL	0.42	0.63	0.73	β, χ	0.11	1.45	0.61	θ, ω
NROPTH	−0.15	1.17	1.02	β, γ	−0.12	1.76	1.14	θ, ω
NROPTL	−0.47	1.56	1.04	β, γ	0.23	0.90	1.53	θ, ω

direction of K status and quantify its equilibrium threshold. It provided a theoretical and practical basis for scientific and effective fertilization in agricultural plants growth. The significance of this study is that we are able to implement investigation on the equilibrium movement of potassium status through path model analysis. It discloses the gradual dynamic process of potassium status, and to decouple the sophisticated interactions between different variables. It is useful to guide the use of potassium fertilizer for cotton crops in practical. The model provides a biogeochemical theory basis of controlling moisture in

cotton plant, applying potash fertilizer and organic fertilizer, and appropriately reducing nitrogen fertilizer.

However, environmental parameters such as soil moisture and temperature are some of the most important factors that influenced crop growth. Observation of dynamic change of soil moisture and temperature in every other time-step is quite sophisticated for us to design the control experiments while it involves in more than 4 times treatments, which means that we have more than 500 pots. By limitation, we had to set the soil moisture into two states, water sufficient (>35%) and deficient (<25%), and temperature were set to indoor temperature.

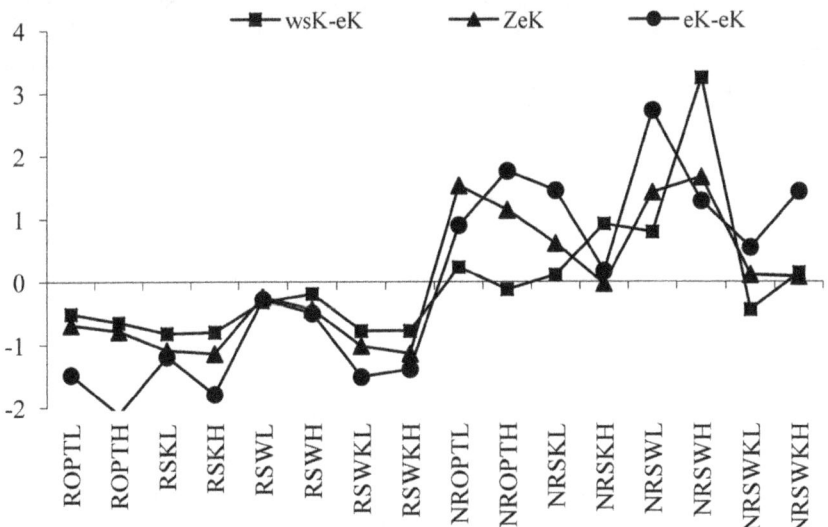

Figure 7. Dynamic equilibrium of exchangeable potassium in rhizosphere/non-rhizosphere soils.

Anyway, we were very aware of the importance of soil moisture and temperature impacts on control experiment potassium status. The next step of our work is to further investigate efficient utilization of potassium and the influence of soil moisture and temperature on exchange of potassium forms. We hope this study is able to gain knowledge of promoting high efficient use of potassium for crops.

Acknowledgments

We are grateful to invaluable review comments of the editor and two anonymous reviewers to make the paper perfect. We would like to express our special thanks to Dr. Feng WU of the Institute of Earth Environment, Chinese Academy of Sciences for his kind technical support.

Author Contributions

Conceived and designed the experiments: WH FC. Performed the experiments: WH. Analyzed the data: WH. Contributed reagents/materials/analysis tools: FC WH. Wrote the paper: WH FC.

References

1. Mahmood-Ul-Hassan M, Rashid M, Rafique E (2011) Nutrients transport through variably structured soils. Soil Science and Plant Nutrition 57: 331–340.
2. Osaki M, Matsumoto M, Shinano T, Tadano T (1996) A root-shoot interaction hypothesis for high productivity of root crops. Soil science and plant nutrition 42: 289–301.
3. Wada S-I, Seki H (1994) Ca-K-Na exchange equilibria on a smectitic soil: Modeling the variation of selectivity coefficient. Soil Science and Plant Nutrition 40: 629–636.
4. Puente M, Li C, Bashan Y (2004) Microbial Populations and Activities in the Rhizoplane of Rock-Weathering Desert Plants. II. Growth Promotion of Cactus Seedlings. Plant Biology 6: 643–650.
5. Wada S-I, Masuda K (1995) Control of salt concentration of soil solution by the addition of synthetic hydrotalcite. Soil Science and Plant Nutrition 41: 377–381.
6. Kobayashi H, Masaoka Y, Takahashi Y, Ide Y, Sato S (2007) Ability of salt glands in Rhodes grass (Chloris gayanaKunth) to secrete Na+and K+. Soil Science and Plant Nutrition 53: 764–771.
7. Han MY, Zhang LX, Fan CH, Liu LH, Zhang LS, et al. (2011) Release of nitrogen, phosphorus, and potassium during the decomposition of apple (Malus domestica) leaf litter under different fertilization regimes in Loess Plateau, China. Soil Science and Plant Nutrition 57: 549–557.
8. He H, Zhou J, Wu Y, Zhang W, Xie X (2008) Modeling the interaction of urbanization and surface water quality environment. Environmental Forensics 9: 215–225.
9. Eick MJ, Sparks DL, Bar-Tal A, Feigenbaum S (1990) Analyses of adsorption kinetics using a stirred-flow chamber: II. Potassium-calcium exchange on clay minerals. Soil Science Society of America Journal 54: 1278–1282.
10. McCune B, Caldwell BA (2009) A single phosphorus treatment doubles growth of cyanobacterial lichen transplants. Ecology 90: 567–570.
11. Schneider A-K, Schröder B (2012) Perspectives in modelling earthworm dynamics and their feedbacks with abiotic soil properties. Applied Soil Ecology 58: 29–36.
12. He H, Jim C (2012) Coupling model of energy consumption with changes in environmental utility. Energy Policy 43: 235–243.
13. He H, Jim C (2010) Simulation of thermodynamic transmission in green roof ecosystem. Ecological Modelling 221: 2949–2958.
14. Brouder S, Cassman K (1994) Evaluation of a mechanistic model of potassium uptake by cotton in vermiculitic soil. Soil Science Society of America Journal 58: 1174–1183.
15. Nakahara O, Wada S-I (1995) Surface Complexation Model of Cation Adsorption on Humic Soils. Soil Science and Plant Nutrition 41: 671–679.
16. Uroz S, Calvaruso C, Turpault M-P, Frey-Klett P (2009) Mineral weathering by bacteria: ecology, actors and mechanisms. Trends in microbiology 17: 378–387.

17. Puente M, Bashan Y, Li C, Lebsky V (2004) Microbial populations and activities in the rhizoplane of rock-weathering desert plants. I. Root colonization and weathering of igneous rocks. Plant Biology 6: 629–642.
18. Hinsinger P, Fernandes Barros ON, Benedetti MF, Noack Y, Callot G (2001) Plant-induced weathering of a basaltic rock: experimental evidence. Geochimica et Cosmochimica Acta 65: 137–152.
19. Hashimoto Y, Kang J, Matsuyama N, Saigusa M (2012) Path Analysis of Phosphorus Retention Capacity in Allophanic and Non-allophanic Andisols. Soil Science Society of America Journal 76: 441–448.
20. Kang J, Hesterberg D, Osmond DL (2009) Soil Organic Matter Effects on Phosphorus Sorption: A Path Analysis. Soil Science Society of America Journal 73: 360.
21. Lambers H, Mougel C, Jaillard B, Hinsinger P (2009) Plant-microbe-soil interactions in the rhizosphere: an evolutionary perspective. Plant and Soil 321: 83–115.
22. Wright S (1921) Correlation and causation. Journal of agricultural research 20: 557–585.
23. Wright S (1920) The relative importance of heredity and environment in determining the piebald pattern of guinea-pigs. Proceedings of the National Academy of Sciences of the United States of America 6: 320.
24. Wright S (1921) Systems of mating. I. The biometric relations between parent and offspring. Genetics 6: 111.
25. Wright S (1921) Systems of mating. II. The effects of inbreeding on the genetic composition of a population. Genetics 6: 124.
26. Kulahci F (2011) A risk analysis model for radioactive wastes. J Hazard Mater 191: 349–355.
27. Berthrong ST, Jobbágy EG, Jackson RB (2009) A global meta-analysis of soil exchangeable cations, pH, carbon, and nitrogen with afforestation. Ecological Applications 19: 2228–2241.
28. Kocev D, Naumoski A, Mitreski K, Krstić S, Džeroski S (2010) Learning habitat models for the diatom community in Lake Prespa. Ecological Modelling 221: 330–337.
29. He C, Cui K, Duan A, Zeng Y, Zhang J (2012) Genome-wide and molecular evolution analysis of the Poplar KT/HAK/KUP potassium transporter gene family. Ecol Evol 2: 1996–2004.
30. Kamewada K (1996) Application of "Four-Plane Model" to the Adsorption of K+, NO3-, and SO42-from a Mixed Solution of KNO3and K2SO4on Andisols. Soil Science and Plant Nutrition 42: 801–808.
31. Kang J, Hesterberg D, Osmond DL (2009) Soil organic matter effects on phosphorus sorption: A path analysis. Soil Science Society of America Journal 73: 360–366.

21

Changes in the Potential Multiple Cropping System in Response to Climate Change in China from 1960–2010

Luo Liu[1,2,3], **Xinliang Xu**[2]*, **Dafang Zhuang**[2], **Xi Chen**[1], **Shuang Li**[2,3]

1 State Key Laboratory of Desert and Oasis Ecology, Xinjiang Institute of Ecology and Geography, Chinese Academy of Sciences, Urumqi, China, **2** State Key Laboratory of Resources and Environmental Information Systems, Institute of Geographical Sciences and Natural Resources Research, Chinese Academy of Sciences, Beijing, China, **3** University of Chinese Academy of Sciences, Beijing, China

Abstract

The multiple cropping practice is essential to agriculture because it has been shown to significantly increase the grain yield and promote agricultural economic development. In this study, potential multiple cropping systems in China are calculated based on meteorological observation data by using the Agricultural Ecology Zone (AEZ) model. Following this, the changes in the potential cropping systems in response to climate change between the 1960s and the 2010s were subsequently analyzed. The results indicate that the changes of potential multiple cropping systems show tremendous heterogeneity in respect to the spatial pattern in China. A key finding is that the magnitude of change of the potential cropping systems showed a pattern of increase both from northern China to southern China and from western China to eastern China. Furthermore, the area found to be suitable only for single cropping decreased, while the area suitable for triple cropping increased significantly from the 1960s to the 2000s. During the studied period, the potential multiple cropping index (PMCI) gap between rain-fed and irrigated scenarios increased from 18% to 24%, which indicated noticeable growth of water supply limitations under the rain-fed scenario. The most significant finding of this research was that from the 1960s to the 2000s climate change had led to a significant increase of PMCI by 13% under irrigated scenario and 7% under rain-fed scenario across the whole of China. Furthermore, the growth of the annual mean temperature is identified as the main reason underlying the increase of PMCI. It has also been noticed that across China the changes of potential multiple cropping systems under climate change were different from region to region.

Editor: Wengui Yan, National Rice Research Center, United States of America

Funding: This research was supported and funded by National Program on Key Basic Research Project (973 Program) (Grant No. 2010CB950901), Project of Remote sensing investigation and assessment for Ecological environment changes in China(No. STSN-14-00) and project of CAS action-plan for West Development (No. KZCX2-XB3-08-01) The funders had no role in study design, data collection and analysis, decision to publish, or preparation of the manuscript.

Competing Interests: The authors have declared that no competing interests exist.

* E-mail: xuxl@lreis.ac.cn

Introduction

Multiple cropping is a general cropping practice which involves growing two or more crops on the same field in a given year. It is one of the most effective ways to increase grain yield and to promote agricultural economic development. There is a common belief that food production and food security are continuously threatened by the increasing price of crude oil, extreme weather, and the diversion of crop land for bio-fuel production. Therefore, food security has long been an issue which has attracted global attention under a situation of urbanization by which the total area of active farmland continues to decrease. Insufficient attention has been paid to processes that increase crop yield, leading to an increase in the multiple cropping index (MCI; ratio of sown area to cropland area) [1–2]. Due to environmental limitations, increased food production is unlikely to be achieved from farmland expansion in many regions [3]. In this situation, multiple cropping is an effective method for improving crop productivity as it is an efficient use of natural resources (i.e. land, light, heat and so on) and human resources. This practice also helps to alleviate competition for land use between food production and economic crops.

China is one of the largest countries in the world and has succeeded in achieving the highest MCI. Nearly half of the cultivated land in China is subject to multiple cropping practices. This has resulted in a 25% increase in crop yields in the past few decades, which has enabled China to feed 22% of the world's population using only 7% of the world's cultivated land [4]. Thus, an increase in the use of multiple cropping practices has assisted China in successfully addressing the issue of food security. Two factors which further underline the importance of multiple cropping practices are the global trends of climate change and urbanization. Both of these factors have caused cropping systems to undergo significant change during the last several decades [5–10]. In the case of China, this has resulted in a decrease in cropland area by 15% [11] (primarily distributed in developed areas).

The statistical method used for acquiring information on cropping system changes was -based on administrative boundaries from the government, which ignored the spatial heterogeneity within the same administrative area and thus failed to represent the spatial characteristics of the multiple cropping systems [12–14]. These systems can be measured according to the phenological and ecological characteristics of crop growth. This is done by identifying the periods of cropping cycles from time series data collected by remote sensing systems including AVHRR/NDVI, SPOT VEGETATION and MODIS [15–18]. This method is restrained by the temporal resolution of remote sensing data, and therefore cannot capture the changes in the multiple cropping systems over long periods of time. However it is suitable for food security assessment and agricultural development planning, which aims to quantify long term sequenced MCI changes and investigate its response to global climate change.

In this study, the potential multiple cropping systems in China were calculated using the method of AEZ based on meteorological observation data. The spatio-temporal change patterns between 1960 and -2010 were also evaluated, via the spatial analysis techniques of the Geographic Information System (GIS). Finally, the characteristic variations of the potential cropping systems were analyzed, and their responses to climate change in different regions were illustrated.

Data and Methods

Data sources

The input data for this study included terrain elevation, soil, land use and meteorological data (including the monthly maximum air temperature, minimum air temperature, precipitation, relative humidity, wind speed and sunshine hours).

Terrain Elevation Data. The terrain elevation dataset derived from the Shuttle Radar Topography Mission C-band (SRTM) was the first publicly available near-global, high resolution raster Digital Elevation Model (DEM) [19]. The SRTM data was distributed with a 90 m spatial resolution around the Earth, reduced from the original 30 m resolution via averaging and sub-sampling. Numerous worldwide applied studies have used SRTM data for environmental analysis [20–21]. The SRTM Version 2 data (by naming convention SRTM3 for 3 arc sec data) was sourced from the Jet Propulsion Laboratory (http://srtm.csi.cgiar.org/SELECTION/inputCoord.asp).

Soil Data. Soil quality was determined by several parameters, including soil type, effective soil depth and soil water holding capacity. A nation-wide soil dataset at the scale of 1:1,000,000 was provided by the Data Center for Resources and Environmental Sciences at the Chinese Academy of Sciences (RESDC) (http://www.resdc.cn/first.asp). Soil data was used to calculate the soil-water balance, which determined the potential and actual evapotranspiration for a reference crop and the length of its growing period (LGP, days).

Land Use Data. A land-use database with a mapping scale of 1:1,000,000 was developed by CAS (The Chinese Academy of Sciences). This database included five time periods: the late 1980s, the mid-1990s, the late 1990s, the mid-2000s and 2008. The primary data source for the land use database was Landsat MSS/TM/ETM CCD digital images. CBERS (the China-Brazil Earth Resources Satellite) and HJ-1 (small satellite constellation for environment and disaster monitoring) images were also used as a supplement for the areas uncovered by Landsat. The land-use data were classified into 25 categories, which were subsequently grouped into six classes: cropland, woodland, grassland, water body, built-up area and unused land. Detailed information on this database can be found in previous papers [22–26]. In this study, farmland data were extracted from 2008 land use data.

Meteorological Data. Meteorological data, which included the monthly maximum air temperature, minimum air temperature, precipitation, relative humidity, wind speed at 10 m height and sunshine hours from 1960 to 2010, were obtained from the national agro-meteorological stations of China maintained by Chinese Meteorological Administration (CMA) (http://cdc.cma.

gov.cn). Because of the diverse terrain across China, the impact of topographic conditions on the interpolation of the meteorological data was also considered. The ANUSPLIN software [27–28], which was designed for spatial interpolation of climate data, was used to interpolate the meteorological data with the DEM mentioned in Section 2.1.1. These data measured monthly for the above six key factors of plant growth were then interpolated by ANUSPLIN to individual 10 km resolutions based on the digital terrain model of China (excluding Taiwan).

Before further processing, all source data were resampled into a raster dataset with a 10 km spatial resolution, and the data were transformed into the same coordinate system (Krasovsky_1940_Albers projection system, with central_meridian 105°E, double standard parallel of 27°N and 45°N, and D_Krasovsky_1940 datum).

Method

The calculation method for the potential multiple cropping systems were derived from the Global Agro-Ecological Zones Model (AEZ) which was developed in the 1970s, and updated in 2010 by the Food and Agriculture Organization (FAO) and the International Institute for Applied Systems Analysis (IIASA) [29–33]. The crop growth process was primarily affected by the local climate conditions during the growth period. The potential multiple cropping index (PMCI) is directly determined by the light-temperature-water condition. Figure 1 shows the technological framework designed for the calculation of the multiple cropping systems.

Multiple cropping zones. To assess the potential multiple cropping systems, a number of multiple cropping zones were defined by matching both the thermal and water requirements with the time required for crop growth in different latitudinal thermal climates. The following parameters were used to define the cropping zones (Table 1, Table 2): length of the growing period (LGP), number of days with mean daily temperatures above 5°C and 10°C ($LGP_{t=5}$ and $LGP_{t=10}$), accumulated temperature on days with a mean daily temperature $\geq 0°C$ and $\geq 10°C$ ($TS_{t=0}$ and $TS_{t=10}$) and accumulated temperature during the growing period with a mean daily temperature $\geq 5°C$ and 10°C ($TS\text{-}G_{t=5}$ and $TS\text{-}G_{t=10}$).

To reveal the impact of water conditions on the multiple cropping systems, two scenarios (irrigated and rain-fed) were considered for the calculation of the potential multiple cropping systems. The rain-fed scenarios [29] were obtained from the light-temperature-water condition, while the irrigated scenarios [30] were calculated using only the light-temperature conditions, which assumed sufficient water for crop growth. The two scenarios were consistent with two existing agricultural management methods in China (naturally rain-fed farmland and irrigated farmland). The water-limited condition referred to the LGP and TS-G mentioned previously with an additional requirement that indicated moisture was insufficient ($ETa < 0.4\ ET_0$) [31]. According to the above criteria (Table 1, Table 2), the following eight zones were classified and mapped:

A. *Zone of no cropping* (too cold or too dry for rain-fed crops)

B. *Zone of single cropping*

C. *Zone of limited double cropping* (relay cropping; single wetland rice cropping may be possible)

D. *Zone of double cropping* (sequential cropping; double cropping with wetland rice cropping not possible)

E. *Zone of double cropping* (sequential cropping; wetland rice cropping possible)

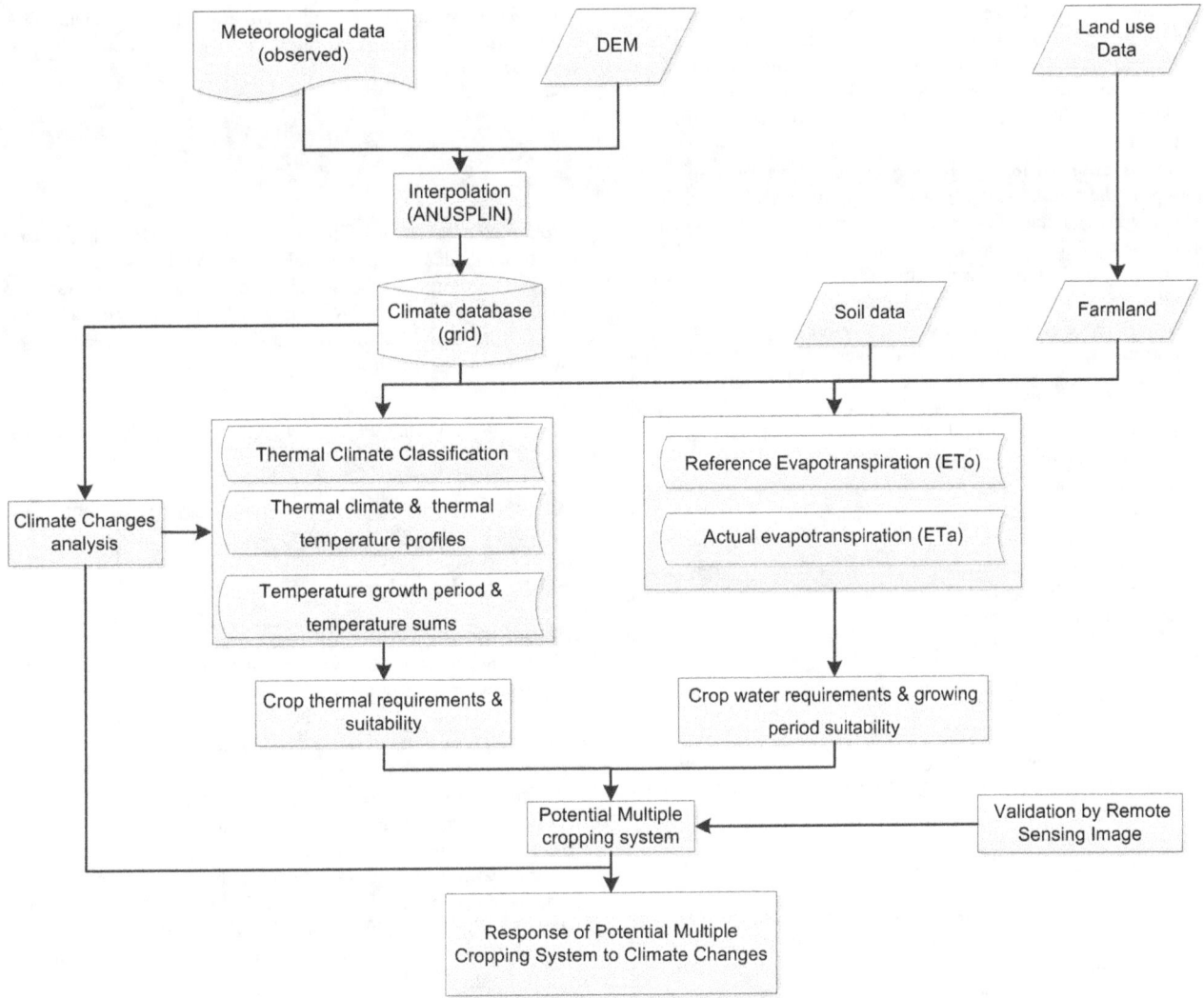

Figure 1. Technological framework designed for potential estimation of multiple cropping systems.

F. *Zone of limited triple cropping* (partial relay cropping; no third cropping possible in case of two wetland rice crops)

G. *Zone of triple cropping* (sequential cropping of three short-cycle crops; two wetland rice crops possible)

H. *Zone of triple rice cropping* (sequential cropping of three wetland rice crops possible)

Latitudinal thermal classification. The latitudinal thermal classification was the basis for the calculation of potential multiple

Table 1. Delineation of multiple cropping zones in tropical areas.

Zone	LGP	$LGP_{t=5}$	$LGP_{t=10}$	$TS_{t=0}$	$TS_{t=10}$	$TS\text{-}G_{t=5}$	$TS\text{-}G_{t=10}$
A	-	-	-	-	-	-	-
B	≥45	≥120	≥90	≥1600	≥1000	-	-
C	≥200	≥200	≥120	≥6400	n.a.	≥3200	≥2700
D	≥240	≥240	≥165	≥6400	n.a.	≥4000	≥3200
E	n.a.	n.a.	n.a.	n.a.	n.a.	n.a.	n.a.
F	≥300	≥300	≥240	≥7200	≥7000	≥5100	≥4800
G	n.a.	n.a.	n.a.	n.a.	n.a.	n.a.	n.a.
H	≥360	≥360	≥360	≥7200	≥7000	-	-

Table 2. Delineation of multiple cropping zones in non-tropical areas.

Zone	LGP	$LGP_{t=5}$	$LGP_{t=10}$	$TS_{t=0}$	$TS_{t=10}$	$TS\text{-}G_{t=5}$	$TS\text{-}G_{t=10}$
A	-	-	-	-	-	-	-
B	≥45	≥120	≥90	≥1600	≥1000	-	-
C	≥180	≥200	≥120	≥3600	≥3000	≥3200	≥2900
D	≥210	≥240	≥165	≥4500	≥3600	≥4000	≥3200
E	≥240	≥270	≥180	≥4800	≥4500	≥4300	≥4000
F	≥300	≥300	≥240	≥5400	≥5100	≥5100	≥4800
G	≥330	≥330	≥270	≥5700	≥5500	-	-
H	≥360	≥360	≥330	≥7200	≥7000	-	-

cropping systems (Table 3). The delineation of a latitudinal thermal regime was based on the total affected accumulation temperature above 10°C (T_{sum10}), which accumulates the daily average temperatures for days in which the daily average temperature is above the appropriate threshold temperature of 10°C [34].

Evapotranspiration. The reference evapotranspiration (ET_0,mm/day) and actual evapotranspiration (ET_a, mm/day) values were calculated using the soil-water balance calculation procedures following the methodologies outlined in "CROP-WAT" [31] and "Crop Evapotranspiration" [35] for crop water requirements and growing period suitability.

The reference evapotranspiration (ET_0, mm/day) represents the evapotranspiration from a defined reference surface which is similar to an extensive surface of green and well-watered grass of uniform height (0.12 meters) that is actively growing and completely shades the ground. The ET_0 was calculated according to the Penman-Monteith equation [31,36,37].

The calculation of ET_m for a 'reference crop' is based on the assumption that sufficient water is available for uptake in the rooting zone. The value of ET_m is related to ET_0 by applying the crop coefficients for the water requirements (kc, fractional) (Equation 1). The kc factors are determined by phenological development and leaf area [35].

$$ET_m = kc \times ET_0 \tag{1}$$

The actual uptake of water for the 'reference' crop was characterized by the actual evapotranspiration (ET_a, mm/day). The calculation of ET_a differentiated between two possible cases depending on the availability of water for plant extraction: (i) adequate soil water availability ($ET_a = ET_m$) and (ii) limited soil-water availability ($ET_a < ET_m$) [35].

Under adequate soil-water conditions, the value of ET_a was assumed to be equal to ET_m as long as the water balance (Wb, mm) was greater than or equal to the "readily" available soil-water (W_{read}, mm). This requirement characterizes a situation in which crops are able to "easily" extract sufficient water and therefore no water stress occurs (Equation 2).

$$ET_a = ET_m \tag{2}$$

when $pre > ET_m$ and $(Wb - W_{read}) > ET_m$

For limited water conditions, the ET_a can be calculated as the product of ET_m and the variable ρ (Equation 3).

$$ET_a = \min[(pre + \rho \times ET_m), ET_m] \tag{3}$$

The *pre* represents the precipitation, and ρ is the quotient of the current water balance (Wb, mm) and the readily available soil water (W_{read}, mm).

$$\rho = \frac{Wb}{W_{read}} \tag{4}$$

Soil-water balance. The volume of water available for plant uptake is calculated using a daily soil-water balance (Wb, mm). The Wb accounts for the accumulated daily water inflow from precipitation (pre, mm/day) or snowmelt (snm, mm/day) and the outflow from actual evapotranspiration (ETa, mm/day) and excess water loss due to runoff and deep percolation.

$$Wb_j = \min(Wb_{j-1} + pre + snm - ET_a, W_{max}) \tag{5}$$

where j is the day of the year, and W_{max} (W_{max}, mm) is the maximum soil water storage capacity.

Results and Analysis

Comparison and Verification

To verify the accuracy of the calculated results, the calculated potential multiple cropping system in 2000 was compared with the actual result from remote sensing monitoring based on MODIS data [11]. Three types of cropping systems were used in the actual situation. The potential multiple cropping system was also classified into three corresponding types: single cropping per year, double cropping per year, and triple cropping per year. Figure 2 demonstrates that the potential multiple cropping system was consistent with the actual system overall. For example, regions that use triple cropping (the area of 93,445 square kilometers) exist in Sichuan Basin, the Middle-Lower Yangtze Plain, the south of the Yangtze River and Hainan Island are completely simulated as triple cropping in the potential cropping system. Similarly, the single cropping in the potential cropping system (the area of 870,093 square kilometers) duplicated the actual single cropping regions. However, as shown in Table 4, the simulated MCI was larger than the actual MCI, which revealed the difference between the calculated potential multiple cropping systems and the actual systems. Farmers tended to select their multiple cropping practices following the traditional pattern which was based on previous climate conditions rather than the optimal multiple cropping systems (which are consistent with the actual local climate conditions). Moreover, many farmers preferred to grow plants that yielded greater economic profits, and the lack of young and middle-aged rural laborers who prefer to work in cities further reduced the MCI.

Spatial patterns of the potential multiple cropping system in China

Based on the method in Section 2.2, the potential multiple cropping systems were calculated using the average climatic conditions at intervals of 10 years from the 1960s to the 2000s. Figure 3 and Figure 4 show the distribution maps of the multiple cropping systems under the rain-fed and irrigated scenarios during each 10-year period.

In the 2000s, the complexity of the potential cropping system pattern and the spatial pattern of the potential multiple cropping systems increased, as manifested from north to south and from west to east (Figure 3, Figure 4). Due to the relatively high

Table 3. Classification of latitudinal thermal climates.

Latitudinal thermal regime	Condition
Tropics	$T_{sum10} > 8000$
Subtropics	$4500 < T_{sum10} < 8000$
Temperate	$1500 < T_{sum10} < 4500$
Boreal	$0 < T_{sum10} < 1500$
Arctic	$T_{sum10} = 0$

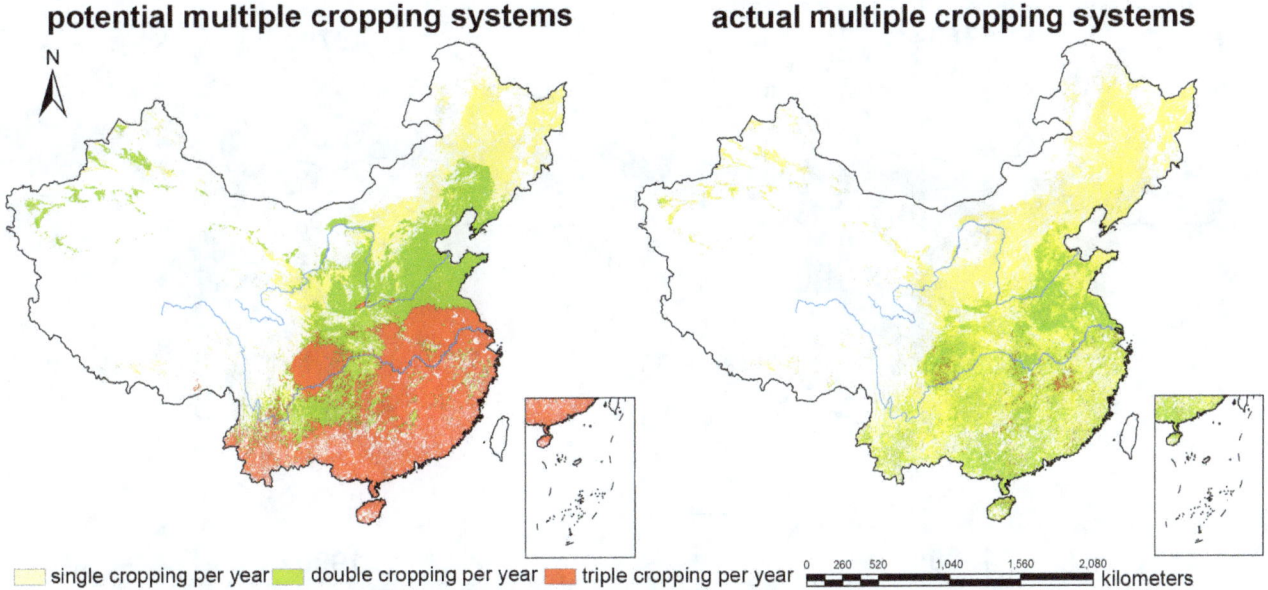

Figure 2. Comparison between potential and actual multiple cropping systems.

temperature and abundant rainfall, the PMCI in the southeastern areas was high. The PMCI declined with increasing distance from the coastal line. In the northwestern portion of the country, potential multiple cropping was constrained by low temperatures and rainfall.

Under the irrigated scenario, the triple cropping zones (including limited triple cropping, triple cropping with rice and triple rice cropping) were primarily distributed over South China and the middle and lower reaches of the Yangtze River, which cover 36.39% of the total land area of China (Figure 5). The double cropping zones (including limited double cropping, double cropping and double cropping with rice) were primarily spread across the Loess Plateau, the Huang-Huai-Hai plain and most of the Gansu-Xinjiang regions, which occupied 32.49% of the total land area of China. The single cropping zones were primarily distributed in the north of Inner Mongolia and along the Great Wall region (denoted as f) and a small portion of the Qinghai-Tibet Plateau, which occupy 31.11% of the total land area of China.

Under the rain-fed scenario, all multiple cropping zones were shifted to the south and the east, and the areas for the double and triple cropping zones were decreased by 7.84% and 8.95% respectively (Figure 5). The limited rainfall led to a decrease of the PMCI from 205% to 181%. The data showed that the irrigated condition was crucial for multiple cropping in China. Therefore,

improving the construction of water facilities for farmland will benefit food productivity in China.

Changes in the potential multiple cropping systems

The potential multiple cropping systems have undergone significant changes between the 1960s -and the 2000s. The fundamental characteristics of the changes in the potential multiple cropping systems were that the single cropping area decreased and the triple cropping area increased. Moreover, the PMCI showed an increasing trend, with the changes primarily occurring after the 1980s.

Under rain-fed scenarios from the 1960s to the 2000s, the single cropping area decreased by 3.23% (approximately 173,900 km^2), whereas the triple cropping area increased by 3.21% (approximately 173,000 km^2) and the PMCI increased by 7% (from 174% to 181%). No significant changes occurred from the 1960s to the 1980s, but shifts occurred in all areas and the PMCI grew rapidly from the 1980s to the 2000s. The major conversion from double cropping to triple cropping and from single cropping to double cropping took place in the time period from the 1980s to the 1990s and from the 1990s to the 2000s respectively. The size of the conversion areas was approximately equal during the two decades (Figure 6).

Under the irrigated scenarios, the single cropping area declined by 7.28% (approximately 391,600 km^2), the triple cropping areas increased by 6.38% (approximately 343,300 km^2), and the PMCI increased by 13% (from 192% to 205%). Major changes occurred from the 1980s to the 2000s, and the PMCI grew much more rapidly. During this period, the increase in the triple cropping area was nearly equivalent to the decrease in single cropping area, whereas the double cropping area remained unchanged as a whole(Figure 6).

Response of potential multiple cropping systems to climate change

As presented in figure 7(a, b, c), the climate change from the 1960s to the 2000s showed that the annual mean temperature, precipitation and radiation data take on different spatial change

Table 4. Comparison table between potential and actual multiple cropping systems (unit: km^2).

		actual		
		single	double	triple
	single	870,093	0	0
potential	double	981,026	301,043	0
	triple	629,786	674,971	93,445

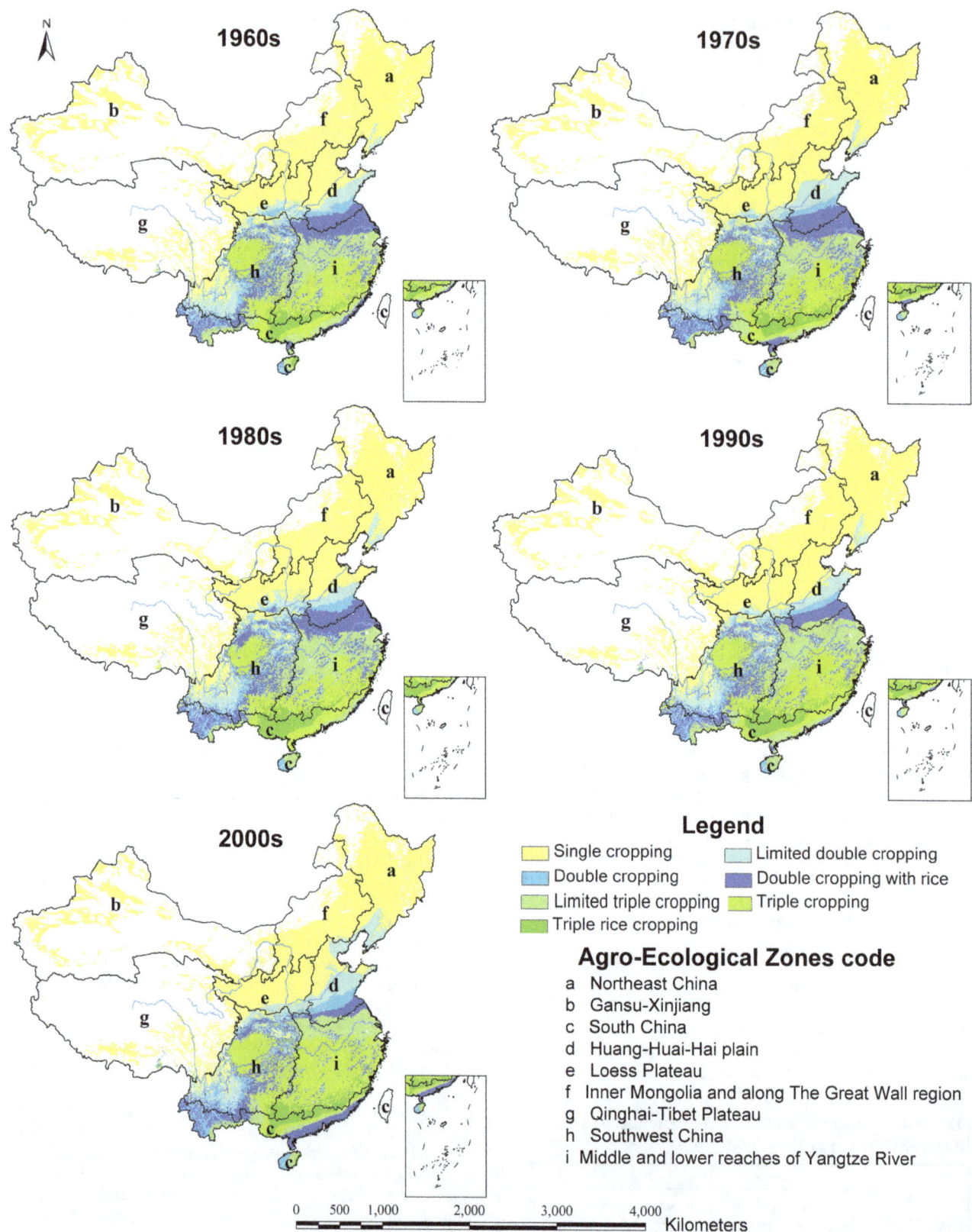

Figure 3. Distribution of multiple cropping systems under the rain-fed scenario.

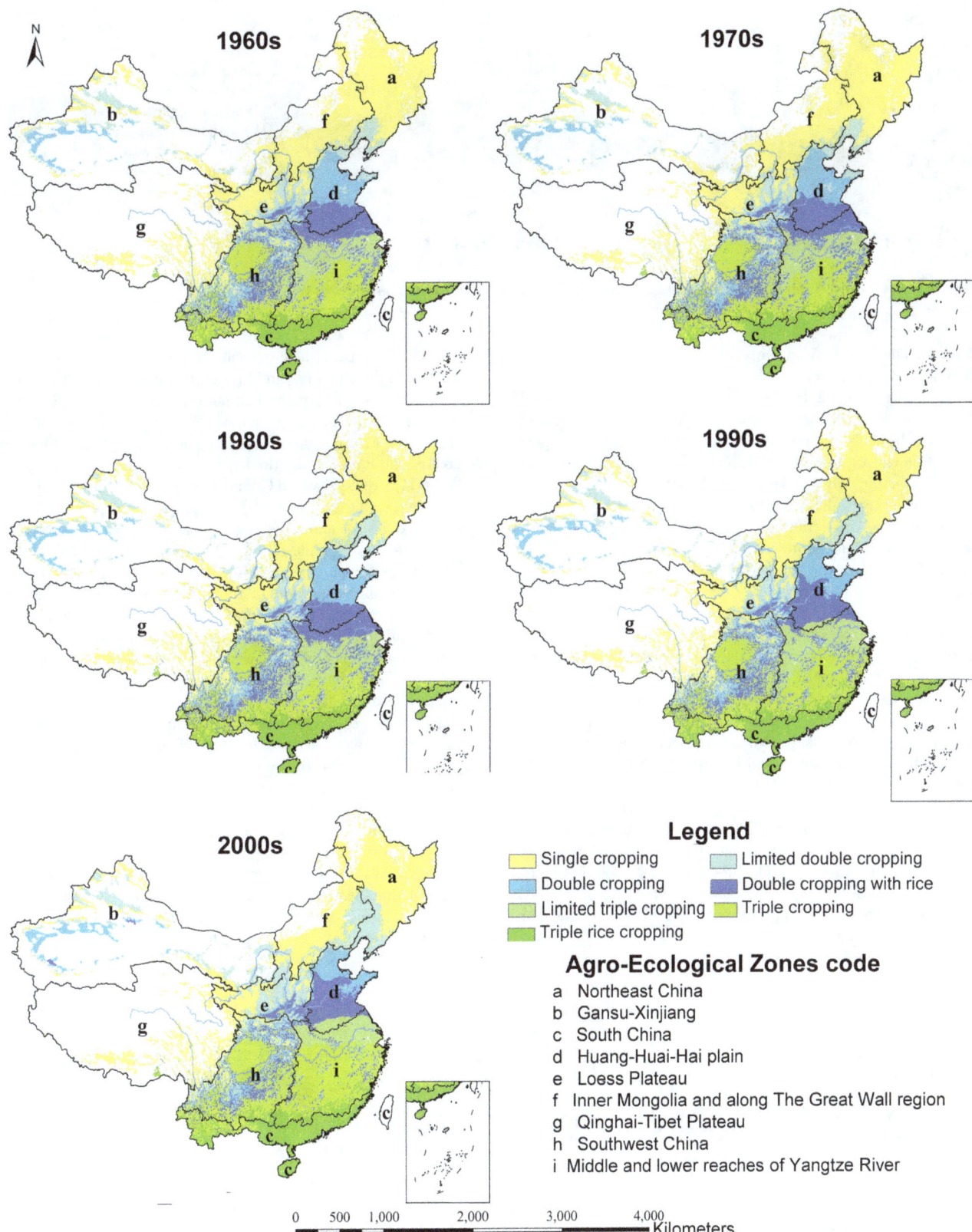

Figure 4. Distribution of multiple cropping systems under the irrigated scenario.

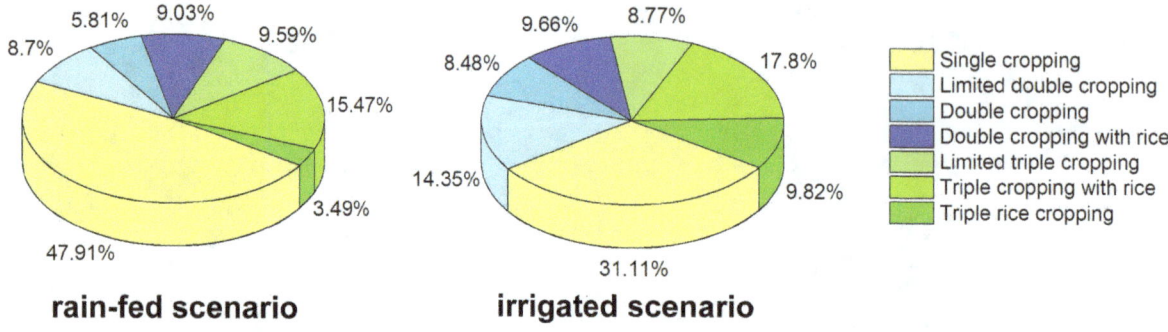

Figure 5. Area ratio difference of the multiple cropping systems under two scenarios in the 2000s.

patterns. The annual mean temperature generally increased across the major cultivation regions. The line chart from the 1960s to the 2000s presented in Figure 8 reflects an obvious upward trend in the temperature change. More specifically, the temperature increased by more than 2°C in parts of the northeast, the north and the northwest of China (Figure 7a). The rate of temperature change after the 1980s (0.41°C/decade) increased more dramatically than prior to the 1980s (0.11°C/decade).

Precipitation was reduced in most regions during the entire period, especially in Southwest China, the Loess Plateau, Inner Mongolia and along the Great Wall region where the precipitation decreased by more than 10%. The areas where precipitation increased were primarily distributed in the northwest of China and the middle and lower reaches of the Yangtze River Region (Figure 7b). As an illustration, the precipitation in the Gansu-Xinjiang and Southern China region increased by 13.7% (approximately 21 mm) and 2% (approximately 31 mm), respectively. In contrast, the temporal change process showed that precipitation did not change significantly before the 1990s but took on an obvious decreasing trend (31 mm) from the 1990s to the 2000s.

Solar radiation decreased in large areas across Huang-Huai-Hai plain and most of southern China but increased in parts of Northeastern China, the Loess Plateau and the western area of Southern China (Figure 7c).The amount of radiation dropped at an average rate of 182 MJ/m^2·decade from the 1960s to the 1980s and at an average rate of approximately 22 MJ/m^2·decade from the 1980s to the 2010s.

The obvious spatial heterogeneity of the climate changes from the 1960s to the 2000s has an important influence on the potential multiple cropping systems in China. Temperature and precipitation were two leading factors that affected the changes of the PMCI, and Figure 8 and Figure 9 show that climate change increased the PMCI under the irrigated and rain-fed scenarios by 13% and 7% respectively in China. The increase in the PMCI was mostly due to the growth of the annual mean temperature. The gap of PMCI between rain-fed scenario and irrigated scenario was 18% in the 1960s, while this gap had increased to 24% in the 2000s.The reason for the increased gap was that the crops could not get enough water due to the reduction in rainfall, which limited the increase of the PMCI under rain-fed conditions. Consequently, in China, moisture was one of the most significant factors for the PMCI. The moisture limit will be the main factor restraining the PMCI under rain-fed scenarios in the present and perhaps also in the future.

In different regions of China, the responses of the potential multiple cropping systems to climate change took on different characteristics. In Northeast China (a), the PMCI continuously increased in two scenarios with rising annual mean temperature and declining radiation throughout the entire period. The gap in the PMCI under the two scenarios remained nearly the same, even when the precipitation sharply dropped by more than 50 mm from the 1980s to the 2000s in the rain-fed scenario. This result indicated that heat is a major factor in limiting the potential multiple cropping systems, whereas precipitation had a lesser influence.

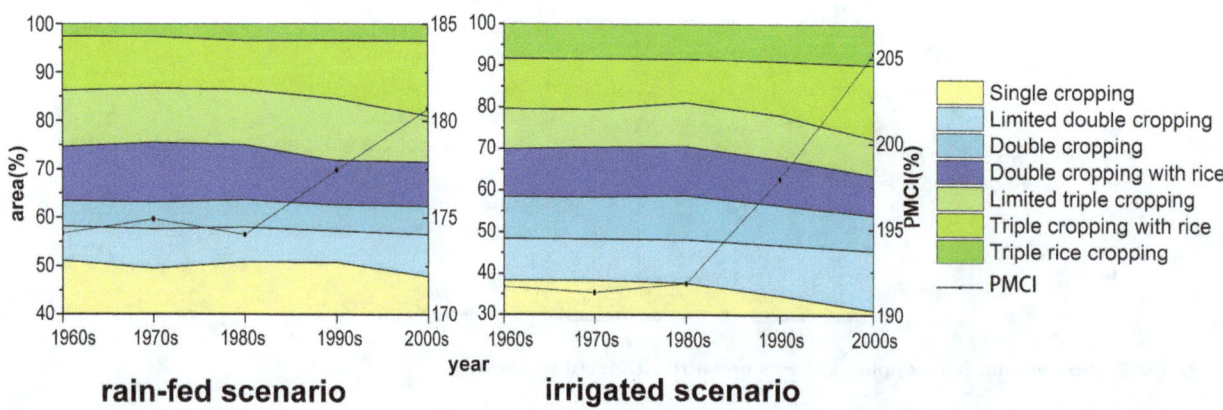

Figure 6. Percentage changes in the multiple cropping area and PMCI under two scenarios from the 1960s to the 2000s.

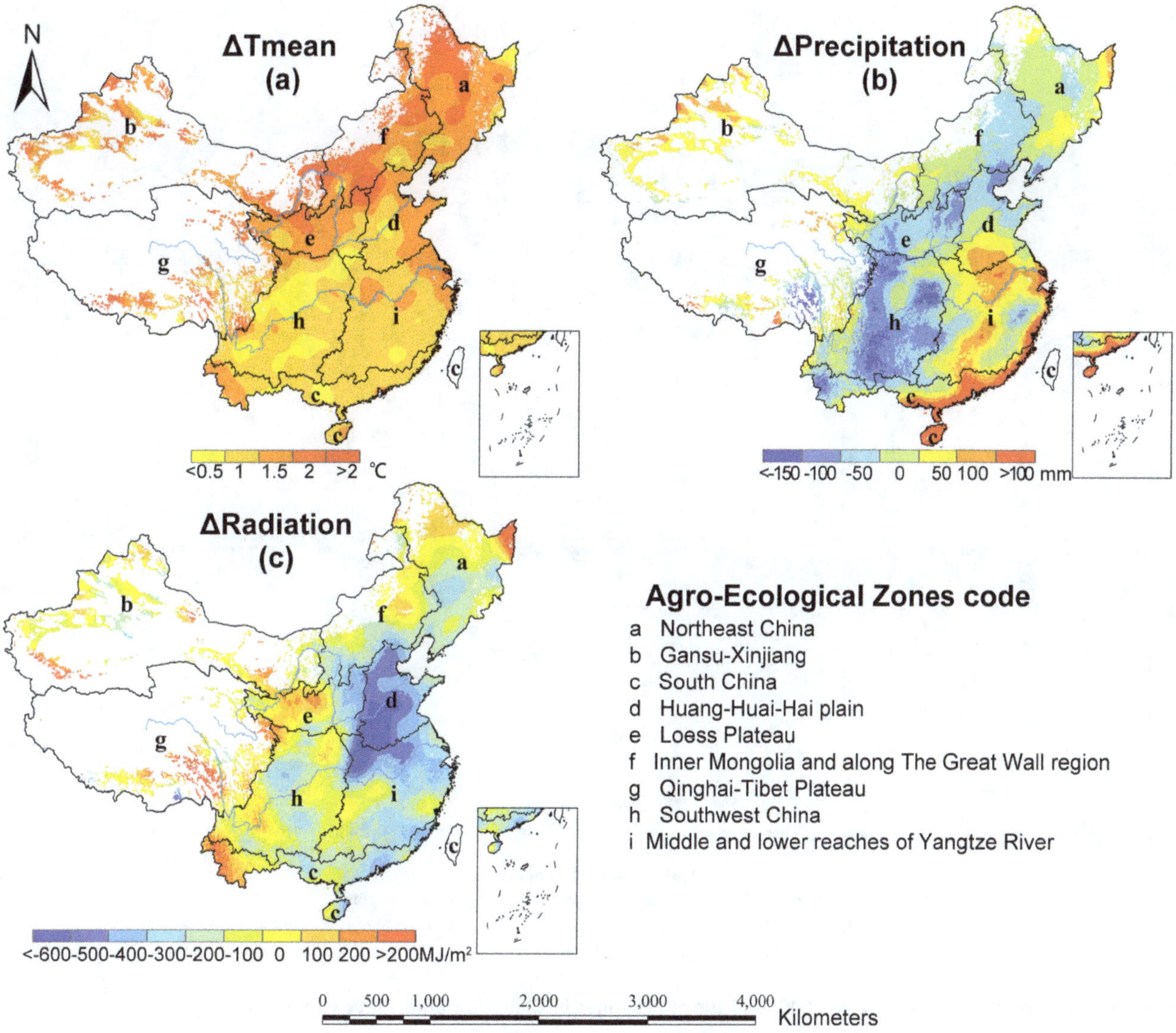

N
ΔTmean
(a)
<0.5 1 1.5 2 >2 ℃
ΔPrecipitation
(b)
<-150 -100 -50 0 50 100 >100 mm
ΔRadiation
(c)
<-600-500-400-300-200-100 0 100 200 >200MJ/m²
Agro-Ecological Zones code
a Northeast China
b Gansu-Xinjiang
c South China
d Huang-Huai-Hai plain
e Loess Plateau
f Inner Mongolia and along The Great Wall region
g Qinghai-Tibet Plateau
h Southwest China
i Middle and lower reaches of Yangtze River
0 500 1,000 2,000 3,000 4,000
Kilometers

Figure 7. Spatial distribution pattern of climate changes from the 1960s to the 2000s in China.

In the Gansu-Xinjiang region (b), the potential multiple cropping systems under the rain-fed scenario was always one crop per year, whereas the PMCI under the irrigated condition rose by 0.178. This result shows that the temperature increase had a positive impact on the PMCI under the irrigated condition. The largest gap of the PMCI under the two scenarios was approx-

11.5
11.0
10.5
900
880
860
5600
5500
5400
5300
Tmean (℃)
Precipitation (mm)
Radiation (MJ/m²)
1960s
1970s
1980s
1990s
2000s
years

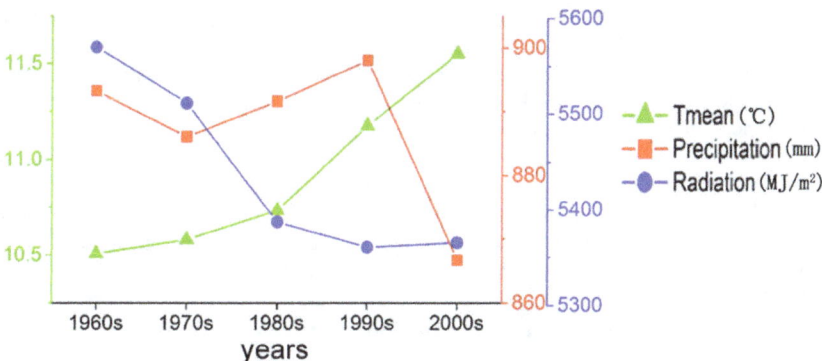

Figure 8. Climate changes from the 1960s to the 2000s in China.

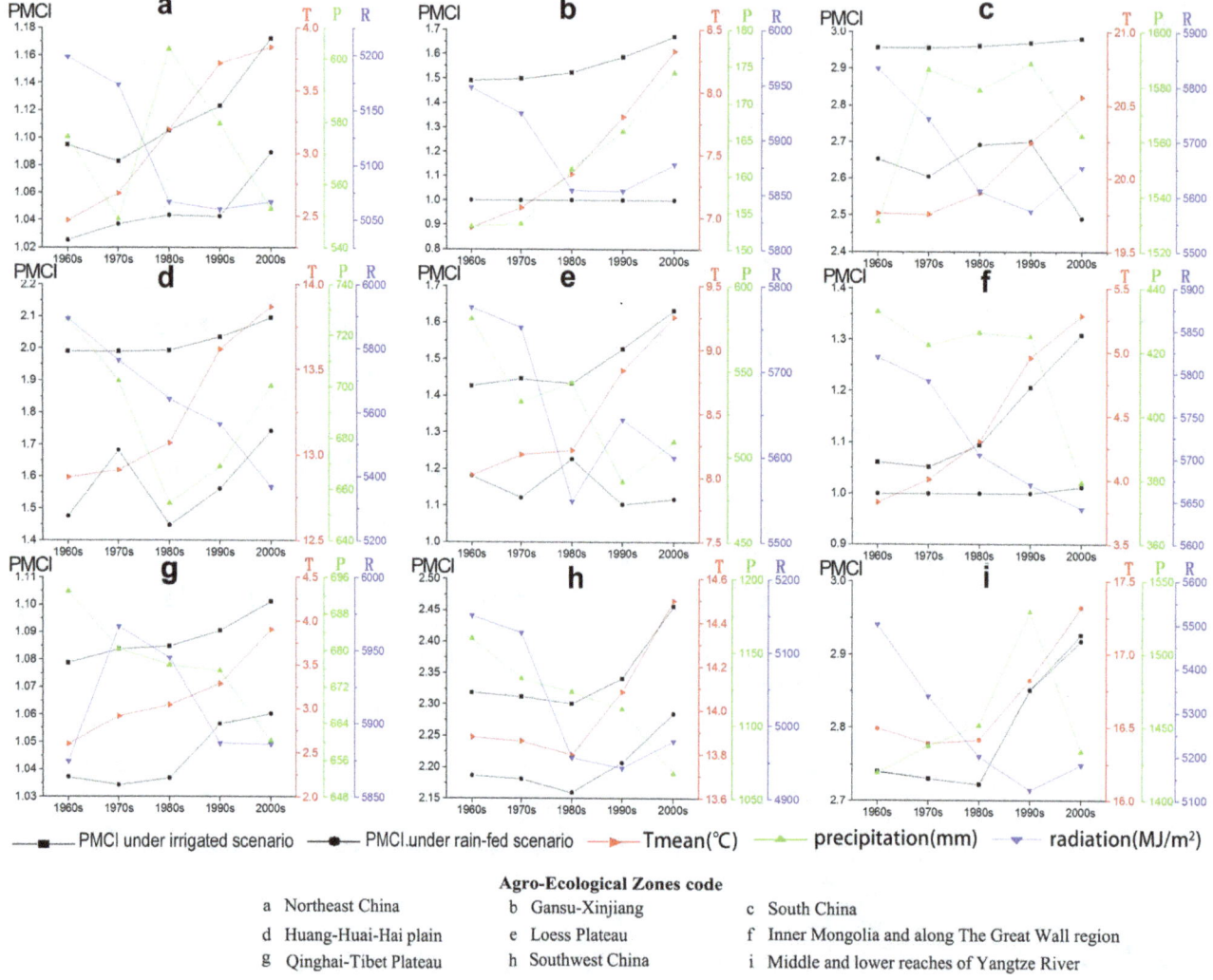

Figure 9. The relationship between the PMCI and climate changes in China from the 1960s to the 2000s.

imately 0.7. However, a small increase in precipitation did not cause the increase in the PMCI under the rain-fed condition due to the extreme deficiency of rainfall.

In Southern China (c), the PMCI under the irrigated scenario was close to 3, which is primarily due to the effect of abundant heat on crop growth. The PMCI under the rain-fed scenario took on an obvious decreasing trend, and the PMCI gap under the two scenarios fluctuated between 0.26 and 0.5. This result was primarily due to the highly uneven rainfall.

In Huang-Huai-Hai plain (d), the annual mean temperature increased by 1.5°C from the 1960s to the 2000s, which led to an increase in the PMCI of 0.103 under the irrigated scenario. Under the rain-fed scenarios, the PMCI increased by 0.266, and displayed a fluctuation that first increased, subsequently decreased, and finally increased again. The change in the PMCI was primarily due to the sharp decrease in radiation by more than 500 MJ/m²·year and the increase in precipitation during certain periods. The gap between the two scenarios was approximately 0.35, which indicates that water restricts crop growth in this region to a certain extent.

In the Loess Plateau (e), Inner Mongolia and along the Great Wall region (f), the annual mean temperature and PMCI under irrigated conditions showed a simultaneous growth trend, whereas

the PMCI under the rain-fed condition changed little, resulting in larger gaps. A potential reason for this finding was that the declining precipitation caused a moisture shortage for the crops.

The PMCI of the Qinghai-Tibet Plateau (h) under the two scenarios were both lower than those of other regions, which rose slightly with increasing temperatures, and the gaps were rarely affected by a reduction in precipitation. The main reason for this appeared to be that the temperature was too low (below 5°C) for normal crop growth.

In Southwest China (i) and the middle and lower reaches of the Yangtze River (j), the annual mean temperature and PMCI under the two scenarios showed a concurrent growth trend. Although the precipitation fluctuated, the PMCI gap under the two scenarios was almost unchanged. This result shows that these areas received sufficient rain for plant growth under the prevailing heat condition, which implies that these areas are the most suitable zones for plant growth in China.

Conclusions and Discussions

The multiple cropping systems demonstrate the efficient use of water, soil, light energy, and other natural resources [38], all of which are important for agricultural production and food security.

Wherever the natural conditions allow, it is necessary to adopt high intensity cultivation to secure efficient food supply in China [2].

This study focused on the calculation and analysis of potential multiple cropping systems in China based on multi-source data. We also analyzed the changes in the potential multiple cropping systems in response to climate change in China from the 1960s to the 2010s. Conclusions from this work are as follows [39–42]:

(1) The spatial pattern of potential multiple cropping systems in China displays tremendous heterogeneity, and an increasing complexity from northern China to southern China, and from western China to eastern China. The decrease in the single cropping area, and the increase in the triple cropping area, are the fundamental characteristics of the potential multiple cropping systems change from the 1960s to the 2000s. The rate of increase of PMCI after the 1980s was far larger than that before the 1980s. During the studied period, the PMCI gap between rain-fed and irrigated scenarios increased from 18% to 24%, which indicated noticeable growth of water supply limitations under the rain-fed scenario. Irrigation conditions were important for multiple cropping practices, underlining the importance of the construction of water facilities would be beneficial for food production in China.

(2) The obvious spatial heterogeneity of climate change from the 1960s to the 2000s had a significant influence on the potential multiple cropping systems in China. The climate change caused the PMCI under the irrigated and rain-fed scenarios to increase by 13% and 7% respectively across the whole of China. Furthermore, the growth of the annual mean temperature was identified as the main reason for the increase of the PMCI.

(3) In the different regions of China, the response of the potential multiple cropping systems to climate change took on different characteristics. In Southern China, the PMCI under the rain-fed scenario took on an obvious decreasing trend because of highly uneven rainfall, although precipitation was abundant. In the Northeast and Qinghai-Tibet Plateau, the heat severely restricted crop growth, and in Gansu-Xinjiang region, Loess Plateau, Inner Mongolia and along the Great Wall region, precipitation was the major factor that limited the potential multiple cropping systems. In Southwest China and the middle and lower reaches of the Yangtze River, the PMCI showed a simultaneous growth trend under both the irrigation and rain-fed scenarios, although precipitation fluctuated.

This study calculated the potential multiple cropping systems based on meteorological data, and subsequently analyzed the relationships between the potential multiple cropping systems and climate change, which assist in dealing with the issue of food security in China. However, certain limitations exist in this research. For instance, (1) due to the lack of irrigation data, we were not able to simulate the actual multiple cropping systems according to real irrigation conditions, instead we simulated two different scenarios (rain-fed and irrigated scenarios); (2) it was assumed that the irrigated scenarios would supply sufficient water for crop growth, but in practice the availability of water under the irrigation scenario still poses limitations to crop growth; (3) extreme weather conditions (i.e., freezing or high temperatures, rainstorms and heavy snow) may have an extreme effect on the multiple cropping systems, but these factors were not taken into consideration in our research.

Author Contributions

Conceived and designed the experiments: XX DZ XC. Performed the experiments: SL. Analyzed the data: XX. Wrote the paper: LL.

References

1. Ziqiang F, Yunlong C, Youxiao Y, Erfu D (2001) Research on the relationship of cultivated land change and food security in China. Journal of Natural Resources 16: 313–319.
2. Hong Y, Xiubin Y (2000) Cultivated land and food supply in China. Land Use Policy 17: 73–88.
3. Von B (2007) Rising world food prices: impact on the poor. The International Journal for Rural Development (Rural 21) 42: 19–21.
4. Guo B (1997) The analysis for the change of multiple cropping index in China. Economic Geography 17: 8–13.
5. Jinshen Y, Limin S (2000) The transition of cropping system in Hebei Province. Acta Agriculture Boreali-sinica 4: 126–130.
6. Shulan J, Lichun H, Lei X (2011a) The Relationship of Multiple Cropping Index of Arableland Change and National Food Security in the Middle and Lower Reaches of Yangtze River. Chinese Agricultural Science Bulletin 27: 208–212.
7. Shulan J, Caiqiu X, Huahua P (2011b) Change Characteristics and Potential Analysis for Multiple Cropping Index of Cropland in Major Grain Producing Areas in China. Guizhou Agricultural Sciences 39: 201–204.
8. Jing L, Zhiyuan R (2011) The Monitoring for cropping index of arable land in northwest region using SPOTNDVI-A case of Shaanxi Province. Journal of Arid Land Resources and Environment 25: 86–91.
9. Dailiang P, Jingfeng H, Huimin J (2007) Monitoring the Sequential Cropping Index of Arable Land in Zhejiang Province of China Using MODIS-NDVI. Agricultural Sciences in China 6: 208–213.
10. Zhiguo Z (2011) Spatial-temporal Characteristics of Multiple Cropping Index and Relationship between Multiple Cropping Index and Grain Yield in Henan Province. Hubei Agricultural Sciences: 3653–3656.
11. Huimin Y, Jiyuan L, Mingkui C (2005) Remotely Sensed Multiple Cropping Index Variations in China during 1981-2000. ACTA Geographica SINICA 60: 559–566.
12. Hongping D (2001) Analysis of the development of Hunan farming system from 1949 to 1998. Cropping and Planting 3: 1–4.
13. Shumin L (2007) Research on the potential of the multiple cropping index and the behavior of multiple cropping of China. Problems of Agricultural Economy 5: 85–90.
14. Mao L (2002) Gray incidence analysis between cultivated area and grain yield and its affecting factors of Henan Province. Progress in Geography 21: 163–172.
15. Xin J, Yua ZR, Van Leeuwen, Driessen PM (2002) Mapping crop key phonological stages in the North China Plain using NOAA time series images. Int. J.Appl. Earth Obs. Geoinf 4: 109–117.
16. Huimin Y, Yuling F, Xiangming X, Heqing H, Honglin H, et al. (2009) Modeling gross primary productivity for winter wheat–maize double cropping system using MODIS time series and CO$_2$ eddy flux tower data. Agriculture Ecosystems and Environment 129: 391–400.
17. Jinlong F, Bingfang W (2004a) A study on cropping index potential based on GIS. Journal of Remote Sensing 8: 637–644.
18. Xiaolin Z, Qiang L, Miaogen S, Jin C, Jin W (2008) A methodology for multiple cropping index extraction based on NDVI time series. Journal of Natural Resources 23: 534–544.
19. Ashton S, Joseph M (2011) Spatial structure and landscape associations of SRTM error. Remote Sensing of Environment 115: 1576–1587.
20. Zandbergen P (2008) Applications of Shuttle Radar Topography Mission Elevation Data. Geography Compass 2: 1404–1431.
21. Farr TG, Rosen PA, Caro E, Crippen R, Duren R, et al. (2007) The Shuttle Radar Topography mission. Reviews of Geophysics 45.
22. Jiyuan L, Mingliang L, Dafang Z, Zengxiang Z, Xiangzheng D, et al. (2003). Study on spatial pattern of land-use change in China during 1995–2000. Science in China (Series D) 46: 373–384.
23. Jiyuan L, Mingliang L, Hanqin T, Dafang Z, Zengxiang Z, et al. (2005a) Spatial and temporal patterns of China's cropland during 1990–2000. Remote Sensing of Environment 98: 442–456.
24. Jiyuan L, Tian H, Mingliang L, Dafang Z, Jerry M, et al. (2005b) China's changing landscape during the 1990s: Large-scale land transformations estimated with satellite data. Geophysical Research Letters 32.
25. Jiyuan L, Zengxiang Z, Xinliang X, Wenhui K, Wancun Z, et al. (2010). Spatial patterns and driving forces of Land use change in China during the early 21st century. Journal of Geographical Sciences 20: 483–494.
26. Jiyuan L, Zhang Q, Hu Y (2012) Regional differences of China's urban expansion from late 20th to early 21st century based on remote sensing information. Chinese Geographical Science 22: 1–14.
27. Hutchinson MF (1998a) Interpolation of rainfall data with thin plate smoothing splines I. Two dimensional smoothing of data with short range correlation. Geographic Information Decision Analysis 2: 153–167.
28. Hutchinson MF (1998b) Interpolation of rainfall data with thin plate smoothing splines II. Analysis of topographic dependence. Geographic Information Decision Analysis 2: 168–185.

29. FAO (1984) Guidelines: Land Evaluation for Rain-fed Agriculture. FAO Soils Bulletin. 52.

30. FAO (1985) Guidelines: Land Evaluation for Irrigated Agriculture. FAO Soils Bulletin. 55.

31. FAO (1992) CROPWAT: A Computer Program for Irrigation Planning and Management. FAO Irrigation and Drainage Paper no 46. Land and Water Development Division, Rome, Italy.

32. Fischer G, M Shah, Van Velthuizen H, Nachtergaele O. (2002a) Global Agro-Ecological Assessment for Agriculture in the 21st Century: Methodology and Results. IIASA RR-02-02, IIASA, Laxenburg, Austria.

33. Fischer G, M Shah, Van Velthuizen H. (2002b) Climate Change and Agricultural Vulnerability. Special Report as contribution to the World Summit on Sustainable Development, Johannesburg 2002. International Institute for Applied Systems Analysis, Laxenburg, Austria: 150–152.

34. Bingwei H (1958) The preliminary draft of comprehensive natural divisions in China. Acta Geographica Sinica 4.

35. FAO (1998) Crop Evapotranspiration. FAO Irrigation and Drainage Paper no. 56 Rome, Italy.

36. Monteith JL (1965) Evapotranspiration and the environment. In The State and Movement of Water in Living Organisms: 205–234.

37. Monteith JL (1981) Evapotranspiration and surface temperature. Quarterly Journal Royal Meteorological Society 107: 1–27

38. Jinlong F, Bingfang W (2004b) A Methodology for Retrieving Cropping Index from NDVI Profile. Journal of Remote Sensing 8: 628–636.

39. Li P, Feng Z, Jiang L, Yujie L, Xiangming X (2012). Changes in rice cropping systems in the Poyang Lake Region, China during 2004–2010. Journal of Geographical Sciences, 22(4), 653–668.

40. Hoque M Z. (1984). Cropping systems in Asia: on-farm research and management. Irri.

41. TianXiang Y, Na Z, Douglas R, ChenLiang W, ZeMeng F, et al. (2013). Climate change trend in China, with improved accuracy. Climatic Change, 1–15.

42. Zhao D S, Wu S H, Yin Y H. (2013). Responses of Terrestrial Ecosystems' Net Primary Productivity to Future Regional Climate Change in China. PloS one, 8(4), e60849.

22

Effect of Subsoiling in Fallow Period on Soil Water Storage and Grain Protein Accumulation of Dryland Wheat and Its Regulatory Effect by Nitrogen Application

Min Sun, ZhiQiang Gao*, WeiFeng Zhao, LianFeng Deng, Yan Deng, HongMei Zhao, AiXia Ren, Gang Li, ZhenPing Yang

College of Crop Science, Shanxi Agricultural University, Taigu, Shanxi Province, China

Abstract

To provide a new way to increase water storage and retention of dryland wheat, a field study was conducted at Wenxi experimental site of Shanxi Agricultural University. The effect of subsoiling in fallow period on soil water storage, accumulation of proline, and formation of grain protein after anthesis were determined. Our results showed that subsoiling in fallow period could increase water storage in the 0–300 cm soil at pre-sowing stage and at anthesis stage with low or medium N application, especially for the 60–160 cm soil. However, the proline content, glutamine synthetase (GS) activity, glutamate dehydrogenase (GDH) activity in flag leaves and grains were all decreased by subsoiling in fallow period. In addition, the content of albumin, gliadin, and total protein in grains were also decreased while globulin content, Glu/Gli, protein yield, and glutelin content were increased. With N application increasing, water storage of soil layers from 20 to 200 cm was decreased at anthesis stage. High N application resulted in the increment of proline content and GS activity in grains. Besides, correlation analysis showed that soil storage in 40–160 cm soil was negatively correlated with proline content in grains; proline content in grains was positively correlated with GS and GDH activity in flag leaves. Contents of albumin, globulin and total protein in grains were positively correlated with proline content in grains and GDH activity in flag leaves. In conclusion, subsoiling in fallow period, together with N application at 150 kg·hm^{-2}, was beneficial to increase the protein yield and Glu/Gli in grains which improve the quality of wheat.

Editor: Tianzhen Zhang, Nanjing Agricultural University, China

Funding: This work was supported by Modern Agriculture Industry Technology System Construction [CARS- 03-01-24]; NSFC (Natural Science Foundation of China) [31101112]; Scientific Research Project of Shanxi Province for Returned Oversea Students [2009037]; Youth Fund of Shanxi Province [2010021028-3]; and the assistance from Shanxi Province for Scientific and Technical Key Projects [20110311001-4]. The funders had no role in study design, data collection and analysis, decision to publish, or preparation of the manuscript.

Competing Interests: The authors have declared that no competing interests exist.

* E-mail: sm_sunmin@126.com

Introduction

Natural precipitation is the only water resource of dryland wheat. Rainfalls always concentrate from July to September at region producing dryland wheat in loess plateau, which is called fallow period. During the fallow period, soil water is replenished and restored. It is the water stored in this period that determines the soil moisture before dryland wheat is sown which has a great influence on the production of dryland wheat [1–4]. In these days, though great progress has been made in the technology of water storage and retention in dryland cultivation [5–8], water storage and retention in fallow period had been ignored. Therefore, finding an effective way to raise the moisture content in fallow period soil is significant to the dryland wheat production.

Many researches showed protein content in wheat grains would increase when the water was deficient in soil [9–13]. For example, Sun et al. (2010) showed that cultivation in dry land could increase the content of albumin, gliadin, glutetin, total protein and Glu/Gli in grains while decrease the content of glublin [11]. Zhao et al. (2007) found that drought in milking stage increased the content of protein and Glu/Gli in grains [9]. Fan et al. (2004) also reported drought in milking stage could increase Glu/Gli [12]. In addition,

Fu et al. (2008) suggested drought in milking stage combined with nitrogen application could increase protein and nitrogen content in wheat grains [10]. Content and component of grain protein, especially for the Glu/Gli, not only influence the nutrition quality of wheat, but also influence the processing quality of wheat [13].

Up to date, problems in production of wheat in dry land were enumerated as follows: 1) Low water-using efficiency caused by poor soil fertility and low ability of soil water storage and retention; 2) Late straw rotting fails to increase the organic matter of soil, and thus, affect the quality of sowing; 3) Although precipitation in fallow period is considerably abundant, a great evaporation loss of surface water fails to store enough water for spring because of the high temperature in summer; 4) Wheat is often sown under enough moisture or rainfall in dry land, so the sowing time is inconsistent. Many researches showed drought can improve the quality of wheat to a certain degree [14–16,10,9,17]. Thus, the wheat quality should be taken into account as the level of water storage raised in the fallow period.

The objective of the present paper was to provide a new way to increase the water storage and retention for dryland wheat. This was realized by subsoiling after wheat harvest and conducting rotary tillage in the late August. The proline content and physiological mechanism about formation of grain protein in dryland wheat were examined. To have a comprehensive understanding for the issue of dryland wheat quality, the effects of fertilizer application were also studied.

Materials and Methods

The field experiment was conducted at Wenxi experimental site (summer fellow) of Shanxi Agricultural University from 2009 to 2011. The soil fertility was measured on 1st July, 2010. Content of organic matter was measured as 8.65 g/kg using potassium dichromate volumetric method. Total nitrogen was measured as 0.74 g/kg using Kjeldahl determination, alkali-hydrolyzable nitrogen was measured as 32.93 mg/kg using alkali-hydrolyzable proliferation method, and rapidly available phosphorus was measured as 20.08 mg/kg using sodium bicarbonate method.

The experimental site is very arid and 60%–70% of the annual precipitation is concentrated in summer and early-fall (July, August, and September). Precipitation conditions at the experimental site in 2005–2011 were shown in Table 1. These data were provided by Agricultural Bureau of Wenxi County. Annual precipitation in 2010–2011 was a little higher than conventional years and the precipitation in fallow period was higher than early growth stage. The precipitation in fallow period averagely accounted for 58% of annual precipitation in 2005–2011 and 75% in 2010–2011 (Table 1).

Design of Field Experiment

The experiment was laid out as split-split plot arrangement and three replications. Two levels (subsoil (SS); and no-tillage (CK)) were allocated to the main plots. Sub plots were three levels of N fertilizer application (75 kg·hm^{-2} (LN), 150 kg·hm^{-2} (MN) and 225 kg·hm^{-2} (HN)). For subsoil plot, the preceding wheat residues were shredded on the fifth day after harvest (i.e. July 1st, 2010), followed by tillage that consisted of disking 30–40 cm deep before planting wheat. Then soil moisture was retained by rotary tillage on 20th August, 2010.

Dryland wheat cultivar Yunhan 20410, which provided by agricultural bureau of wenxi county, was planted on 29th September, 2010 with a row spacing of 20 cm and a seeding rate of 225×10^4/hm^2. The area of experimental site was 300 m^2 (50 m in length and 6 m in width). Phosphate and potash fertilizer had been applied to the field with P$_2$O$_5$ and K$_2$O at a concentration of 150 kg·hm^{-2}.

Sampling and Measurement Methods

Measurement of soil water storage. Soil water storage was measured by drying method. Soil samples in the 0–300 cm depth (20 cm per soil layer) were obtained at pre-sowing stage, pre-wintering stage, seedling establishment stage, elongation stage, booting stage, anthesis stage, and maturation stage using a soil auger. The calculation formula of soil water storage is as follows:

Soil water storage (mm) = [(mass of damp soil − mass of dried soil)/mass of dried soil×100]×soil thickness (mm)×volume weight per layer.

Measurement of proline and activity of nitrogen metabolism enzyme in flag leaves and grains. Wheatears with consistent growth and blooming on the same day were tagged at anthesis stage and fifteen ears were sampled every five days after anthesis. Flag leaves and grains were kept at −40°C after being quickly frozen by liquid nitrogen for measuring the activity of enzyme. Another five fresh samples, i.e. flag leaves and grains, were taken to measure proline content. The measurement of proline content was referred to [18] and activity of glutamine synthetase (GS) and glutamate dehydrogenase (GDH) were measured according to the method of [19].

Measurement of grain protein and its components. Wheatears with consistent growth and blooming on the same day were tagged at anethsis stage and fifteen ears were sampled every five days after anethsis. Grains were separated in the oven where moisture was removed by heating at 105,°C for 30 min and then weighted after drying at 80°C. Grains were used to measure the content of protein and its component after being crushed by micro high-speed universal grinder. Nitrogen content was measured using the method of semi micro Kjeldahl determination and wheat protein content was obtained by multiplying N content (%) by 5.7 [20]. Content of albumin, globulin, gliadin, and glutelin in grains were measured using the method of continuous extraction.

Measurement of grain yield. Number of grains ears per unit area, average number of grains per wheat ear, and 1000-grain weight were investigated at maturation stage. Fifty plants of wheat were taken from each plot to measure biological yield, and twenty square meters of wheat were rope to measure economical yield.

Statistical Analysis

Statistical analysis was conducted using SAS version 8.0 (1999, SAS Institute, Inc., Cary, NC, USA). Analysis of variance (ANOVA) using the General Linear Model procedure and the difference between means using the Duncan test were determined at the α level of 0.05.

Table 1. Precipitation at the experimental site in Wenxi (mm).

Year	Fallow period	Sowing-Pre-wintering	Pre-wintering - Elongation	Elongation - Anthesis	Anthesis-Mature	Total
2005–2011 (average)	291	63	41	34	76	505
2010–2011	402	27	19	22	64	534

Fallow period (The first ten days of Jul. to last ten days of Oct.);
Sowing-Pre-wintering (The first ten days of Oct. to last ten days of Nov.);
Pre-wintering- Elongation (The last ten days of Nov. to first ten days of Apr.);
Elongation -Anthesis (The first ten days of Apr. to first ten days of May);
Anthesis-Mature (The first ten days of May to middle ten days of Jun).

Results

Effect of Subsoiling in Fallow Period on Soil Water Storage and its Regulatory Effect by Nitrogen Application

Effect on soil water storage before sowing. Generally, as the depth of soil increasing, water storage in the 0–300 cm depth soil before sowing increased at first but then decreased to the minimum at 200 cm and finally tended to increase again (Fig. 1). Compared with CK, soil water storage in SS were increased by 16.08 mm, 14.21 mm, and 6.23 mm in the depth of 0–100 cm, 100–200 cm, and 200–300 cm, respectively. The increment in the 0–300 cm depth reached to 36.25 mm in total and the increment of the 60–160 cm depth soil account for more than 10% of the whole storage. Accordingly, subsoiling in fallow period could significantly increase water storage and retention which provided favorable conditions for proper sowing of dryland wheat.

Effect on soil water storage in anthesis. Generally, as the depth of soil increasing, water storage of soil in anthesis in the 0–300 cm depth increased to maximum at 80 cm, and then decreased between 80 cm to 120 cm, and finally tended to increase slowly again (Table 2). Table 2 showed that, under low and medium N application, subsoiling in fallow period could increase water storage at each layer of soil with the depth of 0–200 cm (20 cm per layer). The difference of water storage between SS and CK was statistically significant in the depth of 80, 200, 260, and 300 cm under low N application and in the depth of 20–40 cm and 80–180 cm under medium N application. Compared with CK, subsoiling in fallow period could also increase water storage of soil layers in the 240–300 cm depth (20 cm per layer) under low N application.

Table 2 also indicated that the water storage in the 0–200 cm depth soil under the condition of subsoiling tended to decrease as the nitrogen application increasing and so did the water storage in the depth of 20–300 cm soil in the control. Moreover, soil water storage in the depth of 40 cm, 60 cm and 120–200 cm varied significantly among different treatments (three levels of N application) under the condition of subsoiling. Also, the maxium soil water storage in the 200–300 cm depth was obtained with low nitrogen application while the minimum was obtained with medium nitrogen application under the same condition. And soil water storage in the depth of 40 cm, 100–200 cm and 240–300 cm with low nitrogen application were significantly higher than that with medium and high nitrogen application under control conditions.(Table 2).

Effect of Subsoiling in Fallow Period on Proline Content after Anthesis and its Regulatory Effect by Nitrogen Application

Fig. 2 displays the changes of proline content in flag leaves and grains after anthesis. In all treatments, proline content in flag leaves presented a "M"-type change tendency. Two peaks respectively appeared at 15 d and 20 d after anthesis and the former one was higher. On the other side, proline content in grains displayed a unimodal curve and the only peak appeared at 15d after anthesis.

Fig. 2A indicated that subsoiling in fallow period significantly decreased proline content in flag leaves after anthesis. Under the condition of subsoiling, the increase of nitrogen application led to increase of proline content in flag leaves at 5–15 d and 25 d after anthesis. The proline content was significantly higher at 5 d and 15–25 d after anthesis with high N application. Proline content in flag leaves at 5–25 d in the control was also increased. Proline content in flag leaves varied significantly among different N application treatments at 10–25 d after anthesis. Moreover, the regulatory effect of N application on proline content in flag leaves was more significant in subsoiling plot than that in no-tillage plot at the medium and late milking stage.

Fig. 2B showed that proline content in grains was decreased by subsoiling in fallow period and the difference at 10–20 d after anthesis with low N application, 5–20 d with medium N application, and 5–15 d, 25 d with high N nitrogen application were all statistically significant. In subsoiling treatment, the increase of nitrogen application increased proline content in grains at 5–15 d and 30 d after anthesis. Proline contents in grains at 20 d and 25 d after anthesis were the highest in high N application treatment and the lowest in medium N application treatment, respectively. However, the difference between medium and low N application treatments were not significant. Additionally, the regulatory effect of N application on proline content in grains was more significant in subsoiling plot than that in no-tillage plot at the early and late milking stage.

Effects of Subsoiling in Fallow Period on Activity of Nitrogen Metabolism Enzyme and its Regulatory Effect by Nitrogen Application after Anthesis

Fig. 3 displays the change tendency of GS activity in flag leaves and grains after anthesis. Generally, the GS activity in flag leaves decreased at first and then increased until 15 d after anthesis, and it rapidly decreased again. The GS activity in grains decreased until 20 d after anthesis, and then it tended to level off.

Fig. 3A indicated that GS activity in flag leaves was decreased by subsoiling in fallow period with medium and high N application. The GS activity in subsoiling plot at 5 d, 10 d, and

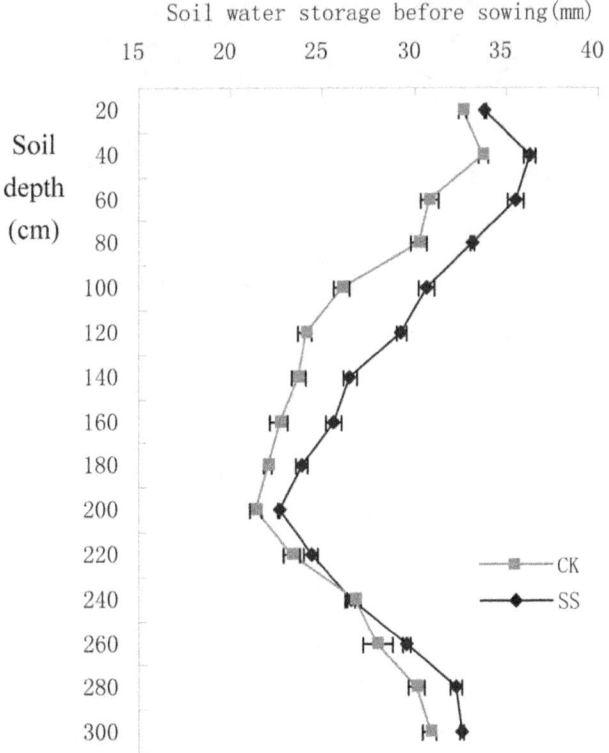

Figure 1. Soil water storage in SS were increased in the depth of 0–300 cm compared with CK. SS, Subsoiling; CK, no-tillage.

Table 2. Effect of subsoiling in fallow period on soil water storage in anthesis and its regulation by nitrogen application(mm).

Tillage	N application amount	soil layer(cm)						
		0–100					0–100	
		20	40	60	80	100		
SS	LN	14.12a	18.14a	20.25a	20.54a	19.45a	92.50a	
	MN	12.58b	17.01b	19.19b	20.31a	18.77a	87.85ab	
	HN	12.18b	15.41c	17.54c	18.24b	15.65b	79.02b	
CK	LN	14.01a	17.68a	19.84a	19.64a	19.02a	90.19a	
	MN	11.08c	15.92b	19.06ab	19.62a	17.63b	83.31b	
	HN	12.21b	15.52b	18.41b	18.98a	16.85c	81.97b	
		100–200					100–200	0–200
		120	140	160	180	200		
SS	LN	19.50a	20.12a	21.05a	21.45a	22.15a	104.27a	196.77a
	MN	18.40b	19.24b	19.98b	20.02b	20.34b	97.97ab	185.83b
	HN	16.52c	17.98c	18.02c	18.95c	18.24c	89.71b	168.73c
CK	LN	19.02a	19.87a	20.85a	21.12a	21.25a	102.11a	192.30a
	MN	17.16b	18.19b	19.24b	19.60b	20.03b	94.21b	177.53b
	HN	17.02b	17.92b	18.96b	19.35b	19.58b	92.83b	174.80b
		200–300					200–300	0–300
		220	240	260	280	300		
SS	LN	21.25a	22.02a	23.25a	24.12a	25.68a	116.32a	313.09a
	MN	20.62a	21.06b	21.20c	22.57b	22.83c	108.28a	294.11b
	HN	20.98a	21.54ab	22.24b	23.75a	23.98b	112.49a	281.22c
CK	LN	21.02a	21.98a	22.57a	23.65a	24.52a	113.74a	306.04a
	MN	20.66a	20.91b	21.13b	22.08b	22.69b	107.46ab	284.99b
	HN	19.85b	20.25b	20.58b	21.68b	22.08b	104.44b	279.24b

The difference between data with different letters from a to c are statistically significant (P<0.05).
SS: subsoiling; CK: no tillage; LN: low nitrogen application; MN: medium nitrogen application; HN: high nitrogen application.

20 d after anthesis with medium N application, and at 10–20 d and 30 d with high N application were significantly decreased compared with that in no-tillage plot. The GS activity in subsoiling plot was decreased at 15–25 d with low N application, but only that at 15 d after anthesis was significantly different. Under subsoiling condition, the increase of N application increased GS activity in flag leaves at 5 d, 10 d and 30 d after anthesis, and the difference between high N application and low N application was significant. In addition, GS activity in flag leaves in no-tillage plot was also increased along with the increase of N application, and the difference between low N application and other treatments at 5–15 d after anthesis were significant. So was the difference between high N application and other treatments at 20–30 d after anthesis. Under the condition of subsoiling, the maxium GS activity in flag leaves at 15–25 d was observed in high N application treatment while the minimum was observed in medium N application treatment. However, the difference among the three N application treatments were not statistical significant. In addition, the regulatory effect of N application on GS activity in flag leaves was more significant in subsoiling plot than that in no-tillage plot at the medium and late milking stage.

Fig. 3B showed that subsoiling in fallow period could increase GS activity in grains in almost all three N application treatments (except at 15 d after anthesis with low N application). And the difference of GS activity between subsoiling plot and no-tillage plot were all statistically significant except for that at 15 d with medium N application and at 25 d with high N application. Increment of N application increased GS activity in grains except for the 15 d sample after anthesis. Besides, the GS activity in low N application treatment was significantly different from other treatments at 15–20 d after anthesis, while the difference between medium and high N treatments were not significant. In the no-tillage plot, the GS activity between high N application and other treatments were significantly different (Fig. 3B).

Fig. 4 displays the change tendency of GS activity in flag leaves and grains after anthesis. Generally, the GDH activity in flag leaves after anthesis increased at first and then decreased, and then increased again. The change of GDH activity influenced by N application was not consistent at 25–30 d after anthesis. In grains, GDH activity kept increasing after anthesis, and had a rapid increment between 15–20 d after anthesis.

Fig. 4A showed subsoiling in fallow period decreased GDH activity in flag leaves after anthesis, and the differences were

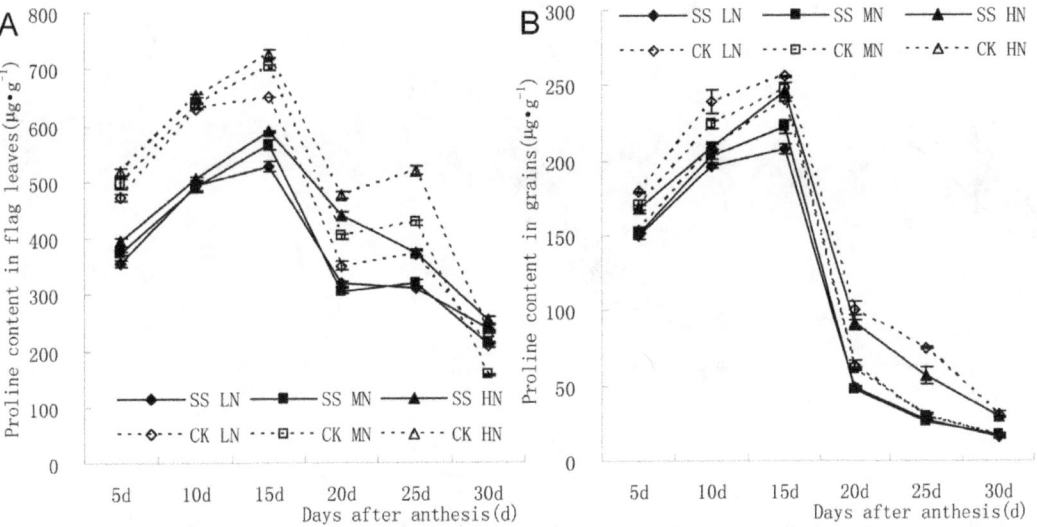

Figure 2. Proline content in flag leaves presented a "M"-type change tendency (A). and proline content in grains displayed a unimodal curve and the only peak appeared at 15d after anthesis (B).

statistically significant at 5–20 d after anthesis. Besides, the increment of N application increased GDH activity in flag leaves at 10–25 d after anthesis, and the difference between low N application and other treatments was significant. In addition, the regulatory effect of N application on GDH activity in flag leaves was more significant in subsoiling plot than that in no-tillage plot.

Fig. 4B showed subsoiling in fallow period could decrease GS activity in grains at 5–20 d with low N application, 5–30 d with medium N application, and 5–15 d with high N application. The differences of GDH activity between subsoiling plots and no-tillage plots were statistically significant at 5–25 d after anethesis. Increment of N application could increase GDH activity in grains under the condition of subsoiling, and difference between high N application treatment and low N application treatment were

significant. In addition, increment of N application could also increase GDH activity in grains at 5–20 d after anthesis in the no-tillage plot, and the difference at 5 d and 15 d after anthesis were significant. In addition, the regulatory effect of N application on GDH activity in grains was more significant in subsoiling plot than that in no-tillage plot at the late milking stage.

Effects of Subsoiling in Fallow Period on Protein Content and its Component in Grains and its Regulatory Effect by Nitrogen Application

Table 3 showed that subsoiling in fallow period significantly decreased the content of albumin and gliadin in grains. The total protein content was also decreased by subsoiling in fallow period and the difference was statistically significant in the low and

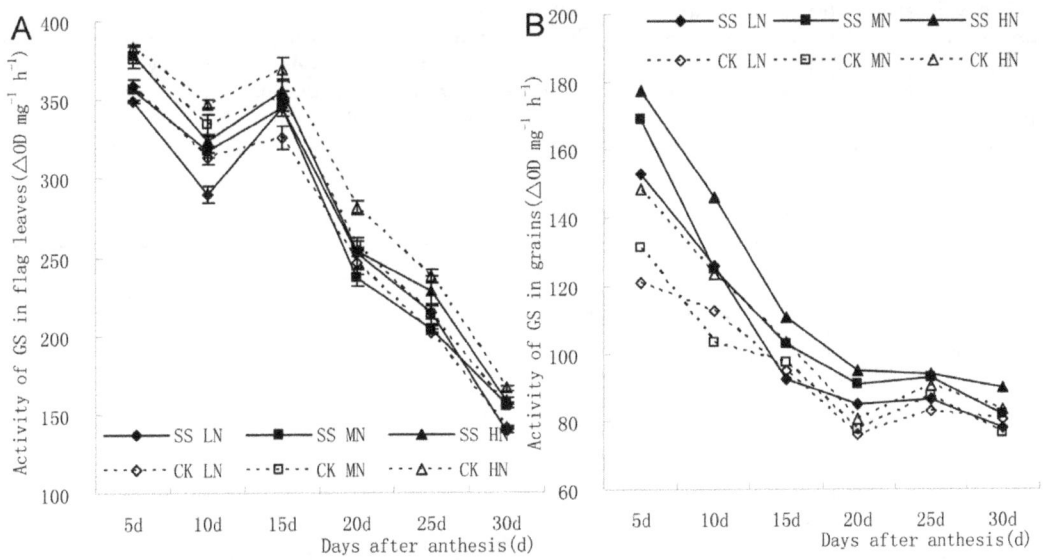

Figure 3. The GS activity in flag leaves decreased at first and then increased until 15 d after anthesis, and it rapidly decreased again (A). The GS activity in grains decreased until 20 d after anthesis, and then it tended to level off (B).

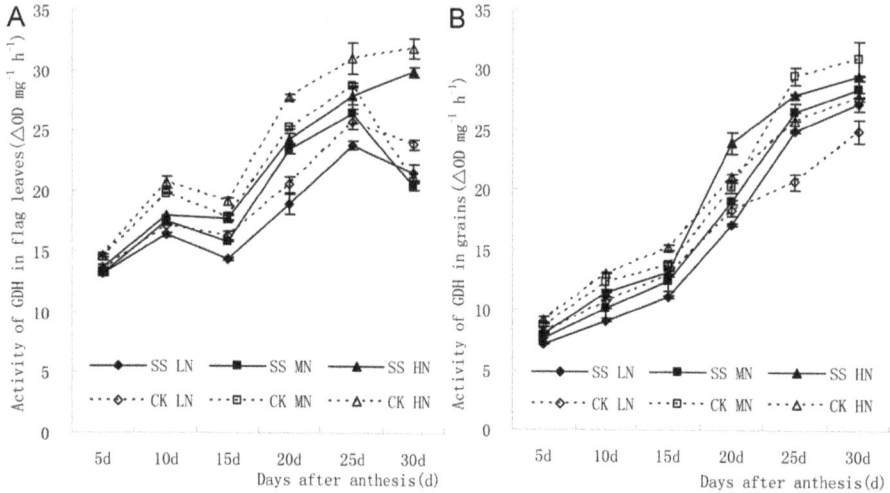

Figure 4. The GDH activity in flag leaves after anthesis increased at first and then decreased, and then increased again (A). In grains, GDH activity kept increasing after anthesis, and had a rapid increment between 15–20 d after anthesis (B).

medium N application treatments. However, globulin content, Glu/Gli (the ratio beween Glutelin/Gliadin), and protein yield were significantly increased. Glutelin content was also increased by subsoiling in fallow period and the difference in high N application were significant. Table 3 also indicated the increment of N application could significantly increase the content of albumin, glutelin and total protein. It also increased globulin content and protein yield, and the difference between low N application and other treatments were significant under subsoiling condition. Gliadin content in high N application treatment was significantly higher than that in the other treatments, while difference between the other treatments was not significant. Glu/Gli in medium N application was significantly higher than that in the other treatments under the condition of subsoiling (Table 3).

Correlation Analysis between Soil Water Storage and Grain Protein Accumulation in Dryland Wheat

Correlation analysis between soil water storage and proline content after anthesis. Table 4 showed soil water storage in the 0–300 cm depth was in negative correlation to proline content in flag leaves and grains. The correlation between soil water storage in the 40–160 cm depth and proline content in grains was significant, while the correlation between water storage

in the 0–300 cm depth soil and proline content in flag leaves was not significant. Table 4 also indicated that proline content in grains was significantly correlated with soil water storage in the two layers of 0–100 cm and 100–200 cm, and the latter correlation was more significant. However, the correlation between water stage in 200–300 cm depth soil and proline content was not significant (Table 4).

Correlation analysis between proline content and activity of nitrogen metabolism enzyme after anthesis. Table 5 showed proline content in flag leaves was in significantly positive correlation to GS and GDH activity in flag leaves after anthesis while its correlation to GS and GDH activity in grains was not significant. Proline content in grains was in significantly positive correlation to GS and GDH activity in flag leaves after anthesis while its correlation to GS and GDH activity in grains was not significant. In addition, the correlation between proline content and GDH activity was a bit higher than that between proline content and GS activity (Table 5).

Correlation analysis between grain protein accumulation and proline content as well as activity of nitrogen metabolism enzyme after anthesis. Table 6 showed contents of albumin, gliadin and total protein in grains were all in positive or significantly positive correlation to proline content in grains and

Table 3. Effect of subsoiling in fallow period on grain protein and its component contents.

Tillage	application amount	Albumin	Globulin	Gliadin	Glutenin	Glu/Gli	Protein	Protein yield
		(%)	(%)	(%)	(%)		(%)	(kg/hm²)
SS	LN	2.43c	1.34b	3.39b	3.97c	1.17b	12.30c	506.95b
	MN	2.53b	1.39a	3.33b	4.21b	1.26a	12.97b	578.60a
	HN	2.69a	1.41a	3.75a	4.56a	1.22b	14.19a	597.81a
CK	LN	2.52c	1.22b	3.65b	3.89c	1.07b	13.08c	423.81b
	MN	2.68b	1.27ab	3.68b	4.14b	1.13a	13.51b	498.33a
	HN	2.86a	1.31a	3.97a	4.38a	1.10a	14.43a	506.64a

The difference between data with different letters from a to c are statistically significant (P<0.05).
SS: subsoiling; CK: no tillage; LN: low nitrogen application; MN: medium nitrogen application; HN: high nitrogen application.

Table 4. Correlation coefficients between soil water storage at different depths and proline content in anthesis.

Soil depth	Flag leaves	Grains
20	−0.4820	−0.5408
40	−0.6424	−0.8550*
60	−0.4059	−0.7612*
80	−0.5719	−0.8299*
100	−0.5087	−0.8225*
0–100	−0.5647	**−0.8224***
120	−0.5935	−0.8182*
140	−0.6591	−0.8494*
160	−0.4522	−0.7615*
180	−0.4847	−0.7504
200	−0.4081	−0.7268
100–200	−0.5212	**−0.7909***
220	−0.7360	−0.7367
240	−0.6631	−0.6878
260	−0.6207	−0.5995
280	−0.6702	−0.5518
300	−0.6296	−0.6225
200–300	−0.6660	**−0.6363**

*$P<0.05$.

Table 5. Correlation coefficients between proline content and activities of the relevant enzymes for protein synthesis.

Proline content	GS activity		GDH activity	
	Flag leaves	Grains	Flag leaves	Grains
Flag leaves	0.7911*	−0.3145	0.8493*	0.5473
Grains	0.9418**	0.1970	0.9845**	0.7130

*$P<0.05$;
**$P<0.01$.

proline content, GS activity and GDH activity in flag leaves. The correlation to proline content in grains was slightly higher than that in flag leaves. The correlation to GDH activity in flag leaves was slightly higher than GS activity. Content of globulin and glutenin together with protein yield were all in positive correlation or significantly positive correlation to GS activity in grains. In addition, content of albumin, glutenin and total protein were in significantly positive correlation to activity of GDH in grains (Table 6).

Discussion

In recent years, dryland cultivators have conducted many researches to study the water storage and retention of dryland wheat, and great progress had been achieved [5–8]. Many results [21–29] showed no-tillage or little tillage had a clear effect on soil water retention. Zhao et al. (2007) stated continuous tillage would form water exclusion in the bottom of harrow which was unfavorable to water storage of soil and the ability of water retention was also limited [9]. However, Zhu and Jia (1997) stated that hydraulic conductivity of soil was improved under protected cultivation, thus water infiltration would increase and massive water loss during plowing would be decreased which result in improving efficiency and availability of water [21]. Besides, Xu et al. (2000) studied effect of different tillage method (conventional tillage, no-tillage, and subsoil tillage) on physical properties and hydraulics of soil. And their results showed subsoiling significantly decreased volume weight of soil and increased porosity of soil and improved the conductivity of soil water. On the other hand, Wang et al. (2006) stated that subsoiling could break the limitation of root system, improve the depth of root system distribution and the efficiency of water use [24].

In agreement with previous studies, our results also showed that subsoiling (depth 30–40 cm) significantly increase water storage

and retention (Fig. 1). Subsoiling can break the plow pan of the farmlands in dry-farming areas which may accelerate the infiltration of rainwater. Additionally, it is easy for plant roots to grow into the deep layer in the loose soil created by subsoiling and the developed root system probably help to hold the rainwater and increase water storage. The different results obtained by researchers might attribute to the varied soil and climate conditions in different area, that is, soil porosity, amount of precipitation, wind-force and temperature, etc. The experimental site which has a relative higher precipitation and lower evaporation rate may result in high water storage and retention after subsoiling.

The present experiment was conducted on a field which was no-tillage in fallow period for many years and performed rotary tillage before sowing. Subsoiling in fallow period broke the bottom of plough layer formed many years which was crucial to water storage and retention, and could increase water storage of 0–300 depth soil at each stage. Especially, the increment in the 60–160 cm depth amounted to more than 10% of the total water storage. It also increased water storage in the 0–200 cm depth soil with low and medium N application until anethesis. However, water storage in the 0–200 cm depth with high N application decreased which was possibly caused by great water consumption due to high biomass of plants. In addition, whether the dryland wheat was subsoiled in fallow period affected the regulatory effect of N application. Under the condition of subsoiling, N application obviously regulated the soil layer with the depth of 40 cm, 60 cm and 120–200 cm. Water storage of soil in the 200–300 cm depth was the least when N application was 150 kg·hm^{-2}. Accordingly, subsoiling in fallow period together with N application at 150 kg·hm^{-2} were beneficial to root downward growth and absorption of water in deep soil.

As subsoiling breaking the plough layer and loosing the soil, fertilizer could also easily reach to the deep soil and used by the plant root in the same layer. And this may heighten the utilization of N to synthesize protein/amino acids and lead to a better wheat productivity. On the other hand, the developed root system would accelerate the absorption of water and nutrition which result in the flourish of wheat leaves. Thus, the photosynthesis is enhanced and more biomass would be accumulated together with a high grain yield.

Osmoregulation is an important mechanism of plant adapting to drought [30], which contributes to continual water adsorption of water from environment with declining water potential and prevents dehydration. Proline plays an important role in osmoregulation [31]. Proline content in the plant is normally low, but it will increase to tenfold under the condition of drought which accounting for more than 30% of total free amino acids [32]. In higher plant, there are two ways to synthesize proline, and both of them are from transformation of glutamic acid which is closely relevant with nitrogen metabolism. Assimilation of

Table 6. Correlation coefficients between content of protein and its components, protein yield and proline content, activity of relevant enzymes for protein synthesis in grains.

protein content and yield	Proline content		GS activity		GDH activity	
	Flag leaves	Grains	Flag leaves	Grains	Flag leaves	Grains
Albumin content	0.8657*	0.9658**	0.9787**	0.1591	0.9918**	0.8239*
Globulin content	−0.4815	−0.0107	0.1366	0.9443**	0.0495	0.2784
Gliadin content	0.9274**	0.9610**	0.8449*	−0.0168	0.9112**	0.6003
Glutenin content	0.2782	0.6925	0.7479	0.7949*	0.7316	0.7945*
Protein content	0.7673*	0.9632**	0.9070**	0.3441	0.9677**	0.7991*
Protein yield	−0.3623	0.0842	0.2466	0.9123**	0.1653	0.4798

*$P<0.05$;
**$P<0.01$.

ammonia is mainly realized by the circulation of GS/GOGAT and an alternate way is the reaction catalyzed by NADH-GDH which is dependent on reductive coenzyme I. This is a compensation to the circulation of GS/GOGAT under certain physiological conditions [33]. Research on tobacco showed that GS played a main role in synthesis of proline under drought while GDH played an important role in providing glutamic acid for synthesis of proline under salty [34]. Study on wheat showed that GS and GDH had different contributions to proline accumulation in wheat seedlings under salty condition. Under low concentration of salt stress, GS played a major role in synthesis of glutamic acid and GDH played a major role in providing precursor for synthesis of proline under high salt stress [35]. The roles of GS in drought stress have been studied in tobacco and rape but the study on the role of GDH in synthesizing proline under drought is lacking. Our study showed that drought in anthesis increased proline content in grains of dryland wheat. GS and GDH in flag leaves both played important roles in synthesis of grain proline and GDH was much more important. In addition, contents of albumin, gliadin and total protein were considerably relative to proline content in grains and GDH activity in flag leaves. Therefore, subsoiling in fallow

period decreased contents of albumin, gliadin and protein in grains.

Conclusion

Our study showed subsoiling could increase glublin content, Glu/Gli, protein yield, glutelin content. Grain protein is an important index to evaluate the quality of wheat. Although subsoiling decreased total protein content, the ratio between glutelin and gliadin was increased. Hence, the quality was improved. And the maximum value was obtained under N application 150 kg·hm^{-2}. In summary, subsoiling in fallow period was proved as an effective technology of water storage and retention. In addition, it was more favorable to improve the quality of dryland wheat grain accompanied with an N application of 150 kg·hm^{-2}.

Author Contributions

Conceived and designed the experiments: MS ZQG. Performed the experiments: MS WFZ LFD YD HMZ AXR GL. Analyzed the data: MS HMZ AXR GL ZPY. Contributed reagents/materials/analysis tools: MS WFZ LFD YD. Wrote the paper: MS ZQG.

References

1. Baumhardt R, Jones O (2002) Residue management and tillage effects on soil-water storage and grain yield of dryland wheat and sorghum for a clay loam in Texas. Soil Tillage Res 68: 71–82.

2. Baumhardt RL, Jones OR (2002) Residue management and paratillage effects on some soil properties and rain infiltration. Soil Tillage Res 65: 19–27.

3. Maiorana M, Castrignanò A, Fornaro F (2001) SW–Soil and Water: Crop Residue Management Effects on Soil Mechanical Impedance. J Agricul Engineering Res 79: 231–237.

4. Bhagat RM, Sharma PK, Verma TS (1994) Tillage and residue management effects on soil physical properties and rice yield in northwestern Himalayan soils. Soil Tillage Res 29: 323–334.

5. Fan T, Stewart B, Yong W, Junjie L, Guangye Z (2005) Long-term fertilization effects on grain yield, water-use efficiency and soil fertility in the dryland of Loess Plateau in China. Agriculture, ecosystems & environment 106: 313–329.

6. Huang Y, Chen L, Fu B, Huang Z, Gong J (2005) The wheat yields and water-use efficiency in the Loess Plateau: straw mulch and irrigation effects. Agricultural water management 72: 209–222.

7. Deng XP, Shan L, Zhang H, Turner NC (2006) Improving agricultural water use efficiency in arid and semiarid areas of China. Agricultural water management 80: 23–40.

8. Huang M, Shao M, Zhang L, Li Y (2003) Water use efficiency and sustainability of different long-term crop rotation systems in the Loess Plateau of China. Soil and Tillage Research 72: 95–104.

9. Zhao H, Jing Q, Dai T, Jiang D, Cao W (2007) Effects of post-anthesis high temperature and water stress on activities of key regulatory enzymes involved in protein formation in two wheat cultivars. Acta Agronomica Sinica 33: 2021.

10. Fu X, Wang C, Guo T, Zhu Y, Ma D et al. (2008) Effects of water-nitrogen interaction on the contents and components of protein and starch in wheat

grains]. Ying yong sheng tai xue bao = The journal of applied ecology/Zhongguo sheng tai xue xue hui, Zhongguo ke xue yuan Shenyang ying yong sheng tai yan jiu suo zhu ban 19: 317.

11. Sun M, Guo PY, Gao ZQ, Wang P, Shi J et al. (2010) Protein Accumulation in Grains of Wheat Cultivars Differing in Drought Tolerance and Its Regulation by Nitrogen Application Amount under Irrigated and Dryland Conditions. Acta Agron Sin 36: 486–495.

12. Fan XM, Jiang D, Dai TB, Jing Q, Wei-xing C (2004) Effects of post-anthesis drought and waterlogging on the quality of grain formation in different wheat varieties. Acta Phytoecol Sin 28: 680–685.

13. Deng ZY, Tian JC, Hu RB, Zhou XF, Yong-xiang Z (2006) Effect of genotype and environment on wheat main quality characteristics. Acta Ecol Sin 26: 2757–2763.

14. Day A, Barmore M (1971) Effects of soil-moisture stress on the yield and quality of flour from wheat (Triticum aestivum L.). Agronomy J 63: 115–116.

15. Jiang D, Xie ZJ, Cao WX, Dai TB, Jing Q (2004) Effects of post-anthesis drought and waterlogging on photosynthetic characteristics, assimilates transportation in winter wheat. Acta Agronomica Sinica 30: 175–182.

16. Johansson E, Prieto-Linde ML, Jönsson JÖ (2001) Effects of wheat cultivar and nitrogen application on storage protein composition and breadmaking quality. Cereal Chemistry 78: 19–25.

17. Zhu J, Khan K, Huang S, O'Brien L (1999) Allelic Variation at Glu-D1 Locus for High Molecular Weight (HMW) Glutenin Subunits: Quantification by Multistacking SDS-PAGE of Wheat Grown Under Nitrogen Fertilization1. Cereal Chem J 76: 915–919.

18. Bates L, Waldren R, Teare I (1973) Rapid determination of free proline for water-stress studies. Plant and soil 39: 205–207.

19. Lu B, Yuan Y, Zhang C, Ou J, Zhou W et al. (2005) Modulation of key enzymes involved in ammonium assimilation and carbon metabolism by low temperature in rice (*Oryza sativa* L.) roots. Plant Sci 169: 295–302.

20. Halvorson A, Nielsen D, Reule C (2004) Nitrogen fertilization and rotation effects on no-till dryland wheat production. Agron J 96: 1196–1201.

21. Zhu W, Jia C (1997) The application and prospect for reduced tillage technology in wheat straw mulching MaiCha field Crops 4: 26–27.

22. Zhao BQL, Li F, Jian X, Li Z (1997) Effect of Different Tillage Methods on Root Growth of Winter Wheat [J]. Acta Agronomica Sinica 5.

23. Xu D, Schmid R, Mermoud A (2000) Effects of Tillage Practices on Temporal Variations of Soil Surface Properties. Journal of Soil and Water Conservation 14: 64–70.

24. Wang XB, Cai DX, Hoogmoed W, Oenema O, Perdok U (2006) Potential effect of conservation tillage on sustainable land use: a review of global long-term studies. Pedosphere 16: 587–595.

25. Sijtsma C, Campbell A, McLaughlin N, Carter M (1998) Comparative tillage costs for crop rotations utilizing minimum tillage on a farm scale. Soil and Tillage Res 49: 223–231.

26. Shipitalo M, Dick W, Edwards W (2000) Conservation tillage and macropore factors that affect water movement and the fate of chemicals. Soil and Tillage Research 53: 167–183.

27. Blevins RL, Frye WW (1993) Conservation Tillage: An Ecological Approach to Soil Management. In: Donald LS (ed) Advances in Agronomy, vol Volume 51. Academic Press, pp. 33–78. doi:10.1016/s0065-2113(08)60590-8.

28. Lal R, Logan TJ, Fausey NR (1989) Long-term tillage and wheel traffic effects on a poorly drained mollic ochraqualf in northwest Ohio. 2. Infiltrability, surface runoff, sub-surface flow and sediment transport. Soil and Tillage Research 14: 359–373.

29. Morin J, Rawitz E, Hoogmoed WB, Benyamini Y (1984) Tillage practices for soil and water conservation in the semi-arid zone III. Runoff modeling as a tool for conservation tillage design. Soil and Tillage Res 4: 215–224.

30. Reynolds M, Fischer R, Sayre K (2005) Osmotic adjustment in wheat in relation to grain yield under water deficit environments. Agronomy J 97: 1062–1071.

31. Verslues PE, Bray EA (2006) Role of abscisic acid (ABA) and Arabidopsis thaliana ABA-insensitive loci in low water potential-induced ABA and proline accumulation. J exp botany 57: 201–212.

32. Blum A (1983) Genetic and physiological relationships in plant breeding for drought resistance. Agricultural water management 7: 195–205.

33. Ireland RJ, Lea PJ (1999) The enzymes of glutamine, glutamate, asparagine, and aspartate metabolism. New York: Marcel Dekker, Inc.

34. Skopelitis DS, Paranychianakis NV, Paschalidis KA, Pliakonis ED, Delis ID et al. (2006) Abiotic stress generates ROS that signal expression of anionic glutamate dehydrogenases to form glutamate for proline synthesis in tobacco and grapevine. The Plant Cell Online 18: 2767–2781.

35. Wang ZQ, Yuan YZ, Ou JQ, Lin QH, Zhang CF (2007) Glutamine synthetase and glutamate dehydrogenase contribute differentially to proline accumulation in leaves of wheat (Triticum aestivum) seedlings exposed to different salinity. J Plant Physiol 164: 695–701.

23

Variation in Yield Gap Induced by Nitrogen, Phosphorus and Potassium Fertilizer in North China Plain

Xiaoqin Dai*, Zhu Ouyang, Yunsheng Li*, Huimin Wang

Key Laboratory of Ecosystem Network Observation and Modeling, Institute of Geographic Sciences and Natural Resources Research, Chinese Academy of Sciences, Beijing, China

Abstract

A field experiment was conducted under a wheat-maize rotation system from 1990 to 2006 in North China Plain (NCP) to determine the effects of N, P and K on yield and yield gap. There were five treatments: NPK, PK, NK, NP and a control. Average wheat and maize yields were the highest in the NPK treatment, followed by those in the NP plots among all treatments. For wheat and maize yield, a significant increasing trend over time was found in the NPK-treated plots and a decreasing trend in the NK-treated plots. In the absence of N or P, wheat and maize yields were significantly lower than those in the NPK treatment. For both crops, the increasing rate of the yield gap was the highest in the P omission plots, i.e., 189.1 kg ha^{-1} yr^{-1} for wheat and 560.6 kg ha^{-1} yr^{-1} for maize. The cumulative omission of P fertilizer induced a deficit in the soil available N and extractable P concentrations for maize. The P fertilizer was more pivotal in long-term wheat and maize growth and soil fertility conservation in NCP, although the N fertilizer input was important for both crops growth. The crop response to K fertilizers was much lower than that to N or P fertilizers, but for maize, the cumulative omission of K fertilizer decreased the yield by 26% and increased the yield gap at a rate of 322.7 kg ha^{-1} yr^{-1}. The soil indigenous K supply was not sufficiently high to meet maize K requirement over a long period. The proper application of K fertilizers is necessary for maize production in the region. Thus, the appropriate application of N and P fertilizers for the growth of both crops, while regularly combining K fertilizers for maize growth, is absolutely necessary for sustainable crop production in the NCP.

Editor: Carl J. Bernacchi, University of Illinois, United States of America

Funding: This study was financially supported by the Strategic Science Plan of Institute of Geographic Sciences and Natural Resources Research (2012ZD004)and National Science and Technology Support Projects of Ministry of Science and Technology of China (2013BAD05B03). The funders had no role in study design, data collection and analysis, decision to publish, or preparation of the manuscript.

Competing Interests: The authors have declared that no competing interests exist.

* E-mail: daixq@igsnrr.ac.cn (XD); liys@igsnrr.ac.cn (YL)

Introduction

The North China Plain (NCP) is the largest and the most important agricultural production region in China. It covers approximately 18.3% of total national farm lands (18 million hectares) and produces 21.6% of the total grain yield of edible crops in the country [1]. The main crops grown are wheat and maize, periodically rotated. The NCP now supplies more than 50% of the nation's wheat and 33% of its maize [2]. Grain production and the maintenance of soil fertility in the NCP are very important for the food security and agricultural sustainability of China.

Nutrient availability is the most yield-limiting factor. To produce higher yields, the over-application of chemical fertilizers has been a common practice in wheat-maize rotation systems and has led to severe environmental problems [3]. Improved nutrient management practices are urgently needed to maximize crop yields and maintain soil fertility while minimizing environmental impacts. The ability to better identify crop response to the application of fertilizers, soil indigenous nutrient supply capability, and the maintenance of soil fertility over time are crucial to the development of improved nutrient management practices. Various long-term experiments have been conducted to test the effects of fertilization on yield or soil fertility throughout the world [4–14], but over long time scales, crop response to the application of N, P and K fertilizers and soil indigenous nutrient supply capability has seldom been clearly understood.

The yield and soil fertility gap between a full NPK fertilizer plot and a fertilizer omission plot was used as a good diagnostic tool to assess the extent of macronutrient limitations. Mussgnug et al (2006) analyzed the yield gap resulting from nutrient limitation on a degraded soil in the Red River Delta (RRD) in Northern Vietnam. They found that in the absence of K application, the yield gap for rice and maize respectively averaged 1.7 Mg ha^{-1} and 3.4 Mg ha^{-1}, while when N or P was omitted the yield gap was less. Potassium was the most yield-limiting macronutrient in RRD of Vietnam [15]. The factors that primarily limit increasing crop yields varied by crop and region [16]. The objectives of the study were to examine the effects of the continuous use of inorganic fertilizers on wheat and maize yield and soil fertility gap to elucidate crop response to fertilizer inputs and the long-term N, P and K supply capability in the NCP.

Materials and Methods

Experimental Site

A long-term field experiment was conducted from 1990 to 2006 at the Yucheng Comprehensive Experimental Station (36°57′ N,

Table 1. Physical and chemical properties of soil at different layers at Yucheng, Shandong Province of China in 1990.

Soil parameters	0–20 cm	20–40 cm
Soil texture	Silt loam	Sandy loam
Clay (%)	17.3	13.7
Silt (%)	61.4	11.4
Sand (%)	21.3	74.9
Bulk density (g cm^{-3})	1.31	1.48
Field capacity (%)	23.3	22.7
Organic C (g kg^{-1})	5.4	4.8
Total N (g kg^{-1})	0.41	0.36
Total P (g kg^{-1})	1.11	1.13
Total K (g kg^{-1})	10.7	14.0
Available N (mg kg^{-1})	46.1	41.4
Available P (mg kg^{-1})	8.3	5.4
Available K (mg kg^{-1})	149.0	125.4

116°36′ E, 28 m a.m.s.l.) of the Chinese Academy of Sciences (CAS) located in Shandong Province, which lies in the NCP. This area represents the moderate- to high-yielding region of the NCP, where the dominant cropping system is double-cropped wheat-maize. The annual mean precipitation is 575 mm, with approximately 70% of the total precipitation falling between June and September. The soil is classified as fluvo-aquic loam soil. Before the initiation of the experiment, soil samples were collected from depths of 0–20 cm and 20–40 cm by stepwise soil augers in the experimental field in 1990. All samples from the same soil layers were mixed thoroughly to create one homogenous sample, and a representative sample was drawn to determine the soil texture, bulk density, organic carbon, total N, P and K, and available N, P and K. The physical and chemical properties of the soil in 1990 are listed in Table 1.

Experimental Design and Treatments

The study consisted of five treatments: no fertilization (control); N, P, and K applied (NPK); N and P applied (NP; K omission); N and K applied (NK; P omission); P and K applied (PK; N omission), with four replications performed for each condition in a randomized complete block design (Table 2). The application rates

Table 2. Treatment and fertilizer nutrient rates (kg ha^{-1}) applied to winter wheat and maize at Yucheng, Shandong Province of China.

Treatments	Winter wheat			Maize		
	N	P$_2$O$_5$	K$_2$O	N	P$_2$O$_5$	K$_2$O
NPK	253	90	171	262	52	298
PK	0	90	171	0	52	298
NK	253	0	171	262	0	298
NP	253	90	0	262	52	0
Control	0	0	0	0	0	0

of N, P and K, if they were applied, were the same in all of the treatments: 253 kg N ha^{-1} as urea (46% N), 90 kg P$_2$O$_5$ ha^{-1} as single superphosphate (16% P$_2$O$_5$), and 171 kg K$_2$O ha^{-1} as potassium sulfate (50% K$_2$O) for winter wheat and 262 kg N ha^{-1} as urea (46% N), 52 kg P$_2$O$_5$ ha^{-1} as single superphosphate (16% P$_2$O$_5$), and 298 kg K$_2$O ha^{-1} as potassium sulfate (50% K$_2$O) for maize (Table 2). Phosphate was applied before sowing and incorporated during land preparation. Winter wheat received 40% of the total N as a basal dressing before sowing, whereas the remainder was top-dressed in two splits (40% at green-up and 20% at flowering). Approximately 50% of the K was applied before sowing and 50% top-dressed at green-up (the first leaf growing in the early of spring growed to 1–2 cm from soil surface and the phenomena was seen on a half of wheat seedling in the field). During the maize season, all of the P and K fertilizers and 35% of the N fertilizer were applied as a basal fertilizer, and 65% of the N was applied at the elongation stage. Each plot area measured 6 m×5 m, and all plots were isolated from one another by a concrete wall to a soil depth of 1 m.

Winter wheat (*Triticum aestivum*) was double-cropped with maize (*Zea mays*), and the crops were grown along the border of the four blocks. Winter wheat was sown in mid-October and harvested in early June of the next year. Maize was sown in mid-June and harvested in early October of the same year. The winter wheat was seeded at a rate of 225 kg ha^{-1}, and the maize was planted at a rate of 75000 seeds ha^{-1}. Surface irrigations were conventionally conducted according to the conditions of crop growth and soil moisture. Wheat and maize were harvested to the level of the soil surface; thus, the stubble left in the field was negligible. However, the roots were left in the soil. All straws were removed from the field. All plots were kept free of weeds by the application of an herbicide. Pesticides were applied in accordance with good practices for crop protection.

Sampling and Chemical Analyses

At harvestable maturity, the grain yields were determined over the whole plot area. Soils samples were collected at a depth of 20 cm at six randomly selected points in every plot after wheat and maize were harvested each year. The soil samples were mixed thoroughly from all cores to obtain a representative soil sample for each plot. The soil samples were air-dried and sieved through a 1-mm mesh to perform available N, Olsen-P and available K measurements and through a 0.149-mm mesh to estimate the organic carbon and total N content. Soil available N was determined using the alkali-hydrolytic diffusion method [17]. Olsen-P was measured by ascorbic acid-molybdate blue colorimetry [18]. Available K was extracted with an ammonium acetate solution (NH$_4$OAc, 1 mol/L) and then determined with a flame photometer [17]. Soil organic carbon was determined using the wet oxidation method of Walkley and Black, and total N was measured following Kjeldahl digestion and distillation [17].

Data Analyses

The results of yield and soil fertility in the first three years of the experiment were not used because in the autumn of 1993 soybean was sown. Therefore, the results reported are from 1994 to 2006. Crop yield losses caused by N, P, K or NPK omission were calculated from the differences between an NPK treatment and the PK, NK, NP or control plot during the same years. The yield differences were called the "yield gap" induced by N (GYG$_N$), P (GYG$_P$), K (GYG$_K$) or NPK (GYG$_{NPK}$) fertilizer omission to assess the extent of macronutrient limitations in the study. The differences in soil available N, Olsen-P and available K between the NPK treatment and the PK, NK, NP or control plots were

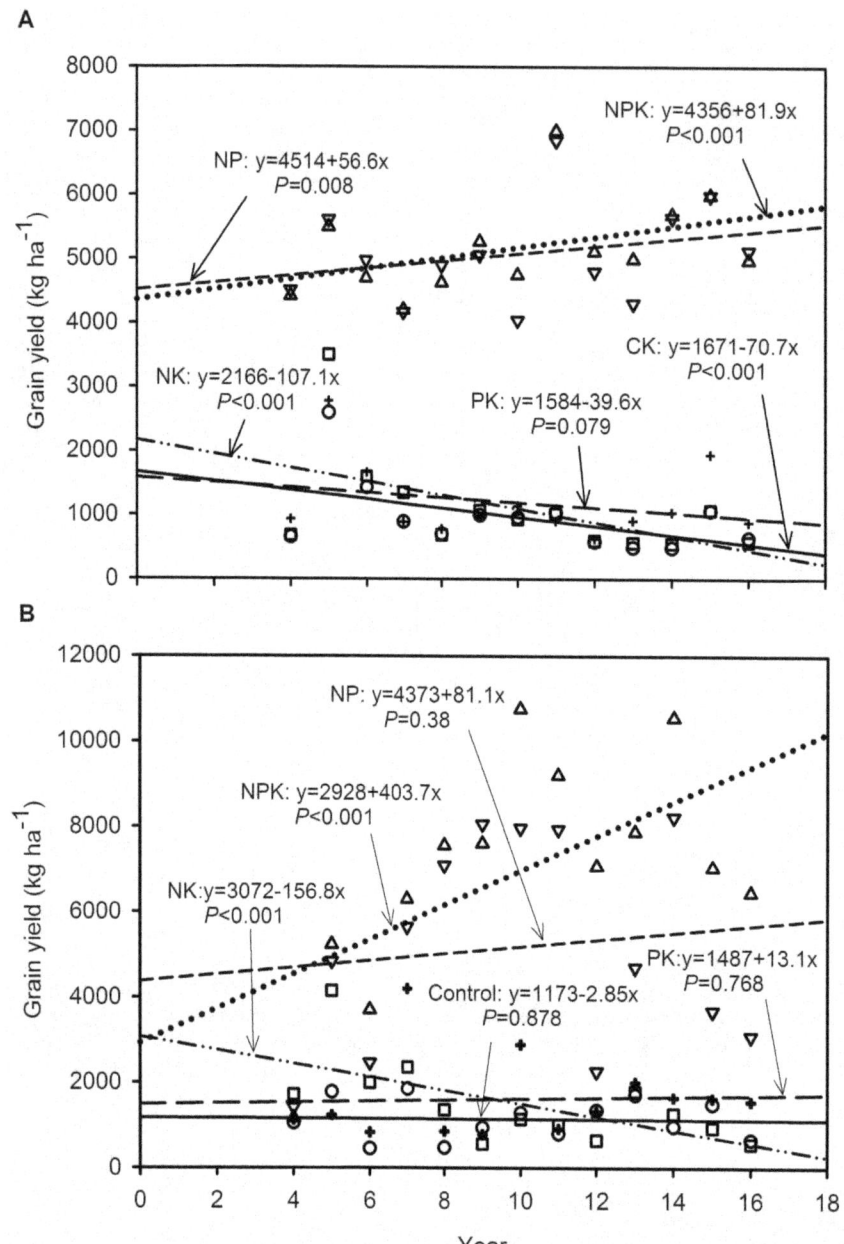

Figure 1. Grain yield of wheat (a) and maize (b) of different fertilizer treatments (Control, circle; NPK, triangle up; NP, triangle down; NK, square; PK, plus) under wheat-maize rotation system from 1994 to 2006 at Yucheng, Shandong Province of China.

calculated to evaluate the variation in soil fertility when nutrients were limited. The soil nutrient differences were called the "soil nutrient gap (SNG)" induced by N (SNG_N), P (SNG_P), K (SNG_P), or NPK (SNG_{NPK}) fertilizer omission.

All statistical analyses including the analysis of variance and regression were conducted using the SPSS package (version 16.0). Considering the data over time was lack of independence, the repeated measures ANOVA between differences in mean through years among treatments for various parameters (grain yield, soil organic C, total N, available N, Olsen-P, available K) were analyzed using a Fisher's protected least significant difference (LSD) test at $P = 0.05$. A mixed models with random block and

block×treatment were done to assess trends (slopes) of grain yield, yield gap and various soil nutrients gap over the years. The P-values of the slopes were used to test whether the observed changes were significantly different from 0.

Results

Grain Yield and Yield Gap

The fertilizer treatments significantly affected the grain yields (Table 3). The wheat and maize yields were highest in the NPK treatment, followed by those in the NP treatment. In the PK or NK treatments, the yields of wheat and maize were significantly

Table 3. Multi-year average grain yield of wheat and maize, as affected by various treatments with fertilizer in a wheat-maize rotation system at Yucheng, Shandong Province of China (mean±standard error).

Treatment	Wheat (kg ha^{-1})	Maize (kg ha^{-1})
Control	963±155 b	1145±133 c
NPK	5174±201 a	6965±724 a
NP	5080±216 a	5184±666 b
NK	1094±218 b	1504±270 c
PK	1188±165 b	1618±269 c

Different letters within a column indicate a significant difference at $P<0.05$ between fertilizer treatments.

lower than those in the NP and NPK treatments. Compared to those in the NPK treatment, the yields of wheat and maize decreased, respectively, by 77% with the N omission, 79% and 78% with the P omission, 2% and 26% with the K omission, and 81% and 84% with no fertilizer. The grain yield losses were much greater in either N or P omission than in plots fertilized combined N and P. Over time wheat yield significantly increased at a rate of 81.9 kg ha^{-1} yr^{-1} in NPK plots ($P=0.008$), and maize yield also significantly increased at a rate of 403.7 kg ha^{-1} yr^{-1} in NPK plots ($P<0.001$, Fig. 1). The yields of wheat and maize in NK treatment decreased significantly at a rate of 107.1 kg ha^{-1} yr^{-1} ($P<0.001$) and 156.8 kg ha^{-1} yr^{-1} ($P<0.001$) over time during the study period, respectively (Fig. 1). The long-term omission of K fertilizer also significantly reduced the maize yield compared to the maize yield of the NPK-treated plot (Table 3). But both crops yields had no significant upward or downward trends when K fertilizer was omitted (Fig. 1).

The variation in the yield gap of wheat and maize between the NPK and PK, NK, NP and control plots is illustrated in Figure 2. For wheat, the yield gap significantly increased at a rate of 152.6 kg ha^{-1} yr^{-1} for the control plots ($P<0.001$), a rate of 189.1 kg ha^{-1} yr^{-1} for the NK plots ($P<0.001$) and a rate of 121.5 kg ha^{-1} yr^{-1} for the PK plots ($P<0.001$). In the NP plots, the yield gap also gradually increased with the cumulative effect of K omission, but the relationship was not significant ($P=0.246$). However, for maize, when the N, P K or NPK fertilizer was omitted, the yield gap significantly increased with the continuous omission of nutrients. The significant increase in the yield gap was 390.6 kg ha^{-1} yr^{-1} for the PK treatments ($P<0.001$), 560.6 kg ha^{-1} yr^{-1} for the NK treatments ($P<0.001$), 322.7 kg ha^{-1} yr^{-1} for the NP treatments ($P<0.001$), and 406.6 kg ha^{-1} yr^{-1} for the control treatments ($P<0.001$). For both crops, the increasing rate of the yield gap was highest in the P omission plots, which indicated that the cumulative effect of P omission was the most significant for wheat and maize in the region. The slopes of the yield gap were the second highest in the N omission plots for both crops, which indicated that N fertilizer inputs were important for wheat and maize growth, especially the latter. It is worth noting that the slopes of the maize yield gap increased significantly with continuous K omission. This result indicates that the soil indigenous K supply was not sufficiently high to meet the maize K requirement over a long period.

Soil Nutrient Status

Long-term continuous treatment with different fertilizers significantly affected the soil organic carbon, available N, Olsen-

P and available K concentrations (Table 4). The multi-year average soil organic carbon content was significantly lower in the control, NK and PK plots than in the NPK and NP plots for wheat and maize. The decrease was 6.7% to 15.6% for wheat and 7.4% to 15.8% for maize. The different treatments except control had no significant influences on soil total N. The available N, Olsen-P and available K concentrations were higher in the plots treated with the corresponding N, P or K fertilizer inputs (Table 4). For wheat, the available N concentration increased significantly in the acquired N fertilizer inputs plots compared to that in the PK plots. The Olsen-P concentration in plots with P fertilizer inputs increased significantly by 7.2 to 12.3 times the concentration in the control plots. Meanwhile, the Olsen-P concentration also increased by 9% and 62% in the NP and PK plots, respectively, compared with that in the NPK plots, indicating that the P uptake of the crops was lower in the NP and PK plots. Similar to the Olsen-P concentration, the available K concentration increased significantly in the plots with K fertilizer inputs, and the increase was as high as 1.2 to 2.3 times the K concentration in the control plots. In the NK and PK plots, the available K concentration increased by 48% and 55%, respectively, compared to that in the NPK plots.

The differences in soil nutrients between the NPK and PK, NK, NP or NPK omission plots are shown in Figure 3&4. During the wheat growth season, the continuous omission of nutrients had no significant effects on the difference in the soil available N concentration (Fig. 3a). The continuous omission of P fertilizer significantly enhanced the differences in the soil Olsen-P concentration in the NPK plots ($P<0.001$), indicating that the long-term absence of P fertilizer resulted in a great deficit in the soil available P concentration. Similarly to the P omission plots, the control plots also possessed a significantly enhanced soil Olsen-P gap ($P<0.001$). However, a surplus concentration of Olsen-P was observed when the continuous N was omitted ($P=0.006$, Fig. 3b). Also a surplus concentration of available K was observed when the continuous P and N inputs were omitted ($P<0.001$) due to a large amount of K fertilizer inputs and lower crop uptake. When the K fertilizer was omitted, the soil available K concentration gap significantly increased ($P=0.001$ for NP and $P=0.002$ for Control), indicating a deficit in soil available K in the NP and Control plots.

During the maize growth season, the soil available N concentration gradually became deficient when the nutrients were omitted, but only when P fertilizer was omitted, the effects were significant (Fig. 4a). Like wheat, P fertilizer omission induced a severe deficit in soil Olsen-P ($P<0.001$). Although the PK plots feature a larger amount of P fertilizer inputs and lower crop uptake, no significant surplus in the soil Olsen-P concentration was observed (Fig. 4b). In the plots fed K fertilizer inputs, a surplus in the soil available K concentration was observed, but the effects were not significant (Fig. 4c). It is worthy to note that the NP plots induced a deficit in soil available K (Table 4). Although the effects were not significant, maize growth was significantly affected due to the shortage of soil available K.

Discussion

The response of wheat and maize growth of combined N and P fertilization (i.e. NP) was much greater than that of K fertilizer application combined with N or P (i.e. NK and PK treatments) (Table 3). The continuous 16-yr omission of N or P fertilizer significantly reduced the wheat and maize yield (Table 3). The grain yield in the N omission plots exhibited no significant decreasing trends over time (Fig. 1a). On the one hand, the

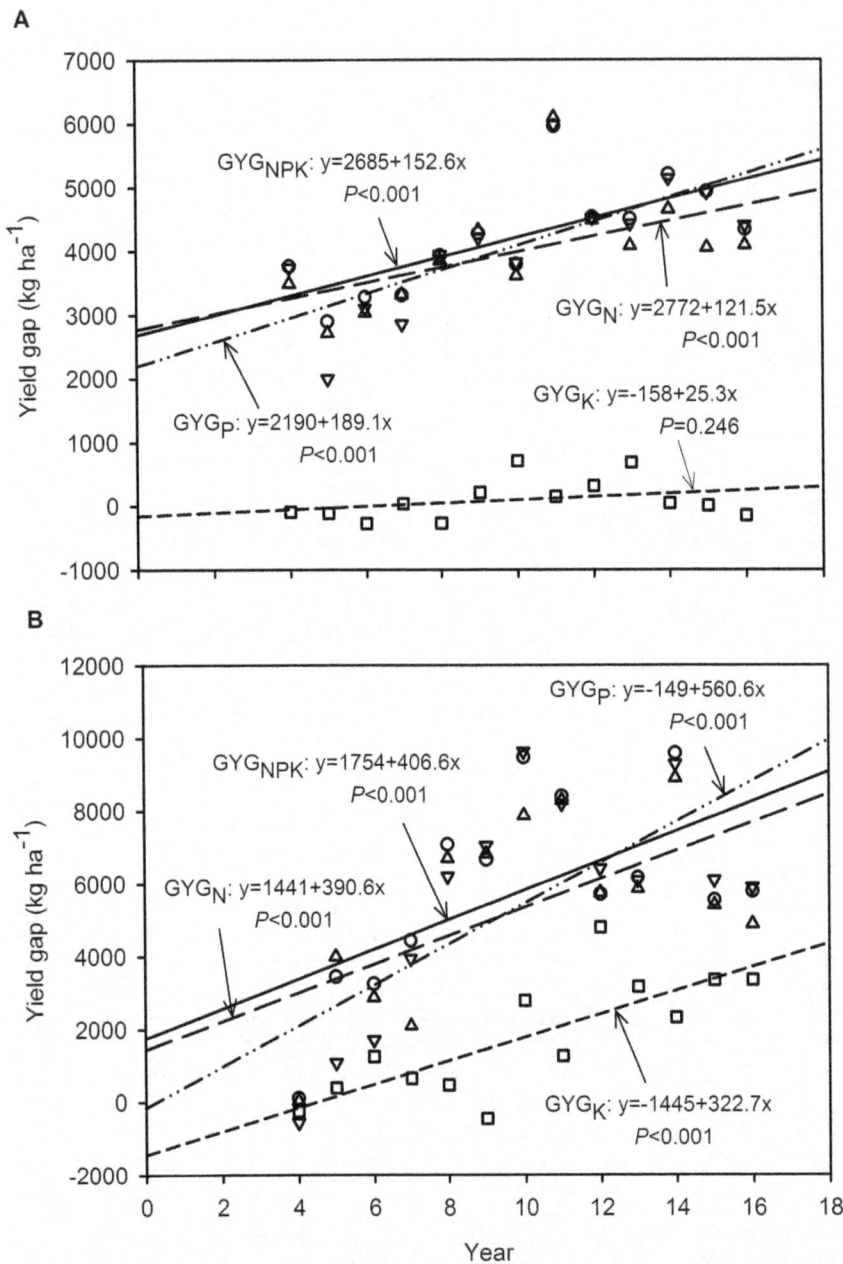

Figure 2. Yield gap variation of wheat (a) and maize (b) by N (GYG$_N$, the yield difference between NPK and PK plots in each year, triangle up), P (GYG$_P$, the difference between NPK and NK plots, triangle down), K (GYG$_K$, the difference between NPK and NP plots, square) or NPK (GYG$_{NPK}$, the difference between NPK and Control plots, circle) fertilizer omission from 1994 to 2006 at Yucheng, Shandong Province of China.

continuous N omission significantly reduced the soil available N concentration, but no significant effects were observed on the total N concentration (Table 4). The differences in soil available N between the NPK and PK plots also exhibited no significant increase over time (Fig. 3a&4a). It might be possible that atmospheric nitrogen deposition in the NCP soil, N mineralization and lower nitrogen depletion may have attributed to the trend in available N over the years in the nitrogen omitted plots. Liu et al. (2006) and Zhang et al. (2008) reported that the bulk deposition of inorganic N in the NCP is approximately 30 kg/ha per year,

which has a significant effect on agricultural systems [19,20]. But yield gap significantly increased over time when the N was omitted (Fig. 2), indicating that the soil indigenous N supply combined with the deposition of N could not meet the N requirement for wheat and maize growth in the NCP.

Both of the wheat and maize yields had significant downward trends in P omission treatment (Fig. 1) and the yield gap of both crops significantly increased with the continuous omission of P fertilizer (Fig. 2), indicating that the cumulative absence of P fertilizers significantly inhibited wheat and maize growth. Tang

Table 4. Multi-year average soil nutrient status at wheat and maize harvest, as affected by various treatments with fertilizer in a wheat-maize rotation system at Yucheng, Shandong Province of China (mean±standard error).

Crops	Treatments	Organic C (g kg^{-1})	Total N (g kg^{-1})	Available N (mg kg^{-1})	Olsen-P (mg kg^{-1})	Available K (mg kg^{-1})
Wheat	Control	5.5±0.19 c	0.56±0.04b	52.5±5.15b	3.1±0.28c	104.5±5.50c
	NPK	6.5±0.21a	0.66±0.05a	63.3±4.35a	25.2±1.95b	226.1±16.0b
	NP	6.3±0.15ab	0.67±0.05a	62.7±4.07a	27.5±2.04b	91.6±8.72c
	NK	5.9±0.17abc	0.64±0.05a	62.5±4.17a	3.2±0.31c	333.4±40.3a
	PK	5.8±0.19bc	0.56±0.04ab	49.1±4.20b	40.8±3.98a	349.3±35.3a
Maize	Control	5.7±0.13c	0.66±0.02ab	56.9±6.11b	2.7±0.32c	106.5±5.22c
	NPK	6.7±0.15a	0.71±0.05a	65.8±6.45a	22.1±2.47b	283.7±24.6b
	NP	6.5±0.15ab	0.69±0.04ab	66.6±6.05a	21.7±1.78b	90.9±4.46c
	NK	6.0±0.13bc	0.66±0.04ab	68.4±6.76a	3.6±0.54c	423.8±43.7a
	PK	5.8±0.14bc	0.62±0.03b	52.1±5.90b	33.5±2.32a	438.8±39.8a

For each crop, different letters within a column indicate a significant difference at $P<0.05$ between fertilizer treatments.

et al. (2008) also reported that wheat and maize yields without P fertilization significantly decreased over time in several other sites in China [21]. The results are different from those reported under European conditions, most likely due to differences in the P supplying abilities of the different soils [22]. It is noted that the continuous omission of P fertilizer significantly reduced the soil Olsen-P concentration for both wheat and maize (Table 4). Meanwhile, the differences in the Olsen-P concentration between the NPK and NK plots significantly increased with continuous P omission (Fig. 3b&4b). The results indicate that the long-term absence of P fertilizer resulted in a great deficit in soil available P. It is interesting that the continuous omission of P fertilizer increased the available N concentration gap between the NPK and NK plots, and the effects were significant for maize (Fig. 4a), which indicates that a large amount of available N was depleted or lost, although the plots were fed N fertilizer inputs. The cumulative absence of P fertilizer intensified the shortage of N for crop growth. Because the treatments fertilized without P had a lower the number of cultivable microorganisms, microbial biomass and community functional diversity than in the treatments with P fertilization [23], which possibly induced the lower N and P mineralization in treatments fertilized without P.

The yield response to K fertilizer was much lower than that to N or P fertilizer (Table 3), most likely due to the high inherent soil K levels, which were in excess of the crop K demands. Most soils of the alluvial floodplain in Asia are high in K, and K is a rare limiting factor [24]. Shen et al. (2004) conducted a 14-yr field trial in Hebei Province of northern China and indicated that the application of N and P enhanced rice yields, while K had no yield-increasing effect due to the large resource of available soil K [25]. In our study, wheat and maize yield in the plots of combined NP and K fertilizer significantly improved, and the yield gap of maize significantly increased with continuous K omission; however, that of wheat yield did not (Fig. 1&2). Compared to that in the NPK plots, the soil available K concentration significantly decreased in the NP plots (Table 4). The differences in soil available K between NPK and NP also improved with the continuous omission of K, and the effects were significant during the wheat season (Fig. 3c). Although soil available K was gradually depleted, the soil indigenous K supply was sufficiently high to meet the requirement

for normal wheat growth over the 16-yr period of the study. If the K fertilizer was continuously omitted for a longer period, the soil available K deficit would possibly inhibit wheat growth, which must be further verified. In addition, the soil available K gap between the NPK and NP plots during maize growth season showed an increasing trend with the omission of K, but the differences were not significant (Fig. 4c). This behavior is because the long-term omission of K fertilizer significantly reduced the maize yield, which resulted in lower K uptake. The results show that long-term maize production is much more sensitive to the absence of K than wheat production in the NCP. The results are supported by those of Tan et al. (2007), who concluded that the effect of K fertilizer on maize was higher than that on wheat under the wheat-maize rotation system of Hebei [26].

In our study, the combination of N and P fertilizers mostly sustained soil organic carbon, total N and available N, P and K levels over time (Table 4). This result is similar to that of Shen et al. (2004), who reported that the soil organic carbon and total N concentrations remained stable over time [25]. Manna et al. (2005) also showed that the recommended NPK plots are adequate for maintaining a constant SOC content under the sub-humid and semi-arid tropical conditions of India over a long period [27]. When N fertilizer was omitted, soil available P and K significantly improved in the plots with P and K fertilizer (PK). Meanwhile, when P fertilizer was omitted, soil available K was significantly accumulated in the plots with N and K fertilizer (NK, Table 4). Because the deficiency of other nutrients inhibited crop production, both crops exhibited lower P or K uptake. The application of inorganic fertilizer allowed the crops to exceed production needs and thus resulted in a substantial build-up of available P or available K, which inevitably induced resource waste and ecological pollution.

Conclusions

The P fertilizer was more pivotal in the long-term growth of wheat and maize and the conservation of soil fertility in the NCP, although the N fertilizer input was important for the growth of both crops as well. Although the crop yield response to K fertilizer was much lower than to N or P fertilizer, the proper application of K fertilizer is also necessary, especially for maize production in the

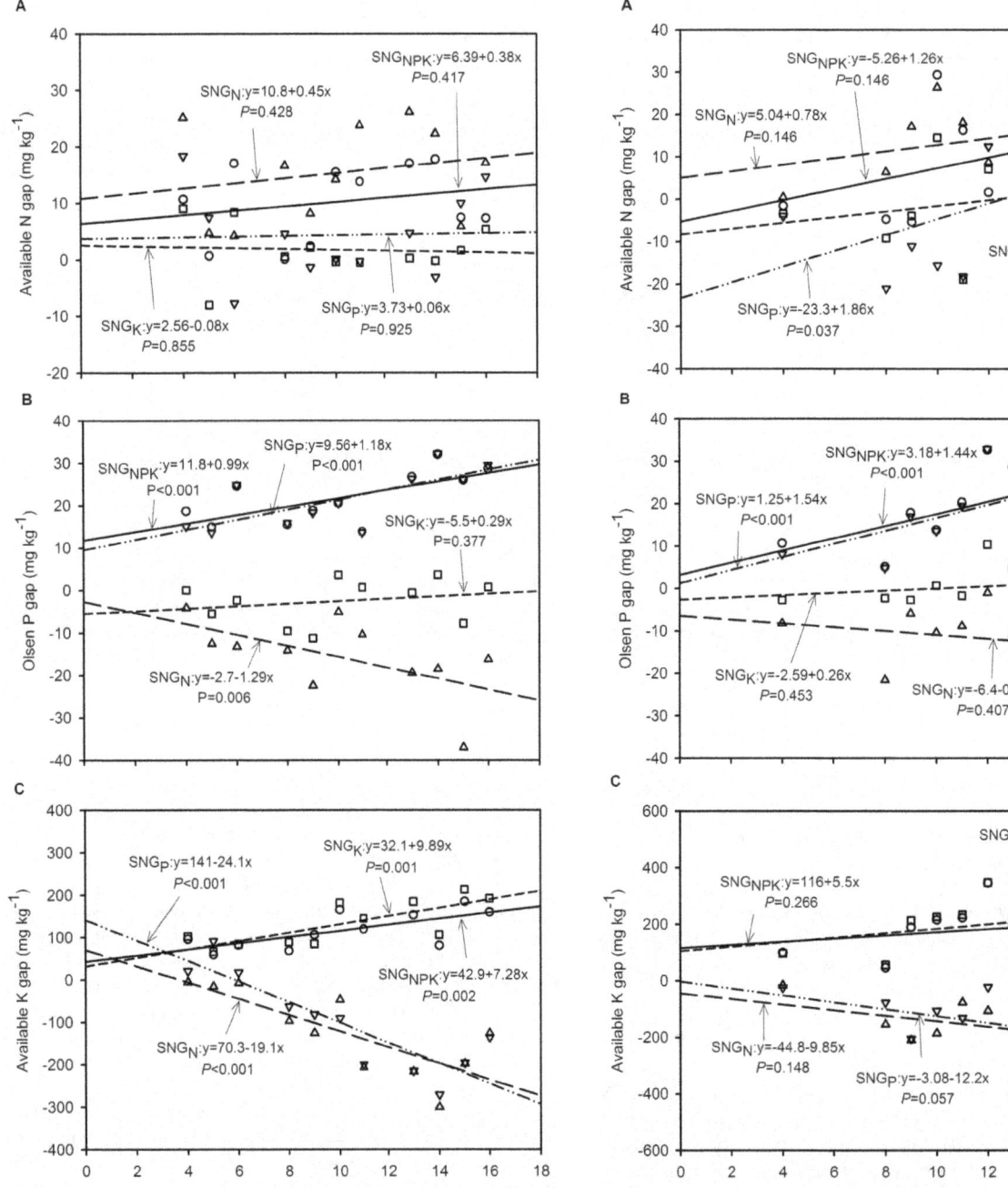

Figure 3. Soil available N (a), Olsen-P (b) and available K (c) gap variation by N (SNG$_N$, soil available nutrient difference between NPK and PK plots, triangle up), P (SNG$_P$, the difference between NPK and NK plots, triangle down), K (SNG$_K$, the difference between NPK and NP plots, square) or NPK (SNG$_{NPK}$, the difference between NPK and Control plots, circle) fertilizer omission from 1994 to 2006 during wheat season at Yucheng, Shandong Province of China.

Figure 4. Soil available N (a), Olsen-P (b) and available K (c) gap variation by N (SNG$_N$, soil available nutrient difference between NPK and PK plots, triangle up), P (SNG$_P$, the difference between NPK and NK plots, triangle down), K (SNG$_K$, the difference between NPK and NP plots, square) or NPK (SNG$_{NPK}$, the difference between NPK and Control plots, circle) fertilizer omission from 1994 to 2006 during maize season at Yucheng, Shandong Province of China.

region. Thus, the appropriate application of N and P fertilizers for both crops, in combination with regular K fertilizers for maize, is absolutely necessary in terms of sustainable crop production in the NCP. However, a longer-range study is required to verify whether

the soil indigenous K supply could continue to meet the requirement for wheat growth over many years.

Acknowledgments

The authors are grateful to Zhenrong Tian in the Yucheng Comprehensive Experimental Station of the Chinese Academy of Sciences (CAS) for many sampling work and Lynn M. Johnson in Cornell University for statistical analysis. Authors also thank the academic editor and anonymous reviewers for their constructive comments, which helped in improving the manuscript.

References

1. Yu Q, Saseendran SA, Ma L, Flerchinger GN, Green TR, et al. (2006) Modeling a wheat-maize double cropping system in China using two plant growth modules in RZWQM. Agr Syst 89: 457–477.

2. China Statistics Bureau (2001) China Statistics Bureau, Annual Agricultural Statistics of Henan Province. Beijing, China Statistics Press. 112–115.

3. Vitousek PM, Naylor R, Crews T, David MB, Drinkwater LE, et al. (2009) Nutrient imbalances in agricultural development. Science 324: 1519–1520.

4. Berzsenyi Z, Győrffy B, Lap D (2000) Effect of crop rotation and fertilization on maize and wheat yields and yield stability in a long-term experiment. Eur J Agron 13: 225–244.

5. Bi LD, Zhang B, Liu GR, Li ZZ, Liu YR, et al. (2009) Long-term effects of organic amendments on the rice yields for double rice cropping systems in subtropical China. Agr Ecosyst Environ 129: 534–541.

6. Cai ZC, Qin SW (2006) Dynamics of crop yields and soil organic carbon in a long-term fertilization experiment in the Huang-Huai-Hai Plain of China. Geoderma 136: 708–715.

7. Fan T, Stewart BA, Wang Y, Luo J, Zhou G.(2005) Long-term fertilization effects on grain yield, water-use efficiency and soil fertility in the dryland of Loess Plateau in China. Agr Ecosyst Environ 106: 313–329.

8. Glendining MJ, Powlson DS, Poulton PR, Bradbury NJ, Palazzo D, et al. (1996) The effects of long-term applications of inorganic nitrogen fertilizer on soil nitrogen in the Broadbalk wheat experiment. J Agr Sci 127: 347–363.

9. Gong W, Yan XY, Wang JY, Hu TX, Gong YB (2009) Long-term manuring and fertilization effects on soil organic carbon pools under a wheat-maize cropping system in North China Plain. Plant Soil 314: 67–76.

10. Jiang D, Hengsdijk H, Dai T, de Boer W, Jing Q, et al. (2006) Long-term effects of manure and inorganic fertilizers on yield and soil fertility for a winter wheat-maize system in Jiangsu, China. Pedosphere16: 25–32.

11. Lu R, Shi Z (1998) Effect of long-term fertilization on soil properties. In: Lu R, Xie J, Cai G, Zhu Q (2010) editors. Soil-plant nutrients principles and fertilizer. Beijing: Chemical Industry Press. 102–110.

12. Wang YC, Wang E, Wang DL, Huang SM, Ma YB, et al. (2010) Crop productivity and nutrient use efficiency as affected by long-term fertilization in North China Plain. Nutr Cycl Agroecosys 86: 105–119.

13. Yan XY, Gong W (2010) The role of chemical and organic fertilizers on yield, yield variability and carbon sequestration – results of a 19-year experiment. Plant Soil 331: 471–480.

14. Zhang H, Xu M, Zhang F (2009) Long-term effects of manure application on grain yield under different cropping systems and ecological conditions in China. J Agr Sci 147: 31–42.

15. Mussgnug F, Becker M, Son TT, Buresh RJ, Vlek PLG (2006) Yield gaps and nutrient balances in intensive, rice-based cropping systems on degraded soils in the Red River Delta of Vietnam. Field Crop Res 98: 127–140.

16. Mueller ND, Gerber JS, Johnston M, Ray DK, Ramankutty N, et al. (2012) Closing yield gaps through nutrient and water management. Nature 490: 254–257.

17. Page AL, Miller RH, Keeney DR (1982) Methods of soil analysis. Part 2. Madison, WI: American Society of Agronomy. 539–871.

18. Olsen SR, Cole CV, Watanabe FS, Dean LA (1954) Estimation of available phosphorus in soils by extraction with sodium bicarbonate. USDA Circular 939.

19. Liu X, Ju X, Zhang Y, He C, Kopsch J, et al. (2006) Nitrogen deposition in agroecosystems in the Beijing area. Agr Ecosyst Environ 113: 370–377.

20. Zhang Y, Liu XJ, Fangmeier A, Goulding KTW, Zhang FS (2008) Nitrogen inputs and isotopes in precipitation in the North China Plain. Atmos Environ 42: 1436–1448.

21. Tang X, Li J, Ma Y, Hao X, Li X (2008) Phosphorus efficiency in long-term (15-years) wheat-maize cropping systems with various soil and climate conditions. Field Crop Res 108: 231–237.

22. Blake L, Mercik S, Koerschens M, Moskal S, Poulton PR, et al. (2000) Phosphorus content in soil, uptake by plants and balance in three European long-term field experiments. Nutr Cycl Agroecosys 56: 263–275.

23. Zhong W, Cai Z (2007) Long-term effects of inorganic fertilizers on microbial biomass and community functional diversity in a paddy soil derived from quaternary red clay. Appl Soil Ecol 36: 84–91.

24. Bajwa MI (1994) Soil K status, K fertilizer usage and recommendation in Pakistan. Potash Review No. 3/1994. Subject 1, 20th Suite. Basel: International Potash Institute. 67.

25. Shen J, Li R, Zhang F, Fan J, Tang C, et al. (2004) Crop yields, soil fertility and phosphorus fractions in response to long-term fertilization under the rice monoculture system on a calcareous soil. Field Crop Res 86: 225–238.

26. Tan D, Jin J, Huang S, Li S, He P (2007) Effect of long-term application of K fertilizer and wheat straw to soil on crop yield and soil K under different planting systems. Agr Sci China 6: 200–207.

27. Manna MC, Swarup A, Wanjari RH, Ravankar HN, Mishra B, et al. (2005) Long-term effect of fertilizer and manure application on soil organic carbon storage, soil quality and yield sustainability under sub-humid and semi-arid tropical India. Field Crop Res 93: 264–280.

Author Contributions

Conceived and designed the experiments: ZO YL. Performed the experiments: XD YL. Analyzed the data: XD HW. Wrote the paper: XD.

24

Comparison of Soil Respiration in Typical Conventional and New Alternative Cereal Cropping Systems on the North China Plain

Bing Gao[1], Xiaotang Ju[1]*, Fang Su[1], Fengbin Gao[1], Qingsen Cao[1], Oene Oenema[2], Peter Christie[1], Xinping Chen[1], Fusuo Zhang[1]

1 College of Resources and Environmental Sciences, China Agricultural University, Beijing, China, 2 Wageningen University and Research Center, Alterra, Wageningen, The Netherlands

Abstract

We monitored soil respiration (Rs), soil temperature (T) and volumetric water content (VWC%) over four years in one typical conventional and four alternative cropping systems to understand Rs in different cropping systems with their respective management practices and environmental conditions. The control was conventional double-cropping system (winter wheat and summer maize in one year - Con.W/M). Four alternative cropping systems were designed with optimum water and N management, i.e. optimized winter wheat and summer maize (Opt.W/M), three harvests every two years (first year, winter wheat and summer maize or soybean; second year, fallow then spring maize - W/M-M and W/S-M), and single spring maize per year (M). Our results show that Rs responded mainly to the seasonal variation in T but was also greatly affected by straw return, root growth and soil moisture changes under different cropping systems. The mean seasonal CO_2 emissions in Con.W/M were 16.8 and 15.1 Mg CO_2 ha^{-1} for summer maize and winter wheat, respectively, without straw return. They increased significantly by 26 and 35% in Opt.W/M, respectively, with straw return. Under the new alternative cropping systems with straw return, W/M-M showed similar Rs to Opt.W/M, but total CO_2 emissions of W/S-M decreased sharply relative to Opt.W/M when soybean was planted to replace summer maize. Total CO_2 emissions expressed as the complete rotation cycles of W/S-M, Con.W/M and M treatments were not significantly different. Seasonal CO_2 emissions were significantly correlated with the sum of carbon inputs of straw return from the previous season and the aboveground biomass in the current season, which explained 60% of seasonal CO_2 emissions. T and VWC% explained up to 65% of Rs using the exponential-power and double exponential models, and the impacts of tillage and straw return must therefore be considered for accurate modeling of Rs in this geographical region.

Editor: Xiujun Wang, University of Maryland, United States of America

Funding: This work was funded by the National Natural Science Foundation of China (41230856, 31172033), the Special Fund for the Agricultural Profession (201103039), the '973' Project (2009CB118606) and the Innovation Group Grant of the National Natural Science Foundation of China (31121062). The funders had no role in study design, data collection and analysis, decision to publish, or preparation of the manuscript.] into [This work was funded by the '973' Project (2012CB417105, 2009CB118606), the National Natural Science Foundation of China (41230856, 31172033), the Special Fund for the Agricultural Profession (201103039), China Postdoctoral Science Foundation (2013M 530778), and the Innovation Group Grant of the National Natural Science Foundation of China (31121062). The funders had no role in study design, data collection and analysis, decision to publish, or preparation of the manuscript.

Competing Interests: The authors have declared that no competing interests exist.

* E-mail: juxt@cau.edu.cn

Introduction

Soils provide a very large sink of carbon (C) in terrestrial ecosystems with C reserves of about 1500 Pg C (1 Pg = 10^{15} g) and make a major contribution to the global carbon equilibrium [1]. Slight changes in soil C might therefore lead to significant changes in the concentration of CO_2 in the atmosphere. Soil respiration is the main terrestrial source of C return to the atmosphere with a flux reaching 98 ± 12 Pg C in 2008 and increasing at a rate of 0.1 Pg C y^{-1} from 1989 [2]. Agricultural soils play a very important role in the global C cycle [3,4] and account for 11% of global anthropogenic CO_2 emissions [5]. It is therefore important to minimize soil respiration and retain more C sequestered in agricultural soils.

Soil respiration comprises mainly autotrophic respiration by plant roots and heterotrophic respiration of plant residues, root

litter and exudates, and soil organic matter by soil microorganisms [6,7]. Its magnitude is affected mainly by soil and climatic conditions [8] such as soil temperature and moisture [1,9,10], vegetation characteristics and management practices [11–14]. Soil respiration therefore shows high spatial and temporal variation [1]. Understanding this variation in different cropping systems in specific region will make a large contribution to the efficient management of C flow in agricultural ecosystems.

Soil respiration in cropland is greatly affected by tillage practices and straw management, with the greatest increase occurring immediately after tillage operations, and cumulative soil CO_2 emissions can be lowered significantly by reducing the intensity of tillage [15,16]. Daily CO_2 fluxes can differ significantly at some sampling dates between conventional moldboard plow tillage and no tillage in continuous corn [17]. Soil CO_2 emission can be enhanced in the short term after crop residues are returned to the field [15,18] but this practice may build up the soil organic carbon (SOC) pools in the long term and may therefore be regarded as a more sustainable way of managing SOC compared to straw burning or other uses for straw [19]. Differences in fertilizer N rates had no significant effect on the CO_2 exchange rates in the same crop rotation and CO_2 fluxes did not differ with crop rotation under no till practices [16]. In addition, crop species and/or other management practices affect soil CO_2 emission as a result of their influence on soil biological and biochemical properties [14,20].

Soil temperature and moisture are two of the most important environmental factors controlling soil respiration [1,21,22]. Soil temperature is significantly positively correlated with soil respiration using linear [7], exponential [1,7], improved Arrhenius [8], power and quadratic [9] and Q_{10} [10] models in different regions. Soil moisture is also a key factor controlling soil respiration, especially in arid or semiarid regions where it can be more important than temperature and become the dominant factor [11]. This shows that when one factor linking soil temperature and moisture is in a higher or lower range, the other might become a major factor controlling soil respiration [13,23,24]. The respiration rate will be limited when soil volumetric water content (VWC%) drops below a threshold of 15% [20]. Soil CO_2 emission increased significantly with increasing temperature up to 40°C, with emissions reduced at the lowest and highest soil moisture contents [20]. Therefore, the single-factor models cannot describe soil CO_2 emission well because they neglect the impacts of interactions between factors. The multiple polynomial models considering both soil temperature and moisture result in a much better description of CO_2 ($r^2 = 0.70$–0.78, $P<0.0001$) emissions than using temperature ($r^2 = 0.27$–0.54, $P<0.01$) or moisture ($r^2 = 0.29$–0.45, $P<0.01$) alone [20].

China has broad climatic regimes and the different ecosystems depend on regional climatic conditions [25]. The North China Plain (NCP) is a major agricultural region. The soil type is Fluvo-aquic soil and the climate is sub-humid temperate monsoon with abundant solar radiation but with cold and dry conditions in winter and spring and warm and wet weather in summer. Evapotranspiration is intense and the spring drought is an important feature. Winter wheat-summer maize is the typical double cropping system and current farming practice involves application of 300 kg N ha^{-1} yr^{-1} for winter wheat and 250 kg N ha^{-1} yr^{-1} for summer maize with a ratio of basal to topdressing applications of 1:1 and 1:1.5, respectively [26,27]. The soil is rotary tilled to 20 cm depth after maize straw removal for sowing winter wheat, and maize is sown directly after removing the wheat straw. Generally, wheat is irrigated three to four times and maize once or twice depending on precipitation. The amount of irrigation water ranges from 60 to 100 mm on each occasion [26]. About 30–60% of N input could be saved without sacrificing yields while significantly reducing environmental risk by adopting optimum N management in the winter wheat-summer maize system as shown by our earlier study [28]. However, over-exploitation of groundwater has become the main factor restricting sustainable agricultural development [29]. There is therefore concern to explore new alternative cropping systems for sustainable use of groundwater and optimum N fertilization to reduce pollution. Winter wheat–summer maize–spring maize with three harvests over two years and a single spring maize system have shown great potential to reduce water use and N use and can achieve balanced use of groundwater [26], and this cropping system may serve as a new alternative system for efficient resource use and sustainable development. However, it is still unclear how these changes will affect soil respiration in the study region.

Low frequency of measurement, lack of data at some growth stages, and failure to consider the interactive effects of soil moisture and temperature on soil respiration may lead to failure to describe the characteristics of soil respiration in this region [18,30,31]. There are indications that the correlation between soil respiration and soil temperature to 5 cm depth is 0.51 but the study that produced this result involved measurement only 21 times over one year [30]. Meng et al. [31] found that soil respiration had a higher correlation with soil temperature to 5 cm depth using the exponential model through weekly measurements of soil respiration under the typical double-cropping system over a whole year. Soil temperature at 5 cm depth explained 63–74% of soil respiration using the exponential model except during the winter, and the application of crop residues had significant positive impacts on soil respiration [18]. The management of N and water, crop residues and tillage practices will change significantly after conversion to new alternative cropping systems [26], an effect closely related to soil respiration. However, no quantitative information is yet available regarding soil respiration in new alternative cropping systems in this region.

In the present study we have compared soil respiration characteristics in different cropping systems with their respective management practices and environmental variables and we explore the factors affecting these differences. We have also analyzed the effects of straw return on variation in seasonal CO_2 emissions on the North China Plain.

Materials and Methods

Site description

A long-term field experiment was set up in October 2007 at Quzhou experimental station (36.87°N, 115.02°E) of China Agricultural University in Hebei province. The site is a sub-humid temperate monsoon area at an altitude of 40 m. The annual mean temperature is 13.2°C. Annual mean precipitation was 494 mm from 1980 to 2010 with a range of 213–840 mm, and 68% of precipitation falls from June to September [26]. The typical double-cropping system is a winter wheat and summer maize rotation which accounts for >80% of agricultural fields in Quzhou county. The soil type is Fluvo-aquic soil and the bulk density of the top 30 cm of the soil profile is 1.37 g cm^{-3}, soil pH is 7.72 (1:2.5, soil:water), SOC content 7.31 g kg^{-1}, total N 0.7 g kg^{-1}, Olsen-P 4.8 mg kg^{-1} and available K 72.7 mg kg^{-1}. Fig. 1 shows the daily mean air temperatures and precipitation during the measurement period (also see Table S1).

Field experiment treatments and management

A completely randomized design was employed with five treatments and four replicates. Each plot is 1800 m^2 (30×60 m). The control is conventional winter wheat and summer maize based on local farming practice (Con.W/M). Four new alternative cropping systems were designed with high-yielding varieties (using optimum planting density and crop management) and optimum water and N fertilizer management compared with conventional practice. They are: optimized two harvests in one year (winter wheat and summer maize - Opt.W/M), three harvests within two years (first year, winter wheat and summer maize or winter wheat and summer soybean; second year fallow then spring maize - W/M-M and W/S-M) and single spring maize per year (M).

Nitrogen input and irrigation for Con.W/M were described in the Introduction above. The basal fertilizer for wheat was surface broadcast before rotary tillage to 20 cm depth after removal of maize straw from the soil and topdressing was broadcast at shooting for wheat followed by irrigation, with both fertilizer

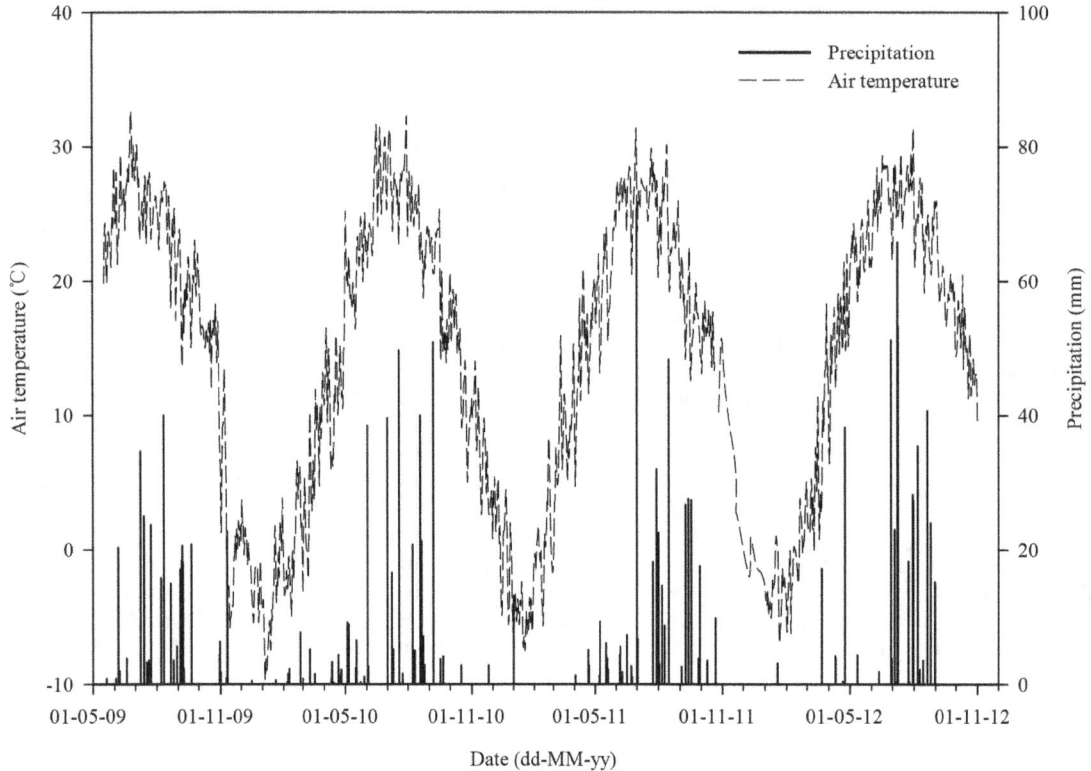

Figure 1. Daily mean air temperature (°C) and precipitation (mm) during the field experiment.

applications at 150 kg N ha^{-1} in the form of urea. The basal application for summer maize comprised 45 kg N ha^{-1} applied to the soil as 15-15-15 compound fertilizer with a seed drill after removing the wheat straw from the soil, and 55 kg N ha^{-1} surface broadcast as urea followed by irrigation and the topdressing 150 kg N ha^{-1} was applied at the ten-leaf stage of summer maize in the form of urea. In the other systems optimized N management was devised according to the N target values minus the soil nitrate-N content in the root zone before side-dressing as described by Cui et al. [27]. For summer maize 45 kg N ha^{-1} was applied as a basal dressing in the same way as for Con.W/M and 80 and 60 kg N ha^{-1} were side-dressed using a soil cover of 0–5 cm after band application at the six- and ten-leaf stages of summer maize, respectively. No other N fertilizer except 45 kg N ha^{-1} was applied as a basal application for soybean as for Con.W/M. Irrigation times and rates were determined by testing the soil water content before the critical growing seasons as described by Meng et al. [26]. The details of nitrogen input and irrigation rate over the whole study are shown in Table 1. Wheat straw was mulched after chopping into 5–10 cm pieces and summer maize or soybean was sown directly. Summer and spring maize and soybean residues were also chopped into 5–10 cm pieces and mechanically ploughed into the top 30 cm of the soil after maize and soybean were harvested and then winter wheat was sown if there was no fallow the following season. The soil was rotavated to 20 cm depth before sowing spring maize.

We measured soil respiration in each plot of the experiment from May 2009 to October 2012. Crops present in the different treatments during gas measurement are shown in Fig. 2.

Soil respiration measurement

Soil respiration was representatively determined in every plot using an automatic soil CO_2 flux system (LI-COR LI-8100, Lincoln, NE). Measurements were carried out daily for 10 days after fertilization events and 3–5 days after irrigation or precipitation events (>10 mm) depending on the size of gas fluxes; for the remaining periods emissions were measured twice per week and once a week when the soil was frozen. Two bases were used in each plot, one on a row and the other in the middle of the row during the maize and soybean seasons. Each base was a PVC tube with an inner diameter of 20 cm and a height of 13 cm inserted 9 cm into the soil for measurement and was removed only before sowing. Soil respiration was measured directly by LI-8100 in units of μmol CO_2 m^{-2} s^{-1} in the field between 08:30 and 11:00 am. Soil respiration is presented as the mean values of four replicated measurements on four different plots. The seasonal amounts of CO_2 emissions were sequentially linearly determined from the emissions between every two adjacent intervals of the measurements.

Auxiliary measurements

Soil temperature to 5 cm depth was measured directly by Li-8100 through a temperature sensing probe during the measurement time. Soil moisture at 0–5 cm is expressed as volumetric water content (VWC%) and was measured directly by Li-8100 through an ECH$_2$O type of EC-10 soil water sensing probe (Decagon Devices, Inc, Pullman, WA). We also measured the top 20 cm depth SOC content in each plot of this field experiment after summer maize harvest in 2011 using the method described by Huang et al. [19]. The daily mean air temperatures and precipitation data during the field experiment were obtained from

Table 1. Nitrogen fertilizer rates and irrigation rates throughout the study period.

Year	N application rate (kg N ha⁻¹)					Irrigation rate (mm)				
	Con.W/M[1]	Opt.W/M	W/M-M	W/S-M	M	Con.W/M	Opt.W/M	W/M-M	W/S-M	M
2009	W[2] 300	W 263	F -[3]	F -	F -	W 250	W 215	F -	F -	F -
	M₁ 250	M₁ 185	M₂ 135	M₂ 210	M₂ 95	M₁ 60	M₁ 60	M₂ 125	M₂ 135	M₂ 125
2010	W 300	W 100	W 140	W 140	F -	W 180	W 120	W 120	W 120	F -
	M₁ 250	M₁ 185	M₁ 185	S 45	M₂ 150	M₁ 60	M₁ 60	M₁ 60	S 60	M₂ 110
2011	W 300	W 139	F -	F -	F -	W 240	W 275	F -	F -	F -
	M₁ 250	M₁ 185	M₂ 162	M₂ 178	M₂ 150	M₁ 70	M₁ 70	M₂ 60	M₂ 60	M₂ 60
2012	W 300	W 140	W 162	W 158	F -	W 180	W 160	W 160	W 170	F -
	M₁ 250	M₁ 185	M₁ 185	S 45	M₂ 266	M₁ 90	M₁ 90	M₁ 90	S 90	M₂ 120
Total	2200	1382	969	776	661	1130	1050	615	635	415

[1]Con.W/M, Opt.W/M, W/M-M, W/S-M and M represent conventional and optimized winter wheat–summer maize, winter wheat–summer maize–spring maize, winter wheat–summer soybean–spring maize and spring maize treatment, respectively.
[2]W, M₁, M₂, S and F represent winter wheat, summer maize, spring maize, summer soybean and fallow.
[3]Denotes no data in the fallow season.

an automatic weather station located 50 m from our experimental site as shown in Fig. 1. Soil respiration and environmental variable data from the present study are presented in the Supplementary Data (Table S1).

Correlations between soil respiration and soil temperature and moisture

The compound factor models of soil respiration with soil temperature and moisture (equations 1–4) were employed as follows:

$$Rs = a + bTs + cWs \qquad (1)$$

$$Rs = aTs^b Ws^c \qquad (2)$$

$$Rs = ae^{bTs} Ws^c \qquad (3)$$

$$Rs = ae^{bTs + cWs} \qquad (4)$$

We established the four compound factor models above among soil respiration (Rs), soil temperature (Ts) and VWC(%) (Ws) using the measured fluxes from May 2009 to May 2012, and compared MAE (the mean absolute error), ME (model efficiency, the ratio of difference in measured and predicted flux in total variation in measured flux, expressed as significant correlation coefficient from −1 to 1), d (the percentage of mean square error and potential error, expressed as significant correlation from 0 to 1) [32,33], RMSE (root mean square error, reflecting the degree of dispersion of one variable), MSEₛ (systematic error) and MSEᵤ (random error) [34] among the models. We comprehensively evaluated the model performances by the sizes of MAE, ME, d, RMSE, MSEₛ and MSEᵤ, and the value of MSEₛ/(MSEᵤ+MSEᵤ). In general, MSEᵤ is close to RMSE in a well fitting model.

Statistical analysis

The primary data were processed using Microsoft Excel 2003 spreadsheets. Total CO_2 emissions in the different treatments were tested by analysis of variance and mean values were compared using SAS statistical software (Version 9.2; SAS Institute, Inc., Cary, NC) to calculate least significant difference (LSD) at the 5% level. Compound factor regression analysis among soil respiration, T and VWC% were performed using Sigmaplot 12.0 (Systat Software Inc., Erkrath, Germany).

Results

Characteristics of soil respiration in the different cropping systems

Over a complete rotation cycle soil respiration gradually increased from March, reached a maximum in July and gradually decreased from August to November, and then remained at the lowest values during winter, in a pattern similar to soil temperature (Figs. 2 and 3A). The mean soil respiration values were 3.35, 4.55, 4.03, 3.35 and 3.25 μmol CO_2 m⁻² s⁻¹ for Con.W/M, Opt.W/M, W/M-M, W/S-M and M throughout the study period, with ranges of 0.02–12.4, 0.26–14.9, 0.31–12.1, 0.34–11.3 and 0.30–11.2 μmol CO_2 m⁻² s⁻¹, respectively. Three peaks per year occurred in the typical double-cropping system, at the shooting stage of winter wheat, six-leaf of summer maize and the period after winter wheat sowing, the first two peaks caused by rapid crop growth and the last by the return of summer maize straw combined with soil tillage. Soil respiration of Opt.W/M was higher than of Con.W/M at the six-leaf stage of summer maize in the middle of July and the period after winter wheat sowing. The maximum peaks of soil respiration in Con.W/M were 8.2, 7.7, 7.8, 12.4 and 4.9, 2.8, 3.6, 2.9 μmol CO_2 m⁻² s⁻¹ during these two periods for four growing seasons, respectively, and they increased to 10.8, 9.6, 10.1, 14.9 and 7.6, 10.3, 6.5, 11.2 μmol CO_2 m⁻² s⁻¹ in Opt.W/M during the corresponding periods. Under the new alternative cropping systems one peak disappeared in the fallow season (season with no winter wheat planted). The highest value of soil respiration was around 7.0 μmol CO_2 m⁻² s⁻¹ in the spring maize season under the new alternative cropping systems, but it increased to more than 10.0 μmol CO_2 m⁻² s⁻¹ for summer maize in Opt.W/M at the corresponding time (Fig. 2).

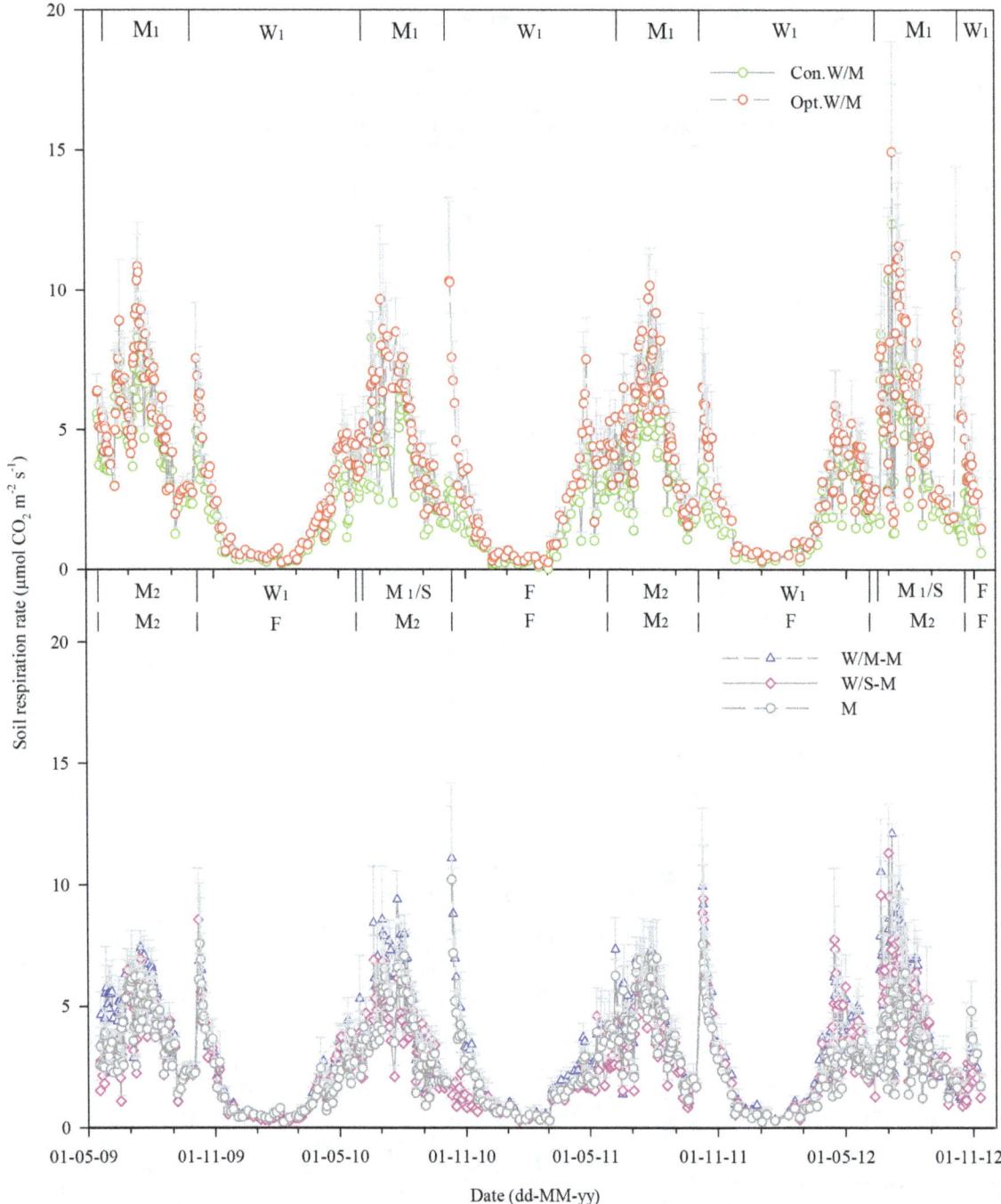

Figure 2. CO₂ emissions of different cropping systems. Con.W/M, Opt.W/M, W/M-M, W/S-M and M represent conventional winter wheat–summer maize in one year, optimized winter wheat–summer maize in one year, winter wheat–summer maize (or summer soybean) –spring maize three harvests in two years and single spring maize system in one year; W, M₁, M₂, S and F represent winter wheat, summer maize, spring maize, soybean and fallow.

Soil respiration was very low even after summer soybean stover return to the field in W/S-M in mid-November 2010 when the soil temperature in the top 5 cm ranged from −2.3 to +4.7°C within a month of soil tillage. A similar phenomenon occurred at the end of October 2012 due to the late spring maize and summer soybean harvests and the soil was tilled when soil temperature to 5 cm

depth was around 10°C, and the peaks of W/M-M, W/S-M and M were only one third of the values of those at the corresponding times in other years. In addition, soil respiration showed large between-year change, so that peaks of soil respiration occurred after irrigation at shooting of winter wheat in other years, but not in winter wheat in 2010.

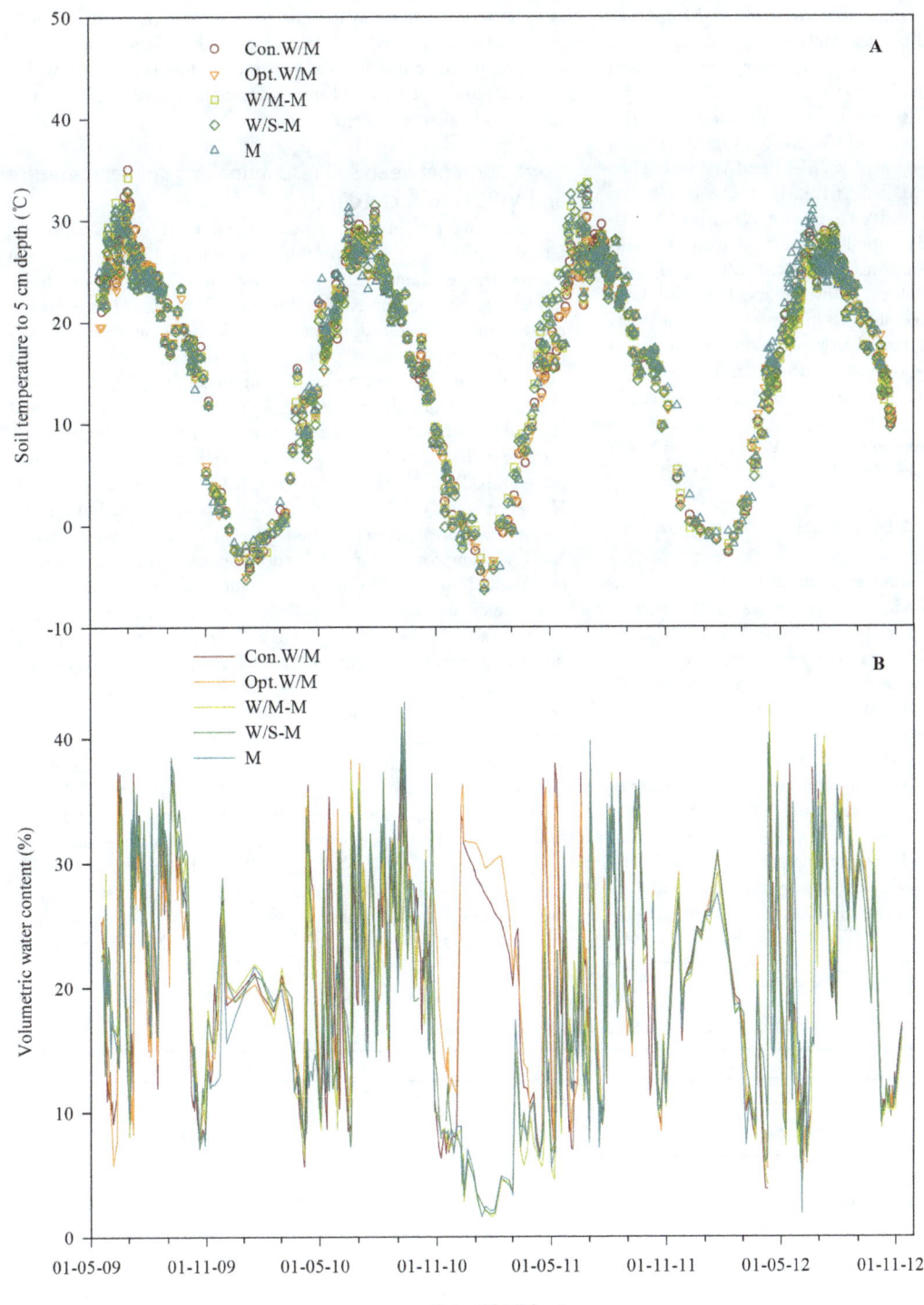

Figure 3. Dynamics of (A) soil temperature and (B) soil VWC% to 5 cm depth.

Total CO$_2$ emissions in each cropping season and each rotation cycle

Total CO$_2$ emissions in each cropping season and each rotation cycle were system dependent (Table 2). The mean seasonal total CO$_2$ emissions of Con.W/M were 16.8 and 15.1 Mg CO$_2$ ha^{-1} for summer maize and winter wheat, respectively. They increased significantly by 26 and 35% in Opt.W/M in the corresponding season. Under the new alternative cropping systems W/M-M showed similar results to Opt.W/M, and the seasonal total CO$_2$ emission of W/M-M was significantly higher than the corresponding season of Con.W/M except spring maize in 2009. However, W/S-M showed no significant difference from Con.W/M in each cropping season and the total CO$_2$ emissions in the fallow season and spring maize of W/S-M were clearly affected by summer soybean planting. Total CO_2 emission of M in each cropping season also showed no clear difference from Con.W/M except spring maize in 2011. In order to compare the impacts of cropping systems on CO_2 emissions of each rotation cycle we calculated the total CO_2 emissions during the period 2011–2012, which included two rotation cycles of Con.W/M, Opt.W/M, and M and a completely rotation cycle of W/M-M and W/S-M. Total CO_2 emission of Con.W/M was 61.9 Mg CO_2 ha^{-1}, and increased significantly by 37 and 29% in Opt.W/M and W/M-M treatment, respectively. The total CO_2 emission of W/M-M was not significantly different from Opt.W/M when there was only one season of winter wheat in two years, but total CO_2 emission of W/S-M decreased sharply in contrast to Opt.W/M when summer soybean was planted to replace summer maize of W/M-M because soil respiration was reduced significantly in the following fallow and spring maize seasons after the low biomass of soybean straw was returned to the soil. Total CO_2 emissions expressed as one complete rotation cycle of W/S-M, Con.W/M and M treatments were not significantly different (Table 2).

Soil respiration as affected by C input in each growth season

The measured soil respiration rates in this study consisted mainly of autotrophic respiration by crop roots in the current season, heterotrophic respiration of root litter and exudates in the current season, and heterotrophic respiration of crop straw return to the soil from the previous season and soil organic matter. As Fig. 2 and Table 2 show, the characteristics and total seasonal cumulative CO_2 emissions were greatly affected by straw return and crop growth status. To further explain soil respiration driven by C input in each growing season, we analyzed the correlation of seasonal cumulative CO_2 emissions with: (1) current-season aboveground biomass only; (2) the sum of C input of straw return from the previous season and the aboveground biomass in the current season. The relationship is improved significantly by inclusion of straw inputs (Fig. 4, equation A) compared to current-season aboveground biomass only (Fig. 4, equation B). Carbon input of straw return from the previous season and the

aboveground biomass in the current season explains up to 60% of seasonal cumulative CO_2 emissions, much higher than that of 27% with current-season aboveground biomass only, which demonstrates that straw C inputs from the previous season can significantly affect soil respiration.

Correlation between soil respiration and soil temperature and VWC% to 5 cm depth

Large changes in soil respiration followed the variation in temperature over a complete year (Figs. 2 and 3A). Soil temperature explained 45% of soil respiration using the quadratic model (Fig. 5, equation A). In addition, soil moisture exerted some impacts on soil respiration under our climatic conditions such as inhibition within a short period after irrigation at shooting and grain filling stages of winter wheat and then a sharp increase which was derived from the effects of drying and wetting cycles. Soil respiration showed significant correlations with soil VWC% using the linear ($R_s = 2.7535 + 0.0447V$, $R^2 = 0.04$, n = 2282) and power ($R_s = 1.7708V^{0.2488}$, $R^2 = 0.06$, n = 2282) models at $P < 0.001$. However, soil VWC% explained only 4–6% of soil respiration. We further examined the combined effects of soil temperature and VWC% using four compound models, namely the linear, power, exponential-power and double exponential models (Table 3). The results indicate that the R^2 values combining temperature and VWC% are significantly higher than using the quadratic model only considering soil temperature. Soil temperature and VWC% explained up to 65% of soil respiration using the exponential-power and double exponential models.

The exponential-power and double exponential models (Table 3) gave significant improvements compared to the linear and power models. We again compared MAE, ME and d, RMSE, MSE_s and MSE_u among the four models and comprehensively evaluated the model performances by the sizes of these indicators and the value of $MSE_s/(MSE_u + MSE_u)$. The exponential-power model was much better for description of soil respiration in response to soil temperature and VWC% (both in the top 5 cm) in our study because it had lower MAE, RMSE, MSE_u, and higher

Table 2. Total CO_2 emissions in each cropping season and each rotation cycle (Mg CO_2 ha^{-1}).

Year	Con.W/M		Opt.W/M	W/M-M		W/S-M		M	
	Crop	CO_2	CO_2	Crop	CO_2	Crop	CO_2	Crop	CO_2
2009	M_1	19.1±0.8bc[1]	22.0±1.5a	M_2	21.4±1.9ab	M_2	17.5±1.3c	M_2	17.4±1.4c
2010	W	15.5±1.2bc	17.2±0.6ab	W	17.8±1.2a	W	17.0±0.9abc	F	14.9±0.5c
	M_1	16.0±0.9b	21.4±1.7a	M_1	20.6±1.8a	S	15.7±0.9b	M_2	18.6±2.5ab
2011	W	13.4±0.7bc	22.4±0.4a	F	19.6±0.4a	F	11.1±1.6c	F	15.9±1.6b
	M_1	15.5±1.4b	20.4±1.9a	M_2	19.6±2.1a	M_2	16.0±1.1b	M_2	18.7±1.2a
2012	W	16.5±0.9bc	21.7±3.5a	W	21.3±2.4a	W	19.3±1.6ab	F	15.0±1.8c
	M_1	16.4±0.6c	20.4±1.9a	M_1	19.4±1.4ab	S	17.0±2.0bc	M_2	18.0±0.7abc
2009-2012 Mean	M_1	16.8	21.1	M_1	20.0	S	16.4	-	
	W	15.1	20.4	W	19.6	W	18.2	-	
	-	-	-	M_2	20.5	M_2	16.8	M_2	18.2
	-	-	-	F	19.6	F	11.1	F	15.3
2011-2012	2 W-M_1[2]	61.9±1.3b	84.9±8.2a	F-M_2-W-M_1	79.8±5.6a	F-M_2-W-S	64.1±2.7b	2 F-M_2	67.1±2.0b

[1]The same letter in the same line denotes no significant difference in different cropping systems by LSD at $P < 0.05$.
[2]2 W-M_1, F-M_2-W-M_1 (or S) and 2 F-M_2 represent two winter wheat-summer maize rotation cycles, fallow-spring maize-winter wheat-summer maize (or summer soybean) rotation cycle and two fallow-spring maize rotation cycles.

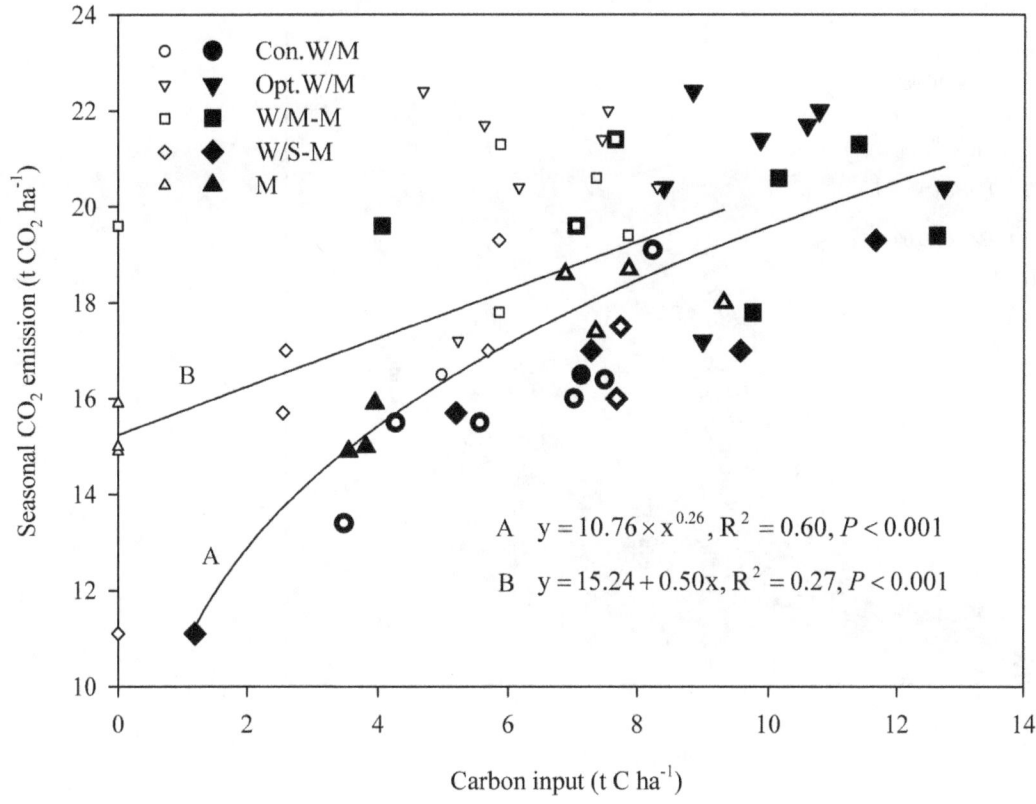

Figure 4. Correlation between seasonal CO$_2$ emission and carbon input. Carbon input was calculated from current-season aboveground biomass only (A); and calculated from straw return of the previous season and the aboveground biomass in the current season (B); the abbreviations of the treatment are shown in the footnotes in Fig. 2.

ME than the double exponential model and the values of MSE$_s$/ (MSE$_s$+MSE$_u$) were similar using both models.

Discussion

Soil respiration in croplands is affected mainly by soil properties, cropping system (which is related to crop species), tillage and straw management, water and nutrient management, and environmental variables (soil temperature, moisture etc.) [1,6,20,35]. There is temporal variation within the same cropping system and spatial variation among different cropping systems [16,17,31]. Changes in soil respiration in our sub-humid temperate monsoon region are largely affected by the seasonal variation in temperature, which is in line with most previous reports [30,31,36]. However, soil respiration responded little to soil temperature as shown in Fig. 5, equation A using the quadratic model because some data points did not fit the model with the impacts of soil tillage before the wheat crop was sown. The R^2 value improved by 18%, and up to 53% when the data within one month after tilling were excluded (Fig. 5, equation B). Moreover, we found that soil respiration tended to follow the variation in temperature from August to the following March when the data after tillage were excluded (Figs. 2 and 3A). Soil temperature explained 74% of soil respiration when only the data from August to March were included (Fig. 6, equation A). Therefore, the impacts of tillage must be considered for modeling soil respiration on the NCP.

The short decline in soil respiration after irrigation might be attributable to blocked diffusion of CO$_2$ with high moisture and limited oxygen concentrations in the soil matrix [37], and the

flushes afterwards may be due to the stimulation of decomposition of plant residues [21], root litter and exudates or autotrophic respiration of rapid root growth, which taken together induced the effects of drying and wetting cycles. Soil respiration would be limited when soil moisture was too high or too low and the maximum range is usually close to field water holding capacity [38]. The disappearance of respiration flushes was due to the low soil temperatures within a week after irrigation at the shooting stage of wheat in 2010 relative to other years (6–10°C in 2010 vs 12–21°C in 2011 and 12–19°C in 2012) (Fig. 3A). Soil moisture was not the key driving factor over the whole study period but did affect soil respiration slightly at particular stages and therefore only explained a very small proportion of the variation in soil respiration in our study area.

Numerous studies have reported that soil respiration is significantly affected by tillage practices combined with straw management [14–16]. Total soil respiration was significantly higher in Opt.W/M than Con.W/M as the latter soil was rotary tilled to 20 cm depth after maize straw removal and Opt.W/M was ploughed into the top 30 cm of the soil after maize straw return to the soil, soil respiration increased sharply after soil disturbance by tillage operations possibly because increased soil aeration accelerated the decomposition rate of crop residues which was associated with higher microbial activity [14,15,39]. However, the impacts of maize straw return and tillage were lowered by delaying tillage until the soil temperature to a depth of 5 cm reached 10°C or lower.

Although seasonal cumulative CO$_2$ emission in Opt.W/M and W/M-M increased significantly relative to Con.W/M as a result of

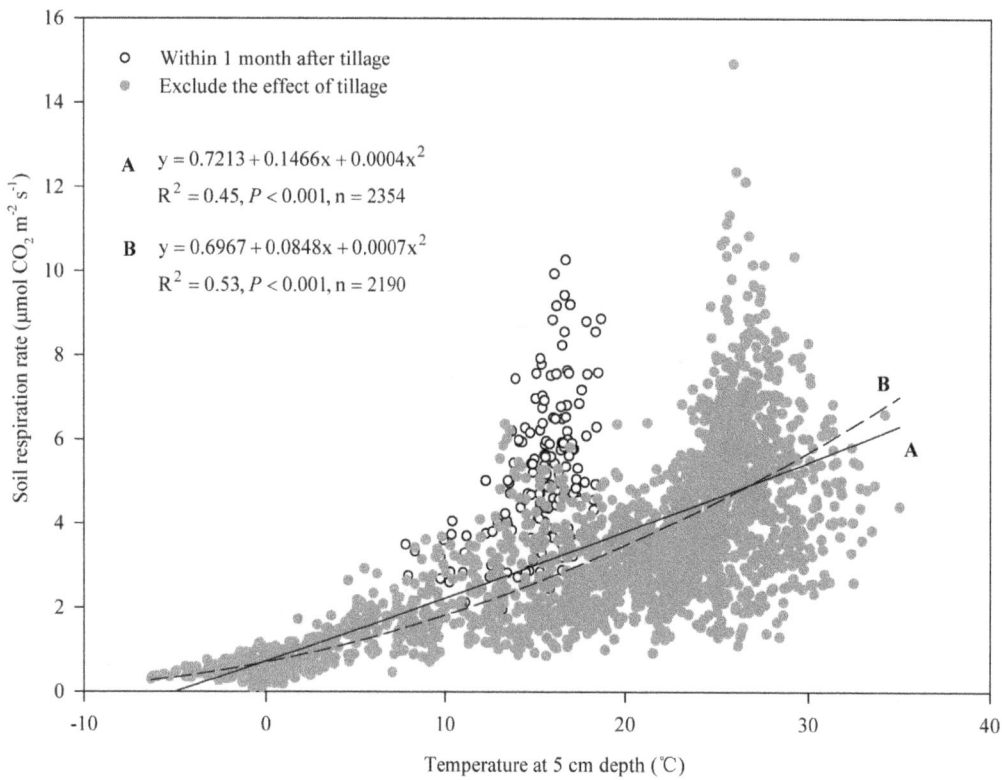

Figure 5. Impacts of soil tillage combined with straw return on soil respiration. Correlation between soil respiration and soil temperature at 5 cm depth over the whole year (equation A) and the correlation between soil respiration and soil temperature at 5 cm depth excluding the data within one month of tillage (equation B).

straw return [14,15], this practice also increases the SOC content over the long term [16,17,40]. The SOC content in the top 20 cm of the soil profile in straw return treatments increased by 3.9–16.5% relative to the straw removal treatments in winter wheat–summer maize double-cropping systems on the NCP, with a mean increase in rate of 0.04 to 1.44 t C ha^{-1} y^{-1} over a six-year period as shown by Huang et al. [19]. We also measured SOC to a depth of 20 cm in the present field experiment after summer maize harvest in 2011 and all values increased to 8.07, 8.71, 7.93 and 7.52 g kg^{-1} in Con.W/M, Opt.W/M, W/M-M and W/S-M, respectively, with the sole exception of a slight decrease to 7.18 g kg^{-1} in M (from 7.31 g kg^{-1} at the start of the field experiment in 2007). Although there was no crop straw return, Con.W/M also showed a clear increment relative to the initial value in line with Huang et al. [19], and this may have been due to the large amounts of crop roots and rhizo-deposited carbon. Con.W/M

showed a greater increase in SOC than W/M-M, W/S-M and M, possibly due to the lower intensity of tillage in Con.W/M than in W/M-M, W/S-M and M. Our results show that soil respiration responded mainly to the seasonal variation in soil temperature but was also greatly affected by straw return, root growth and soil moisture changes under the different cropping systems.

Supporting Information

Table S1 Measured soil respiration fluxes, soil temperature and soil volumetric water content to 5 cm depth, daily mean air temperature and precipitation in the field experiments from 18th May 2009 to 11th November 2012.

Table 3. Correlation between soil respiration and soil temperature and VWC(%) to 5 cm depth.

Model	Fitting equation	n	R^2	MAE	ME	RMSE	MSE$_s$	MSE$_u$	d
Linear	$R_s{}^1 = 0.7712 + 0.1581T - 0.0030V$	1905	0.47*^{2}	1.11	0.48	1.48	1.28	1.03	0.80
Power (T>0)	$R_s = 0.6291T^{0.5929}V^{-0.0096}$	1811	0.56*	1.17	0.36	1.57	2.13	0.59	0.70
Exponential-power	$R_s = 0.9347e^{0.069T}V^{-0.0464}$	1905	0.65*	1.14	0.38	1.61	1.41	1.67	0.80
Double exponential	$R_s = 0.8924e^{0.0693T-0.0045V}$	1905	0.65*	1.20	0.31	1.69	1.36	1.81	0.78

^{1}R$_s$, T and V represent soil respiration, soil temperature and VWC% to 5 cm depth, respectively.
^{2}* represents highly significant correlation at P<0.001.

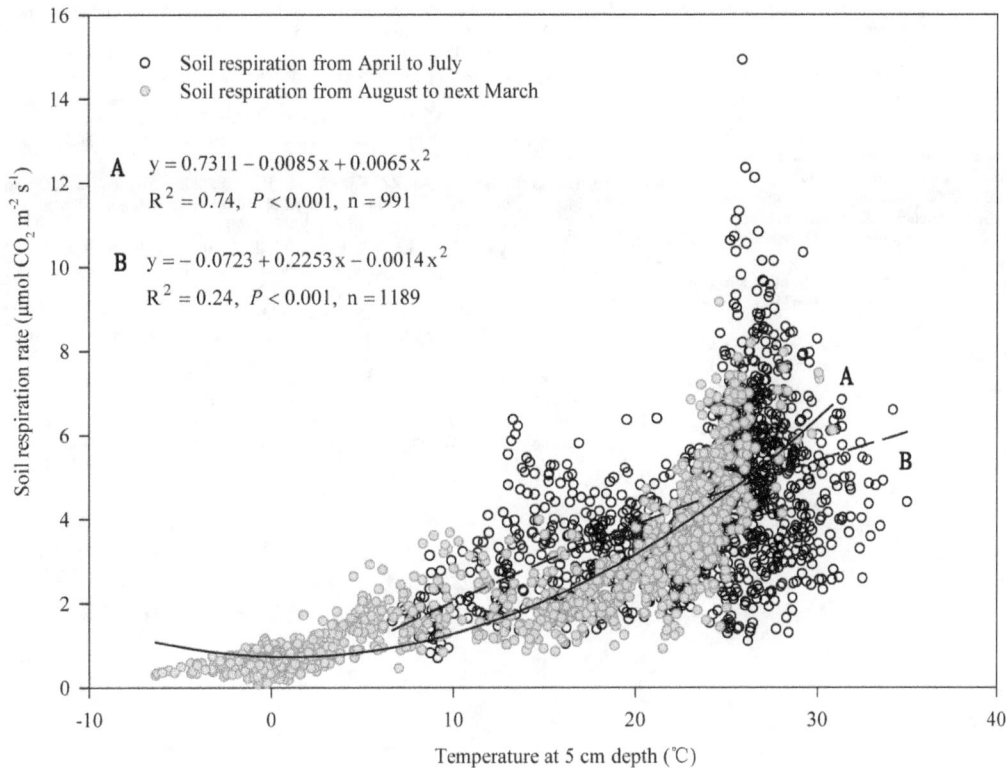

Figure 6. Correlation between soil respiration and soil temperature at 5 cm depth. Equations A and B represent the correlations between soil respiration and soil temperature at 5 cm depth from August to March after removing the impacts of tillage and from April to July, respectively.

Author Contributions

Conceived and designed the experiments: XJ XC FZ. Performed the experiments: BG FS FG QC. Analyzed the data: BG XJ. Contributed reagents/materials/analysis tools: XJ. Wrote the paper: BG XJ. Given some suggestion and modified the language for the manuscript OO PC.

References

1. Davidson EA, Belk E, Boone RD (1998) Soil water content and temperature as independent or confounded factors controlling soil respiration in temperate mixed hardwood forest. Global Change Biol 4: 217–227.
2. Bond-Lamberty B, Thomson A (2010) Temperature-associated increases in the global soil respiration record. Nature 464: 579–582.
3. Robertson GP, Paul EA, Harwood RR (2000) Greenhouse gases in intensive agriculture: contributions of individual gases to the radiative forcing of the atmosphere. Science 289: 1922–1926.
4. Mahecha MD, Reichstein M, Carvalhais N, Lasslop G, Lange H, et al. (2010) Global convergence in the temperature sensitivity of respiration at ecosystem level. Science 329: 838–840.
5. Grace J, Rayment M (2000) Respiration in the balance. Nature 404: 819–820.
6. Raich JW, Schlesinger WH (1992) The global carbon dioxide flux in soil respiration and its relationship to vegetation and climate. Tellus B 44: 81–99.
7. Melling L, Hatano R, Goh KJ (2005) Soil CO₂ flux from three ecosystems in tropical peatland of Sarawak, Malaysia. Tellus B 57: 1–11.
8. Lloyd J, Taylor JA (1994) On the temperature dependence of soil respiration. Funct Ecol 8: 315–323.
9. Fang C, Moncrieff JB (2001) The dependence of soil efflux on temperature. Soil Biol Biochem 33: 155–165.
10. Li HJ (2008) Studies on soil respiration and its relations to environmental factors in different ecosystems. Shanxi: PhD thesis, Shanxi University (in Chinese).
11. Wang WJ, Dalal RC, Moody PW, Smith CJ (2003) Relationships of soil respiration to microbial biomass, substrate availability and clay content. Soil Biol Biochem 35: 273–284.
12. Carlisle EA, Steenwerth KL, Smart DR (2006) Effects of land use on soil respiration: Conversion of Oak woodlands to vineyards. J Environ Qual 35: 1396–1404.
13. Rey A, Pegoraro E, Tedeschi V, De Parri I, Jarvis PG, et al. (2002) Annual variation in soil respiration and its components in a coppice oak forest in central Italy. Global Change Biol 8: 851–866.
14. Bavin TK, Griffis TJ, Baker JM, Venterea RT (2009) Impact of reduced tillage and cover cropping on the greenhouse gas budget of a maize/soybean rotation ecosystem. Agric Ecosyst Environ 134: 234–242.
15. Al-Kaisi MM, Yin XH (2005) Tillage and crop residue effects on soil carbon and carbon dioxide emission in corn-soybean rotations. J Environ Qual 34: 437–445.
16. Mosier AR, Halvorson AD, Reule CA, Liu XJ (2006) Net global warming potential and greenhouse gas intensity in irrigated cropping systems in Northeastern Colorado. J Environ Qual 35: 1584–1598.
17. Alluvione F, Halvorson AD, Del Grosso SJ (2009) Nitrogen, tillage, and crop rotation effects on carbon dioxide and methane fluxes from irrigated cropping systems. J Environ Qual 38: 2023–2033.
18. Zhang QZ, Wu WL, Wang MX, Zhou ZR, Chen SF (2005) The effects of crop residue amendment and N rate on soil respiration. Acta Ecologica Sinica 25: 2883–2887 (in Chinese with English abstract).
19. Huang T, Gao B, Christie P, Ju XT (2013) Net global warming potential and greenhouse gas intensity in a double cropping cereal rotation as affected by nitrogen and straw management. Biogeosciences Discuss 10: 13191–13229.
20. Iqbal J, Hu RG, Lin S, Ahamadou B, Feng ML (2009) Carbon dioxide emissions from Ultisol under different land uses in mid-subtropical China. Geoderma 152: 63–73.
21. Xu LK, Baldocchi DD, Tang JW (2004) How soil moisture, rain pulses, and growth alter the response of ecosystem respiration to temperature. Global Biogeochem Cycl 18: GB4002.
22. Conant RT, Dalla-Bett P, Klopatek CC, Klopatek JM (2004) Controls on soil respiration in semiarid soils. Soil Biol Biochem 36: 945–951.
23. Howard DM, Howard PJA (1993) Relationships between CO₂ evolution, moisture content, and temperature for a range of soil types. Soil Biol Biochem 25: 1537–1546.
24. Lellei-Kovács E, Kovács-Láng E, Botta-Dukát Z, Kalapos T, Emmett B, et al. (2011) Thresholds and interactive effects of soil moisture on the temperature response of soil respiration. Eur J Soil Biol 47: 247–255.
25. Fang JY, Chen AP, Peng CH, Zhao SQ, Ci LJ (2001) Changes in forest biomass carbon storage in China between 1949 and 1998. Science 292: 2320–2322.

26. Meng QF, Sun QP, Chen XP, Cui ZL, Yue SC, et al. (2012) Alternative cropping systems for sustainable water and nitrogen use in the North China Plain. Agric Ecosyst Environ 146: 93–102.

27. Cui ZL, Chen XP, Zhang FS (2010) Current Nitrogen management status and measures to improve the intensive wheat-maize system in China. Ambio 39: 376–384.

28. Ju XT, Xing GX, Chen XP, Zhang SL, Zhang LJ, et al. (2009) Reducing environmental risk by improving N management in intensive Chinese agricultural systems. Proc Natl Acad Sci USA 106: 3041–3046.

29. Hu C, Delgado JA, Zhang X, Ma L (2005) Assessment of groundwater use by wheat (*Triticum aestivum* L.) in the Luancheng Xian Region and potential implications for water conservation in the Northwestern North China Plain. J Soil Water Conserv 60: 80–88.

30. Niu LA, Hao JM, Zhang BZ, Niu XS, Lu ZY (2009) Soil respiration and carbon balance in farmland ecosystems on North China Plains. Ecol Environ Sci 18: 1054–1060 (in Chinese with English abstract).

31. Meng FQ, Guan GH, Zhang QZ, Si YJ, Qu B, et al. (2006) Seasonal variation in soil respiration under different long-term cultivation practices on high yield farm land in the North China Plain. Acta Scientiae Circumstantiae 26: 992–999 (in Chinese with English abstract).

32. Guo RY, Nendel C, Rahn C, Jiang CG, Chen Q (2010) Tracking nitrogen losses in a greenhouse crop rotation experiment in North China using the EU-Rotate_N simulation model. Environ Pollut 158: 2218–2229.

33. Willmott CJ (1982) Some comments on the evaluation of model performance. Am Meteorological Soc 63: 1309–1313.

34. Xiao M (2006) Effect of long-term fertilization on soil carbon and nitrogen storages and dynamics of nitrogen mineralization. Beijing: China Agricultural University (in Chinese).

35. Lohila A, Aurela M, Regina K, Laurila T (2003) Soil and total ecosystem respiration in agricultural fields: effect of soil and crop type. Plant Soil 251: 303–317.

36. Han GX, Zhou GS, Xu ZZ, Yang Y, Liu JL, et al. (2007) Soil temperature and biotic factors drive the seasonal variation of soil respiration in a maize (*Zea mays* L.) agricultural ecosystem. Plant Soil 291: 15–26.

37. Gaumont-Guay D, Black TA, Griffis TJ, Barr AG, Jassal RS, et al. (2006) Interpreting the dependence of soil respiration on soil temperature and water content in a boreal aspen stand. Agric For Mete 140: 220–235.

38. Davidson EA, Verchot LV, Cattanio JH, Ackerman IL, Carvalho JEM (2000) Effects of soil water on soil respiration in forests and cattle pastures of eastern Amazonia. Biogeochemistry 48: 53–69.

39. Jackson LE, Calderon FJ, Steenwerth KL, Scow KM, Rolston DE (2003) Responses of soil microbial processes and community structure to tillage events and implications for soil quality. Geoderma 114: 305–317.

40. Huang Y, Sun WJ (2006) Changes in topsoil organic carbon of croplands in mainland China over the last two decades, Chinese Sci Bull 51: 1785–1803.

Permissions

All chapters in this book were first published in PLOS ONE, by The Public Library of Science; hereby published with permission under the Creative Commons Attribution License or equivalent. Every chapter published in this book has been scrutinized by our experts. Their significance has been extensively debated. The topics covered herein carry significant findings which will fuel the growth of the discipline. They may even be implemented as practical applications or may be referred to as a beginning point for another development.

The contributors of this book come from diverse backgrounds, making this book a truly international effort. This book will bring forth new frontiers with its revolutionizing research information and detailed analysis of the nascent developments around the world.

We would like to thank all the contributing authors for lending their expertise to make the book truly unique. They have played a crucial role in the development of this book. Without their invaluable contributions this book wouldn't have been possible. They have made vital efforts to compile up to date information on the varied aspects of this subject to make this book a valuable addition to the collection of many professionals and students.

This book was conceptualized with the vision of imparting up-to-date information and advanced data in this field. To ensure the same, a matchless editorial board was set up. Every individual on the board went through rigorous rounds of assessment to prove their worth. After which they invested a large part of their time researching and compiling the most relevant data for our readers.

The editorial board has been involved in producing this book since its inception. They have spent rigorous hours researching and exploring the diverse topics which have resulted in the successful publishing of this book. They have passed on their knowledge of decades through this book. To expedite this challenging task, the publisher supported the team at every step. A small team of assistant editors was also appointed to further simplify the editing procedure and attain best results for the readers.

Apart from the editorial board, the designing team has also invested a significant amount of their time in understanding the subject and creating the most relevant covers. They scrutinized every image to scout for the most suitable representation of the subject and create an appropriate cover for the book.

The publishing team has been an ardent support to the editorial, designing and production team. Their endless efforts to recruit the best for this project, has resulted in the accomplishment of this book. They are a veteran in the field of academics and their pool of knowledge is as vast as their experience in printing. Their expertise and guidance has proved useful at every step. Their uncompromising quality standards have made this book an exceptional effort. Their encouragement from time to time has been an inspiration for everyone.

The publisher and the editorial board hope that this book will prove to be a valuable piece of knowledge for researchers, students, practitioners and scholars across the globe.

List of Contributors

Jayne Green, Dong Wang, Catherine J. Lilley, Peter E. Urwin and Howard J. Atkinson
Centre for Plant Sciences, University of Leeds, Leeds, United Kingdom

Xiu-liang Jin
Institute of Crop Science, Chinese Academy of Agricultural Sciences/Key Laboratory of Crop Physiology and Production Ministry of Agriculture, Beijing, China
Beijing Research Center for Information Technology in Agriculture, Beijing, China

Wan-ying Diao and Chun-hua Xiao
Key Laboratory of Oasis Ecology Agriculture of Xinjiang Construction Crops, Shihezi, China

Fang-yong Wang and Bing Chen
Institute of Cotton, Xinjiang Academy of Agricultural Reclamation Sciences, Shihezi, China

Ke-ru Wang and Shaokun Li
Institute of Crop Science, Chinese Academy of Agricultural Sciences/Key Laboratory of Crop Physiology and Production Ministry of Agriculture, Beijing, China
Key Laboratory of Oasis Ecology Agriculture of Xinjiang Construction Crops, Shihezi, China

Eiko E. Kuramae
Department of Microbial Ecology, Netherlands Institute of Ecology (NIOO-KNAW), Wageningen, The Netherlands
Department of Ecological Science, VU University Amsterdam, Amsterdam, The Netherlands

Erik Verbruggen
Department of Ecological Science, VU University Amsterdam, Amsterdam, The Netherlands

Remy Hillekens and Mattias de Hollander
Department of Microbial Ecology, Netherlands Institute of Ecology (NIOO-KNAW), Wageningen, The Netherlands

Wilfred F. M. Röling
Department of Molecular Cell Physiology, VU University Amsterdam, Amsterdam, The Netherlands

Marcel G. A. van der Heijden
Research Station ART, Agroscope Reckenholz Tänikon, Zurich, Switzerland
Department of Biology, Utrecht University, Utrecht, The Netherlands

George A. Kowalchuk
Department of Microbial Ecology, Netherlands Institute of Ecology (NIOO-KNAW), Wageningen, The Netherlands
Department of Biology, Utrecht University, Utrecht, The Netherlands

Zhenyao Shen, Lei Chen, Qian Hong, Hui Xie, Jiali Qiu and Ruimin Liu
State Key Laboratory of Water Environment Simulation, School of Environment, Beijing Normal University, Beijing, P.R. China

Benjamin D. Duval
Energy Biosciences Institute, University of Illinois at Urbana-Champaign, Urbana, Illinois, United States of America
Global Change Solutions, Urbana, Illinois, United States of America

Kristina J. Anderson-Teixeira and Sarah C. Davis
Energy Biosciences Institute, University of Illinois at Urbana-Champaign, Urbana, Illinois, United States of America

Cindy Keogh
Natural Resource Ecology Laboratory, Fort Collins, Colorado, United States of America

Stephen P. Long and Evan H. DeLucia
Energy Biosciences Institute, University of Illinois at Urbana-Champaign, Urbana, Illinois, United States of America
Global Change Solutions, Urbana, Illinois, United States of America
Department of Plant Biology, University of Illinois at Urbana-Champaign, Urbana, Illinois, United States of America

William J. Parton
Natural Resource Ecology Laboratory, Fort Collins, Colorado, United States of America

Eva L. M. Figuerola, Leandro D. Guerrero, Silvina M. Rosa, Leandro and Simonetti
Instituto de Investigaciones en Ingeniería Genética y Biología Molecular (INGEBI-CONICET) Vuelta de Obligado 2490, Buenos Aires, Argentina

Matías E. Duval and Juan A. Galantini
CERZOS-CONICET Departamento de Agronomía, Universidad Nacional del Sur, Bahía Blanca, Argentina

José C. Bedano
Departamento de Geología, Universidad Nacional de Río Cuarto, Río Cuarto, Córdoba, Argentina

Luis G. Wall
Departamento de Ciencia y Tecnología, Universidad Nacional de Quilmes, Roque Sáenz Peña 352, Bernal, Argentina

Leonardo Erijman
Instituto de Investigaciones en Ingeniería Genética y Biología Molecular (INGEBI-CONICET) Vuelta de Obligado 2490, Buenos Aires, Argentina
Facultad de Ciencias Exactas y Naturales, Universidad de Buenos Aires, Ciudad Universitaria, Pabellón 2, Buenos Aires, Argentina

Amita Mohan, William F. Schillinger and Kulvinder S. Gill
Department of Crop and Soil Sciences, Washington State University, Pullman, Washington, United States of America

Lei Deng and Zhou-Ping Shangguan
State Key Laboratory of Soil Erosion and Dryland Farming on the Loess Plateau, Northwest A&F University, Yangling, Shaanxi, China

Sandra Sweeney
Institute of Environmental Sciences, University of the Bosphorus, Istanbul, Turkey

Zhaoliang Song
Zhejiang Provincial Key Laboratory of C Cycling in Forest Ecosystems and C Sequestration, Zhejiang Agricultural and Forestry University, Lin'an, Zhejiang, China, School of Environment and Resources, Zhejiang Agricultural and Forestry University, Lin'an, Zhejiang, China
State Key Laboratory of Environmental Geochemistry, Institute of Geochemistry, Chinese Academy of Sciences, Guiyang, Guizhou, China

Jeffrey F. Parr
Southern Cross GeoScience, Southern Cross University, Lismore, New South Wales, Australia

Fengshan Guo
School of Environment and Resources, Zhejiang Agricultural and Forestry University, Lin'an, Zhejiang, China

Bin Zhao, Shuting Dong, Jiwang Zhang and Peng Liu
State Key Laboratory of Crop Biology, College of Agronomy, Shandong Agricultural University, Tai'an, China

Raymon S. Shange and Ramble O. Ankumah
Department of Agricultural and Environmental Science, Tuskegee University, Tuskegee, Alabama, United States of America

Abasiofiok M. Ibekwe
United States Department of Agriculture-Agricultural Research Service-United States Salinity Lab, Riverside, California, United States of America

Robert Zabawa
George Washington Carver Agricultural Experiment Station, Tuskegee University Tuskegee, Alabama, United States of America

Scot E. Dowd
Molecular Research LP, Shallowater, Texas, United States of America

Ling Zhang and Evan Siemann
College of Resources & Environmental Sciences, Nanjing Agricultural University, Nanjing, China
Department of Ecology & Evolutionary Biology, Rice University, Houston, Texas, United States of America

Yaojun Zhang, Hong Wang and Jianwen Zou
College of Resources & Environmental Sciences, Nanjing Agricultural University, Nanjing, China

Lubo Gao, Huasen Xu, Biao Bao, Xiaoyan Wang, Chao Bi and Yifang Chang
College of Water and Soil Conservation, Beijing Forestry University, Beijing, P.R. China

Huaxing Bi
College of Water and Soil Conservation, Beijing Forestry University, Beijing, P.R. China
Key Laboratory of Soil and Water Conservation, Ministry of Education, Beijing, P.R. China

Weimin Xi
Department of Biological and Health Sciences, Texas A&M University-Kingsville, Kingsville, Texas, United States of America

Michele C. Pereira e Silva, Jan Dirk van Elsas and Joana Falcão Salles
Department of Microbial Ecology, Centre for Life Sciences, University of Groningen, Groningen, The Netherlands

Armando Cavalcante Franco Dias
Department of Soil Science, "Luiz de Queiroz" College of Agriculture, University of São Paulo, Piracicaba, São Paulo, Brazil

Sikander Khan Tanveer, Xiaoxia Wen, Xing Li Lu, Junli Zhang and Yuncheng Liao
College of Agronomy, Northwest A&F University Yangling, Shaanxi, P.R. China

Hisashi Morise, Erika Miyazaki, Shoko Yoshimitsu and Toshihiko Eki
Molecular Genetics Laboratory, Division of Bioscience and Biotechnology, Department of Environmental and Life Sciences, Toyohashi University of Technology, Tempakucho, Toyohashi, Aichi, Japan

Xinjian Zhang, Yujie Huang, Hongmei Li, Yan Ren, Jishun Li, Jianing Wang and Hetong Yang
Shandong Provincial Key Laboratory of Applied Microbiology, Biotechnology Center of Shandong Academy of Sciences, Jinan, Shandong Province, People's Republic of China

Paul R. Harvey
Shandong Provincial Key Laboratory of Applied Microbiology, Biotechnology Center of Shandong Academy of Sciences, Jinan, Shandong Province, People's Republic of China
CSIRO Sustainable Agriculture National Research Flagship and CSIRO Ecosystem Sciences, Glen Osmond, South Australia, Australia

Taoxiang Zhang, Haizhen Wang, Laosheng Wu, Jun Lou, Jianjun Wu, Philip C. Brookes and Jianming Xu
College of Environmental and Natural Resource Sciences, Zhejiang Provincial Key Laboratory of Subtropical Soil and Plant Nutrition, Zhejiang University, Hangzhou, China

Marie-Hélène Robin and Célia Cholez
Institut National de la Recherche Agronomique, Unité Mixte de Recherche 1248 Agrosystèmes et agricultures, Gestion des ressources, Innovations et Ruralités, Castanet-Tolosan, France
Université de Toulouse, Institut National Polytechnique de Toulouse, Ecole d'Ingénieurs de Purpan, Toulouse, France

Nathalie Colbach
Institut National de la Recherche Agronomique, Unité Mixte de Recherche 1347 Agroécologie, Dijon, France

Philippe Lucas and Françoise Montfort
Institut National de la Recherche Agronomique, Unité Mixte de Recherche 1099 Biologie des Organismes et des Populations appliquée à la Protection des Plantes. Le Rheu, France

Philippe Debaeke and Jean-Noël Aubertot
Institut National de la Recherche Agronomique, Unité Mixte de Recherche 1248 Agrosystèmes et agricultures, Gestion des ressources, Innovations et Ruralités, Castanet-Tolosan, France
Université Toulouse, Institut National Polytechnique de Toulouse, Unité Mixte de Recherche 1248 Agrosystèmes et agricultures, Gestion des Ressources, Innovations et Ruralités, Castanet-Tolosan, France

Wenming He
Key Laboratory of Aquatic Botany and Watershed Ecology, Wuhan Botanical Garden, Chinese Academy of Sciences, Moshan, Wuchang, Wuhan, Hubei Province, China
Graduate University of Chinese Academy of Sciences, Beijing, China

Fang Chen
Key Laboratory of Aquatic Botany and Watershed Ecology, Wuhan Botanical Garden, Chinese Academy of Sciences, Moshan, Wuchang, Wuhan, Hubei Province, China
International Plant Nutrition Institute, Wuhan, China

Luo Liu
State Key Laboratory of Desert and Oasis Ecology, Xinjiang Institute of Ecology and Geography, Chinese Academy of Sciences, Urumqi, China

State Key Laboratory of Resources and Environmental Information Systems, Institute of Geographical Sciences and Natural Resources Research, Chinese Academy of Sciences, Beijing, China
University of Chinese Academy of Sciences, Beijing, China

Xinliang Xu and Dafang Zhuang
State Key Laboratory of Resources and Environmental Information Systems, Institute of Geographical Sciences and Natural Resources Research, Chinese Academy of Sciences, Beijing, China

Xi Chen
State Key Laboratory of Desert and Oasis Ecology, Xinjiang Institute of Ecology and Geography, Chinese Academy of Sciences, Urumqi, China

Shuang Li
State Key Laboratory of Resources and Environmental Information Systems, Institute of Geographical Sciences and Natural Resources Research, Chinese Academy of Sciences, Beijing, China

University of Chinese Academy of Sciences, Beijing, China

Min Sun, ZhiQiang Gao, WeiFeng Zhao, LianFeng Deng, Yan Deng, HongMei Zhao, AiXia Ren, Gang Li and ZhenPing Yang
College of Crop Science, Shanxi Agricultural University, Taigu, Shanxi Province, China

Xiaoqin Dai, Zhu Ouyang, Yunsheng Li and Huimin Wang
Key Laboratory of Ecosystem Network Observation and Modeling, Institute of Geographic Sciences and Natural Resources Research, Chinese Academy of Sciences, Beijing, China

Bing Gao, Xiaotang Ju, Fang Su, Fengbin Gao, Qingsen Cao, Peter Christie,
Xinping Chen and Fusuo Zhang
College of Resources and Environmental Sciences, China Agricultural University, Beijing, China

Oene Oenema
Wageningen University and Research Center, Alterra, Wageningen, The Netherlands

Index

www.ingramcontent.com/pod-product-compliance
Lightning Source LLC
Chambersburg PA
CBHW080409190526
45161CB00003B/175